BEAMED ENERGY PROPULSION

Related Titles from AIP Conference Proceedings

654   Space Technology and Applications International Forum - STAIF 2003:
Conference on Thermophysics in Microgravity;
Conference on Commercial/Civil Next Generation Space Transportation;
20th Symposium on Space Nuclear Power and Propulsion;
Conference on Human Space Exploration;
1st Symposium on Space Colonization
Edited by Mohamed S. El-Genk, February 2003, 0-7354-0114-4
CD-ROM: 0-7354-0115-2

608   Space Technology and Applications International Forum - STAIF 2002:
Conference on Thermophysics in Microgravity;
Conference on Innovative Transportation Systems for Exploration of the Solar System and
Beyond;
19th Symposium on Space Nuclear Power and Propulsion;
Conference on Commercial/Civil Next Generation Space Transportation
Edited by Mohamed S. El-Genk, February 2002, 0-7354-0052-0
CD-ROM: 0-7354-0053-9

552   Space Technology and Applications International Forum - 2001:
Conference on Space Exploration Technology;
Conference on Thermophysics in Microgravity;
Conference on Innovative Transportation Systems for Exploration of the Solar System and
Beyond;
Conference on Commercial/Civil Next Generation Space Transportation;
18th Symposium on Space Nuclear Power and Propulsion; Space Radiation and Environment
Effects Track
Edited by Mohamed S. El-Genk, February 2001, 1-56396-980-7
CD-ROM: 1-56396-981-5

To learn more about these titles, or the AIP Conference Proceedings Series, please visit the
webpage **http://proceedings.aip.org/proceedings**

BEAMED ENERGY PROPULSION

First International Symposium on
Beamed Energy Propulsion

Huntsville, Alabama   5–7 November 2002

EDITOR
Andrew V. Pakhomov
The University of Alabama in Huntsville
Huntsville, Alabama

SPONSORING ORGANIZATIONS
NASA George C. Marshall Space Flight Center
The University of Alabama in Huntsville (UAH)
NASA Experimental Program to Stimulate
   Competitive Research (EPSCoR)
Alabama Space Grant Consortium
American Institute of Aeronautics and Astronautics
Propulsion Research Center, UAH
Seoul National University, BK-21 Mechanical
   Engineering Research Division
Rensselaer Polytechnic Institute

Melville, New York, 2003
AIP CONFERENCE PROCEEDINGS ■ VOLUME 664

Editor:

Andrew V. Pakhomov
Department of Physics
University of Alabama in Huntsville
Huntsville, AL 35899
USA

E-mail: pakhomov@email.uah.edu

L.C. Catalog Card No. 2003104537
ISBN 0-7354-0126-8
ISSN 0094-243X
Printed in the United States of America

CONTENTS

MISCELLANEOUS APPLICATIONS

BEAM-PROPELLED MICROAIRCRAFT

BEAM GENERATION, PROPAGATION, AND RECEPTION

Preface

Dear Reader,

The First International Symposium on Beamed Energy Propulsion (ISBEP) was held on the campus of the University of Alabama in Huntsville, on November 5 - 7, 2002. The preparations of ISBEP took over a year. When it was over, I received a lot of replies from participants, from which I can see that the symposium was a success. ISBEP attracted 114 scientists from 8 countries and 14 states, and 70 talks were presented overall. As a first international conference of this kind, ISBEP demonstrated that Beamed Energy Propulsion is an established worldwide field of research and engineering, and an extremely active field of new technology. Four countries have already launched their lightcrafts, and the day when they will reach space is looming in the near future. A major demonstration of microwave propulsion *in space* will be conducted this year. Laser-based microthrusters are becoming off-the-shelf technology. Microairplanes and dirigibles are now riding on laser beams. Even planetary defense systems left sci-fi thrillers to become a matter of feasibility discussion. But enough about space at large, the microspace may turn into a domain for beamed propulsion, too: One can read about the experiments on x-ray systems, intended to drive vehicles inside human blood vessels. You can find all of these in this book.

The composition of the Proceedings is close to the original ISBEP program, *i.e.* plenary talks are arranged in order of presentation, followed by eleven regular sessions. There are only a few exceptions to this rule. One is made for the keynote speech of Prof. Arthur Kantrowitz, which is placed ahead of all other talks. The Proceedings also include the paper of Prof. Myrabo on the concept of the Mercury lightcraft. This paper was not listed in the original program, but Prof. Myrabo kindly agreed to present this work at the Plenary session as a substitute for a cancelled talk. You can find this paper in the section on System Analysis. Finally, Prof. Yabe has combined his talks given at Ablative Laser Propulsion (ALP) and BEP Microaircraft sessions into one paper. This paper is placed in the ALP section. The titles of papers included in the Program, but not presented at the symposium, are listed in the Addendum. The Addendum also includes four presented papers, which were unavailable for the print.

ISBEP resulted from a great collective effort, and I am honored to acknowledge everyone (if that would be possible) whose personal commitment and support made the conference a success. I would like to start with those who worked with me side-by-side on the organization of this event: the Co-Chairman of the symposium, Edward "Sandy" Montgomery (NASA MSFC), and symposium coordinators Donald J. Larson (NASA MSFC) and Cecil G. Stokes, Jr. (Propulsion Research Center, UAH). We worked on ISBEP together from February 2002, and a list of what we did is too long to be reproduced here. Prof. Leik N. Myrabo (Rensselaer Polytechnic Institute) was involved in this work from the beginning as my co-author on the original ISBEP statement, along with being an informal, yet very active and helpful co-organizer. In particular, his organizational energy was "beamed" into the assembly of the International Program Committee, which he chaired. The Committee helped us greatly, and the names of its members are listed after this text. Also, I would like to acknowledge a great job done by Janine Roskowski (NASA MSFC) for preparing

ISBEP Program and fliers for print, Phil Gentry (UAH), ISBEP media advisor; Cynthia K. Brasher and Hanan Badreldin (UAH), the conference secretaries, Trenette Brown (UAH, Accounting) and Jennifer G. Middleton, sales director of the Bevill Center. Finally, I think that ISBEP would not have been the same without the keynote dinner and the excellent music performed by Microwave Dave (David Gallaher) and the Nukes (Rick Godfrey and Skip Skipworth).

Although the complete list of supporting organizations is presented within the following pages, I would like to give special recognition to those who supervised this support. First, to John Cole (NASA MSFC), whose letter from 12/05/01 announced the initial (and major) support of ISBEP. The cause was followed by Dr. Ron Greenwood (UAH), Prof. John Gregory (Alabama Space Grant, UAH), Prof. Clark Hawk (Propulsion Research Center, UAH), Prof. In-Seuck Jeung (Seoul National University, Korea), and Prof. William Baeslack (Rensselaer Polytechnic Institute).

I am expressing my deepest gratitude to Prof. Arthur Kantrowitz (Dartmouth College, former Chairman of Avco-Everett Research Labs) for coming to ISBEP and giving us his historical keynote talk. The founding role of Dr. Kantrowitz in the field of beamed-energy propulsion was honored at the symposium with the First ISBEP Award (see the photo in the Keynote Session). Prof. Kantrowitz also received an award from Rensselaer Polytechnic Institute, presented by Prof. Myrabo.

I also would like to acknowledge those who made their contribution to this volume. First, our plenary speakers, especially those who undertook the burden of presenting and writing papers on works performed, in some cases, a quarter of a century ago: Dr. Anthony Pirri (Northeastern University), Dr. Jordin T. Kare (Kare Technical Consulting), Prof. John D.G. Rather (Wayne State University), Prof. Leik N. Myrabo, and Lee W. Jones (B.G. Smith and Assoc.), and for the reviews on research in Japan, given to us by Dr. Masayuki Niino (National Aerospace Laboratory) and Germany, by Dipl.-Ing. Wolfgang O. Shall. I would like to thank those who opened ISBEP with their warm welcomes: Mayor Loretta Spencer (City of Huntsville), Prof. Lewis J. Radonovich (UAH), John Cole, and Robert L. Sackheim (NASA MSFC). I would like to give special credit to Dr. James Benford (Microwave Sciences, Inc.) for both his kind *in memoriam* speech in honor of Dr. Robert Forward, and his good advice on the preparation of ISBEP. Also I would like to acknowledge the help of Dr. Claude R. Phipps (Photonic Associates), who graciously allowed us to conduct a preparatory panel at the High-Power Laser Ablation Symposium IV (Taos, NM), and also Prof. Brian Landrum (UAH) for similar help at the 33rd AIAA Lasers and Plasmadynamics Conference (Maui, HI). A great credit must also be given to our session chairs: Prof. Leik N. Myrabo, Prof. Akihiro Sasoh (Tohoku University, Japan), Dr. Claude R. Phipps, John Cole, Dr. James Benford, Dr. C. Les Johnson (NASA MSFC), Dr. Jordin T. Kare, Prof. Takashi Yabe (Tokyo Institute of Technology), Dr. Jonathan Campbell (NASA MSFC), Dr. Hal Bennett (Bennett Optical Research Inc.), Gordon Woodcock (Gray Research Inc.), Ron Litchford (NASA MSFC), and Dr. Michael Lander (Anteon Corporation). Finally, I would like to say to all participants: thank you for coming to Huntsville and being a part of the First ISBEP.

On a personal note, I would like to express my gratitude to my colleagues who helped with my work on ISBEP: Prof. Don Gregory, Prof. Lloyd Hillman, Kenneth A. Herren (NASA MSFC), graduate students M. Shane Thompson and Jun Lin, and last, but not least, my wife, Yelena N. Zakin, for allowing me to stay long hours away from home.

The First International Symposium on Beamed Energy Propulsion is over, and this book is the last material product of the event. We all are aware of the "time-sensitive" value of conference proceedings, but I believe that this particular book will have a longer lifetime than an average one of its kind. Why? Because this collection of papers goes beyond the tradition of narrow-focused scientific writing, though still written by scientists. If you have received this book as a participant of the First ISBEP, I hope that you enjoyed the meeting, and that you are now busy with your travel preparations for Sendai, Japan, where ISBEP 2 will take place this year. If you are an interested reader, I hope you find the answers to your questions in this rich collection of works. Finally, if you are a student, I invite you: Join us, go into BEP research. There is a wealth of work still to be done in this new field, where plenty of new heights are ahead of us.

Andrew V. Pakhomov,
Department of Physics, University of Alabama in Huntsville, 2002

First International Symposium on Beamed Energy Propulsion (ISBEP)
Huntsville, Alabama, November 5-7, 2002

Symposium Co-Chairs:
Prof. Andrew V. Pakhomov, The University of Alabama in Huntsville (UAH)
Edward E. Montgomery IV, NASA Marshall Space Flight Center (MSFC)

Symposium Coordinators:
Donald J. Larson, NASA MSFC (Huntsville, Alabama)
Cecil G. Stokes, Jr., Propulsion Research Center, UAH

ISBEP International Program Committee:
Dr. Adrian Alden, Communications Research Centre (Ottawa, Canada)
Prof. Dr. Willy L. Bohn, Deutsches Zentrum fur Luft- und Raumfahrt (DLR) - Institute of Technical Physics (Stuttgart, Germany)
John Cole, NASA MSFC (Huntsville, Alabama)
Prof. Andrei Ionin, Lebedev Physical Institute (Moscow, Russia)
Prof. In-Seuck Jeung, Seoul National University (Korea)
Dr. Jordin T. Kare, Kare Technical Consulting (San Ramon, California)
Prof. Kimiya Komurasaki, The University of Tokyo (Japan)
Dr. Mike Lander, Anteon Corp. (Wright-Patterson Air Force Base, Ohio)
Dr. Alain Lebéhot, Laboratoire d'Aérothermique (Orléans, France)
Dr. Marco A.S. Minucci, Centro Technico Aerospacial (CTA), Instituto De Estudos Avancados (Sao Jose dos Campos, Brazil)
Prof. Leik N. Myrabo, Committee Chair, Rensselaer Polytechnic Institute (Troy, New York)
Dr. Masayuki Niino, Kakuda Space Propulsion, National Aerospace Laboratory of Japan (Kakuda, Japan)
Prof. Andrew V. Pakhomov, The University of Alabama in Huntsville
Dr. Claude R. Phipps, Photonic Associates (Santa Fe, New Mexico)
Dr. Yuri A. Rezunkov, Research Institute for Complex Testing of Optoelectronic Devices (Sosnovy Bor, Russia)
Prof. Akihiro Sasoh, Tohoku University (Sendai, Japan)
Dipl.-Ing. Wolfgang Schall, (DLR) - Institute of Technical Physics (Stuttgart, Germany)
Dr. Shigeaki Uchida, Institute for Laser Technology (Osaka, Japan)
Prof. Takashi Yabe, Tokyo Institute of Technology (Japan)

Robert L. Forward, 1932 - 2002.

Photo courtesy of Dr. James Benford.

Forward Thinking
Memories of Robert L. Forward

Physicist and science fiction author Dr. Robert L. Forward passed away six weeks ago in Seattle, Washington, at the age of 70 years.

Few are perfectly named. To answer to "Forward!" from youth must have shaped Robert's remarkable mind. He always kept moving—forward, of course.

Everyone who knew Bob was inspired by him. He used physics like a sculptor uses a chisel, shaping fascinating ideas and possibilities.

Most importantly for this Symposium, he pioneered the idea of driving sails with beams—laser or microwave—with a savvy eye cocked toward the truly grand goal of interstellar travel. His two-stage interstellar sail vehicle, and his ideas for harnessing light and microwaves made interstellar flight seem realizable. There are two sessions on this tomorrow. He would be here if he could.

He knew his physics, but refused to be blinkered by its conventional wisdom. Starting with his doctoral thesis, designing a new class of gravitational wave antenna under Joe Weber, he always thought long range.

Elegance of concept marked all his many inventions. Orbital tethers will be both graceful and useful. Antimatter beckoned to him as the most fundamental method of containing energy. His best summary work is "Indistinguishable From Magic", taken from Clarke's Third Law, "Any Sufficiently Advanced Technology is Indistinguishable From Magic."

He was playful with ideas and wasn't constrained by formality: Titles from his 157 technical publications that I like are "General Relativity for the Experimentalist", "Guidelines to Anti-Gravity" and (my favorite) "Laser Weapon Target Practice with Gee-Whiz Targets."

Beyond these areas, any of which could make a full career, he also wrote novels "about ideas I can't get into the physics journals". Life on a neutron star. Two planets orbiting so close they shared an atmosphere. Life forming at near absolute zero. Bob made it all seem quite possible.

But he didn't have time to explore all of his ideas. When he learned unexpectedly that he had only months left, he spent part of it writing down the ideas he hadn't had time to work up and reveal to us. He did live to see sails fly on beams. He did not live to see our attempt at the first solar sail flight, an audacious privately funded attempt to fly a solar sail, this coming spring. In our last conversation I told him we're going to propel that sail with a microwave beam from Earth, and he loved the audacity of it. He didn't live to see it, but he did know that realizations of his beam/sail propulsion concepts are starting to come true.

He was a true Renaissance man, cruelly struck down when he still had his darting imagination.

The best memorial we can give him is to keep moving—forward, of course.

James Benford
November 2002

KEYNOTE SESSION

THE WORLD, THE FLESH and THE DEVIL

Thayer School of Engineering, Dartmouth College, Hanover, New Hampshire 03755

I have taken the title of this talk from the title of a wonderful book of prophesy written in 1929 by the great x-ray crystallographer J. D. Bernal. His scientific work laid the basis for the discovery of the double helix structure of DNA. His prophesy in this tiny 81 page book has had a great impact on my thinking. My remarks are an update divided into the same three parts, which constitute the main chapters of his book.

THE WORLD

The first chapter, The World, is a prediction that a substantial portion of mankind will leave the earth to live in space. Bernal imagined that they would build "globes", miles in diameter perhaps manufactured from a convenient asteroid. The obvious barrier to these opportunities is overcoming the earth's gravitation. The energy to reach low earth orbit (LEO) is about 8.75 kWh/kg or about 20 cents/kg at current wholesale electricity rates. I attach great significance to this number because it shows that, given enough invention and development, fundamental energy considerations will not prevent the migration of substantial fractions of humanity into space.

Current technology costs about $10,000/kg to LEO and five times that to get to geosynchronous orbit. It is most striking to me that these costs have not declined in the 48 years since Sputnik. Commercial launchers of communication and surveillance satellites are now engaged in a competitive market and their costs may decline somewhat. However chemical rocketry is a mature discipline and like other mature disciplines its vision is limited. This limitation was highlighted for me by a 1998 National Academy of Sciences report entitled "Breakthrough Technologies to Meet Future Air and Space Transportation Needs and Goals". After devoting one short paragraph to "Novel Launch System Concepts" they went on to recommend only the search for "Advanced Propellants."

Let's list some of the possibilities which were not on their list.

Several attempts to use nuclear energy were started in the fifties and sixties and abandoned. In my opinion, nuclear propulsion is hostage to the irrational terror of radiation. For example, there was the suggestion by Stan Ulam that a series of very

CP664, *Beamed Energy Propulsion: First International Symposium on Beamed Energy Propulsion*,
edited by A. V. Pakhomov
© 2003 American Institute of Physics 0-7354-0126-8/03/$20.00

small nuclear explosions could be used to exert thrust on a pusher plate. George Dyson (Freeman's son) has written "Orion" which is a fine description of that effort.

The possibility of using a large satellite in a stationary orbit to hoist payloads into stationary orbits has suffered from the lack of materials with sufficient strength to weight ratio. Dennis Reilly suggested to me that perhaps that difficulty will be overcome with the nanotubes we have recently learned to produce in quantity.

30 years ago I was engaged in an effort at the Avco-Everett Research Laboratory to increase the power available from lasers. After we had increased the power of lasers by 4 orders of magnitude it occurred to me that we could use a very powerful laser to ablate material from a vehicle going into orbit and to heat that vapor to temperatures much higher than can be achieved from chemical energy. This perhaps could produce specific impulse high enough to reach LEO more cheaply than chemically powered rockets.

In these 30 years laser propulsion has been blessed by brilliant advances made by people who probably never even heard of laser propulsion.

The astronomical observing community in recent decades has been strikingly successful in advancing the resolving power of telescopes. Atmospheric density fluctuations had blurred the images obtainable so that the resolving power of large telescopes was not much better than that of a good pair of binoculars. Introducing flexible mirrors controlled by many actuators enabled compensation in milliseconds for atmospheric distortions. It has been possible to sharpen the images of fixed stars and at the same time other objects in the same field of view with a tenfold increase in resolving power.

Using a 20 watt dye laser whose frequency is tuned to excite sodium atoms in a layer about 100 kilometers above the earth, artificial guide stars have been created and the resolving power of telescopes has been dramatically improved. For example, the Keck telescope with 349 actuators is reported to have a resolving power of .025 microradians. This resolving power is better than I thought was required for laser propulsion to LEO. Imaginative utilization of this new capability provides a new opportunity for invention.

We have heard several interesting papers this afternoon on the use of beamed microwaves for propulsion. The use of multiple phase locked dishes such as are used for radio astronomy in the Very Large Array in New Mexico. The array of phase locked dishes gives a resolving power of .02 microradians. It will be exciting to follow this development.

Today's world provides a variety of possibilities for dramatically reducing the cost of ascent to LEO by orders of magnitude. I will hazard the prediction that this group of possibilities and perhaps many more will be taken seriously in this century. Nations aspiring to share the kind of technological leadership that the US now enjoys will make large scale access to space one of their prominent goals.

THE FLESH

When expansion to new worlds becomes feasible I am convinced that there will be many who will grasp that opportunity. Engineers and scientists have a great role to play in enabling these advances.

To my mind the greatest opportunity which space affords is the opportunity to innovate beyond what is socially acceptable to your neighbors. The migration of the Pilgrims to America to freely practice their religion was a bold move in this sense. The revolutionary ideas of the 1960's show that the capability and the motivation for social innovation is alive and well. Space colonies will provide more room for the bold social experiments which will advance mankind.

We also must remember that most earlier migrations were motivated by a felt need to escape the intolerable conditions that rulers impose on large fractions of humanity. In today's world - contemplating war to decelerate the spread of nuclear weapons - the need to expand our habitat will, I fear, become much more compelling. Migration to space will provide humanity with new opportunities for the fresh starts which are so important for progress.

Bernal writing before the breathtaking advances in 20th century biology predicted great changes in anatomy and physiology that would evolve in the portion of humanity that migrated into space. He discussed the striking possibility that mankind would split into two species. I don't feel competent to update Bernal's biological predictions in view of the revolutionary changes which have occurred in biology, so I will move on to the devil with whom I have more experience.

THE DEVIL

40 years ago I had a personal experience with one of the devil's most powerful weapons - secrecy - which dramatically changed my life.

The Soviet Union had scored a great propaganda coup with Sputnik on 4 Oct. 1957. In the fall of 1960 the Air Force was still competing with NASA to acquire a larger role in the rapidly growing space program. As part of that competition Lt. Gen. Bernard A. Schriever, who had led the successful development of the US Intercontinental Ballistic Missile, created a committee led by a former Asst. Secretary of the Air Force, the late Trevor Gardiner, to aid the Air Force in advising the incoming President on a space program for the U.S. My appointment to the Gardiner committee began a decisive part of my education on the devilish relationship between science and politics.

Soviet Rockets were larger than US rockets and could launch larger payloads into orbit. The media came to believe that matching the Soviet's bigger boosters was the key to reestablishing US prestige. The Apollo mission, designed for this purpose, required the injection into earth orbit of rocket propellant and equipment weighing about 250,000 lb. The Atlas and Titan rockets which had been developed for the ICBM program could launch only a few thousand pounds. One obvious answer was to build a much bigger booster (e.g. the Saturn V).

An alternative approach which, Von Braun had described to the committee, was to assemble payloads in earth orbit. The Atlas and Titan boosters which the Air Force then had coming off production lines in California and Maryland could in this way launch any load. By assembling payloads in earth orbit we could not only go to the moon but we could utilize America's tremendous production capabilities to outmatch any booster the Soviets could build. Further we estimated that earth orbital assembly (EOA) would be 10 times cheaper than the big booster approach. But EOA required that we depend on the successful development of the technologies of rendezvous and assembly in orbit. After much argument and after hearing many presentations the Gardiner committee was still divided on which approach to recommend. I was convinced that EOA was the better strategy. Our unclassified report (Report of the Air Force Space Study Committee 20 March 1961) recommended that both approaches be pursued for the time being.

The next thing I heard was not from the Air Force but from the National Space Council which was headed by Vice President Lyndon Johnson. I was informed by telephone that our report had been classified TOP SECRET that only 25 copies would be printed, and that all of them would be stored in Lyndon Johnson's safe. Since all our meetings and the information we used had been unclassified, I was flabbergasted. The report was Top Secret until 1996 when it was released by a Freedom of Information Request still bearing the word "unclassified" on its title page.

I sought out those members of the President's Science Advisory Committee, whom I knew. I persuaded many of them that suppression of the EOA option was an important mistake. A panel under Prof. Frank Long of Cornell was tasked with examining the matter. The panel held a meeting at which a NASA representative laid out the bigger booster plan. I then put forward evidence that EOA would be at least ten times cheaper. Long turned to the NASA representative and asked what was wrong with what I had said. His total reply was - "If you decide in favor of EOA it will be very embarrassing to us."

Some weeks later Jerome Wiesner, Pres. Kennedy's Science Advisor, asked me to come and discuss the matter with him. For about two hours we went over the arithmetic and he too was convinced that we were making a big mistake. Several years afterward he told me that during his tenure as Science Advisor nobody had caused him as much trouble as I did. He had taken the matter to Kennedy. The President told him that it was none of his business and that he should stay out of it.

The national policy would be to build the biggest booster, regardless of its cost. The Apollo mission was successful and restored US prestige. However it was the most expensive mission the US had ever undertaken and the culture of disregard for costs would mean that for a long time space operations would be frustratingly expensive. The Saturn V booster developed for Apollo was simply discarded a few years after the Moon mission was completed. The expensive new start allowed the center of space activity to be chosen by Lyndon Johnson. However a key measure of progress in space - the cost of launching a pound into LEO has not fallen as would be expected in a healthy young industry. It now costs at least twice as much in inflation adjusted dollars as it did in 1960. The suppression of scientific information for political purposes has delayed the start of what one day will be an important part of the world economy. As far as I know this consideration of Earth Orbital Assembly is not mentioned in historical accounts of NASA's early years. It is of course in use for the assembly of the International Space Station.

(Note that three years later LBJ employed secrecy again to manipulate facts to suit political purposes and to secure the passage of the notorious Tonkin Gulf Resolution. This cleared the way for our disastrous Vietnam war.)

The transmission to the public of information needed for the democratic control of technology is essential to the faith in progress. I had witnessed the suppression of scientific information by powerful people for political purposes. This abuse of authority, if sufficiently widespread, would destroy that faith. Since that experience, I have devoted a considerable fraction of my time to means for improving the ability of science to communicate what it knows, and especially what it does not know, to the public.

From left to right: Leik N. Myrabo, Arthur Kantrowitz (with ISBEP Award) and Andrew V. Pakhomov, at the dinner and award ceremony, Bevill Center, November 6, 2002. Photo by Phil Gentry, UAH, 2002.

PLENARY SESSION

Laser Propulsion: The Early Years

Peter E. Nebolsine* and Anthony N. Pirri**

*Physical Sciences Inc., Andover, MA 01810

**Division of Technology Transfer, Northeastern University, Boston, MA 02115

Abstract. The early years of laser propulsion were driven by the vision of Arthur Kantrowitz and his team of researchers at the Avco Everett Research Laboratory. The pioneering research continued during the 1970s both at Avco and Physical Sciences Inc. (PSI), a company founded by Avco scientists. This paper presents the results of the experimental and theoretical research that was the precursor for much of the work that was done in the 1980s and 1990s. The successes are presented as well as the failures, and the enthusiasm of the early days is discussed, as well as the matching realization that the limiting factors for progress in the 1970s were the progress in laser development and the identification of missions for such systems. All this set the stage for research activities of the 1980s that were carried out by subsequent believers of the Kantrowitz vision.

INTRODUCTION

Laser propulsion was first introduced by Arthur Kantrowitz in 1971 [1,2] and at the time he presented a remarkable vision driven by his optimism over the future of laser technology. The laser at the time had been invented only 13 years earlier. Kantrowitz realized that beaming energy to a chemically inert propellant would provide a mechanism for propulsion that would overcome the low specific impulse limitations of existing and planned chemical propulsion systems. By using laser absorption instead of requiring a chemical reaction, rocket propulsion parameters could be controlled by the rocket designer and not by the limitations of propellant chemistry. The specific impulse of a rocket would depend simply on the ability to focus and thus, concentrate the beamed energy. The thrust/mass ration would be determined by the available laser power.

Theoretically, any choice of specific impulse and thrust/mass ratio can be achieved with laser propulsion. Therefore, mission requirements would control these choices. Kantrowitz envisioned the use of lasers to launch low mass payloads into earth orbit at a high rate using the optimum specific impulse of 800 seconds. In addition, the simplicity of propellant choices would allow the use of environmentally friendly propellants (such as water or inert gases) to perform this mission.

An artist's conception of a laser propulsion system launching a rocket into earth orbit was presented in a paper summarizing the status of research at the Avco Everett Research Laboratory in 1974 [3] and is reproduced as Fig. 1. In this figure a ground based laser, which could be an assembly of multiple laser cavities to get the high power, would beam energy directly into the exhaust section of a rocket nozzle where it is focused by the nozzle into the propellant flow. This simplified schematic was the basis for many of the early rocket thruster concepts. Beaming directly into the

CP664, *Beamed Energy Propulsion: First International Symposium on Beamed Energy Propulsion,*
edited by A. V. Pakhomov
© 2003 American Institute of Physics 0-7354-0126-8/03/$20.00

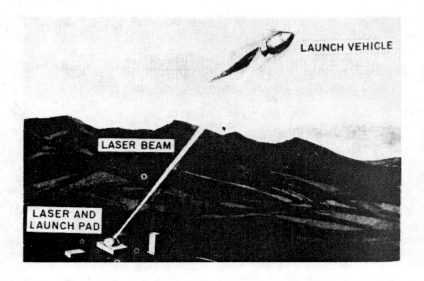

FIGURE 1. Artist's Conception of a Laser-Powered Rocket

exhaust plane of the thruster would mean a simple thruster design. However, it was recognized that the mission profile might not permit this simplicity and other concepts are discussing in the following section.

EARLY LASER-POWERED THRUSTER CONCEPTS

The early years were spent primarily on thruster designs and configurations for demonstration of concept feasibility. Differences in designs were a result of laser specification and the location of laser energy absorption in the propellant. At the time it was not clear whether future lasers would be repetitively pulsed or continuous wave devices. Therefore, thruster concepts were proposed for both laser types.

There are three features of a laser-powered thruster that control its performance and design requirements. The first is the optics used to concentrate the laser power. The second is the location of the absorption zone for stable laser absorption, and the third is the propellant feed rate [3,4]. Thruster concepts for continuous wave (CW) lasers that use nozzle focusing were believed to be simple in design but potentially faced with absorption zone instabililities [5]. Repetitively pulsed laser thruster concepts with nozzle focusing would optimize performance by pulsing the laser in sync with the frequency of the absorption zone instabilitiy. Thruster design concepts with external optics to focus the laser energy into a conventional plenum chamber could alleviate stability issues but at the expense of increased thruster design complexity.

Early research at the Avco Everett Research Laboratory (AERL) and at Physical Sciences Inc. (PSI) was with simple thruster designs using nozzle focusing while at NASA early research concentrated on external focusing and plenum chamber designs.

A nozzle design used in early simulation experiments for continuous wave laser propulsion is presented in Fig. 2 [3]. The nozzle was sized such that by utilizing a long pulse laser, the steady state operation of a CW laser powered thruster would be duplicated. The nozzle serves as the focusing optic concentrating the laser energy into the region where the propellant is injected. Although gaseous propellants could by used, for simplicity solid propellant rods were incorporated into the design.

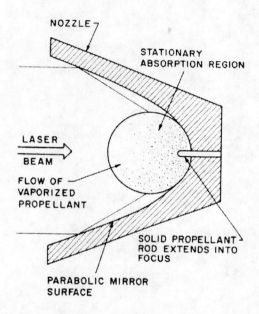

FIGURE 2. Steady-State Simulation Nozzle Design

The solid propellant is vaporized and the vapor is heated in a region downstream of the focal point. Specific impulse is dependent upon the propellant choice and the amount of heating that occurs in the laser energy absorption zone.

Figure 3 presents the early thruster concept for a repetitively pulsed laser powered thruster. The nozzle wall again focuses the laser radiation to a point downstream of the nozzle throat. At the focal point a laser-induced breakdown of the gas occurs and either a laser-induced blast wave or detonation wave explosively heats the propellant that exhausts the nozzle to provide thrust.

A more complex thruster design that uses external optics and upstream laser heating is shown in Fig. 4. This concept was presented in a paper by Avco scientists in 1977 [6]. Heating takes place upstream of the nozzle throat so from the absorption chamber back to the exit plane of the nozzle, this rocket engine resembles a conventional chemical rocket. The photo in Fig.5 appeared in an early Lockheed report [7] and the external focusing optic for this type of engine design can be seen positioned on an orbital transfer vehicle.

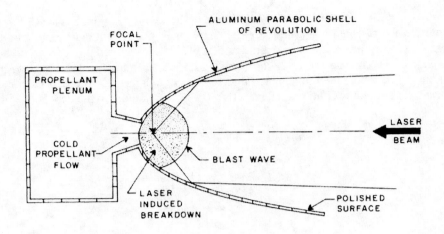

FIGURE 3. Repetitively Pulsed Laser Powered Thruster Design

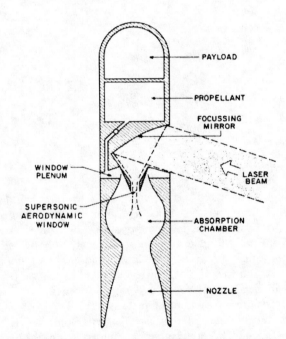

FIGURE 4. Two Port CW Laser Powered Thruster Design

FIGURE 5. Lockheed/NASA Concept Photo

TECHNICAL ISSUES ADDRESSED BY EARLY EXPERIMENTS AND ANALYSES

Early analyses and experiments that went along with the above thruster concepts were aimed at addressing the following technical issues that needed to be resolved:

1) The fundamentals of propulsion by external energy absorption needed to be understood. Since laser propulsion is not photon propulsion, an understanding of the techniques for efficiently heating a gas with single wavelength radiation and transforming the heated propellant into thrust was desired. In conventional chemical rocket propulsion, the governing parameter is the specific impulse. The specific impulse is governed by the stagnation temperature, a result of the chemical reaction, and the thrust is dependent upon the mass flow of the heated gas than can be forced through the nozzle. In laser propulsion a new parameter appears: the coupling coefficient (the ratio of thrust to laser power), and it can be shown that the product of the coupling coefficient and the specific impulse is proportional to the overall efficiency of the process.

2) Optics is now coupled to fluid mechanics for those laser powered thruster concepts where the nozzle has to both focus the radiation and expand the heated gas. The trade-off between optical requirements and gas expansion for maximum thrust is dependent upon the type of laser utilized. Since it was not known at the time which laser system would be the "winner", concepts coupling optics and fluid mechanics were explored for both CW and repetitively pulsed (RP) laser configurations.

3) The choice of propellant and its laser radiation absorption characteristics plays a key role in the performance of the thruster because of stability requirements in the heating zone. An understanding of absorption zone stability and its dependence upon absorption coefficient variations in the heated plasma was needed to verify that simplicity of propellant choice was a major feature of laser propulsion concepts.

Although the above issues occupied the attention of early researchers, it is safe to safe to say that much of the time was spent during the early years of laser propulsion on simple concept feasibility and performance, ie. just demonstrating that it works. In the following section, results of the early analyses and experiments performed at AERL and PSI are presented.

RESULTS OF EARLY EXPERIMENTS AND ANALYSES

A photograph is presented in Fig. 6 of the steady state simulation thruster experiment that is described along with Fig.2. A long pulse (100 microsecond) 10.6 micron laser was utilized to simulate CW laser propulsion by scaling down the nozzle size such that the exhaust gases would traverse the nozzle several times during the duration of the laser pulse [3].

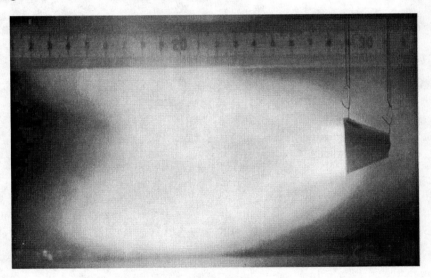

FIGURE 6. Steady-State Simulation Thruster Experiment of Ref. 3.

Experimental data obtained from this experiment are presented in Fig. 7. Shown plotted are the coupling coefficient and exhaust gas specific impulse measured for two solid rod propellants. Lines of constant efficiency of converting laser power to exhaust power (proportional to the product of the coupling coefficient and the specific impulse) are also presented. It can be seen that the efficiency in these early experiments was quite low. It is believed that minimal laser energy absorption was occurring in the vapor and the laser energy was simply vaporizing the solid material. The exhaust velocity and corresponding specific impulse observed was characteristic

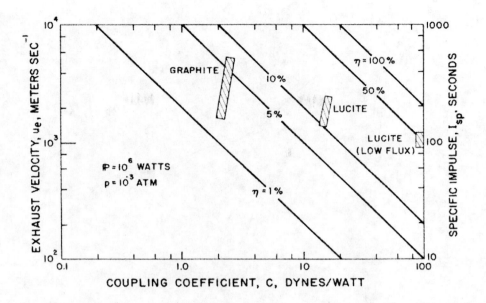

FIGURE 7: Coupling Coefficient and Specific Impulse Results Obtained in Steady-State Simulation Experiments

of the vaporization process plus the gas dynamic expansion of the vapor from the nozzle throat to the nozzle exhaust plane. At the conclusion of these experiments, the researchers believed that high specific impulse that was desired could only be achieved with CW lasers by heating the gas in a stable manner upstream of the throat. This was the approach being pursued by NASA contractors [8]. At PSI a computational model that predicted the performance of a such a thruster with upstream laser heating was developed [9]. Example results of this computational model are presented in Fig. 8.

The results presented in Fig. 8 are for a hydrogen propellant. The plasma in the plenum is at tens of thousands of degrees Kelvin and the calculated exhaust velocity will be greater than ten kilometers per second. The corresponding specific impulse will be greater than 2000 sec.

Early experiments to simulate repetitively pulsed laser propulsion were performed at PSI with a series of 10.6 micron single pulse lasers that were triggered sequentially [10]. The photograph presented in Fig. 9 is a conical nozzle version of the concept presented in Fig. 3. By using a conical nozzle the performance could be better understood without the added complexity of the nozzle focusing optic. Later versions of the thruster shown in Fig. 9 used a parabolic nozzle as shown schematically in Fig. 3. The thruster is approximately eight inches long.

Results of these experiments for the single pulse coupling coefficient with air as the propellant are presented in Fig. 10 as a function of single pulse laser parameters.

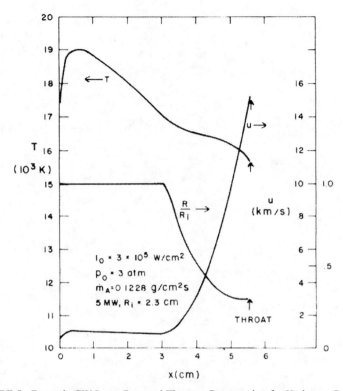

FIGURE 8. Example CW Laser Powered Thruster Computation for Hydrogen Propellant

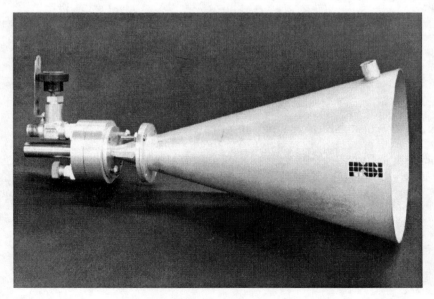

FIGURE 9. Conical Nozzle Configuration for Simulation of Repetitively Pulsed Laser Propulsion

18

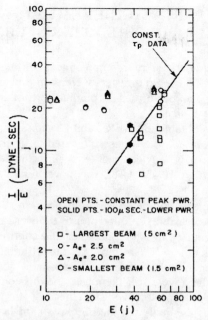

FIGURE 10. Single Pulse Coupling Coefficient Results with Air Propellant

Data obtained with multiple pulses for the specific impulse as a function of the time between pulses are presented in Fig. 11.

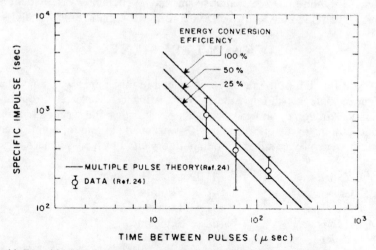

Figure 11. Repetitively Pulsed Thruster Results for Specific Impulse vs. Time Between Pulses. Argon and Hydrogen Propellants

It can be seen in Fig. 11 that these experiments achieved an energy conversion efficiency of approximately 50% when the data is analyzed using the multiple pulse theory of Ref. [11]

SUMMARY OF EARLY ACHIEVEMENTS AND LIMITING FACTORS IN LASER-POWERED THRUSTER DEVELOPMENT

The above sections summarize the accomplishments of the '70s at AERL and PSI. Several concepts for laser-powered thrusters were proposed and a significant amount of theoretical research was performed to predict performance from these devices and optimum propellant choices.

Experimental data was obtained at AERL and PSI for simulated CW and RP laser propulsion. During this period a CW laser powered thruster was also designed and built by Rocketdyne under NASA contract [8]. Unfortunately, this thruster, which was designed for the NASA 10 kW laser, was never tested because of laser device limitations.

Although performance data and demonstrations performed during the '70s verified that laser propulsion was feasible, optimization of thruster designs was not possible. Insufficient time was spent on mission, guidance and control constraints, and optics and propagation considerations. The most serious limitation on progress was a result of the lack of lasers of sufficient power and reliability. Because of this lack of required laser devices, there were still skeptics. The mission driver was a search for a less expensive space transportation system because the shuttle was expensive. Single stage to orbit was the goal, and the big payoff for laser propulsion was with rapid launch of many small objects into orbit. In addition, NASA continued to explore other missions that would not require the large laser powers necessary for earth launch [7].

In spite of the limitations of insufficient laser power during this time period, and the existence of non-believers, enthusiasm and excitement was evident as more researchers signed on to the Kantrowitz vision at the beginning of the '80s.

ACKNOWLEDGMENTS

A significant fraction of the research described in this paper was supported by the Department of Defense (through the Defense Advanced Research Projects Agency and the U. S. Air Force) and the National Aeronautics and Space Administration. Major contributions to the effort in the '70s were made by the following members of the staff at the Avco Everett Research Laboratory and Physical Sciences Inc:

G. E. Caledonia	D. Douglas Hamilton
J. S, Goela	A. R. Kantrowitz
N. H. Kemp	H. H. Legner
M. J. Monsler	P. E. Nebolsine
A. N. Pirri	D. A. Reilly
R. G. Root	D. I. Rosen
D. I. Rosen	G. A. Simons
V. H. Shui	R. F. Weiss
G. M. Weyl	P. K. S. Wu

REFERENCES

1. Kantrowitz, A., *Astronautics & Aeronautics (A/A)*, **9**, No.3 ,34-35 (1971).
2. Kantrowitz, A., *Astronautics & Aeronautics (A/A)*, **10**, No. 5, 74 (1972).
3. Pirri, A.N., Monsler, M.J.,and Nebolsine, P.E., *AIAA Journal, 12,* No. 9, 1254-1261 (1974).
4. Weiss, R.F., Pirri, A.N., and Kemp, N.H., *Aeronautics & Astronautics (A/A),* 50-58 (March 1979).
5. Wu, P.K.S., and Pirri, A.N., *AIAA Journal,* **14,** No.3, 390-392 (1976).
6. Legner, H.H. and Douglas-Hamilton, D.H., *Journal of Energy,* **2,** No.2, 85-94 (1978)
7. Jones, W.S., "Laser-Powered Aircraft and Rocket Systems with Laser Eneergy Relay Units", *Radiation Energy Conversion in Space,* Progress in Astronautics and Aeronautics, Vol. 61, pp. 264-270, 1978.
8. Shoji, J.M. and Larson, V.R., "Performance and Heat Transfer Characterisitics of the Laser-Heated Rocket-A Future Space Transportation System" AIAA Preprint 76-1044, International Electric Propulsion Conference, Nov. 1976.
9. Kemp, N.H., Root, R.G., Wu, P.K.S., Caledonia, G.E. and Pirri, A.N., "Laser-Heated Rocket Studies," Physical Sciences Inc. TR-53, NASA CR-135127, May 1976.
10. Nebolsine, P.E., Pirri, A.N., Goela, J.S., Simons, G.A. and Rosen, D.I., "Pulsed Laser Propulsion", Physical Sciences Inc. TR-142, Presented at AIAA Conference on Fluid Dynamics of High Power Lasers, Oct. 1978.
11. Simons, G.A. and Pirri, A.N., *AIAA Journal,15,* No.6, 835-842 (1977).

Laser Launch—The Second Wave

Jordin T. Kare

Kare Technical Consulting, 222 Canyon Lakes Pl., San Ramon CA 94583 jtkare@attglobal.net

ABSTRACT

In the spring of 1986, a Workshop on Laser Propulsion was held to discuss the feasibility of using large free electron lasers to launch payloads into Earth orbit. This workshop kicked off a four-year program, supported by the U.S. Strategic Defense Initiative Organization (SDIO), to develop laser propulsion technology. This talk reviews the concepts addressed by the SDIO Laser Propulsion Program, and the results of both modeling and experiments on double-pulse planar ablative thrusters, which remain a promising approach to laser launch. Other program topics to be discussed include trajectory and system modeling, and air-breathing thruster and vehicle concepts.

BACKGROUND

The concept of laser launch was suggested by Arthur Kantrowitz in 1972 [1] nearly simultaneously with the development of the first high-average-power lasers. Various approaches to laser propulsion and laser launch were explored in the mid to late 1970's. However, by the late 1970's military interest in large lasers had declined, and NASA was deeply committed to the Space Shuttle, so interest in laser propulsion waned.

In the mid-1980's, three factors favored a renewed interest in laser launch: the development of the Free Electron Laser (FEL), and two aspects of the increased political support for ballistic missile defense, as reflected in Ronald Reagan's Strategic Defense Initiative (SDI). One aspect of SDI was increased development of "directed energy" weapons, including ground-based laser weapons, leading to progress in atmospheric correction, large beam directors, and other relevant technologies. The other was support for space-based weaponry, especially space-based kinetic-kill vehicles (SBKKV's), with the implied need to launch large masses of hardware and propellants into Earth orbit at low cost.

CP664, *Beamed Energy Propulsion: First International Symposium on Beamed Energy Propulsion,*
edited by A. V. Pakhomov
© 2003 American Institute of Physics 0-7354-0126-8/03/$20.00

THE 1986 WORKSHOP

In 1986, Lawrence Livermore National Laboratory (LLNL) was developing the Induction Linac Free Electron Laser (ILFEL) for strategic defense applications [2]. Work was underway on a large 10 μm ILFEL demonstration, and proposals were being circulated for building high power lasers (10's to 100's of megawatts) within a few years. At the same time, members of LLNL's O-Group had been exploring possible solutions to the long-standing problem of improving access to space, including laser launch [3]. At the suggestion of Dr. Lowell Wood, a Workshop, jointly sponsored by the Strategic Defense Initiative Organization (SDIO) and the Defense Advanced Research Projects Agency (DARPA), was planned for the summer of 1986. The original Workshop announcement is reproduced in Appendix A.

The express charter of the workshop, and of the subsequent SDIO Laser Propulsion Program, was to develop concepts and technology for near-term ground-to-orbit launch capability. As such, technologies for orbital maneuvering were of secondary interest, and priority was given to approaches likely to scale well for small, near-term launchers.

Invitations were distributed early in 1986 to researchers and organizations known to have an interest in laser propulsion, notably those referenced in [4]. Somewhat optimistically, the workshop was scheduled for two full weeks. Some 30 people from 18 organizations attended the first few days of the Workshop, and presentations were given covering past and current studies of laser propulsion, FEL's and other laser technologies, adaptive optics and beam propagation, and several related topics. The Workshop Proceedings [5] were published as three volumes: Executive Summary, Contributed Papers, and Viewgraphs.

During the second week of the Workshop, a smaller set of dedicated (or less time-pressured) attendees, notably including Prof. Arthur Kantrowitz, reviewed the Workshop discussions and attempted to define key characteristics for a laser launch system. While the Workshop as a whole did not select any preferred technologies, this smaller group did arrive at a set of concepts which became the basis for the SDIO Laser Propulsion Program.

Results of the Workshop

Three primary conclusions were drawn from the 1986 Workshop:

1. Laser launch was potentially feasible in the near term, due to progress in lasers, large optics, and adaptive optics.

2. The most promising technical approach was the double-pulse ablative thruster, originally proposed by Reilly [6].

3. The most promising thruster and vehicle geometry would be a planar thruster occupying the full base area of a conical vehicle.

The double-pulse thrust cycle is illustrated in Figure 1. Preliminary calculations [7] suggested that a double-pulse thruster could achieve a thruster efficiency (ratio of exhaust kinetic energy to total laser pulse energy, $\delta m <v_{exh}>^2/2E$) of 30% or better at a specific impulse of 600 - 1000 seconds.

A) "Metering" pulse evaporates a thin layer of propellant
$$\tau_1 = 2 - 5 \ \mu s$$

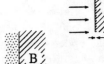

B) Gas expands to the desired density
$$\tau_{1-2} = 0 - 5 \ \mu s$$

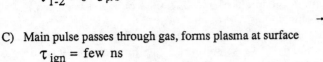

C) Main pulse passes through gas, forms plasma at surface
$$\tau_{ign} = \text{few ns}$$

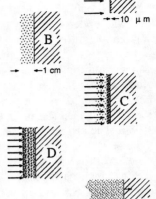

D) Plasma absorbs beam by inverse bremsstrahlung;
absorbing layer (LSD wave) propagates through gas
$$\tau_2 = 1 \ \mu s$$

E) Uniformly hot gas expands in 1-D, producing thrust
$$\tau_{exp} = 3 - 10 \ \mu s$$

F) Exhaust dissipates; cycle repeats at 100 Hz - few kHz

FIGURE 1. Double-pulse thrust cycle

The double pulse thruster was favored over alternatives based on several factors:

- CW plasma thrusters required more complex vehicles, and appeared to scale poorly for launch applications (as opposed to on-orbit maneuvering). At launch-system scales, the thrust-chamber window requirements exceeded known technology. Also, CW thrusters were incompatible with the ILFEL.

- Single-pulse ablation was expected to be inefficient in the range of I_{sp}'s optimum for ground-to-LEO launch, although potentially better at higher I_{sp}'s (>2000 s). Also, efficient single-pulse plasma formation was predicted to require significantly higher flux and fluence than the double-pulse LSD-wave required, forcing the vehicle to carry a beam-concentrating optic.

- More complex technologies (notably the Apollo Lightcraft) required both more development and much more sophisticated infrastructure (multi-gigawatt lasers in orbit) than ablative thrusters.

The planar thruster concept -- using a flat solid propellant surface with no skirt or focusing nozzle, and generating efficient thrust by 1-D expansion of a thin hot gas layer over the entire surface -- appears to have originated at the Workshop. Prior publications (e.g., [8]) assumed the use of an exhaust cone to focus the incoming light and constrain the expanding gases.

The planar thruster offered three key advantages:

- Extreme simplicity

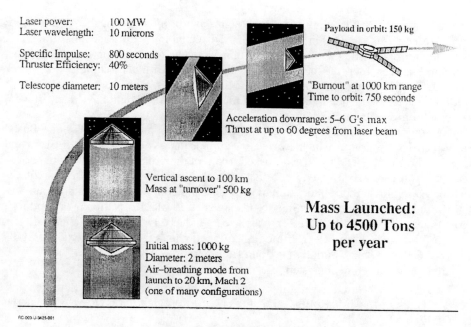

Laser power: 100 MW
Laser wavelength: 10 microns

Specific Impulse: 800 seconds
Thruster Efficiency: 40%

Telescope diameter: 10 meters

Payload in orbit: 150 kg

"Burnout" at 1000 km range
Time to orbit: 750 seconds

Acceleration downrange: 5–6 G's max
Thrust at up to 60 degrees from laser beam

Vertical ascent to 100 km
Mass at "turnover" 500 kg

**Mass Launched:
Up to 4500 Tons
per year**

Initial mass: 1000 kg
Diameter: 2 meters
Air–breathing mode from
launch to 20 km, Mach 2
(one of many configurations)

RC-003-U-3425-001

FIGURE 2. Planar Thruster Vehicle and Launch System Concept

- Acceleration vector independent of the laser beam vector, allowing acceleration at a large angle to the laser beam, and

- Ground-based control: steering the vehicle by changing the laser beam profile

Accelerating at an angle to the beam is of great importance in a practical launcher, since "beam riding" designs constrained to accelerate along the laser beam vector are severely limited in the orbits they can reach.

The double-pulse planar-thruster vehicle concept consists of little more than a block of propellant with a payload on top. A conical shroud permits transverse acceleration without exposing anything except propellant to the laser. Figure 2 illustrates the launch system concept, and conveys the characteristic trajectory and mass scales under consideration.

The simplicity of the planar thruster vehicle, plus the possibility of steering the vehicle from the ground, led Arthur Kantrowitz to formulate his "4P" principle [9]:

"A laser should launch only Payload, Propellant, and Photons, Period."

THE SDIO LASER PROPULSION PROGRAM

Based on the Workshop output, a program was defined to develop technology leading to a high-volume laser launch capability with payloads of 20 - 100 kg in as little as 5 years. At that time, similarly aggressive programs were being proposed, and in some cases funded, in a variety of SDI technology areas, so this timescale, while highly optimistic, was not wholly unreasonable. In addition, SDIO was just beginning to consider the concept later known as Brilliant Pebbles: very large constellations of very small SBKKV's. Brilliant Pebbles, with masses of a few 10's of kilograms and launched in quantities of many thousands, would have been an ideal payload for a laser launch system.

The proposed program was funded by SDIO as a component of the Ground-Based Laser (GBL) program, which was also supporting work on large FEL's and associated technologies. LLNL was asked to direct the program, but requested not to keep the entire effort in-house. To that end, a call for proposals for research was distributed in November 1986. An advisory steering committee for the program was established, consisting of: Dr. Gregory Canavan (Los Alamos National Laboratory), Prof. Freeman Dyson (Institute for Advanced Studies), Dr. Edward Teller (LLNL), and Prof. Arthur Kantrowitz (Dartmouth College), plus a representative of the SDIO Directed Energy Office, initially Dr. John Hammond.

Initially, the program consisted of: the following activities:

- Two theoretical studies of double-pulse propulsion, at Physical Sciences Inc. (PSI) under Mr. Chris Rollins, and at the Naval Research Laboratory (NRL) under Dr. Elaine Oran.

- Three small-scale experimental studies of double-pulse propulsion, using 10 - 100 J lasers with 50-100 ns pulse lengths, at PSI, Avco Everett Research Laboratory (AERL) under Mr. Dennis Reilly, and Spectra Technology, Inc (STI) under Dr. Michael Hale.

- Work at LLNL under Dr. Jordin Kare on air-breathing propulsion and trajectory and system modeling.

- Work at Rensselaer Polytechnic Institute (RPI) under Prof. Leik Myrabo, on a nearer-term, scaled-down version of the multi-propulsive-mode Lightcraft, and

- A computational study of blast wave expansion and thrust generation, at Stanford University under Prof. I-Dee Chang.

As the program evolved, some of these efforts were suspended or modified, and new studies were added:

- Lehigh University (Prof. Yong W. Kim): measuring ablation and plasma initiation characteristics of propellants at 1.06 microns

- University of Washington (Prof. Robert Brooks): High-energy (500 J) single- and double-pulse ablation experiments at 10.6 µm.

1987 Workshop

A second Workshop was held at Los Alamos National Labs in September, 1987. At this workshop, the results of the first year of the SDIO program were presented. A

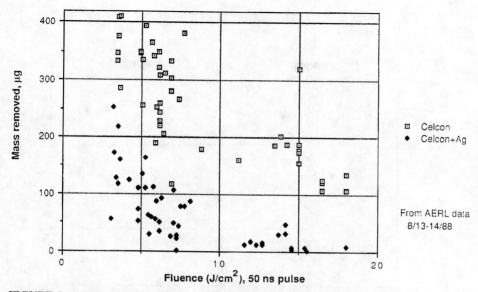

FIGURE 3. Adding approx. 20% silver-coated glass microspheres to Celcon plastic reduces mass loss from a single laser pulse in air by up to 10-fold.

Proceedings volume was published [10] but not widely distributed. The Table of Contents of the 1987 Workshop Proceedings is reproduced in Appendix B.

A major topic at the 1987 Workshop was the need to optimize the double-pulse propellant for multiple characteristics: short absorption depth, prompt plasma ignition, good LSD-wave propagation, and (ideally) low frozen-flow losses. Ignition, propagation, and chemical recombination issues are critical in the range of pulse widths and fluxes appropriate to a launcher; in particular, slight delays in ignition and formation of a surface-shielding plasma could result in large "dribbling losses" -- propellant ablated at low velocity and not contributing significant thrust.

To enhance plasma ignition, PSI [11] suggested embedding sub-wavelength-scale aluminum flakes in the propellant. Reilly [12] proposed creating a "tuned ignition array" propellant, containing embedded half-wavelength-wide metallic stripes.

Delrin plastic [polyacetal, $(CH_2O)_n$] was identified as a good candidate, having strong C-O bond absorption at 10.6 microns (absorption depth later measured at <2.5 µm) and containing only light elements. However, modeling suggested that adding an easily-ionized component to the propellant would improve LSD-wave characteristics. Efforts to add sodium in the form of sodium valerate $(CH_3(CH_2)_3COONa)$ to Delrin were unsuccessful, but PSI succeeded in loading a similar plastic, Celcon, with up to 4% sodium valerate. (It is worth noting that sodium valerate has an extremely strong odor, and is best handled under oil, with disposable gloves!). Also, various forms of wavelength-scale metallic additives were tested, including 5 µm diameter x 2 µm aluminum flakes and silver-coated microspheres, at densities of ~1 per 1000 cubic microns. These were found to be

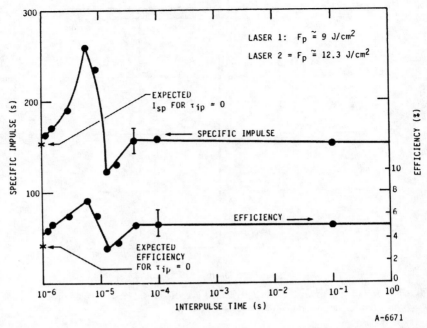

FIGURE 4. Specific impulse and efficiency for double 50 ns pulses on glass, showing enhanced performance at optimum interpulse time.

effective in triggering rapid plasma formation and surface shielding for targets in air, as shown in Figure 3 [13]. Other sodium-loaded propellants were under development when the Program ended; e.g. a promising, but untested, material was a mix of poly(sodium methacrylate) (50%), poly(methyl methacrylate) (30%), carbon powder (15%) for optical absorption, and aluminum flakes (5%).

The 1987 Proceedings contained at least two other interesting firsts: the first explicit observation of enhanced I_{sp} and efficiency using double pulses, from PSI experiments [14], shown in Figure 4, and the first publication of the now-familiar Myrabo spike-tailed vehicle configuration [15].

ScaleUp Experiment

After the 1987 Workshop, the two key priorities for the experimental program were 1) a larger scale experiment, with sufficient pulse length to get past the plasma ignition transient, and sufficient energy to approximate 1-D expansion conditions, and 2) a repetitively-pulsed (rep-pulse) experiment to reach steady-state propellant surface conditions and measure dribbling losses. Unfortunately, no laser suitable for a double-pulse rep-pulse experiment existed, and rebuilding an older Avco CO_2 laser such as HUMDINGER was determined to be impractical. The Program efforts therefore focused on setting up a high-energy double-pulse experiment.

For characteristic LSD-wave propagation velocities of 1 cm/μs, initial estimates were that a true 1-D expansion experiment with a 1 μs main pulse would require a target radius of 20 - 50 cm and a pulse energy of up to 30 kJ -- too much for a practical experiment. By mid-1987, however, it was realized that, because the 2-dimensional loss effects propagate inward as a rarefaction wave from the edge of the target spot, accurate measurements could be obtained over a small area at the center of a relatively small target. The resulting "centerline" experiment was designed to provide accurate impulse data over a ~1 cm^2 area at the center of an 8 cm target.

Even at the lower pulse energy requirement (1-2 kJ) the Program could not locate an available laser with the desired pulse properties. While building a dedicated new laser was considered, the Program accepted an Avco proposal to modify an existing e-beam-pumped excimer laser, the ScaleUp laser, into a 4-cavity 2-kJ CO_2 laser with a ~1 μs pulse length. A smaller CO_2 TEA laser, loaned by NRL, would provide the evaporation pulse.

The ScaleUp experiment was, unfortunately, plagued with problems, starting with the laser cavities themselves: Avco was unable to obtain the 4 10" diameter NaCl windows needed for the ScaleUp conversion. While such windows had been readily available at the peak of CO_2 laser research in the 1970's, two attempts to fabricate windows in 1989 yielded only two usable windows, reducing the ScaleUp output power by half. Other problems included a factor-of-two error in calibrating the pulse energy sensor, discovered late in the program, and the poor condition of the NRL TEA laser. A 100-J Lumonics 602 laser from LLNL was substituted, but lacked the power and pulse length to optimize the evaporation process.

Despite these difficulties, the ScaleUp experiment succeeded in collecting data from 69 shots. Unfortunately, the combination of lower-than-design pulse energy and limited trials led to no unambiguously-successful double-pulse shots.

One exceptionally high-performance result was obtained using Celcon targets in air: ~130 dyne-s/J with an apparent efficiency of ~500% (~800 s I_{sp}). This result was apparently due to combustion of Celcon breakdown products in air, as tests in a nitrogen atmosphere gave ~25 dyne-s/J and ~30% efficiency.

(There is also evidence in the ScaleUp data for a significant target mass loss delayed approximately 100 us from the laser pulse. This mass loss was noted but not studied in detail at the time. It may be similar to the delayed mass loss observed by Pakhomov in recent short-pulse experiments [16], believed to be due to superheating of a liquid surface layer.)

Other Activities

Recognizing that a ground-launched vehicle would expend considerable time and energy traveling through the atmosphere, the SDIO Program supported two investigations of air-breathing propulsion. RPI, as noted above, pursued the design of a focusing-nozzle air-breathing vehicle, which has since been successfully flight tested at the 10-kW scale and widely reported upon elsewhere.

LLNL [17] demonstrated a unique approach to creating controlled air breakdown over large areas, using a dimpled plate: a reflective surface (copper plate) machined with a spherical tool into a pattern of overlapping concave reflectors, typically a few mm in diameter and focal length. Each dimple created a breakdown point above the reflector surface, with the resulting plasmas expanding and (ideally) merging to form a continuous LSD wave, shielding the surface from laser damage. Experiments showed that dimpled plates did generate plasma as expected, and produced respectable coupling coefficients, but the individual plasmas did not merge into a uniform LSD wave in the ~100 ns pulse duration available.

The Stanford group modeled the effect of the vehicle base area and shape on delivered impulse [18]. For nominally-planar thrusters, thruster efficiency falls off if the detonation-wave thickness is comparable to the thruster diameter. However, even a short, large-opening-angle "skirt" can substantially increase the delivered impulse and thruster efficiency, without interfering with the incoming beam.

Finally, Sandia Laboratories [19] explored the potential synergy between laser launch and nuclear reactor-pumped lasers. Reactor-pumped lasers could potentially be more compact and cheaper to build and operate than electrically-pumped lasers plus the required electric generating and transmission infrastructure.

Additional laser propulsion workshops were held, at Stanford University (1988), Lehigh University (1989), Physical Sciences, Inc. (1990) and at Dartmouth College (1991), but formal proceedings for those workshops (which were primarily program reviews) were not published, and as of this writing, copies of presentations and submitted papers could not be located.

The Air Force Office of Scientific Research (AFOSR) developed an interest in laser propulsion paralleling that of SDIO, and an AFOSR Workshop was organized in 1988 by Dr. Mitat Birkan [20], but the Air Force did not at that time fund a research program.

The End of the Program

The SDIO Laser Propulsion Program took an abrupt turn with the 1989 decision [21] of the SDIO GBL Program to pursue the RF-Linac FEL proposed by Los Alamos and Boeing, rather than the ILFEL proposed by LLNL. The RF-Linac FEL pulse format (a continuous chain of picosecond-length micropulses at gigahertz rates) was wholly incompatible with pulsed ablative thrusters. The Laser Propulsion Program was thus left with no prospect of a usable demonstration laser being funded by any other program, and, given the difficulties of the ScaleUp conversion, little confidence in the prospects for large CO_2 lasers..

One route out of this impasse was to make a high performance thruster that would retain the desirable properties of the double-pulse planar thruster, but work with the CW beam format of the RF-linac FEL. This line of attack led, in the summer of 1990, to the invention of a new approach to laser launch: the heat exchanger (HX) thruster [22, 23]. The HX thruster uses a flat, solid heat exchanger to heat hydrogen propellant, and would thus work with almost any type of laser.

The HX thruster concept arrived too late, however, to affect the fate of the SDIO Laser Propulsion Program. Due to reductions in the SDIO Directed Energy budget, no funding was made available to the Laser Propulsion Program for Fiscal Year 1991; the entire SDIO ground-based laser effort was terminated shortly thereafter. Funding for laser propulsion returned to the modest level available through NASA and Air Force advanced technology efforts.

RESULTS OF THE SDIO PROGRAM

The most significant results of the SDIO Laser Propulsion Program can be summarized as follows:

1) Double pulse thrusters have a theoretical and computer-modeled efficiency of approximately 30% for the I_{sp} range of 600 - 1000 seconds.

Note electric propulsion 50%

2) Double pulses can significantly reduce mass ablation and increase I_{sp} and efficiency relative to equivalent total flux and fluence in a single pulse; however, demonstrated I_{sp}'s and efficiencies were well below the goals of 600-800 s and 25%.
PSI demonstrated 300 s and 8-10% efficiency, while the Avco ScaleUp experiment demonstrated only ~220 s but 20-30% efficiency. However, the ScaleUp experiments did demonstrate ~80% shielding of the target from the second pulse energy
Spectra Technology, working with LiH propellant observed ~800 s I_{sp} at 10% efficiency, but with a limited data set. They also observed the onset of LSD-wave formation (a sharp increase in ablated gas velocity) in LiH and Delrin, using a Schlieren technique.

3) Ignition thresholds for surface plasmas can be lowered significantly by appropriate seeding of dielectric propellant with metallic structures.

4) Dribbling losses may be significant for ablative thrusters in steady-state repetitive-pulse operation at plasma temperatures above approximately 10,000 K.

5) If double pulses are used, rapid plasma ignition is critical; a steep leading-edge spike on the second pulse is desirable, as is a sharp trailing-edge cutoff to minimize dribbling losses.

Lessons Learned

As an effort to develop technology for laser launch, the SDIO Laser Propulsion Program had mixed success. Despite encouraging theoretical results and some suggestive experimental data, it failed to conclusively demonstrate the expected performance of the double-pulse thruster.
On the other hand, the Program has left several valuable legacies. The current Myrabo lightcraft design derives directly from work done under the SDIO Program, as does the choice of Delrin as an ablative propellant. The HX thruster remains a viable alternative approach to laser launch.

Relatively little of the SDIO program's work was published in refereed journals. This has led, inevitably, to some valuable results being effectively lost. Several of the concepts and techniques explored, including seeded propellants and centerline impulse measurements, have not been further pursued. Many of the concepts and calculations developed at the 1986 workshop and the subsequent SDIO program have also been repeated by later investigators, for example the calculation of thruster and trajectory efficiency and optimum I_{sp} by Kare [24] echoed by Phipps et al. [25].

From a programmatic point of view, the SDIO Program deliberately tried to create a community of laser propulsion researchers broad and diverse enough to support a large development effort. Unfortunately, when that large effort failed to materialize, the community proved to be broad, but insufficiently deep; except for Prof. Myrabo at RPI no one organization possessed a critical mass of researchers or an organizational commitment to laser propulsion, and the Program, regrettably, left no significant legacy of research teams or funding channels.

Perhaps the most fundamental lesson to be learned from the SDIO laser propulsion program, though, is that laser launch is still a stepchild of both high-average-power lasers and large-scale activity in space. As such, it may need to wait for the next generation of interest in both its step-parents before it achieves its place in the Sun.

REFERENCES

NOTE: Copies of the 1986 and 1987 Workshop proceedings are no longer available through Lawrence Livermore National Laboratory. Requests for copies may be directed to the author at jtkare@attglobal.net

1. Kantrowitz, A., "Propulsion to Orbit by Ground-Based Lasers," *Astronautics and Aeronautics* **10**(5) 74 (1972).
2. Briggs, R. J., "Free Electron Laser Work at LLNL," in *Proc. SDIO/DARPA Workshop on Laser Propulsion, 7-18 July 1986*, J. T. Kare, ed., LLNL CONF-860778, Vol. 3, pp. 23-41, LLNL 1987.
3. Wood, L. and Hyde, R., "Science and Technology In Space During The Coming Decade," invited presentation at the U.S. Space Foundation Symposium, *Space, The Next Ten Years* (Colorado Springs, CO, 26-28 Nov. 1984). LLNL UCRL-91783 (1984).
4. Glumb, R. J. and Krier, H., "Concepts and Status of Laser-Supported Rocket Propulsion," *Journal of Spacecraft and Rockets*, January, 1984, 70-79 (1984).
5. *Proc. SDIO/DARPA Workshop on Laser Propulsion, 7-18 July 1986*, J.T. Kare, ed., LLNL CONF-860778, Vol. 1-3, LLNL 1987.
6. Douglas-Hamilton, D. H., Kantrowitz, A., and Reilly, D. A., "Laser Assisted Propulsion Research," in *Progress in Astronautics and Aeronautics*, Vol. 61, ed. by K. W. Billman, AIAA New York, 1978, pp. 271-278.
7. R. A. Hyde, "One-Dimensional Modeling of a Two-Pulse LSD Thruster," in *Proc. SDIO/DARPA Workshop on Laser Propulsion, 7-18 July 1986*, J. T. Kare, Ed., LLNL CONF-860778 LLNL 1987, Vol. 2, pp. 79-88.
8. Chapman, P. K., Douglas-Hamilton, D. H., and Reilly, D. A., "Investigation of Laser Propulsion," Vol. II, Avco Everett Research Laboratory, DARPA Order 3138, (Nov. 1977).
9. Kantrowitz, A., private communication (1986).
10. *Proceedings, 1987 SDIO Workshop on Laser Propulsion*, J. T. Kare, ed., LLNL CONF-8710452, LLNL 1990.

11. Rollins, C., Bailey, A., Gelb, A., Rosen, D., Weyl, G., and Wu, P., "Issues in Laser Propulsion," in *Proc. 1987 SDIO Workshop on Laser Propulsion*, J. T. Kare, ed., LLNL CONF-8710452, 57-102 LLNL 1990, pp. 57-102.

12. Reilly, D. A., "Advanced Propellants for Laser Propulsion," in *Proc. 1987 SDIO Workshop on Laser Propulsion*, J. T. Kare, ed., LLNL CONF-8710452, LLNL 1990, pp. 145-156.

13. Reilly, D. A., "Laser Propulsion Experiments, Final Report," Avco Research Laboratory Inc., Everett, MA, USA, 1991.

14. Rollins, C., et al. *op cit.* (1990), pp. 95-98.

15. Myrabo, L. M., "Airbreathing Laser Propulsion for Transatmospheric Vehicles," in *Proc. 1987 SDIO Workshop on Laser Propulsion*, J. T. Kare, ed., LLNL CONF-8710452, LLNL 1990, pp. 173-209.

16 Pakhomov, A. V., Thompson, M. S., and Gregory, D. A., "Ablative Laser Propulsion Efficiency," AIAA 2002-2157, presented at 33rd AIAA Plasmadynamics and Lasers Conference, Maui, HI, USA, 2002.

17 Kare, J. T., "Laser Supported Detonation Waves and Pulsed Laser Propulsion", J. T. Kare, **Current Topics in Shock Waves**, Y. W. Kim, ed., Amer. Inst. of Physics Conference Proceedings 208 (17th Int. Symp. on Shock Waves and Shock Tubes, Lehigh U., Bethlehem, PA, July 1989), AIP, 1990. (Also UCRL 101677, LLNL, 1989.)

18. Chang, I. and Mulroy, J. R., "The Effect of Lateral Propellant Expansion on the Impulse Received by a Pulsed Laser Thruster," Final Report, LLNL Subcontract 1871603, Dept. of Aeronautics and Astronautics, Standford University. (See also, Mulroy, J. R., "Numerical Investigation Of The Thrust Efficiency Of A Laser Propelled Vehicle," Ph.D. Thesis, Stanford University Dept. of Aeronautics and Astronautics, 1990.)

19 Lawrence, R. J., Kare, J. T, Monroe, D. K., and Zaworsky, R. M., "System Requirements for Low-Earth-Orbit Launch Using Laser Propulsion", Proc. 6th International Conference on Emerging Nuclear Energy Systems, Monterey, CA, USA June 1991. (Also SAND 91-1687C, Sandia National Laboratory, 1991.)

20 Proceedings of the AFOSR Workshop on Laser Propulsion, M. Birkan, ed., AFOSR-TR-88-1430, February 1988.

21. Gilmartin, P., "Boeing Aerospace Wins Strategic Defense Initiative Contract for RF-Driven Free Electron Laser," in *Aviation Week and Space Technology*, **131**(17), 21 (1989).

22. Kare, J.T., "Laser Powered Heat Exchanger Rocket for Ground-to-Orbit Launch," *J. Propulsion and Power* **11**(3), 535-543 (1995).

23. Kare, J. T., "Near-Term Laser Launch Capability: The Heat Exchanger Thruster," in these proceedings.

24. Kare, J. T., "Trajectory Simulation for Laser Launching," in *Proc. SDIO/DARPA Workshop on Laser Propulsion, 7-18 July 1986*, J. T. Kare, Ed., LLNL CONF-860778, LLNL 1987, Vol. 2, pp. 61-78.

25. Phipps, C. R., Reilly, J. P., and Campbell, J. W., "Optimum Parameters for Laser-Launching Objects Into Low Earth Orbit," *Lasers and Particle Beams*, **18**(1) 1-35 (2001).

ANNOUNCING A

LASER PROPULSION WORKSHOP
7 JULY - 18 JULY 1986
At The
LAWRENCE LIVERMORE NATIONAL LABORATORY
Livermore, California

" ... to extend understanding of the key technical issues in the design of a laser-energized, ground launch-to-orbit system, and specifically to document answers to three questions:

- *What are the potential capabilities of ground-based lasers being developed by SDIO for launching interesting payloads into various Earth orbits?*

- *What R&D must be done to validate this capability, and what is a reasonable program for accomplishing it?*

- *If relatively near-term ground-based laser systems can provide useful launch capability, what steps should be taken, in the creation of these and successor systems, to maximize this capability?"*

Co-Sponsored By

The Strategic Defense Initiative Organization
And

The Defense Advanced Research Projects Agency
U.S. Department of Defense

All U.S. citizens having pertinent technical background and interest are invited to contact Dr. Jordin T. Kare, Workshop Coordinator, at (415)423-8300 regarding attendance. This is a Gordon Research Conference-format Workshop: attendance is by invitation only and all attendees are expected to contribute actively, in prepared presentations and in technical discussions, as well as in the documentation of Workshop Proceedings, which will be published. Some Workshop sessions will require DoD SECRET clearances to attend.

APPENDIX B
1987 Workshop Proceedings Table of Contents

Contents

APPENDIX C
PSI Single Pulse Data

In the course of an unrelated program (the Hardening Data Base Program), PSI in 1988 performed repetitive single-pulse ablation tests on carbon-carbon materials, using the NRL REP III laser. A few of these tests had high enough flux to reliably light plasmas at the target. In PSI's final Laser Propulsion Program report [1], these results were evaluated in laser propulsion terms, and we reproduce them here because the calculated performance (shown in Table C1) was remarkably good.

TABLEC1. PSI HDBP RP-3 Test Results [2] 4/17/88

Run No.	Sample	No. of Pulses	J/Pulse	Q* kJ/gm	Impulse dyne-s	Isp sec	Efficiency
409	CB-07-18	100	192	548	692	2025	0.359
410	CB-10-17	100	183	383	722	1545	0.598
411	CB-06-19	79	208	675	811	2680	0.511
Pulse 10.6 μm, 25 μs, PRF 105 Hz, spot size 1.407 cm^2,							

These results may have been affected by the test chamber pressure, which was at <1 Torr but not a good vacuum. Very small amounts of background gas entrained in the laser-produced plume could increase the apparent delivered impulse and Isp, and even more strongly affect the apparent efficiency, which varies as (impulse)^2 for fixed propellant mass. However, the results are consistent with PSI's modeling of single-pulse effects. Repetition of these tests would be very valuable, since if the results were confirmed, a single-pulse thruster using carbon-carbon could be very attractive for ground-to-space launch, especially for high-delta-V launches to geosynchronous transfer orbits or Earth escape.

1. Bailey, A., Lo, E., Gelb, A, Goldey, C., Miller, M., Pirri, A., Pugh, E., Rollins, C., Rosen, D., and Weyl, G., "Laser Propulsion Activities: Final Report," LLNL Subcontract Number B076277, PSI-1076/TR-1123, Physical Sciences Inc., Andover, MA, June 1991

2. Pugh, E., et al., "Hardening Data Base Program Final Report Volume II: Repetitively Pulsed Laser Studies Part II: Technical Report," Contract No. F33615-83-C-5074, Wright Research and Development Laboratories, Materials Laboratory Rpt. Doc. No. 8400-89-0015, April 1989.

Ground to Space Laser Power Beaming: Missions, Technologies, and Economic Advantages

John D. G. Rather

University Professor of Physics, Wayne State University, Detroit, MI 48202, USA

Abstract. Systems studies initiated by NASA HQ in 1991 confirmed that important new missions can be implemented by ground-to-space laser power beaming. Large near-term space development opportunities can be enabled by fast, efficient and relatively inexpensive transportation of massive payloads from low earth orbit to any location in cislunar space. Early development of a permanent lunar base would be made feasible by this capability. Near-earth objects up to 20 meters in diameter could be captured and maneuvered into useful orbits. All key technology requirements were quantified by our studies, and several research and development efforts were funded. The present paper will review the overall rationale, discuss tradeoff issues and quantify system requirements.

INTRODUCTION

Large scale development of space is severely curtailed by two fundamental problems: (1) the very high cost, low efficiency, and low versatility of space transportation systems, and (2) the high cost and low efficiency of systems for providing large amounts of electric power in space. Indeed, the demise of NASA's hopes for returning to the moon "to stay" and proceeding with manned exploration of Mars is directly traceable to extremely high costs derived from "business as usual" in space transportation and power. The clear need for radical innovations led me create a technology development effort at NASA HQ[*] in 1991 that sought new ways to attack these problems. Quite a lot of progress was made in the subsequent decade, much of it accomplished by people attending this meeting. I would be remiss not to acknowledge particularly the many important activities of Hal Bennett, Sandy Montgomery and Glenn Zeiders, and very significant contributions from a host of other collaborators too numerous to mention.

It is well known that electromagnetic power can be beamed losslessly through space, efficiency limitations being associated only with the generation, beam control, and reception processes. Much has been written about concepts for microwave power beaming, including the very extensive literature on Space Power Satellites. Unfortunately, however, basic physics impinges heavily upon the feasibility of long distance power transmission because of practical limits upon the transmitter and

---

[*] The author was Assistant Director for Space Technology (Program Development), (1990-95); and Manager, Advanced Concepts Systems Integration, (1995-'97).

CP664, *Beamed Energy Propulsion: First International Symposium on Beamed Energy Propulsion*, edited by A. V. Pakhomov

receiver diameters. The wavelength of the emissions is the critical parameter. For example, if a microwave system operates at a wavelength of 10 centimeters, beaming power with low loss over a range of 40,000 kilometers from geostationary earth orbit (GEO) to Earth (or *vice versa*) will require transmitters and receivers having diameters approximately 2 kilometers in diameter. (Reciprocal tradeoffs can be made, but the aperture product remains very large.) Alternatively, if a laser system operates at a near-infrared wavelength of 1 micrometer, the required telescope diameters for low-loss power transmission at GEO range are less than 10 meters. Thus the mass in space of the necessary hardware will be the order of a million times less for a laser system than for a microwave system intended to perform the equivalent function. Figure 1 shows receiver size versus range for a variety of wavelengths, where the laser transmitter beam expander is assumed to have a diameter of 12 meters.

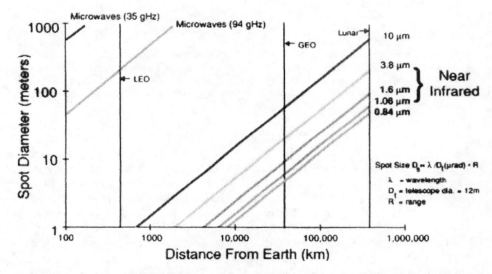

FIGURE 1. A 12meter diameter ground-based beam expander can deliver 0.84μm wavelength laser power to a receiver 7 meters in diameter at Geosynchronous Earth Orbit (GEO) or 80 meters in diameter on the moon.

The advantages of laser power transmission are particularly notable in the Earth-to-space case, where the laser, its prime power source, and all supporting complications remain on the ground in a "hands-on" environment. In our efforts, NASA Marshall Space Flight Center, supported by Lewis (now Glenn) Research Center, Jet Propulsion Laboratory, the U.S. Navy China Lake Laboratory, and several private industry participants, conducted detailed systems studies to determine the potential mission capabilities of such systems. These studies validated that ground-to-space laser power beaming can engender large cost reductions and much enhanced performance over existing systems and methods. I will discuss some interesting examples below.

To be reasonably credible about costs and time required to achieve initial operational capability, it is important to quantify the magnitude of the contemplated endeavors. Back in 1979, I created the chart in Figure 2 to show the vast range of parameters associated with many conceptual laser applications, some of which will be debated in the present symposium. (This chart was prepared for the U.S. Senate laser hearings in 1979-'80 that I instigated with the interest and encouragement of the great Senator from Alabama, Howell Heflin.) The important lesson from Figure 2 is that orbit-raising and maneuvering are orders of magnitude easier to accomplish than earth to space propulsion. The relatively modest technologies supporting in-space propulsion can become near-term catalysts to bootstrap much grander-scale future capabilities in space. Therefore I will focus upon practical limitations, especially in the context of presently reachable technologies.

By far, the two most prominent technical challenges are economical construction and operation at suitable ground sites of (1) adaptive optical transmitter telescopes with diameters up to 12 meters, and (2) continuously operating lasers producing up to 10 megawatts output power (either pulsed or CW) at near infrared wavelengths. A powerful lesson learned from nearly forty years of Department of Defense (DoD) laser system development programs is that costs and difficulties exponentiate if the existing state-of-the-art is pushed too far beyond levels of reasonable scalability. Accordingly, we undertook in 1991 an R&D effort to develop relatively low cost, highly reliable new technologies. The work proceeded under three acronyms: SELENE (Space Laser Energy) which addressed overall system optimization, PAMELA (Phased Array Mirror, Extendable Large Aperture) which addressed the design of high power adaptive optics telescopes, and NAOMI (National Advanced Optics Mission Initiative) which sought to initiate site development at the U.S. Navy China Lake test range. I will now describe key findings of each of these efforts.

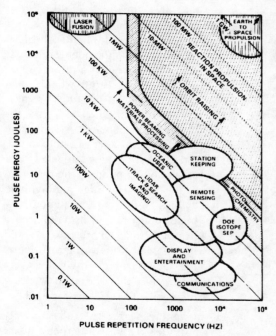

FIGURE 2. How laser characteristics combine to satisfy various applications. Average powers are indicated on diagonal lines. 1-10 MW enables many missions.

SELENE MISSIONS AND SYSTEMS

Both civilian and defense space activities would benefit greatly from enhanced electric power and propulsion. Since the present symposium is focused upon propulsion, the material below will skip discussion of many of the electric power applications considered in the SELENE effort. The important hybrid application of laser-electric plasma propulsion, however, drives technology development to levels that are quite adequate for many space-electric needs. Laser-electric and laser thermal propulsion would both be made highly feasible by a SELENE system.

Propulsion Thrust and Power Requirements

For the past forty years, orbit changing and maneuvering in space have depended entirely upon either chemical reaction propulsion or low-thrust ion propulsion. Moderately high thrust solar-thermal or solar-electric propulsion are technically feasible, but the modest cost-efficiency benefits have not warranted development of these options. A significant problem with solar energy as a prime source is that the spacecraft would be in eclipse much of the time when in low earth orbit (LEO). Also, for high performance missions, the solar concentrators or photovoltaic arrays must be quite large and must continuously point in the direction of the sun. The huge photovoltaic "wings" on the Space Station now under construction provide a case in point: much more power could be obtained from a single 10 meter diameter laser photovoltaic array than from the entire solar array complex. The reasons for the laser

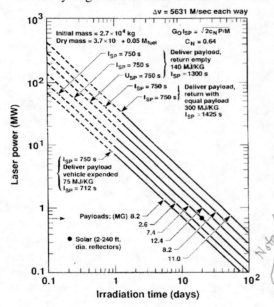

FIGURE 3. Power requirement for payload delivery from LEO to GEO. Laser powers of 1 to 10 MW could perform major missions quickly.

advantages are that the sun only provides 1290 watts per square meter of broad-band irradiation, while a laser can supply ten or twenty "suns" of narrow-band energy tailored to minimize power management and overall system weight. Thus a solar energized propulsion system would achieve much less performance per pound of system weight, and the mission performance would be proportionately reduced. Figure 3 compares performance for various examples of LEO to geosynchronous earth orbit (GEO) transfer missions with specific impulse optimizations from 750 to 1425 seconds.

Grant Logan at Lawrence Livermore National Laboratory did a comprehensive system optimization study that emphasized the case for specific impulses (I_{sp}) in the range from 750-2500 seconds. These values are well matched to the demonstrated performance of electrodynamic plasma thrusters. Figure 4 shows the reason for this performance optimization: Ion propulsion requires much more power per unit thrust although fuel consumption is very low. Chemical rockets, on the other hand, require relatively huge amounts of fuel.

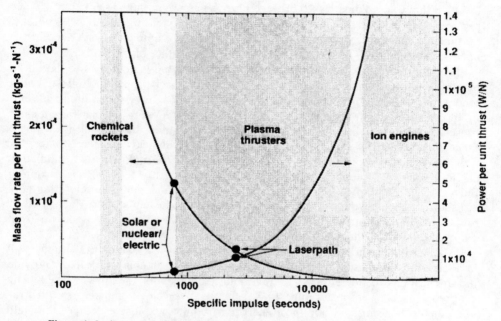

Figure 4. Optimum Specific impulse for laser missions ranges from 750 – 2500 seconds

Laser Propelled Tugboat for Orbital Maneuvering

A particularly interesting family of missions could be performed by an I_{sp}=1500 "tugboat" having a dry mass of 4000 kg and an equal fuel mass tank volume. Every Space Shuttle mission reaches LEO with about 4000 kg of residual hydrogen remaining in the main fuel tank. By utilizing that fuel with 10 MW of laser power input, the entire 30,000 kg tank could be moved to GEO in five days, thus salvaging this valuable, air-tight resource that is wasted back into the earth's atmosphere on every current Shuttle flight. *The salvaged tanks in high orbit could become the basis for many valuable things including a manned station in GEO, a circulating manned "hotel" from LEO to low lunar orbit (LLO), shielding against ionizing radiation and meteoroids, a near-term defense against asteroid impacts, and even a plausible link to manned exploration of Mars. An essentially free, high quality mass accumulation of 180,000 kg per year (current Space Shuttle average) is not to be taken lightly (pun intended) as a major space development opportunity!*

Capturing Small Near-Earth Objects

When I chaired the Near Earth Object (NEO) Impact Mitigation Study for the U.S. Congress in 1992, I was impressed by the data indicating a very large flux of asteroids 10 to 20 meters in diameter. A sphere having the same radius as the moon's orbit will have approximately 20 of these objects passing through at any time. (Once per year there is an impact somewhere on Earth yielding about 10 kilotons. Fortunately, most of these small impacting objects are completely vaporized in the upper atmosphere with little collateral damage on the surface of the Earth.)

It is interesting that some NEO's have quite low relative velocities to the Earth, making capture maneuvers a real possibility. A laser-propelled tugboat could rendezvous with an object of opportunity and impart enough ΔV to put it into a useful orbit. The remarkable thing about ground-to-space laser energy is that megatons of useful momentum transfer can be accomplished by running flat-out for a few days! Since an estimated 3% of the NEO population will contain heavy metals, 20% ices for drinking water and reaction mass, and the remaining 77% rocky debris for radiation shielding and other applications, the prospect of using these resources is exciting indeed. I particularly like the idea of large and comfortable "hotels" routinely transporting people from near the Earth to the moon and eventually to Mars. Lots of radiation shielding and useful natural resources change the whole scenario from cramped misery to hedonistic comfort!

Another important point is that the ability to maneuver a small NEO into the path of a large incoming asteroid or comet threatening the Earth could very likely provide the simplest possible defense. A useful number to ponder is 3.5 km/sec. The kinetic energy released by any mass impacting at this relative velocity happens to equal the yield of the same mass of the most energetic chemical explosives. Since the yield increases as the square of the relative collision velocity, the disruptive energy imparted to fast incoming objects by a planned collision can be tremendous.

Building the Initial Lunar Base

A mission extensively studied in the SELENE effort is the construction of a permanent lunar base. A major problem with the moon is that its polar axis is tipped only five degrees with respect to the ecliptic pole. The result is that practically every location on the moon endures about two weeks of day temperatures above 100 C followed by two weeks of night below −100 C. Hence, solar cells with massive recyclable fuel cell energy storage or nuclear reactors with massive heat radiators have been the only options considered for large-scale lunar development. With present transportation costs to the lunar surface approaching a million dollars per kilogram, the costs of developing such a base are unthinkable.

Since the moon always presents the same face to the Earth, beaming laser power from the Earth provides a remarkable new option. A 1-watt beacon on the moon would provide the reference signal to ensure that the 80 meter diameter laser spot remains stably focused on the low-mass photovoltaic array. Only a small amount of energy storage would be needed for contingencies when one of several Earth-based transmitters might be out of commission. Figure 5 compares solar/fuel cell, nuclear, and laser systems.

As discussed above, the same laser system would also power the tugboat fleet providing logistics from low earth orbit to low lunar orbit. The final leg from LLO to the lunar surface would be handled by a shuttle craft fueled by hydrogen and oxygen recovered from the regolith. One megawatt of electric power could recover per year of operation more than 250 tonnes of oxygen, 50 tonnes of nitrogen and 10 tonnes of hydrogen from the lunar soil.

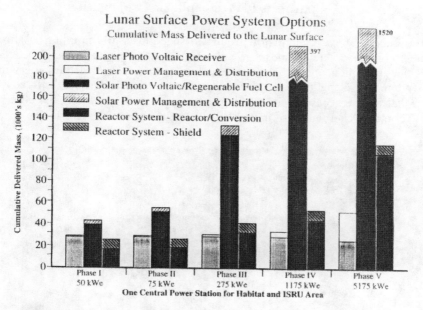

FIGURE 5. Comparison of solar / fuel cell, nuclear, and laser power systems on the moon.

PAMELA ADAPTIVE OPTICS

New technologies are the key to success. For any kind of laser propulsion to be feasible, exquisitely precise beam control is the *sine qua non*. The only way to achieve this is with adaptive optics, because atmospheric disturbances, thermal distortions induced by the laser, wind, and gimbal pointing uncertainties render active wavefront control the only reasonable option. In the 1980's, I proposed a demonstration project to prove the feasibility of precise wavefront control pointing and tracking. It grew into a huge project of the Strategic Defense Initiative Organization (SDIO) known as the STARLAB Wavefront Control Experiment (WCE). Although the space components were never launched, precise hardware tests in the lab demonstrated WCE success: pointing error <50 nanoradians r.m.s. was achieved.

It was clear from the WCE work that innovations were needed to reduce the cost of large laser beam projector telescopes while increasing their reliability. This led my team to create the PAMELA project, which was initiated by SDIO and ultimately taken over by NASA at Marshall Space Flight Center (MSFC). Impressive results were achieved at MSFC. To understand the properties of PAMELA, it is necessary to consider how several different types of constraints affect the design.

Constraints of Atmospheric Propagation

The choice of wavelength for ground-to-space laser transmission is strongly affected by three different atmospheric physical phenomena, *viz* Rayleigh scattering, molecular absorption, and turbulence scale. Figure 6 shows how scattering is a major loss factor at wavelengths shorter than 0.7 micrometer, particularly at the large zenith angles that will be necessary for tracking space objects. Strong molecular absorption bands form a "picket fence" of unusable wavelengths at wavelengths longer than 0.7 μm.

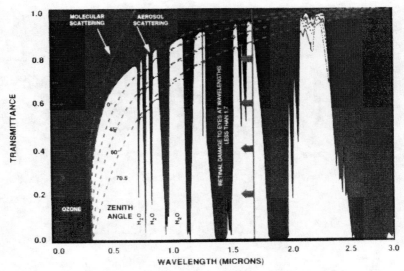

FIGURE 6. Atmospheric transmittance to space with Laser located at 2 km above sea level.

Figure 7 gives typical turbulence scales for five wavelength examples: The coherence scale, r_0, can be thought of as referred to the size of a patch on the laser projector primary mirror wherein phase-coherent wavefront corrections can be accomplished if the adaptive optics can resolve that size. It can be seen from Figure 7 how, even for wavelengths longer than 1.0 μm, large zenith angles drive the scale below 4 centimeters. This means that a 12 meter diameter projector must have adaptive optics capable of precisely controlling the phase of about 200,000 patches on the primary aperture.

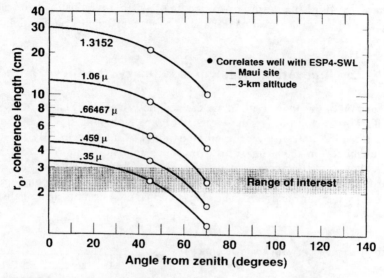

FIGURE 7. Turbulence coherence length as a function of wavelength and zenith angle.

Constraints of Beam Control

The most fundamental requirement for any optical system is to approach as closely as possible to diffraction limited performance permitted by basic physics. Success is usually expressed by the Strehl ratio, which remains favorable (>0.8) if total wavefront errors over the beam do not exceed ~0.1 wavelength. To accomplish this, a beacon near the receiver in space must send down a low power (~1 watt) reference beam to a wavefront analyzer that senses the entire optical path through the atmosphere and telescope optics. Then the adaptive elements must be adjusted in a millisecond to the exact positions that will cancel all disturbances to the outgoing beam so that it will arrive at the receiver as near to the diffraction limited spot as possible. The resulting control problem for the adaptive surface is profound but not insurmountable for modern digital technology. Each adaptive element must be positioned to an accuracy of ~50 nanometers r.m.s. over a range of ~5 mm at a frequency >100 Hz. This performance will enable successful operation of the power beaming system described below.

Constraints upon Beam Expander Telescope Design

The beam expander telescope diameter optimizes at 12 meters for two principal reasons: it is large enough so that the laser power within the telescope is distributed over areas sufficiently large that the optical elements will not be destroyed, and it is also large enough to deliver a near-diffraction-limited spot less than 10 meters in diameter to GEO range (40,000 km). Conversely, it is small enough that its cost and maintenance are affordable. A laser wavelength of ~1 μm is compatible with all of the foregoing constraints and is also well matched to high efficiency photovoltaic power conversion.

Key Innovation: Segmented Primary Adaptive Optics

To satisfy all of the above criteria, we created the PAMELA (Phased Array Mirror, Extendable Large Aperture technologies. Figure 8 depicts how the concept is optimized to address all of the coupled constraints of the system. By locating the adaptive elements on the primary mirror, some amazing simplifications become feasible. Notably, the telescope structure can become lighter weight and lower cost – more like a radio telescope than an optical telescope – because the structural errors can be corrected by the same adaptive control system that corrects atmospheric distortions and pointing errors.

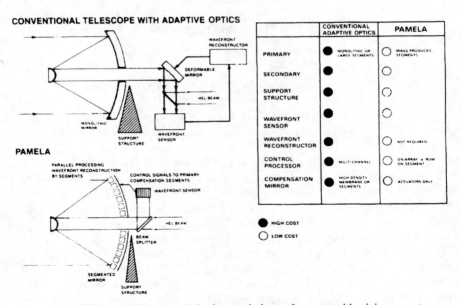

FIGURE 8. PAMELA is optimized to maximize performance with minimum cost.

NASA Marshall Space Flight Center carried PAMELA segment development through three generations as shown in Figure 9a. The integrated test telescope that successfully proved the PAMELA principle at MSFC is shown in Figure 9b.

FIGURE 9. (a) Three generations of segment development. **(b)** PAMELA Proof of principle segmented telescope.

NAOMI

In 1995, a collaborative effort between NASA and the U.S. Navy was undertaken under the acronym NAOMI (National Advanced Optics Mission Initiative.) Considerable progress was made toward actual design of a ground-based power beaming site at the U.S. Navy's China Lake test range. Dr. Hal Bennett, former Director of optical research at the China Lake laboratory and former President of the SPIE, has continued to pursue the SELENE / NAOMI objectives in his private company, Bennett Optical Research since his retirement.

Advantages of Free Electron Lasers

Free Electron Lasers are likely to be the lasers most suitable for sustained high power operation. Different designs can yield very high peak power pulses suitable for thermal reaction propulsion, or modest microplulses at very high rep rates suitable for electrodynamic plasma propulsion. FEL's can operate efficiently at any chosen wavelength suitable for SELENE, particularly in the near infrared. They can produce light with diffraction limited quality. Great progress has been made in FEL design, particularly by the innovative group at the Budker Institute of Nuclear Physics in Novosibirsk, Russia. A 200 kilowatt design exists, and a test machine may already be operating. Russia hosted a meeting in 1993 to try to initiate collaboration with the U.S. Subsequently, in 1996 a Russian delegation visited China Lake and the Lawrence Berkeley Laboratory. Although the door remains open, it has not been possible to obtain funding in the U.S. for this important joint effort.

Birchum Mesa Site Study

A very detailed site study was performed at China Lake, establishing Birchum Mesa as an ideal location. It has an average of 350 days per year of acceptable weather, a high site with good access, cooling water and electric power availability, and no danger of eye damage from adverse beam scattering. Preliminary engineering studies showed that two high power FEL's with 2 Km beam expander vacuum pipes could easily be accommodated.

EFFICIENCY AND ECONOMICS

Marshall Space Flight Center (MSFC) performed a comprehensive cost-benefit study of SELENE. Figure 10 summarizes some exemplary findings of the study.

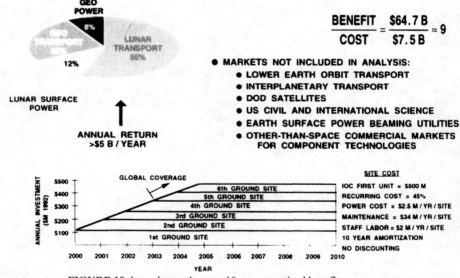

$$\frac{\text{BENEFIT}}{\text{COST}} = \frac{\$64.7 \text{ B}}{\$7.5 \text{ B}} \approx 9$$

- MARKETS NOT INCLUDED IN ANALYSIS:
  - LOWER EARTH ORBIT TRANSPORT
  - INTERPLANETARY TRANSPORT
  - DOD SATELLITES
  - US CIVIL AND INTERNATIONAL SCIENCE
  - EARTH SURFACE POWER BEAMING UTILITIES
  - OTHER-THAN-SPACE COMMERCIAL MARKETS FOR COMPONENT TECHNOLOGIES

FIGURE 10. Laser beamed power 10 year amortized benefits versus cost.

ACKNOWLEDGMENTS

All of the work cited in this paper was performed with U.S. Government funding by the National Aeronautics and Space Administration, the Department of Energy, and the Department of Defense. The author pays homage to the many talented individuals who participated in the work over three decades.

Brief History of the Lightcraft Technology Demonstrator (LTD) Project

Leik N. Myrabo

Department of Mechanical, Aerospace, and Nuclear Engineering
Rensselaer Polytechnic Institute
Troy, New York 12180

Abstract. This paper presents a brief history of the laser lightcraft research and development project that lead to the first outdoor free-flight demonstrations for the U.S. government on November 1997 - using the 10 kW PLVTS $CO2$ laser at the High Energy Laser Systems Test Facility (HELSTF), White Sands Missile Range (WSMR), NM. Summarized herein are some of the technological milestones that brought the Lightcraft Technology Demonstrator (LTD) concept vehicle to its present state of development: 1) from the initial LTD conceptual design, systems integration, and performance analysis for the Strategic Defense Initiative Organization (SDIO); 2) through laboratory engine bench testing, indoor wire-guided and vertical free-flight testing phases under contract to NASA and the USAF; and 3) finally resulting in outdoor flights to the current record altitude of 71 meters (233 ft) on 2 October 2000, sponsored by the Foundation for International Non-governmental Development of Space (FINDS).

INTRODUCTION

Several of the author's earliest works on beamed energy propulsion are given in given in [1-4]. Kantrowitz [5-6] was largely responsible for inspiring the author's personal quest for an advanced launch system that could cut the cost of space access by a factor of 1000 below chemical rockets. The author and Dean Ing, were the first to coin the word "lightcraft" in their 1985 book The Future of Flight [4]. A "lightcraft" is defined as any flight platform, airborne vehicle, or spacecraft designed for propulsion by a beam of light – be it microwave or laser.

Note that the earliest lightcraft design efforts at Rensselaer Polytechnic Institute (RPI) were focused on manned, "tractor-beam" vehicles energized from above, by space solar power stations [7-9]. These studies were sponsored largely by NASA - through the Universities Space Research Association (USRA), under its Advanced Design Program. The next RPI design effort focused on a near-term, unmanned "pusher-beam" Lightcraft Technology Demonstrator (LTD) designed as a microsatellite launcher [10]. This three year LTD study (~1985 to1988) was sponsored by the Strategic Defense Initiative Office (SDIO), under its Laser Propulsion Program – through the Lawrence Livermore National Laboratory.

The LTD study was carried out with the participation of several RPI graduate and undergraduate students. The 1.4-m diameter, 100 kilogram LTD spacecraft was designed for the reconnaissance or telecommunications roles - to be boosted into low Earth orbit by a powerful 100-MW ground-based laser. The initial atmospheric portion

CP664, *Beamed Energy Propulsion: First International Symposium on Beamed Energy Propulsion,*
edited by A. V. Pakhomov
© 2003 American Institute of Physics 0-7354-0126-8/03/$20.00

of the boost was to employ an airbreathing pulsed detonation engine (PDE) mode, followed by the rocket mode (using onboard liquid nitrogen, or perhaps water) – for when the air becomes too thin; the proposed transition point was Mach 5 and 30-km altitude. RPI's final report on the LTD concept vehicle was published on 30 June 1989 [10].

Figure 1 shows the relative sizes of the tractor-beam Apollo (5-place) lightcraft, Mercury (1-place) lightcraft, and Gemini (2-place) lightcraft, in comparison with the LTD vehicle. The dashed lines on the three tractor-beam craft indicate the approximate contours of the associated U.S. space capsules. Note that the fuel specific impulse of these very advanced laser-electric lightcraft engines (e.g., with MHD slipstream accelerators) can theoretically reach 6000 to 20,000 seconds, so the propellant tanks shrink to insignificant dimensions.

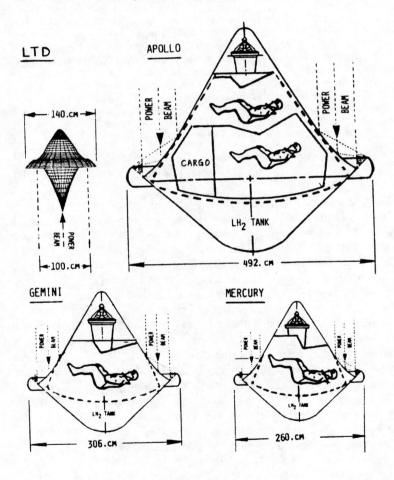

FIGURE 1. Family of RPI laser-boosted lightcraft designs.

Two distinct categories of laser propulsion technology are in evidence for lightcraft engines: 1) simple laser-thermal engines, and, 2) more complex laser-electric engines. The laser-thermal' engines must only absorb the beamed electromagnetic power directly into the working propellant. The more sophisticated laser-electric engines must first convert the captured beam power into electricity, before airbreathing, magnetohydrodynamic (MHD) engines can use it. The latter category holds the greatest promise for evolving hyper-energetic manned Lightcraft in the next 15-20 years. Most beamed-energy propulsion systems will likely exploit the atmosphere for momentum exchange to improve engine efficiency radically beyond that of today's chemical rockets. The matrix of feasible engine/optics/vehicle configurations is near infinite [3], but to pull off a new lightcraft design successfully will require the close integration of widely different engineering disciplines. Lightcraft engine design is a formidable exercise in interdisciplinary design.

Over the decade from 1985 to 1995, the author collaborated with hundreds of RPI students in developing the foundation for this laser propulsion technology, supported largely under NASA and USAF research contracts totaling just over $1 million. In the process, laser lightcraft technology was advanced to a critical threshold level where real success with building actual flight engines for the U.S. government was eminent. The author doesn't claim all the credit, and owes a much to these student collaborators - whose young creative minds significantly advanced the state-of-the-art. Without such preparation, the prospects for success in the subsequent 4-year "hardware" phase - funded under the USAF/NASA Laser Lightcraft Program - would have been slim indeed.

In 1996, the Air Force Research Laboratory (AFRL) initiated a contract with RPI to transform this LTD vehicle concept into flight hardware. Note that within one year NASA Marshall Space Flight Center joined on as an equal sponsor of the endeavor. With much creative work, the LTD configuration was quickly transformed into the present laser lightcraft geometry (Series #200) that is now flying at WSMR [11-18]. The development work took place over a 4-year period involving 23 laser propulsion tests at WSMR. A patent covering the current laser lightcraft geometry was finally awarded on 2 December 2002 (U.S. Patent # 6488233 – "Laser Propelled Vehicle." The author is credited as the sole inventor on the patent.

TRANSFORMATION INTO HARDWARE

The design and construction of the first lightcraft engine and test equipment began at RPI in 1995 with a $45K contract from the Phillips Laboratory (commonly known as the Rocket Lab) at Edwards AFB. One year later, the author was invited to join the Rocket Lab in what became a 3-yr IPA Fellowship (i.e., sabbatical from RPI - 1 Sept. '96 through 31 Aug. '99), to transform his 1989 LTD lightcraft concept into hardware reality.

Throughout those 4 years of collaboration with the Rocket Lab (which later merged into AFRL), the author effectively served as principal investigator for this Laser Lightcraft Program. To be entirely factual, the author personally designed every piece of lightcraft test hardware, flightweight and laboratory engines, test stands, launch stands, and auxiliary equipment during this 4-year period. The author also supervised the fabrication of this equipment by the RPI Central Machine Shop, and wrote/conceived all but two of the 23 WSMR test plans. The author's former RPI Ph.D. student (Don Messitt) assembled the computer-based data acquisition and schlieren photography systems. Several other RPI undergraduate and graduate students were also involved in the 4-year R&D activity.

In summary, it took a combined USAF/NASA investment of just under $1M over the four-year period to advance the Lightcraft Technology Program. The author personally solicited (and secured) nearly half of this from the NASA Marshall Space Flight Center - to grow it into a joint-sponsored program. Note also that only 10% of the $1M was spent for RPI Central Machine Shop services and the purchase (through RPI) of essential test equipment/ instrumentation for that 4-year period.

LIGHTCRAFT FLIGHT TEST HISTORY

Perhaps the simplest 'yardstick' for measuring the progress of laser lightcraft technology is the sequence of altitude records set along the way (see Table 1), starting with the first 1-foot vertical free-flight on 23 April 1997. The information contained in Table 1 has been quite thoroughly disclosed in numerous television documentaries, newspapers and magazines produced over the past seven years. Additional details have been revealed in several recent technical papers [19-21] from AFRL scientists. The design of the author's best-flying, spin-stabilized, beam-riding lightcraft, designated the Series 200 or Model 200, is revealed in [15, 17-19]. The Series 200 engine has claimed all the altitude records of 50-ft and beyond. These records were attained through a sequence of subtle improvements in the lightcraft engine geometry, PLVTS laser, telescope, launcher, and launch procedures.

TABLE 1. History of Lightcraft Vertical Free-Flight Records

Altitude (feet)	Date	Lightcraft Model #	Miscellaneous Details
1.0	23 Apr. '97	early #100	Non-spinning; truncated optic; conical forebody & shroud.
6.5	26 Aug. '97	early #150	First gyro-stabilized flight, inside lab; flat-plate forebody.
14.	1 Oct. '97	early #150	Inside lab; flat-plate forebody.
50.	5 Nov. '97	#200	First outdoor record flight.
73.	4 Dec.' 97	#200	Last record flight by a 14.7-cm Lightcraft
91.	18 Apr. '98	#200-3/4	First record flight by an 11-cm Lightcraft.
99.	22 Apr. '98	#200-3/4	Last airbreathing engine record flight.
127.	9 July '99	#200-3/4SAR	First ablative rocket record flight; 23rd WSMR test.
233.	2 Oct. '00	#200-5/6SAR	FINDS sponsored record flights; author's 24th WSMR test.

The Model #200 engine contours were designed to provide an autonomous, 'beam-riding' function that causes the exhaust flow to automatically vector in flight - thereby pulling the vehicle back into the center of the laser beam during flight. This critical "beam-riding" feature is the principal ingredient made all these record-breaking flights feasible in the first place. The 99-ft flight on 22 August 1998 was the last altitude record to be demonstrated with an airbreathing, pulsed detonation engine (PDE) - which has a propellant specific impulse of infinity because no propellant is carried on-board. The 9 July 1999 record flights were the first to employ an ablative rocket propellant. The 127-ft record flight occurred at the 23rd and last WSMR test for which the author functioned as Principal Investigator of the USAF/NASA (i.e., government-sponsored) Laser Lightcraft Program.

Transition to Solid Ablative Rockets

The LTD's air breathing engine mode had a predicted theoretical impulse Coupling Coefficient (CC) performance of 580 Newton's (thrust) per megawatt (laser power) at the launch altitude of 3-km (mountain top), but required a laser pulse width of 0.3 microseconds [10]. This pulse width is 50 to 100 times shorter than the 18 to 30-μsec pulses available from the PLVTS laser. At the 18-μsec pulse width, the #200-3/4 airbreathing engine had experimentally demonstrated (in the laboratory) a peak performance of only 160 N/MW, which is a factor of 3.6X less than that predicted for the LTD. Since PLVTS laser cannot be converted for such short pulse widths, other solutions to improving the thrust performance of the #200-3/4 engine had to be identified. Furthermore, attempts at powered flight times longer than 4 seconds with this bare aluminum engine were abruptly terminated - when the "absorption chamber" (i.e., the shroud) melted and blew apart.

To circumvent both problems (i.e., for the short term), the author adapted the #200-series lightcraft engine to use solid ablative rocket propellants that absorbed the infrared energy volumetrically (i.e., in depth). This solution came out of conversations with Dennis Reilly and Claude Phipps; both scientists had had experimentally confirmed that superior laser impulse enhancements were indeed available from such solids. Reilly suggested Delrin, and it indeed worked very well. The ablative rocket approach revealed a dramatic increase in thrust for the basic model #200-series lightcraft engine (i.e., linked to PLVTS), at the expense of having to carry propellant (see below for more details). Otherwise, the 99-ft. record flight of 22 August 1998 might still be standing today.

July '99 Flights to 127-ft

As mentioned earlier, July '99 was the 23rd and last WSMR laser propulsion test (under the USAF/NASA Laser Lightcraft Program) that the author effectively served in the capacity of principal investigator. The following provides more details on this 9 July 1999 test, the essentials of which were disclosed in [19 and 20].

In two consecutive launches on 9 July 1999, lasting ~3.5 seconds each, two different laser-propelled ablative rockets climbed rapidly to an altitude of 128-ft (39-m) and impacted a 4-ft x 8-ft plywood "beam-stop" suspended by a crane. These lightcraft hit the beam-stop with a velocity sufficient to crush in their aluminum noses. The spin-stabilized, beam-riding rocket design was a logical derivative of an earlier airbreathing engine design that established the former free-flight altitude record of 99-ft on 22 August 1998. The new lightcraft were designated model #200-3/4SAR, for which the 'SAR' stands for Solid Ablative Rocket.

These 127-ft record flights (as usual) took place in New Mexico on the White Sands Missile Range (WSMR) at the High Energy Laser Systems Test Facility (HELSTF) – using the 10-kW pulsed infrared laser called PLVTS. This US Army laser is operated by the Directorate for Applied Test, Training and Simulation (DATTS). The historic videos of the record 127-ft flights were included in a National Geographic Television production (Explorer-Series) entitled "The Quest for Space" - which was first aired on 6 July 2000.

The two record-breaking rocket lightcraft were made at RPI from 6061-T6 aluminum to a diameter of 11-cm (4.3 inches), and had launch masses of 26.3 and 29 grams. The second rocket to be flown carried slightly more ablative propellant aloft than the first. An especially important design feature of this engine was to pack the inert ablator in a place that would increase lightcraft stability, rather than to reduce it. By wrapping the plastic propellant in a thin band about the thrust chamber's perimeter, gyroscopic stability was indeed improved - at least until the propellant was all ablated. The rapidly evaporating plastic probably reduced the engine's operating temperature, thereby extending engine life.

Post flight inspection of the two rocket lightcaft and their rear reflectors (i.e., thin aluminum mirrors) revealed that the first engine to fly, showed no signs of stress. However, the second engine's optic (on the heavier lightcraft) had indeed sustained visible thermo-mechanical damage; it was somewhat warped, but still in one piece. Neither rear optic appeared to be contaminated (i.e., coated or blackened) by the ablated propellant; i.e., no obvious evidence of such condensation could be found on the mirrored surfaces. Also, when videos of the two flights were carefully reviewed, no indication of particulate-induced air breakdown could be observed in the beam transmission path.

Advantages of SAR-Lightcraft

For the earlier airbreathing-engine versions, peak air plasma temperatures can reach 20,000-K to 30,000-K, which is 3 to 5 times hotter than the sun's surface. Hence, run times with the 11-cm bare aluminum engines are severely limited to about 100 laser pulses, or less than 4 seconds of operation, before the chamber walls melted and blew apart. This of course abruptly ended the flight. In sharp contrast, the aluminum shroud on both ablative rocket flights survived well, sustaining minimal damage. This suggests that the craft could have flown well beyond the 127-ft plywood beam stop. Note that the PLVTS infrared laser was set up to deliver 18μsec, 450-joule pulses at repetition rate of 26 to 28 pulses per second.

Not only does the ablative propellant increase vehicle gyroscopic stability (i.e., through its proper placement – as mentioned above) and reduce engine temperatures, but it also more than doubles the thrust available from PLVTS. Whereas the pure airbreathing engine version developed only 0.36-lb (1.6 newtons) of time-averaged thrust, the ablative rocket gives better than 0.81-lbs (3.6 N) - on the same 10 kilowatts of laser power. At the 18μsec pulse width, the #200-3/4 airbreathing engine has experimentally demonstrated (in the laboratory) a peak impulse Coupling Coefficient (CC) performance of only 160 to 170 N/MW. In contrast, the #200-3/4 SAR engine has exhibited 360 N/MW at low pulse energy of ~95 Joules, at which energy the airbreathing engine develops only 80 N/MW. This was the principal ingredient that enabled rocket-propelled lightcraft to easily exceed the previous year's 99-ft record.

Finally, by packing this ring of ablative solid propellant into the annular focal region of the rear concentrating optic, the lightcraft's propulsion system may also be transformed into a kind of rocket-based, combined-cycle (RBCC) engine - at least within the dense atmosphere at subsonic speeds. Under these conditions an auxiliary airbreathing process, wherein the local ambient air is expelled ahead of the hypervelocity ablating rocket exhaust, might augment the engine's thrust.

Record Flights to 233-ft

Early in the morning of 2 October 2000 on the High Energy Systems Test Facility (HELSTF), a new altitude record of 233 feet (71 meters) was set with a 4.8 inch (12.2 cm) diameter laser boosted rocket (see Fig. 2) - in a flight lasting 12.7 seconds. Although most of the 8:35 am flight was spent hovering at 230+ ft., the lightcraft sustained no real damage, and will fly again. Besides setting a new altitude record, the craft simultaneously demonstrated the longest ever laser-powered free-flight, and the greatest 'air time' (i.e., launch-to-landing/ recovery). The improved #200-series lightcraft employed plastic ablative propellant, and were spin-stabilized to at least 10,000 RPM just prior to launch.

FIGURE 2. Lightcraft and Launcher (WSMR, 2 Oct. '00).

All three lightcraft were an improved version of the basic model #200-5/6SAR configuration. Table 3 gives the initial gross liftoff mass for the vehicles, and the flight Run #'s for each. The ablative rocket propellant ring added 4.12 grams to the initial launch mass, and it was not replaced with "fresh propellant" between flights. Note that lightcraft #1 was flown thrice, and accumulated 71.6-ft + 159-ft + 184-ft = a total of 415 feet of altitude in the process, with quite a lot of propellant still remaining. Clearly, with the proper telescope arrangement PLVTS could have boosted this #200-series craft to 500-ft, which indeed is now the objective for the next flight tests planned by the author.

TABLE 2. Lightcraft Flight Details (2 October 2000 Tests)

Flight (Run #)	Theodolite (degrees)	Maximum Altitude (meters & feet)		Time to Peak Altitude (sec)	Time to 23-m (sec)	Time to 46-m (sec)
1	30	21.8-m	(71.6-ft)	3.5	--	--
2	62	71.1-m	(233-ft)	12.73	2.57	5.17
3	14	9.4-m	(30.9-ft)	--	--	--
4	52	48.4-m	(159-ft)	5.16	2.03	4.8
5	18	12.3-m	(40.3-ft)	--	--	--
6	56	56.0-m	(184-ft)	5.57	1.6	4.5
7	22	15.3-m	(50.1-ft)	--	--	--

TABLE 3. Lightcraft Gross Liftoff Mass (Recorded before first flight)

Lightcraft (vehicle #)	Launch Mass (grams)	Diameter (cm)	Flight #'s	Accumulated Altitude (meters & feet)
#1	49.02	12.2	Runs 1, 4, & 6	124-m (415-ft.)
#2	50.62	12.2	Run 2	71-m (233-ft.)
#3	51.05	12.2	Runs 3, 5, & 7	37-m (121-ft.)

The record flights were (as usual) performed with the 10 kW pulsed carbon dioxide laser named "PLVTS" by the Directorate for Applied Training, Test and Simulation (DATTS). The beam profile at the launch pad is given below in Fig. 3. Even though the laser was suffering from a grounding or arcing problem in the sustainer that caused it to run erratically, the time-average beam power was still adequate to propel the craft to record altitudes. (Incidentally, this laser problem was identified and fixed in the following day.) Note that vehicle #3 did not fly well, but this might be blamed on unlucky timing with a rough-running laser.

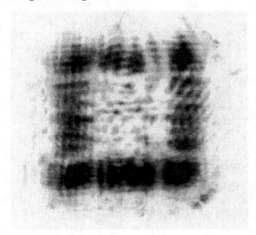

FIGURE 3. PLVTS beam profile at the launch pad.

These were the first ever vertical, free-flight tests to be performed without a 4-ft X 8-ft plywood "beam-stop," suspended by a crane - to intercept stray laser energy that sometimes 'spills' around the vehicle in flight. The author's flights were carried out with the cooperation of NORAD and WSMR range control to avoid the irradiation of both low Earth orbital satellites and low flying aircraft. Twelve launch "windows" varying from 2.56 to 41.25 minutes in length were secured from NORAD. With the 233 ft. flight on Oct. 2, the author attained his objective of nearly doubling his previous altitude record of 128 ft. - set on July 9, 1999 (with a 11 cm #200-series lightcraft) under prior joint USAF/NASA funding.

One major concern with longer lightcraft flights was that ablating rocket propellant might contaminate either: a) the engine's rear parabolic optic, or, b) the laser beam's atmospheric propagation path. In the former case, any vaporized propellant that condenses and sticks to the rear optic, could eventually 'darken' the brightly polished aluminum surface, thereby reducing its reflectivity. This may cause the thin metal reflector to overheat and warp, especially in the forward portion of the optic where high over-pressures are created from reflecting laser-induced detonations [15, 17, 18]. However, these fears were unfounded.

Post flight inspection of the three lightcraft's rear reflectors revealed no signs of either mechanical stress or contamination. Also when the videos of the seven flights were carefully reviewed, no indication of particulate-induced air breakdown could be observed. No attempt was made to measure laser transmission losses through the rocket's plume.

A detailed computer-based motion analysis study of the videos taken of Flights #2, #4, and #6 was carried out. Note that the camera used to film these flights was aligned with the DATTS "man-lifter" base (a horizontal reference plane that was centered in the video frame), to indicate an altitude of 75 feet from the launch pad. Hence the maximum altitude recorded by the camera was 150 feet. Fortunately, theodolite readings were also taken to identify the maximum altitude of each flight. Two other cameras were employed to record additional flight footage: a vertical "look-up" camera positioned roughly 1-foot from the launch pad, and a mobile hand held camcorder.

In the record flight (Run #2) results shown in Figs. 4 and 5, the lightcraft accelerated over the first two seconds, thereafter maintaining a quasi-constant velocity of 27 ft/sec for the next 3.5 seconds. Most of the 12.73 second flight was spent hovering above200 feet, before setting the new World's altitude record (at least for laser lightcraft) of 233 feet. During Run #4, the craft reached a top velocity of 35 ft/sec on the way to a peak altitude of 159 feet. The lightcraft in Run #6 reached the 140-ft point in 3.5 seconds, which was 1.5 seconds quicker than Run #2, and 1.0 second faster than Run #4. Run #6 attained a maximum velocity of 49 ft/sec on the way to a peak altitude of 184 feet. If the PLVTS laser had not been suffering that morning from an electrical grounding problem, no doubt all peak velocities would have been in this range, or higher.

Note that the 2 October 2000 record-breaking lightcraft flights at WSMR were sponsored by the Foundation for International Non-governmental Development of Space (FINDS), a non-profit organization dedicated to promoting low cost access to space. With FINDS funding, laser launch technology has been decidedly moved out of the exclusive realm of government-sponsored research. The next objective is to double the current altitude record, again—to attain altitudes of 500 ft. or beyond

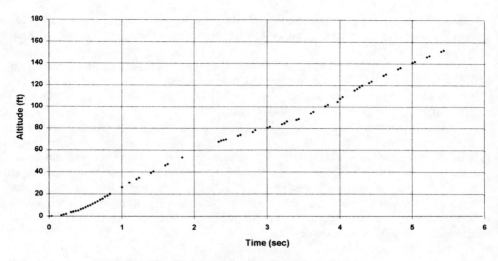

FIGURE 4. Altitude vs. time for Flight 2 (233-ft record).

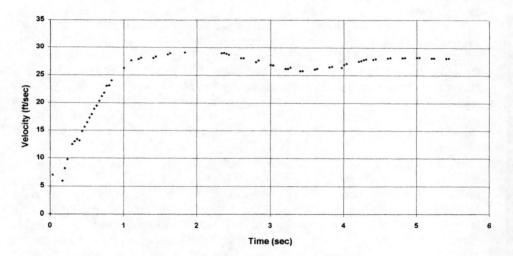

FIGURE 5. Average velocity vs. time for Flight 2 (233-ft record).

within the foreseeable future. Clearly, ten more doublings in altitude are necessary to reach the edge of space (e.g., 256-kft, or 78-km).

Scaling Lightcraft Technology

The July 1999 and October 2000 WSMR tests have conclusively proven that the requisite lightcraft engine technology for flights to 'extreme altitudes' is certainly close at hand. What remains to be done is to scale the engine size to 23-cm (or larger) for a reasonable near-term objective of 10-km to 30-km altitudes. Note that two 23-cm lightcraft have been constructed at RPI under the author's supervision, prior to September 1999. The first 23-cm complete lightcraft was designed and built as a rotating wind tunnel model as the central focus of a graduate student thesis project by Andrew Panetta (M.S. student supervised by the author). The rotating model has been run under several campaigns in RPI's 4-ft X 6-ft subsonic wind tunnel, to obtain aerodynamic performance data needed for flight dynamics studies [14,22].

Note that a second 23-cm diameter lightcraft engine had been fabricated at RPI - designed for laboratory experiments with and without ablative propellants. The engine was finally tested with solid ablative inserts in Test Cell 3 at HELSTF on 9-11 July 1999, with PLVTS. The performance data revealed coupling coefficients as high as 460 newtons per megawatt (N/MW) using laser pulse energies up to 667 joules with a 30-μsec pulse duration. Further increases in CC performance beyond 500 N/MW may be feasible - at higher laser pulse energies. Fully developed, the 23-cm laser propulsion engine could conceivably produce a thrust of at least 100 newtons (22.5 pounds) with a 150-kW infrared laser that delivers, say perhaps, 10 kJ pulses @ 150 Hz. For a projected 23-cm lightcraft vehicle empty weight (i.e., no propellant or payload) of just 1.5 newtons (5.3 ounces), thrust/ weight ratio would be nearly 67. This scale of lightcraft could certainly enable flights to the edge of space, but the discussion is academic until a 10X more powerful CO_2 electric discharge laser is built.

SUMMARY AND CONCLUSIONS

The feasibility of ambitious flight demonstrations for laser lightcraft to extreme altitudes (e.g., 10 to 30 kilometers) in the near future, depends upon three key steps: 1) scaling the launch laser power to at least 100-150 kilowatts; 2) linking this laser to a 1.4 meter telescope; and finally, 3) scaling the #200-series lightcraft design to diameter of 23-cm or more. Clearly, without at least a 10-fold upgrade in beam power beyond the present 10-kilowatt carbon dioxide laser at WSMR, everyone's ability to set future lightcraft altitude records will be severely curtailed. One thousand feet might be the best that PLVTS can do.

In the author's opinion, accomplishing the notable milestone of launching the first micro-satellite into orbit with a megawatt-class laser is simply a matter of will, and finances. The scientific knowledge and engineering expertise to pull this off is already here. Let's get it done!

ACKNOWLEDGEMENTS

The author gratefully acknowledges the important contributions to this project by the RPI Central Machine Shop; Stephen Squires, Chris Beairsto and Mike Thurston of DATTS; Ron Grimes for recording the peak altitudes for the 2 October 2000 lightcraft flights with a theodolite; and FINDS for sponsoring the author's most recent flight tests. In addition, the author would like to thank the AFRL and NASA Marshall Space Flight Center for the opportunity to guide their Laser Lightcraft Program through the first 23 WSMR tests from 1995-1999.

REFERENCES

1. Myrabo, L.N., "Solar-Powered Global Air Transportation," AIAA Paper 78-698, Washington, D.C., April 1978.
2. Myrabo, L.N., "A Concept for Light-Powered Flight," AIAA Paper 82-1214, Washington, D.C., June 1982.
3. Myrabo, L.N., "Advanced Beamed-Energy and Field Propulsion Concepts," BDM Corp. publication BDM/W-83-225-TR, Final Report for the California Institute of Technology and Jet Propulsion Laboratory under NASA Contract NAS7-100, Task Order No. RE-156, dated 31 May 1983.
4. Myrabo, L. and Ing, D., *The Future of Flight*, Baen Books, publisher, distributed by Simon & Schuster, 1985.
5. Kantrowitz, A. "The Relevance of Space," *Astronautics & Aeronautics*, V. 9, 34-35 (March 1971).
6. Kantrowitz, A., "Propulsion to Orbit by Ground-Based Lasers," *Astronautics and Aeronautics*, V. 10, 74-76 (May 1972).
7. Myrabo, L.N., et al., "Apollo Lightcraft Project," Final Report, prepared for the NASA/USRA Advanced Design Program, 3rd Annual Summer Conference, Washington, D.C., 17-19 June 1987.
8. Myrabo, L.N., et al, "Apollo Lightcraft Project, Final Report, prepared for the NASA/USRA Advanced Design Program, 4th Annual Summer Conference, Kennedy SFC, FL, 13-17 June 1988.
9. Walton, D., List, G., Myrabo, L. et al., "Investigations into a Potential 'Laser-NASP' Transport Technology", Proceedings of the 6th Annual Summer Conference, NASA/USRA University Advanced Design Program, 11-15 June 1990; See also, D. Walton, G. List, and L. Myrabo, "Economic Analysis of a Beam-Powered, Personalized Global Aerospace Transport System," Rensselaer Polytechnic Institute, December 1989.
10. Myrabo, L.N., et al., "Lightcraft Technology Demonstrator," Final Technical Report, prepared under Contract No. 2073803 for Lawrence Livermore National Laboratory and the SDIO Laser Propulsion Program, dated 30 June 1989.
11. Myrabo, L.N., Messitt, D.G., and Mead, F.B., Jr., "Ground and Flight Tests of a Laser Propelled Vehicle," AIAA Paper 98-1001, Jan. 1998.
12. Mead, F.B., Jr., Myrabo, L.N., and Messitt, D.G., "Flight Experiments and Evolutionary Development of a Laser Propelled, Transatmospheric Vehicle," Space technology & Applications International Forum (STAIF-98), Albuquerque, NM, 25-29 Jan. 1998.
13. Mead, F.B., Jr., Myrabo, L.N., and Messitt, D.G., "Flight and Ground Tests of a Laser-Boosted Vehicle," AIAA Paper 98-3735, July 1998.
14. Panetta, A.D., Nagamatsu, H.T., Myrabo, L.N., Minucci, M.A.S. and Mead, F.B., Jr., "Low Speed Wind Tunnel Testing of a Laser Propelled Vehicle," Paper 1999-5577, 1999 World Aviation Conference 19-21 October 1999, San Francisco, CA
15. Wang, T.-S., Cheng, Y.-S., Liu, J., Myrabo, L.N., and Mead, F.B., Jr., "Performance Modeling of an Experimental Laser Propelled Lightcraft," AIAA 2000-2347, June 2000.
16. Messitt, D.G. , Myrabo, L.N., and Mead, F.B., Jr., "Laser Initiated Blast Wave for Launch Vehicle Propulsion," AIAA 2000-3848, Washington, D.C., July 2000.
17. Wang, T.-S., Chen, Y.-S, Liu, J. Myrabo, L.N., and Mead, F.B., Jr., "Advanced Performance Modeling of Experimental Laser Lightcrafts," AIAA Paper 2001-0648, January 2001.
18. Wang, T.-S., Chen, Y.-S, Liu, J. Myrabo, L.N., and Mead, F.B., Jr., "Advanced Performance Modeling of Experimental Laser Lightcrafts," *Journal of Propulsion and Power*, Nov.-Dec. 2002, pp. 1129-1138.
19. Mead, F.B., Jr., Squires, S., Bearisto, C. and Thurston, M., "Flights of a Laser-Powered Lightcraft During Laser Beam Hand-off Experiments," AIAA Paper 2000-4384, July 2000.
20. Larson, C.W., and Mead, F.B., Jr., "Energy Conversion in Laser Propulsion," AIAA Paper 2001-0646, January 2001.
21. Mead, F.B., Jr., "Laser-Powered, Vertical Flight Experiments at the High Energy Laser System Test Facility," AIAA Paper 2001-3661, July 2001.
22. Libeau, M.A., Myrabo, L.N., and Filippelli, M., "Combined Theoretical & Experimental Flight Dynamics Investigation of a Laser-Propelled Vehicle," AIAA Paper 2002-3781, July 2002.

A Brief History of Laser Propulsion at the Marshall Space Flight Center

Lee W. Jones

B.G. Smith & Associates, Huntsville, AL 35801

Abstract. A laser propulsion technology program was conducted between 1977 and about 1987 at the NASA Marshall Space Flight Center. The project ended in about 1987, when the high power laser failed. In addition to the experimental activity, several contracted efforts, mostly aimed at modeling the physics of laser-plasma interactions, are briefly described.

INTRODUCTION

This paper describes the technology program in laser propulsion that was conducted between 1977 and about 1986 at the Marshall Space Flight Center. It is not intended to be a technical account, although a few data are presented, but rather a historical summary of the work. It is written from the perspective of the author, who was involved totally at the beginning, then much less as time passed and others with appropriate technical skills, such as Dwayne McCay and Richard Eskridge, were brought in to perform the experimental work. The project ended in about 1987, when the high power laser could no longer be made to operate. The author became involved once more, in an attempt to secure funds to rebuild it, but that effort was unsuccessful.

Beginning of the MSFC Effort

Sometime in 1977, NASA's Office of Aeronautics & Space Technology (OAST) decided to move the laser propulsion work that was being done by the Lewis Research Center to Marshall. The informal reason given was that Lewis had become bogged down in doing high-power laser development, and they were spending a lot of money in that area, rather than looking hard at the feasibility of using lasers for propulsion. The official reason, as spelled out in Jerry Mullins' charts to the HELRG on July 8, 1980, was that the agency was de-emphasizing long-term space research and technology (R&T) work at Lewis and shifting manpower to aeronautics and terrestrial energy R&T. Both were probably accurate.

The principal researchers at Lewis, Steve Cohen and Dick Lancashire, had already been told that the work would be transferred to Marshall. They wanted to ensure that

CP664, *Beamed Energy Propulsion: First International Symposium on Beamed Energy Propulsion*,
edited by A. V. Pakhomov
© 2003 American Institute of Physics 0-7354-0126-8/03/$20.00

funding for Physical Sciences Inc. (PSI), in Woburn, MA, would be continued – the investigators there were very capable and were doing good work. The principals at PSI were Tony Pirri and Nelson Kemp, and they were performing an analytical study of the physics of coupling of laser energy into a working fluid, and developing the equations for modeling it. A detailed account of PSI's work is presented in Dr. Pirri's paper, [1] and in their reports [2], [3], [4].

Lewis had also funded Wayne Jones at Lockheed/Palo Alto to perform a study of laser-propelled aircraft and OTVs [5], and Jim Shoji at Rocketdyne for studies and for the design of a thruster to be tested in Lewis's high power laser facility. [6]

One of the things Lancashire related was that they had become heavily involved in developing a suitable high power laser for their experiments, because there weren't any available that were high enough in power, had good beam quality, and were reliable. He warned us about getting drawn into a laser development program; it would eat up the budget, and divert you from the primary objective, which is to develop a propulsion system.

Scope of the MSFC Work

Very early in the transition from Lewis, it became clear that we were going to have to decide on the scope of our activities, and restrict it in order to get anything done. The field of laser propulsion was broad – there was ablative propulsion for launch vehicles, the idea first proposed by Arthur Kantrowitz,[7],[8], continuous-wave (CW) and repetitively-pulsed (RP) versions of propulsion systems for upper stages and orbital spacecraft, laser-electric propulsion, and others. The mission class we selected was orbit transfer vehicles or spacecraft in low Earth orbit (Figure 1). The Lewis work had been aimed at this mission class also, and it made sense to continue in that direction. A further selection was the use of CW lasers, either space or ground-based, rather than pulsed lasers. It is noted, however, that very rapidly pulsed lasers, such as the free electron laser (FEL), will sustain an LSC wave. Keefer, in experiments with a large FEL at Los Alamos, demonstrated this in 1987.

At that point, it seemed that the biggest barrier to using high power lasers for propulsion was converting the photon energy of the laser into kinetic energy of a hot gas leaving a thruster nozzle, e.g., the means of heating the working fluid. Nelson Kemp of PSI had explained that there were three processes that could be used: one is the inverse bremsstrahlung absorption process, producing a laser-supported combustion (LSC) wave; the second is molecular absorption of the laser radiation, and the third is particulate absorption. In the first case, the high operating temperatures (~20,000 K), lead to high radiant energy losses from the hydrogen plasma and large heat transfer rates to the walls. Consequently, PSI chose to focus on the molecular absorption approach using hydrogen/alkali metal mixtures.
However, we were attracted to the higher temperature method, because that is, after all, the rationale for pursuing research in laser propulsion – it offers higher specific impulse than chemical rocket engines because it heats a working fluid to very high

temperatures. That is also the reason that we pursued hydrogen as the working fluid; a simple inspection of the Isp equation, where Isp is directly proportional to the square root of the working fluid and inversely proportional to the square root of its molecular weight, will drive the search to high temperature hydrogen. The laser rocket engine concept is shown in Figure 2.

Figure 1. Laser Orbit Transfer Vehicle

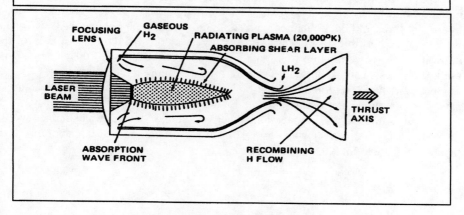

Figure 2. Laser Rocket Engine

The rest of 1977 and most of 1978 were spent in learning more about high power lasers, and laser propulsion, and getting to know the technical community – those who were doing work in these areas. During this period, and until we began the experimental program, there was only person at Marshall working on this project – the funding from OAST was barely sufficient to keep PSI funded.

COLLABORATION WITH THE ARMY

Sometime during this period, contact was made with personnel of the Army's High Energy Laser (HEL) Laboratory, notably Dr. Raymond Conrad. Conrad had previously observed the effects of (LSC) waves, when he was doing lethality studies using the 150 kW Tri-Services Laser (TSL) in the 1972-1975 time period. The TSL was a large continuous-wave gas dynamic laser that was subsequently removed from service. Conrad had reported on this work in classified venues [9], but he gave us an unclassified version of the major results. Our early discussions were mostly about the availability of very high power lasers for the experimental program that we had decided to undertake. While existing lasers had power levels of as much as 300 kW, none of them were available to us. Conrad also told us about a laser that had been part of a weapon demonstration, called the Mobile Test Unit (MTU). Its rated power level was classified still, but the laser had been removed from the tracked vehicle containing the rest of the weapons system, and set up in a nearby facility as a research laser. In that configuration, it had a nominal power level of 30 kW. Early in 1979, we arranged for Conrad and his associates to conduct some experiments for us using the MTU laser. [10]

The object of these experiments was to produce LSC waves in pure hydrogen, determine the power and irradiance required for wave generation, measure the absorption of the plasma, and obtain information on the character of the plasma radiation. These objectives were only partially met, but LSC waves were generated in hydrogen, threshold data obtained, and some radiometric data obtained. LSC waves were obtained 8 times; with laser power levels at the focal plane of 7.5 to 9.2 kilowatts. LSC wave ignition was sporadic, because of the nature of the electric spark used. (Later tests using a pulsed laser spark produced more reliable ignition). The peak irradiance required for ignition was about 5×10^5 W/cm^2 in hydrogen at 45 psia. The experimental setup is shown in Figure 3.

An estimate of the propagation velocity of the LSC waves was obtained with a fast silicon photodiode. It was 4.6×10^3 cm/sec, which compared favorably to a theoretical velocity of 3.8×10^3 cm/sec.

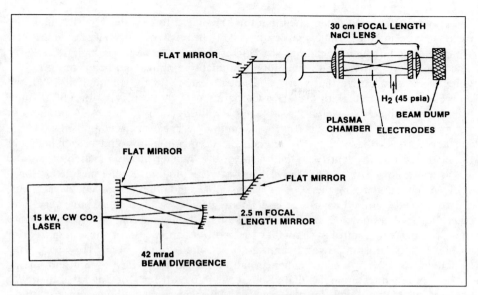

Figure 3. Experimental Configuration for Plasma Ignition Tests

CONTRACTED ACTIVITIES

The University of Tennessee Space Institute (UTSI) was involved from the earliest days of the MSFC program. Dr. Dennis Keefer, a professor in the Engineering Science & Mechanics Department, had done some work on laser instrumentation with the Army, and was highly regarded. They were awarded a small contract to participate in the Marshall activity, and guided the experimental program to a great degree [11]. They proposed a two-phase experimental approach. Phase 1 would develop reliable ignition techniques, obtain preliminary spectroscopic data, and establish a stable hydrogen plasma. Phase 2 would determine the influence of mass flow rate and pressure on the plasma characteristics. Measurements of hydrogen bulk temperature would be obtained in Phase 2. Keefer, with support from graduate students and other UTSI colleagues, constructed a series of models of the laser supported plasma that had increasing fidelity. [12-17]

Dr. Owen Hofer of the Lockheed Huntsville Research and Engineering Center, and Dr. Juergen Thoenes and Dr. Francis Wang of the same organization later won a contract for support of the Marshall activity. The contract also called for them to evaluate the relative merits of various lasers, assess several modeling techniques for modeling the physical phenomena in a laser rocket engine, and evaluate chamber window materials, and cooling methods for the absorption chamber [18].

Earlier, we had determined that the Marshall effort should focus on the laser rocket engine and the technical issues that had to be resolved before we could proceed with the design of such an engine. The DoD was heavily invested in high power laser research, as well as needed technologies in adaptive optics and beam propagation. Our idea was to utilize that work, and not repeat it. In the summer of 1980, we entered into a contract with the BDM Corporation of McLean, VA, for a brief study effort. Their objectives were to assess the overall direction of the NASA program in laser propulsion, to compare laser propulsion to other means of accomplishing NASA missions, assess the possible synergism between propulsion and power beaming applications, and to identify an approach for the transfer of DoD technology to the NASA program. The principal investigators were Dr. John Rather and Peter Borgo [19]. After summarizing the SOA in lasers, optics, and pointing and control systems, they observed that the high power lasers being developed for defense have very short run times, and are not applicable to propulsion missions, and that the precision optics and pointing and control systems have capabilities in excess of those required for laser propulsion and power beaming for civilian applications. However, they concluded that NASA could enhance national security by developing and testing civilian oriented technologies that also have high payoff for DoD, and that there is a logical development path for major civilian laser missions leading from ground-to-space power beaming and propulsion to space-based systems.

CONGRESSIONAL INTEREST

During 1979, a strong advocate for high power laser research & technology took a very active interest in the Marshall laser propulsion work. He had access to the Senator from Alabama, Howell Heflin, and got Sen. Heflin interested in the broad topic of high power laser technology, and in our work. The result of this interest was one or more letters to the Administrator recommending more NASA involvement in laser propulsion, and asking some specific questions the agency program. One result of that was a call to headquarters to brief Dr. Walter Olstad, the AA for OAST, and to meet with the Administrator, Dr. Frosch. Frosch did not appear to be pleased about having to respond to Heflin's questions. The Senate Subcommittee on Science, Technology and Space, of which Sen. Heflin was a member, scheduled hearings on laser research and technology for December 12-14, 1979 in Washington, January 8, 1980 in Huntsville, AL, and January 12 in Albuquerque, NM.

In a February, 1980 call from a subcommittee staffer, we learned that, as a result of the hearings, Sen. Heflin had ascertained that there was a high level of interest within the DOD for high power laser work, but very little interest within NASA. He had scheduled the hearings because he was looking for a major new job for Marshall, and our advocate had convinced him that an enhanced laser propulsion program and a joint Army/NASA/TVA energy transmission demonstration program would be such a "plum". After learning of NASA's lack of interest, Heflin decided to look elsewhere for a new program for Marshall. The staffer also quoted this exchange between Heflin and Dr. Olstad during the hearings in Huntsville: In response to Heflin's comment that the subcommittee had heard testimony that laser propulsion could be a near-term program, and that NASA should re-evaluate its stated position, Olstad said that there were differences of opinion about the readiness of laser propulsion, and added, "We are excited about the potential, and if we become more excited, we plan to re-assess the funding levels at Marshall".

A funding augmentation never materialized as a result of the interest by Sen. Heflin. In fact, because of the time and effort required to respond to his questions, and to attend and testify at the hearings, OAST management became less inclined to fund the program at Marshall.

OAST HIGH POWER LASER WORKING GROUP

Following the March 1979 OAST High-Energy Laser Workshop, which was held to review state-of-the-art and identify the most promising applications for high power lasers, the NASA advisory committees showed heightened interest in this technology area. At the Woods Hole Conference on Innovation, held in June 1980, Dr. Arthur Kantrowitz once again advocated laser propulsion for launch vehicles. In July 1980, the NASA program was reviewed for the High Energy Laser Review Group (HELRG) of the Defense Advanced Research Projects Agency (DARPA). In August 1980, Dr. John Rather of the BDM Corporation briefed the NASA Administrator, Dr. Robert Frosch, on ground-to-space high power laser applications, and in September 1980, as a result of strong congressional interest, Dr. Frosch requested a review and reformulation of the NASA program on high power lasers. This led to the formation in October 1980 of the NASA High Power Laser Working Group, with the following membership: Dr. Frank Hohl, Chairman (LaRC), Dr. Lynwood Randolph (HQ), Dr. Gerald Walberg (LaRC), Dr. Russell DeYoung (LaRC), Dr. Edmund Conway (LaRC), Dr. James Morris (LaRC), Mr. Robert English (LeRC), Dr. Joseph Randall (MSFC), Mr. Lee Jones (MSFC), and Dr. Joseph Mangano (DARPA).

The group's charter was to "assess the current NASA program in the context of potential civilian and military applications and recommend a balanced program, complementing the DoD and DoE efforts that would provide a strong technological base for future civilian space applications".

Over the next few months, the Working Group reviewed the existing activities within the Agency and recommended changes necessary to establish a research & technology base for civilian space power applications of laser power generation, transmission, and conversion. It recommended an augmented NASA program which included work at Langley in blackbody and direct solar pumped lasers, at Lewis in laser energy conversion and power transmission, and at Marshall in laser propulsion [20]. It was a modest program, with funding levels of $1.6 to $3.6 M and manpower levels of 25 – 30 direct man-years, which was never fully funded by the agency.

OUR VERY OWN LASER!

In October, 1980, we learned that the Army's HEL Lab was being reorganized. Col. Woody DeLeuil, was now the Director of the new Directed Energy Directorate. The CO_2 laser we had been using in the experiments would be moved from Bldg. 4762 to the HEL area. Later, Col. DeLeuil told us that once the laser was moved, he anticipated that we could continue our work with Conrad.

A few days later, on Feb. 8, we learned that the Army had decided to divest itself of the MTU, and the Air Force was interested in it. Naturally, that was disturbing news, inasmuch as we did not have access to another laser. So, after discussions with management, a letter was drafted from Dr. Lucas, the Marshall Center Director, to General Rachmeler, the MICOM Commander, asking that the Army transfer the MTU laser to Marshall. It was signed on 2/14/80, and Gen. Rachmeler responded affirmatively on 2/26/80. This was to be the easiest part of getting our own high-power laser. We still had to get the laser disassembled and moved to a Marshall facility that had yet to be built. The process of selecting a suitable location, having it modified, and installing the laser took most of the next year.

EXPERIMENTAL WORK AT MSFC'S LASER FACILITY

The Marshall laser propulsion facility was operational in August 1981. Our goal was to establish proof of concept of continuous-wave (CW) laser propulsion, and our objectives were to:
- Demonstrate stable LSC waves in hydrogen gas
- Demonstrate that all of the incident laser energy is absorbed
- Maintain the LSC wave stationary
- Transfer energy from the LSC wave to the working fluid
- Demonstrate the ability to cool the chamber walls

The experimental program had two distinct efforts; hydrogen plasma ignition experiments (Figure 4) and steady-state hydrogen plasma experiments. [21, 22]

Ignition Experiments

We decided to use a pulsed CO_2 laser to achieve plasma ignition. It was the cleanest, least intrusive, and most reliable of the options that had been demonstrated. We used a Lumonics 103 TEA laser, which delivered 7 Joules at 150 nanoseconds per pulse. The output beam was introduced into the test cell through the "hole" in the center of the main beam, which made it easier to align the two focal volumes. This was possible because the MTU was a CO_2 electric discharge laser with unstable resonator optics, and its output profile resembled a torus, with near zero power at the center of the beam. The pulsed laser produced a spark within the test cell gas (hydrogen), a region of high electron density, which could ignite a sustained plasma (or LSC wave).

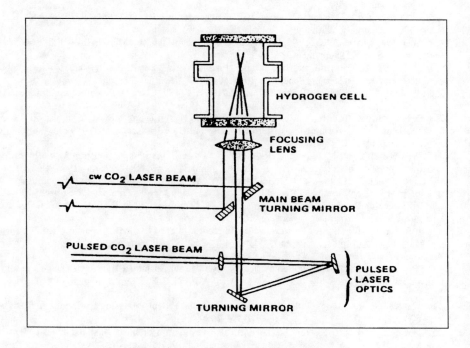

Figure 4. Pulsed Laser Ignition Experimental Setup

SUMMARY

It can be said that the Marshall program demonstrated that a laser-supported plasma could be initiated and maintained in hydrogen, and that the laser energy was absorbed by the plasma. Further experimentation was blocked by the failure of the laser, and lack of funds to repair it. The program was plagued by lack of funds, no clearly-defined NASA mission to which it could be applied, and the necessity of maintaining an expensive, high power laser that had been designed to accomplish a specific Army demonstration, not to be used as a research tool. A good analytical base was generated during the program, upon which future investigators can build.

REFERENCES

1. Peter E. Nebolsine and Anthony N. Pirri, "Laser Propulsion – The Early Years", in *Proceedings of the First International Symposium on Beamed Energy Propulsion*, Edited by A. Pakhomov, et al, AIP Conference Proceedings, Melville, New York, 2002.

2. Robert F. Weiss, Anthony N. Pirri, and Nelson H. Kemp, "Laser Propulsion", <u>Astronautics & Aeronautics</u>, March 1979.

3. N. H. Kemp and R.G. Root, "Laser-Heated Thruster Interim Report", PSI TR-205, NASA Contract NAS8-33097, February 1980.

4. N.H. Kemp and R.H. Krech, "Laser-Heated Thruster Final Report", NASA CR-161666, NASA Contract NAS8-33097, September 1980.

5. W.S. Jones, J.B. Forsythe, and J.P. Skratt, "Final Report, Laser'Rocket Systems Analysis", NASA Contractor Report CR-159521, September 1978. (Contract NAS3-20372)

6. J.M. Shoji, "Laser-Heated Rocket Thruster", NASA CR-135, NASA Lewis Research Center, May, 1977.

7. A.R. Kantrowitz, "The Relevance of Space", March 1971, <u>Astronautics & Astronautics</u>, pp. 34-35.

8. A.R. Kantrowitz, "Propulsion to Orbit by Ground-Based Lasers", May, 1972, <u>Astronautics & Aeronautics</u>, p.74

9. R.W. Conrad, D.W. Mangum and R.G. Polk, " Laser-Supported Combustion Wave Investigation with the Army Tri-Service Laser", Army Missile Command Report RR-75-2, July 1975 (CONF).

10. R.W. Conrad, E.L. Roy, C.E. Pyles, and D.W. Mangum, "Laser-Supported Combustion Wave Ignition in Hydrogen" Technical Report RH-80-1, U.S. Army Missile Command, October 1979.

11. D.R. Keefer, H. Crowder, R. Elkins, and R. Eskridge, "Final Report: Laser Heated Rocket Analytical and Experimental Support" July 1981

12. D. R. Keefer, R. Elkins, C. Peters, and L.W. Jones, "Laser Thermal Propulsion", in <u>Orbit-Raising and Maneuvering Propulsion: Research Status and Needs</u>, edited by L. Caveny, Vol. 89 of *Progress in Astronautics and Aeronautics*, 1984.

13. S-M Jeng and D.R. Keefer, "Numerical Study of Laser-Sustained Hydrogen Plasmas in a Forced Convective Flow" AIAA-86-1524, AIAA/ASME/SAE/ASEE 22nd Joint Propulsion Conference, June 16-18, 1986, Huntsville, AL.

14. D.R. Keefer, S-M Jeng and R. Welle, "Laser Thermal Propulsion Using Laser-Sustained Plasmas", <u>Acta Astronautica, Vol. 15, No. 6/7</u>, pp. 367-376, 1987.

15. S-M Jeng and D.R. Keefer, "A Theoretical Investigation of Laser-Sustained Plasma Thruster", AIAA-87-0383, AIAA Aerospace Sciences Meeting, Jan. 1987, Reno, NV.

16. S-M Jeng, D.R. Keefer, R. Welle and C.E. Peters, "Laser-Sustained Plasmas in Forced Convective Argon Flow, Part II: Comparison of Numerical Flow with Experiment", AIAA Journal, Vol.25, No.9, pp.1224-1230, Sept. 1987.

17. San-Mou Jeng, Ronald Litchford and Dennis.R. Keefer, Computational Design of an Experimental Laser-Powered Thruster", Final Report, Contract NAS*-36220, March 24, 1988.

18. Owen C. Hofer, "Final Report: Study of Laser-Heated Propulsion Devices" LMSC-HREC TR D784744, April, 1982

19. J. Rather and P. Borgo, "Laser Propulsion Support Program Final Report", BDM/W-80-652-TR, November 1, 1980

20. R.J. DeYoung, G.D. Walberg, E.J. Conway and L.W. Jones, "A NASA High-Power Space-Based Laser Research and Applications Program", NASA Special Publication 464, 1983.

21. R.H. Eskridge, T.D. McCay and D.M. Van Zandt, "An Experimental Study of Laser-Supported Plasmas for Laser Propulsion – Final Report", NASA TM-86583, January, 1987.

22. L.W. Jones and D. R. Keefer, "NASA's Laser Propulsion Project", <u>Astronautics & Aeronautics</u>, pp. 66-73, September, 1982.

Activities of Laser Propulsion in Japan

Masayuki Niino

Kakuda Space propulsion Laboratory, National Aerospace Laboratory

Abstract. This paper describes the status of laser propulsion research and related studies in Japan. Japan covers wide range of laser propulsion research including from ground launching to orbital transfer applications. The current status of Japanese laser propulsion is introduced and the elemental studies of Tohoku University, National Aerospace Laboratory, University of Tokyo, Tokai University and Institute for Laser Technology are presented.

TOHOKU UNIVERSITY[1,2]

At Institute of Fluid Science, Tohoku University, experimental studies on the Laser-driven In-Tube Accelerator (LITA, Fig. 1) have been intensively conducted. A projectile of up to 3 grams has been lifted off in the acceleration tube. A coupling coefficient can be larger than 300 N/MW. As the propellant, monatomic gases, argon, krypton and xenon, have been tested. In the experiments, the impulse scales with the reciprocal of the speed of sound. This implies that in order to increase the impulse the molecular weight of the propellant species should be increase. Sasoh et al. gave a simple explanation of this impulse characteristics: Assuming a constant volume energy addition from a pulsed laser irradiation, the post-shock overpressure and the shock Mach number are unchanged. Yet, the duration time of the force is in inverse proportion to the speed of sound, thereby resulting in the above-mentioned impulse characteristics. The peak overpressure measured on the LITA acceleration tube wall is independent of the species, though it depends on the fill pressure (see Fig. 2). Various extensions of the LITA technology are currently explored. Also, numerical study has been conducted as a collaborative work with Seoul National University and Pusan National University, Korea.

CP664, Beamed Energy Propulsion: First International Symposium on Beamed Energy Propulsion,
edited by A. V. Pakhomov
© 2003 American Institute of Physics 0-7354-0126-8/03/$20.00

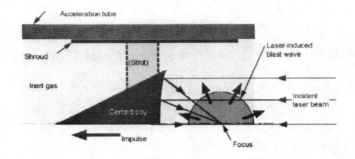

FIGURE 1. LITA operation principle.

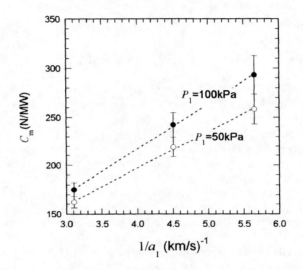

FIGURE 2. Measured impulse vs. the reciprocal of speed of sound.

NATIONAL AEROSPACE LABORATORY

A thrust impulse generation from laser-illuminated targets in vacuum has been investigated. The laser source was Nd: YAG operated in Q-switch mode with a fundamental 1.064 µm output. The laser beam was focused into a propellant placed in a vacuum chamber. The illuminated propellant became high temperature expanding gas, which was received by a nozzle to produce thrust impulse. A series of thrust impulse generation experiments are conducted using water as propellant to realize low-cost reusable propulsion system (Figs. 3 and 4).

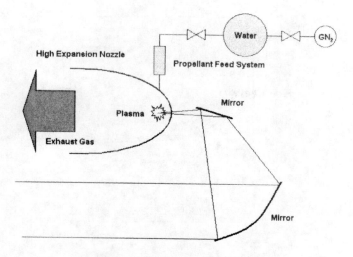

FIGURE 3. Laser-Propelled OTV engine.

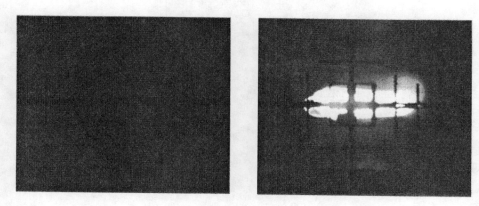

FIGURE 4. Laser Irradiation into Water Droplet.

THE UNIVERSITY OF TOKYO

RP Laser Thruster

Thrust and drag produced in a Lightcraft type vehicle were computed.[3,4] The thrust history is plotted in Fig. 5. Figure 6 shows the computed velocity-time diagrams of a laser SSTO. The vehicle could not reach the orbital velocity at $E_i \leqq 30MW$. The fraction of laser pulse energy that is converted to blast wave energy η_s was experimentally examined.[5,6] It was found a function of ambient pressure. (Fig. 7)

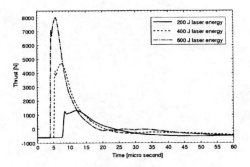

FIGURE 5. Computed thrust history.

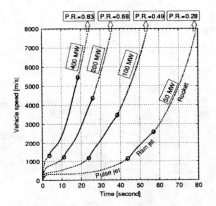

FIGURE 6. Velocity-time diagram of a laser SSTO

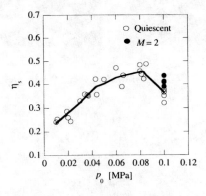

FIGURE 7. Blast wave energy conversion efficiency η_s

CW Laser Thruster

Figure 8 shows a CW laser thruster. Thrust efficiency is estimated by measuring thrust.[7,8] Temperature distribution is computed by CFD.7) The maximum temperature is around 15,000K. (Fig. 9)

FIGURE 8. The University of Tokyo CW laser thruster

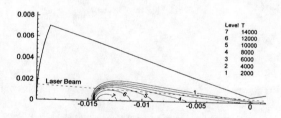

FIGURE 9. Temperature contours. $P=700W$, Ar.

TOKAI UNIVERISTY

Forward Plasma Acceleration by Intense Laser Pulses for Space Propulsion

A feasibility study of the utilization of a forward laser plasma accelerator for space propulsion applications has been conducted at Tokai University.[10-12] A part of experiments is also being conducted at Advanced Photon Research Center of Japan Atomic Energy Research Institute (JAERI). The interaction of the ultraintense laser pulses with solid targets leads to the generation of fast particles, from x- and g-ray photons to high energy ions, electrons, and positrons.[13-17] In particular, here, an interest has developed in ion acceleration by compact high-intensity lasers with potential applications for the space propulsion. The laser light terminates at the target surface and drives high-energy electrons generated in front of it deep inside the target. Because of planar charge separation these electrons produce a strong electrostatic field accelerating ions in a forward direction. In this case, high energy electrons expand faster and the ions form a well collimated relativistic beam confined in the transverse direction by the pinching in the self-generated magnetic field (Fig.10).[17] Maximum ion energy of order several MeV through the acceleration has been reported in previous studeis.[10-17] Assuming a proton accelerated up to 1 MeV, its speed corresponds to 3 ~ 4 % of the speed of light and the specific impulse Isp of order 106 sec.

At Tokai University, propulsive performance tests for the plasma acceleration with laser pulses irradiated on various targets are conducted as a preliminary study.[12] The experiment includes, (1) impulse measurement using time variation of a cantilever displacement, (2) plasma speed measurement through plasma wave visualization with a highspeed camera, and (3) plasma speed measurement with TOF analysis (Fig.11). In addition to the experiment, a PIC simulation on the fast ion acceleration mechanism is also conducted.

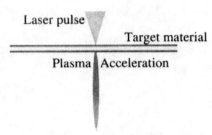

FIGURE 10. Forward Plasma acceleration through a solid target with high power laser irradiation

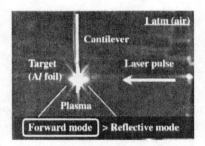

FIGURE 11. Photo of laser-induced plasma (Ti:S laser, pulse energy: 10mJ/pulse, pulse width: 70 fsec).

INSTITUTE FOR LASER TECHNOLOGY[18]

The research unit of Institute for Laser Technology, National Aerospace Laboratory and Kinki Universty is studying laser propulsion mechanism using a double layered target. Laser ablation is used to generate thrust. Momentum coupling coefficient has been enhanced by at least an order of magnitude. The momentum and temporal profile of force exerted on the targets are measured using pendulum and piezoelectric sensors. It was found that momentum generation from the double layered target exhibits two components (Fig.12). High momentum coupling coefficient, Cm is due to the second component which carries large amount of mass with very low velocity. This method can be used not only for generating high Cm but also for controlling Cm in wide range of parameter (Fig.13). When applied to LOTV (Laser Orbital Transfer Vehicle) or other on-orbit thrusting applications, this method can optimise the efficiency of energy, propellant mass or pay-load-propellant mass ratio by tailoring the operating parameters to specific use.

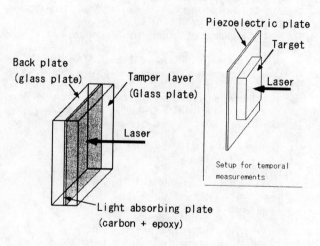

FIGURE 12. Double layered target used for the experiments

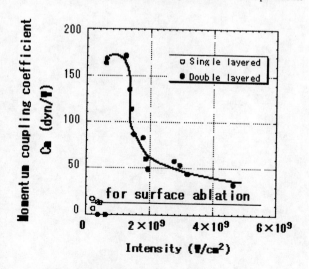

FIGURE 13. Cm strongly depends on laser intensity.

ACKNOWLEDGMENTS

The author is deeply indebted to the colleagues who provided the updated information on laser propulsion research activities of each organization.

REFERENCES

1. Sasoh A (2001) Laser-driven in-tube accelerator. Review of Scientific Instruments, 72:1893-1898
2. Sasoh A, Choi J-Y, Jeung I-S, Urabe N, Kleine H, Takayama K (2001) Impulse Enhancement of Laser Propulsion in Tube. Postepy Astronautyki, 27:40-50
3. Katsurayama, H., et al.: Numerical Analyses on Pressure Wave Propagation in Repetitive Pulse Laser Propulsion, AIAA Paper 2001-3665, 2001.
4. Katsurayama, H., et al.: Computational Performance Estimation of Laser Ramjet Vehicle, AIAA Paper 2002-3778, 2002.
5. Mori, K, et al.: A Far-Field Repetitive Pulse Laser Thruster, AIAA Paper 2001-0649, 2001.
6. Mori, K., et al.: Laser Plasma Production and Expansion in a Supersonic Flow, AIAA Paper 2002-0634, 2002.
7. Toyoda, K., et al.: Continuous-wave Laser Thruster Experiment, Vacuum, Vol. 59, 2000, pp. 63-72.
8. Hosoda, S., et al.: The Observation of the Laser-Sustained Plasma in a CW Laser Thruster, Proc. 22nd Int'l Symp. Space Tech. and Sci., ISTS-2000-b-5, Morioka, 2000.
9. Komurasaki, et al.: Numerical Analysis of CW Laser Propulsion, *Trans. Japan Soc. Aero. Space Sci.*, Vol. 44, No. 144, 2001, pp. 65-72.
10. Horisawa, H., et al., AIAA Paper 2000-3487 (2000).
11. Horisawa, H., et al., AIAA Paper 2001-3662 (2001).
12. Horisawa, H., et al., IEPC 01-206 (2001).
13. Campbell, P.M., et al., Physical Review Letters, 39 (5), 1977, pp.274-277.
14. Decoste, R., et al., Physical Review Letters, 40 (1), 1978, pp.34-37.
15. Fews, A.P., et al., Physical Review Letters, 73 (13), 1994, pp.1801-1804.
16. Roth, M., et al., First Intl. Conf. Inertial Fusion Sci. Appl., 1999.
17. Sentoku, Y., et al., Physical Review, E62 (5), 2000, pp.7271 – 7281.
18. Shigeaki Uchida, S., et al.: Comparison of Surface and Internal Laser Irradiation of Solid Propellant for Propulsion, AIAA Paper 2002-992.

Laser Propulsion Activities in Germany

Willy L. Bohn and Wolfgang O. Schall

DLR-Institute of Technical Physics, D-70503 Stuttgart, Postfach 80 03 20, Germany

Abstract. Activities related to laser supported propulsion concentrate on investigations of the fundamental phenomena arising from the interaction of high-power CO_2 laser pulses with a simple bell engine. Breakdown dynamics, plasma, and lightcraft accelerations are carefully measured using optical and laser diagnostics. In a first-order approach the expanding plasma can be described by the point explosion model with counterpressure. Wire guided vertical flight experiments in the laboratory have been undertaken and analyzed. Comparative impulse measurements in ambient air and ablated material are presented for a series of experiments performed either at atmospheric pressure or at reduced pressures down to vacuum. Impulse coupling coefficients, average exhaust velocities, specific impulse, and jet efficiencies are derived from the experimental data. The repetitively-pulsed CO_2 laser device used in all experiments shows a potential of achieving 50 kW average power. Finally, long-term perspectives of laser propulsion will be addressed.

INTRODUCTION

Laser propulsion activities at the DLR-Institute of Technical Physics are aimed at the investigation and understanding of fundamental phenomena in pulsed laser propulsion. The application envisioned is the launching of so-called lightcrafts with comparably small payload masses in the order of several kilograms either as high-altitude sounding rockets or nanosatellites in low Earth orbit (LEO). Going back to an idea of Kantrowitz [1], published in 1972, pulsed laser propulsion promises single stage to orbit flights (SSTO) with high thrust to weight ratios and low launch costs at a rapid turn around. After Myrabo has carried out free flights of a lightcraft in cooperation with the US Air Force in 1997 [2, 3], DLR has started its research at the end of 1998 and was soon able to launch a small wire-guided lightcraft to the ceiling of the laboratory. After demonstrating the feasibility of laser propulsion the research was motivated to understand, optimize and scale the laser propulsion mechanism by using a simple lightcraft geometry and the available laser source [4]. In 2000 and 2001 a cooperation with the US Air Force Research Laboratory, Edwards AFB, CA allowed a rigorous comparison of the performance of two very different lightcrafts for many parameters in ambient air, with and without solid propellant as well as in

CP664, *Beamed Energy Propulsion: First International Symposium on Beamed Energy Propulsion*,
edited by A. V. Pakhomov

vacuum and in low pressure surroundings. These experiments have been performed at the DLR test stand [5]. The performance investigations will be backed in the future by the investigation of tracking schemes. This paper provides a survey over the work at DLR and summarizes the major results. At the end, an outlook and a brief discussion of the open questions will be given.

EXPERIMENTAL FACILITIES

The major parts of the DLR lightcraft test facility are schematically summarized in Figure 1. The two main constituents are a pulsed, electron beam sustained electric discharge laser and a vacuum test chamber with a pendulum for impulse measurements. The laser can be operated with several different laser gases at their laser wavelengths, such as CO_2, CO and CO overtone and ArXe. For CO_2 the laser is rated at an average power of 50 kW [6], although the present status of the power supply does not allow for such a high power. Maximum obtained pulse energy is 450 J and maximum pulse repetition rate is 100 Hz. The laser beam can be coupled out by either a stable resonator, producing a nearly homogeneous flat top near field profile, or an unstable resonator with a ring shaped profile as indicated in Figure 1. The pulse length for the experiments was selected to 11 µs, although a few tests were run at a shorter pulse length, too. The laser beam energy is measured online and fed into a PC, where the signal is translated into Joules. The laser beam is transmitted to the vacuum tank via a reflecting telescope for beam size adaptation to the lightcraft aperture. The tank inside is 800 mm in diameter and 1150 mm high. The beam enters the tank through a 120 mm diameter window. The lightcraft is suspended as a

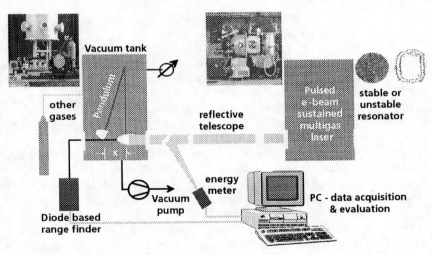

FIGURE 1. Schematic of the test stand setup. The photographic inserts show the open tank on the left and the laser in the center, as well as the two resonator modes on the left.

pendulum with a length of 645 mm. A laser diode based range finder is aimed at the backside of the lightcraft and measures the displacement of the pendulum, x, with a submillimeter resolution as a function of time. These data are also fed into the PC and allow the determination of the impulse, I, and together with the laser pulse energy, E, the impulse couling coefficient, $c_m = I / E$. The tank can be evacuated to a pressure of < 1 mbar within a few minutes. It can also be filled with other gases like N_2. The pressure is measured by a conventional manometer.

The tank can be substituted by a simple launch pad with a bending mirror to redirect the laser light to enter the lightcraft from below. The lightcraft slides along a thin wire up to the laboratory ceiling, which is 6 m above the launch pad.

EXPERIMENTS

Analysis of the Expanding Plasma

Figure 2 shows a photograph of a parabolic lightcraft with a diameter of 100 mm and a height of 62.5 mm. A sector of the wall is cut away to allow a view into the thrust chamber. The wall is polished on the inside and focuses the laser beam onto a spot in 10 mm distance from the apex of the paraboloid. A metal needle extends 20 mm from the apex along the axis of symmetry and ensures a proper plasma breakdown even for low pulse energies. The plasma ball is seen in the photograph.

The measurement of the duration of the visible light that is emitted from the plasma has produced a curve with a double peak. The first maximum is reached, when the laser pulse is terminated (approximately after 10 µs). This peak is followed by a second, even higher maximum after another 14 µs, before the emission decays during the following 35 µs. This unexpected behavior of the plasma discharge is explained by the ingnition of a spherical blast wave at the focal point. The blast wave

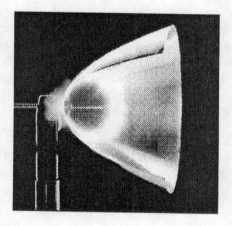

FIGURE 2. Air breakdown in bell-shaped lightcraft.

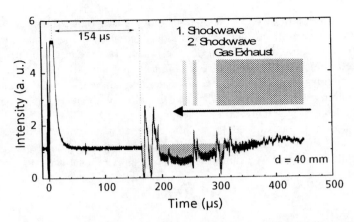

FIGURE 3. Time dependent density change 40 mm behind thruster exit after a laser induced air breakdown in the lightcraft

expands into the surrounding gas and hits the wall of the thrust chamber. There it is reflected and runs a second time through the still hot gas, recompressing it and raising its temperature again. As a consequence of this mechanism two succesive (shock) waves should be observable emanating from the exhaust of the thruster, followed by a bulk of hot gas. This expectation has been proven to be correct by a time dependent probing of the density of the gas at some distance from the exit. For this purpose a HeNe laser beam was sent across the exhaust and recorded by a photodiode with a small slit in front. Any density distortion would sweep the HeNe beam across the slit. As Figure 3 shows, the trace of the HeNe light signal at a distance of 40 mm from the thruster exit exhibits two sharp signals 154 µs after the plasma ignition, followed by a long lasting density distortion attributed to the bulk of hot gas. Measurement of the arrival time of the first wave at various distances from the exit allows the determination of the velocity, which for a pulse energy of 165 J reaches Mach no. 1.82. So it is clearly a shock wave.

In an attempt to understand the experimental findings the measured velocities are traced back towards the initial location of the air breakdown. The formation of a laser supported detonation wave following the air breakdown cannot be brought into coincidence with the measured data. However, if the model of a strong explosion in an atmosphere is employed with a time dependence of the shock location then $x \propto (E/\rho_0)^{1/5} t^{2/5}$ [7]. At some distance from the explosion center the calculated blast wave velocity goes over into that of the measured shock wave with constant propagation velocity. E is the instantaneous energy deposit into the undisturbed gas of density ρ_0. This is shown in Figure 4. With x = 5 cm the transition is still inside the lightcraft.

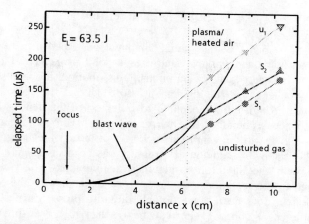

FIGURE 4. Evolution of the shock and plasma waves in the time-space domain.

Wire-Guided Flight

Wire-guided flights to the ceiling of the laboratory have been untertaken with different lightcraft masses up to 53 g and various pulse energies and repetition rates. The geometry was the same as for the diagnostic investigations. The flight height as a function of time was recorded using the laser diode based range finder. Figure 5 is an example of a flight and its analysis. With 40 pulses of 175 J at a repetition rate of 45 Hz a 53 g lightcraft reached a height of 6 m within 1.5 s. The propelled distance was about 4 m with a final velocity of 6.5 m/s after 890 ms. The maximum acceleration of 10.5 m/s^2 corresponds to a thrust of 1.05 N and a coupling coefficient of 137 N/MW. In addition, it has been noted that the small aluminum lightcraft structure undergoes consideral mechanical and thermal stress at laser pulse energies greater than 100 J. About 1/3 of the laser pulse energy is found as heat in the structure.

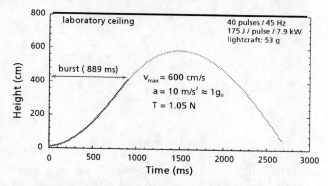

FIGURE 5. History of a vertical laboratory flight experiment.

Impulse Measurements

Impulse measurements with a pendulum have been performed for a variety of parameters in air with and without Delrin as a solid propellant. Parameters tested were the laser pulse energy, the beam intensity distribution, the intensity at the target for the solid propellant, and the ambient pressure. During the ascent of a spacecraft from the ground through the atmosphere to outer space the density of the air drops exponentially to zero. Therefore, the surrounding air eventually ceases to be useful for propulsion and, as in conventional rocketry, on-board fuel must replace it. The decline of the thrust with reducing air pressure has been investigated by suspending the pendulum with the lightcraft inside the vacuum tank. The result is shown in Figure 6 and led to the remarkable discovery that, depending on the pulse energy, the coupling coefficient does not change notably down to pressures of 200 to 500 mbar. Therefore, altitudes of about 11 km can be reached in the air-breathing operation mode without loss in performance (Fig. 7) and more than 20 km, if a certain loss in thrust is admitted. The result is of significance, since it allows a substantial reduction of necessary on-board fuel in exchange for increased payload. Otherwise, the flight through the atmosphere would be particularly fuel consuming due to the added force of air drag, which increases with the square of the flight velocity.

Additional propellant may increase the performance during atmospheric flight, however. It is indispensable during propelled flight at high altitudes and in the vacuum of space. Delrin, a plastic material, has been selected as a first choice. Cylindrical Delrin pins of 8 and 10 mm diameter and of 8.5 and 17 mm in length have been inserted into the lightcraft. The laser light irradiates the pin either on the

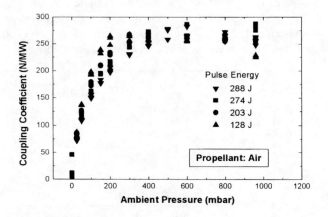

FIGURE 6. Impulse coupling coefficient for various pulse energies in ambient air at reduced ambient pressure.

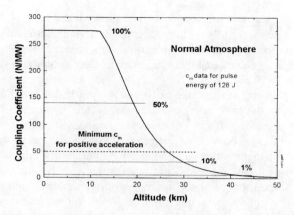

FIGURE 7. Change of the coupling coefficient with altitude in the normal atmosphere.

circumference and evaporates material in radial direction or on the front surface. The intensity distribution on the pin surface varies with the diameter of the pin.

Figure 8 shows the variation of the coupling coefficient with pressure for a fixed pulse energy of 250 J and for the following 3 cases: 1) Air as the only propellant, like in the previous experiments, 2) Delrin as additional propellant, and 3) Delrin in a surrounding atmosphere of pure nitrogen. It is seen that Delrin substantially increases the c_m-value at all pressures. In vacuum the coupling coefficient amounts to about 250 N/MW. As the air pressure rises c_m increases steadily, displaying no leveling off. In contrast, the operation with surrounding pure nitrogen exhibits a simple displacement of the air curve by a practically constant value. The difference between the two curves with Delrin in air and nitrogen is attributed to a release of combustion energy from a reaction of Delrin vapor with the oxygen in the air. In this case and at

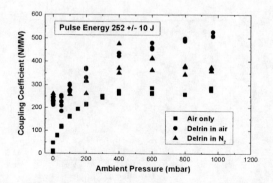

FIGURE 8. Variation of the coupling coefficient with changing pressures of ambient air or nitrogen using Delrin as a solid propellant.

the proper pulse energy a maximum coupling coefficient of 610 N/MW has been found, compared to 400 N/MW in vacuum. The combustion effect improves the propulsive conditions in a hybrid operation where either the performance of a chemical propellant is enhanced by laser energy or laser propulsion is augmented by a chemical reaction.

As Figure 9 shows, the diameter of the Delrin pin had a significant effect on the achieved impulse as the laser energy is changed. While in vacuum for the 8 mm pin there is almost no increase if the pulse energy is doubled, an almost linear increase also by a factor of two is observed with the 10 mm pin. If the short pin is irradiated on the front side just behind the focal point only a very weak impulse is found. Along with these experiments the mass loss has been determined by weighing the Delrin pin before and after three shots of equal parameter setting. It turned out that the mass loss showed exactly the same behavior as the impulse. If the impulse is plotted versus the mass loss a linear one to one dependence is found. Two significant conclusion can be drawn from these experiments:

1) The impulse of the lightcraft in vacuum, $\Delta m \cdot v_e$, is a function of the evaporated amount of Delrin Δm only (v_e: average exhaust velocity). It is not possible to raise the inner energy in the Delrin vapor by a higher pulse energy so as to produce a higher exhaust velocity. More energy produces only more mass. The specific impulse is therefore a constant.

2) As the experiments with different target intensities show, the intensities for the thinner pin were too high to fully absorb the enrgy in the target. It is assumed that in this case a plasma wave front (LSD wave) is formed that absorbes part of the pulse energy, in particular during the later phase of the pulse duration. This energy fraction cannot be made available anymore as a pressure increase inside the thrust chamber. For optimum deposition of energy it is necessary to tailor the pulse intensity to lower values and accept a corresponding loss in impulse. To achieve a high thrust nevertheless, the repetion rate of the pulses has to be increased.

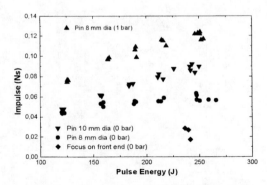

FIGURE 9. Behavior of impulse exerted to the lightcraft for different propellant pin diameters and irradiations as a function of the pulse energy.

FIGURE 10. a) German bell-type lightcraft,   and b) US plug nozzle type lightcraft.

Within the cooperation with the Air Force Research Laboratory (AFRL), Edwards AFB, CA it became possible to directly compare the impulse measurements of the German bell type lightcraft (G-Lightcraft) (Fig. 10a) under identical conditions with a lightcraft of the same diameter but with an entirely different geometry. This US-lightcraft has the shape of a plug nozzle with a central spike and a parabolic contour directed radially outward [8] (Fig. 10b). The highly reflective contour focusses the incident light beam onto a line focus along the outer rim of a shroud where a ring of solid propellant may be placed. Comparative measurements of the impulse and the mass loss of Delrin have been performed only with the stable resonator beam configuration and its flat beam intensity profile.

Figure 11 compares the derived coupling coefficient of the two lightscrafts for a fixed pulse energy and as a function of the ambient air. For the G-Lightcraft the result with the 8 mm diameter pin is shown. Quite in contrast to the ever increasing

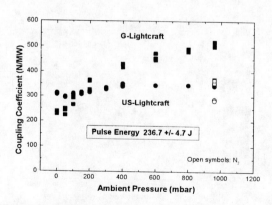

FIGURE 11. Comparison of the coupling coefficients as a function of the ambient pressure in air and nitrogen with Delrin propellant.

coupling coefficient with increasing pressure of the G-Lightcraft, the coupling coefficient for the plug type US-lightcraft starts out with higher values at full vacuum. Up to a pressure of 400 mbar a marginal increase is observed, but for pressures > 400 mbar it remains constant and agrees well with the value for the G-Lightcraft in an inert nitrogen atmosphere of 1 bar. The comparative nitrogen value for the US-Lightcraft again corresponds to the vacuum value. This behavior, in combination with the already described hybrid combustion process for the G-Lightcraft at higher pressures, suggests that the observed energetic interaction with air is of minor importance for the geometry of the US-Lightcraft. The combustion occurs apparently outside of the realm of the plug nozzle and can thus not contribute to the thrust.

The impulse for the US-Lightcraft is found to be a linearly increasing function with the pulse energy. For infinite pulse energy the coupling coefficient therefore approaches a limiting value that is calculated from the measured data as 350 N/MW. In Figure 12 the experimental data are plotted versus the pulse energy for the US-Lightcraft as well as for the G-Lightcraft with the pin diameters of 8 and 10 mm (side-on irradiance). The general behavior is drastically different for each of the three cases. Starting from about 275 N/MW at the lowest pulse energy of 120 J the coupling coefficient for the US-Lightcraft increases with increasing energy asymptotically towards its limiting value. In contrast, the G-Lightcraft displays the highest values (up to 400 N/MW) for the lowest energy and decreases towards higher pulse energies. This decrease is small for the 10 mm diameter pin but quite substantial for the pin with 8 mm diameter. Since the intensity on the Delrin ring of the US-Lightcraft is almost an order of magnitude lower than on the G-Lightcraft pin the postulated shielding effect does not yet occur for the US-Lightcraft, even at the highest pulse energy.   An effective exhaust velocity that is averaged over the exhaust cross-section and the pulse time can be determined from the impulse and the loss of mass. The velocity and depending quantities like the jet efficiency (ratio of the kinetic jet energy to the laser pulse energy), are only meaningful if no substantial other mass

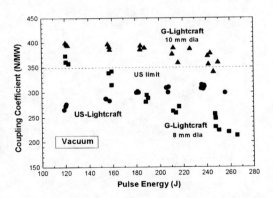

FIGURE 12. Comparison of the coupling coefficient in vacuum as a function of the pulse energy.

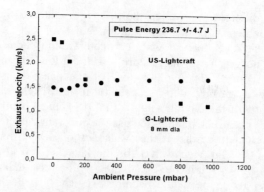

FIGURE 13. Average exhaust velocity vs. ambient
pressure of air with Delrin propellant.

like air is exhausted from the interior of the lightcraft. This holds only for vacuum.
Since the air fraction is not known a reasonable assumption has to be made in order to
derive the air fraction and the velocity from the balance equations of momentum and
energy. Assuming a constant energy deposition (not suitable, if combustion is
significant) the dependency of the velocity on the ambient pressure is found as shown
in Figure 13. Because a rapidly increasing mass fraction of air is ejected from the G-
Lightcraft as the ambient pressure goes up, the exhaust velocity drops from 2.5 km/s
to less than half. However, the specific impulse remains that of vacuum of 255 s
because the ablated Delrin mass does not change with the ambient pressure. In
contrast to findings at the AFRL [9] with an unstable resonator beam the ablated mass
for the US-Lightcraft was substantially higher with the homogenous stable laser beam
of the DLR laser. Therefore, the resulting exhaust velocity is lower and even rises
slightly with the air pressure. The lower exhaust velocity also results in a lower jet
efficiency. With an efficiency of 40% at the lowest pulse energy the efficiency for the
G-Lightcraft is relatively high. However, as the incident energy on the propellant
increases, the efficiency decreases, while for the US-Lightcraft it goes up from 15 %
to nearly 25 %.

OPEN QUESTIONS

Inspite of the discovery of several fundamental effects for pulsed laser propulsion
there remain some open questions. As propellants only air and nitrogen gas and the
solid Delrin have been investigated so far. For Delrin the existence of a flux
limitation has been found. However, it is quite certain, that a proper change in the
pulse length of the laser, as well as a different wavelength can improve the situation.
Also, a different propellant would require a different set of intensity, pulse length and
wavelength. Some quick experiments in atmospheric air have indicated an
improvement in the coupling coefficient for a pulse length reduced to 7 μs. It is

89

further known that a different opening of the bell type lightcraft thruster leads to a different coupling coefficient [10]. A specific impulse of 600 s and higher, as required for SSTO, has not yet been shown for a lightcraft until today.

Another problem not yet rigorously addressed at DLR is the question of flight stability and attitude control. To avoid a loss of the beam in a free flight it is important to achieve a "beam riding" capability for the lightcraft, as has been demonstrated already with the Myrabo lightcraft [3]. The problem of tracking and guidance of a lightcraft is currently addressed using a quadrant detector approach.

Finally, the most crucial question is the availability of an appropriate pulsed laser with sufficient power (several Megawatts) for the launch of nanosatellites. Since a focal range of more than 500 km has to be covered with a focal diameter of the order of 1 m the transmitter telescope would become excessively large for the CO_2 laser with a wavelength of 10.6 μm (>15 m mirror diameter). Unfortunately, this is the only laser that is mature enough to be scaled to the requested power level. There are other lasers with shorter wavelengths that have already achieved MW power levels, however only in the continuous operation mode (cw): The HF/DF laser and the chemical oxygen-iodine laser (COIL). While the DF laser (wavelength 3.7 to 4.5 μm) in the pulsed chain reaction mode yields very high pulse energies, it is probably not practical in long term operation with high rep rates. The COIL (1.3 μm) could not yet be operated at high average pulse powers. Other suitable lasers are the CO-overtone (3-4 μm) and the ArXe laser (1.7 μm). Both lasers have shown single pulse energies in excess of 100 J, but not in repetitive operation. It is very doubtful if a solid state laser can ever be scaled to the required power level, unless it comes to a fusion type laser facility of considerable extension. Excimer lasers like KrF could well meet the goal, but their wavelength is too short and would require Raman shifting. Finally, the free electron laser (FEL) could in principle fulfill all requirements, although the required operation conditions have not yet been demonstrated either and the investment costs appear excessive for commercial laser propulsion applications.

CONCLUSIONS

A small test stand has been made available at DLR, comprising a vacuum tank with a pendulum for impulse measurements and a pulsed multigas laser with pulse energies up to 450 J for the wavelength of CO_2. The laser operates with a stable or an unstable resonator with different intensity distributions and beam divergences. The pulse length can be adjusted between 3 and 12 μs by a change in the pulse forming network. An indoor launch to a height of 6 m is possible. With this capability measurements of the impulse, coupling coefficient and specific impulse have been carried out for different pulse energies, ambient pressures and target intensities.

Some of the more remarkable findings are: The dependency of the impulse in air at various pressure levels; the effect of hybrid propulsion by a laser induced combustion of Delrin vapor in air; the significance of the target intensity for maximum energy deposition into the target; the dependence of the exerted impulse to the lightcraft on the evaporated Delrin mass only; and the significantly different performance behavior of a bell-type thruster and plug nozzle-type thruster.

ACKNOWLEDGMENTS

Part of this research has been supported by the EOARD, London under the contract numbers F61775-00-WE033 (2001) and FA8655-02-M4017 (2002). The authors thank Dr. F.B. Mead Jr. and Dr. C.W. Larson for the fruitful cooperation. We also appreciate the contributions of Dr. H.-A. Eckel and S. Walther, who are an indispensible part of the experimental crew.

REFERENCES

1. Kantrowitz, A., "Propulsion to Orbit by Ground-Based Lasers," *Astronautics and Aeronautics*, Vol. 10, No. 5, pp.74-76, 1972.
2. Myrabo, L. N., Messitt, D. G., Mead Jr., F. B., "Ground and Fight Tests of a Laser Propelled Vehicle", *36th Aerospace Science Meeting & Exhibit*, paper AIAA 98-1001.
3. Myrabo, L. N., "World record flights of beam-riding rocket lightcraft: Demonstration of "disruptive" propulsion technology", *37th AIAA/ASME/SAE/ASEE Joint Propulsion Conference*, paper AIAA 2001-3798.
4. Bohn, W. L., "Laser lightcraft performance," *High-Power Laser Ablation II,"* Proceedings of SPIE Vol. 3885 (1999), pp. 48-53.
5. Schall, W. O., Eckel, H.-A., Mayerhofer, W., Riede, W., Zeyfang E., "Comparative lightcraft impulse measurements", *International Symposium on High-Power Laser Ablation IV*, C. R. Phipps, Editor, Proceedings of SPIE Vol. 4760, 908-917, 2002.
6. Mayerhofer, W., Zeyfang, E., Riede, W., "Design data of a repetitively pulsed 50 kW Multigas Laser and recent experimental results," *XII Intern. Symp. on Gas Flow and Chemical Lasers and High-Power Laser Conf.*, Proceedings of SPIE, Vol. 3574, pp. 644-648, 1998.
7. Zel'dovich, Ya. B. and Raizer, Yu. P., *Physics of Shock Waves and High-Temperature Hydrodynamic Phenomena*, Academic Press, New York and London, 1966, p. 94.
8. Larson, C. W., Mead Jr., F. B., and Kalliomaa, W. M., "Energy conversion in laser propulsion III" in *High-Power Laser Ablation IV*, edited by C. R. Phipps, Proceedings of SPIE Vol.4760 (2002), pp. 887-898.
9. Larson, C. W., Mead Jr., F. B., "Energy conversion in laser propulsion", *39th AIAA Aerospace Sciences Meeting & Exhibit*, paper AIAA 2001-0646.
10. Myrabo, L. N., Libeau, M. A., Meloney, E. D., Bracken, R. L., Knowles, T. B., "Pulsed Laser propulsion Performance of 11-cm Parabolic 'Bell' Engines Within the Atmosphere," *AIAA 33rd Plasmadynamics and Lasers Conference*, paper AIAA 2002-2206.

FLUID DYNAMICS IN
LASER PROPULSION

Conversion of Blast Wave to Impulse in a Pulsed-laser Thruster

K.Mori*, H. Katsurayama*, Y. Hirooka**, K. Komurasaki**, and Y. Arakawa**

*Department of Advanced Energy, University of Tokyo
**Department of Aeronautics and Astronautics, University of Tokyo
Hongo 7-3-1, Bunkyo-ku, 113-8656, Tokyo, JAPAN

Abstract. Energy conversion efficiency in the Repetitive Pulse Laser Thrusters has been studied. Whole conversion processes are divided into two processes; Energy conversion from the laser energy to the blast wave energy and conversion from the blast wave to impulse. The fraction of laser energy that is converted into the blast wave energy was deduced to be 50% by visualizing the blast wave motion. A simple model has been proposed to estimate the impulse from the blast wave energy, and it has been validated by the experimental results.

INTRODUCTION

Laser propulsion is one of the futuristic candidates for satellite launching. An air-breathing Repetitively Pulse (RP) laser thruster, which is illustrated in Fig.1, utilizes the atmospheric air as a working fluid introduced through an intake. In our laboratory, experiments and numerical computations are performed to clarify the influence of ambient pressure and flow velocity on the thruster performance. High-power laser beam, which is transmitted remotely from the ground, is focused in a nozzle, and a breakdown process is initiated in the vicinity of focus.

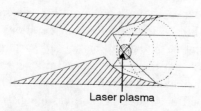

Laser plasma

FIGURE 1. Air-breathing RP Laser thruster

CP664, *Beamed Energy Propulsion: First International Symposium on Beamed Energy Propulsion*,
edited by A. V. Pakhomov

The development stage of the breakdown is accompanied by propagation of laser-absorption region along the laser light channel [1]. At the laser power density greater than 10^7 W/cm^2, absorption occurs in the Laser Supported Detonation (LSD) regime. In this regime, a shock wave and plasma front propagate together at supersonic speed absorbing the laser beam.

At the power density lower than 10^6 W/cm^2, the absorption occurs in the Laser Supported Combustion (LSC) regime. The LSC wave can be defined as the localized plasma region that propagates at subsonic speed to absorb the laser beam. In this regime, the influence of shock compression on the absorption is expected slight since the shock wave propagates apart from the LSC wave. If the input power is low and the absorption length is long, the absorption occurs mostly in the isobaric way.

Since the shock wave compresses the absorption region, the absorption coefficient in the LSD wave is higher than that in the LSC wave, the absorption occurs quite efficiently. LSD regime terminates in the transient range of the power density between 10^6 and 10^7 W/cm^2.

Plasma expands to drive a blast wave. The absorbed energy is converted into the blast wave energy E_{bw}, radiation loss, and chemical potential energy. Here, the blast wave energy is defined as the sum of translational and kinetic energy within the blast wave. After the blast wave has gone away, a high-temperature region, "fire-ball" remains in the vicinity of the explosion center. The chemical potential of the fire-ball is frozen partly.

Impulse is imparted on the nozzle by reflecting the blast wave as shown in Fig.1. Since E_{bw} uniquely determines the pressure distribution within the blast wave, it is useful to characterize the energy conversion from the laser energy E_i to the blast wave energy E_{bw} by the blast wave energy conversion efficiency η_{bw} defined as

$$\eta_{bw} \equiv E_{bw}/E_i \qquad (1).$$

In addition, the process of the conversion from the blast wave energy to the impulse must be understood to predict the performance. This process is greatly influenced by the size and shape of nozzle since it is dominated by the expansion of the blast wave under a boundary condition determined by the nozzle configurations. The influence of nozzle configurations was investigated for nozzles with simple shapes (Conical and parabolic) systematically by Ageev et.al [2], and it was concluded that there was an optimum nozzle length for a certain input energy. On the other hand, the influence of the input laser energy on the momentum coupling coefficient has been investigated using a parabola (Bell) nozzle and Myrabo type Lightcraft [3-5]. These studies also suggest that there would be an optimum input energy for a certain nozzle size. Although many experimental and numerical investigations on the momentum coupling coefficient are presented, it is not clearly presented how efficiently the input energy can be converted into the fluid dynamic energy (how much is the blast wave energy conversion efficiency?) and how efficiently the nozzle can convert the fluid dynamic energy to the impulsive thrust (how can we define the nozzle efficiency and how much is it?). Such a situation makes it difficult to predict the performance and to construct the design rules. In order to define and predict the nozzle efficiency, analytical expressions on the relation between nozzle scale and input energy is very useful.

In this study, η_{bw} has been deduced experimentally using a shadowgraph method, and a simple model to estimate the impulsive thrust has been proposed. The simple model agreed reasonably well with an experimental result.

BLAST WAVE ENERGY

Blast wave energy was obtained from the shadowgraph images. The experimental apparatus is presented in Ref. [6]. A TEA CO_2 laser was used and the input laser energy was changed from 2.3 to 12.8 J. The laser beam Plasma was produced under an atmospheric air. Shadowgraphs were taken using an ICCD camera with a delay circuit.

The theory of blast wave energy is presented in APPENDIX since it is not the main point of this paper to present how the blast wave energy was deduced. Figure 2 shows the relation between E_i and E_{bw}. E_{bw} was proportional to E_i, and η_{bw} was found 0.48 ± 0.05. The error was mostly originated from the laser energy fluctuation. The efficiency was insensitive to the input laser energy within the tested range.

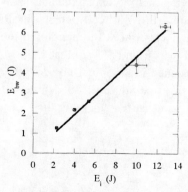

FIGURE 2. E_{bw} v.s. E_i. η_{bw} was found 0.48.

IMPULSIVE THRUST

Measurement Apparatus and data processing

A ballistic pendulum was used to measure the impulsive thrust. Its characteristics are shown in Table. I. The arm was supported by a ball-bearing. The pendulum movement was monitored using a laser displacement meter. After an impulse is imparted at the end of the arm, the pendulum oscillates slowly, and the displacement decays due to the friction of the bearing. Theoretically, the motion is described by an equation:

$$I\ddot{\theta} + c\dot{\theta} + k\theta = 0 \qquad (2)$$

where θ is the angle of the pendulum, I, the moment of inertia, c, the decay constant, and k is connected with the frequency of the oscillation ω as

$$\omega = \sqrt{\frac{k}{I} - \left(\frac{c}{2I}\right)^2} \qquad (3).$$

From the solution of Eq. (2), the displacement of the pendulum can be described as

$$\Delta X(t) = P\left[\frac{l^2}{I\omega}\exp\left(-\frac{c}{2I}t\right)\sin \omega t\right] \qquad (4)$$

where P and l is impulse imparted at the end of the pendulum and the length of the pendulum, respectively. The impulse was deduced by least-square fitting of Eq. (4) to the displacement curve of the pendulum.

Calibration was also performed using a load cell on an impulse hammer to validate the impulse deduction method explained above. When the hammer hit the end of the ballistic pendulum, the load cell detects the force reacted on the end, and the laser displacement meter traces the pendulum motion. Consequently, the relative error of the impulse was found 5% at most.

The laser beam was focused in a conical nozzle, which was placed at the end of pendulum. It has a divergence angle of 30 degree, and length of 37 mm. As shown in Fig. 3, impulse was measured changing the focus position Z_f.

Table I Characteristics of pendulum

Moment of inertia	I	0.85	(kg m^2)
Length	l	89	(cm)
Frequency	ω	~ 1	(Hz)
Decay constant	c	~ 0.09	(kg m^2 / s)

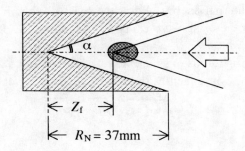

FIGURE 3. Schema of the conical nozzle

Impulse Model

A simple model has been constructed to estimate the impulse. Considering the case in which the laser focus is fixed at the apex of cone, the expression becomes quite simple.

Impulsive thrust imparted on a conical nozzle is described as

$$
I = \int_{o}^{t_{arr}} \int_{A} \left(p(t) - p_a \right) dA\,dt
$$

$$
= I^+ - I^-
$$

$$
I^+ \equiv \int_{o}^{t_{arr}} \int_{A} p(t)\,dA\,dt \qquad (5)
$$

$$
I^- \equiv \int_{o}^{t_{arr}} \int_{A} p_a\,dA\,dt
$$

where t_{arr} is the time when the shock wave arrives at the exit of the cone, $p(t)$ is the pressure within the blast wave. After the shock wave arrives at the exit, the pressure within the cone is assumed to recover to the atmospheric one soon.

The simple model is based on the Kompaneets approximation [7], in which the blast wave is regarded to have homogeneous static pressure $p(t)$, which is connected with the blast wave energy E_{bw} and the blast wave volume $V_{bw}(t)$ by a relation:

$$p(t) = (\gamma - 1)\frac{E_T}{V_{bw}(t)} \qquad (6)$$

$$E_T = \beta_T E_{bw}$$

where E_T is the internal energy within the blast wave, and β_T is the ratio of the internal energy to the blast wave energy. According to Sedov solution[8], $\beta_T = 7/9$.

The time varying radius of shock wave, and blast wave volume is described by the Sedov solution. Consequently, the impulse imparted to the cone is given as

$$I = \sqrt{2ME_{bw}}\left(\sqrt{\eta_{ec}^+} - \sqrt{\eta_{ec}^-}\right) \qquad (7)$$

where

$$
\begin{aligned}
\eta_{ec}^+ &\equiv \frac{\left(I^+\right)^2}{2ME_{bw}} \\
&= \frac{75\pi(\gamma-1)^2}{128\pi\xi_0^5}\left(\frac{\sin\alpha}{\sin(\alpha/2)}\right)^4 \beta_T^2 \\
\eta_{ec}^- &\equiv \frac{\left(I^-\right)^2}{2ME_a} \\
&= \frac{25(\gamma-1)^2}{384\pi\xi_0^5}\left(\frac{\sin\alpha}{\sin(\alpha/2)}\right)^4\left(\frac{E_a}{E_{bw}}\right)
\end{aligned}
\qquad (8)
$$

Here, M and E_a are defined as the mass and the internal energy filled within the cone before the external energy is deposited, respectively. α is the divergence half-angle of the cone. Impulse can be predicted with η_{ec}^+ and η_{ec}^-, which are the energy conversion efficiency defined in a way conventional to the stationary working rocket engines.

The relation between the impulse and the nozzle length is shown in Fig. 4 for different η_{bw}. In this figure, the result of experiment conducted by Ageev et al. are also plotted in the figure. They used a TEA CO$_2$ laser whose energy was 5 J/pulse, and are quite similar to our experimental conditions. The theoretical curve and the experimental plots are quite close to each other when $\eta_{bw} = 0.6$ in the case of $\alpha = 8.6$ degrees, and when $\eta_{bw} = 0.4$ in the case of $\alpha = 20$ degrees.

In both cases, the theoretical curve and the experimental plots are quite close to each other when $R / R^* < 0.3$. Since this simple model assumes the Sedov solution and hence extremely strong explosion, it is natural that this theory cannot predict the cases of large R/R*. The shock strength, which corresponds to $R/R^* < 0.3$ is $p_s/p_a > 5.7$ where p_s is the pressure soon behind the shock wave. According to Zel'dovitch's book [9], the Sedov solution is valid when $p_s / p_a \gg (\gamma+1) / (\gamma-1)$, and $p_s / p_a \gg 6$ when $\gamma = 1.4$. Hence, the limit of this theory: R/R* < 0.3 is reasonable.

η_{ec}^+ was 0.25. η_{ec}^- increased with the nozzle length, and it was 0.03 when $R/R^* = 0.3$.

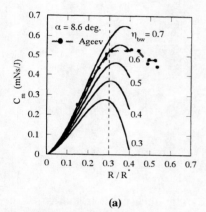 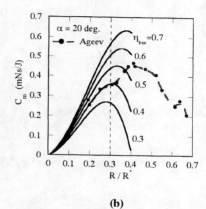

(a)   (b)

FIGURE 4. Theoretically deduced relation between C_m and R/R^* a) in the case of $\alpha = 8.6$ degrees, and b) $\alpha = 20$ degrees. Experimental results by Ageev, which were taken from Ref. [2], are also plotted, and are represented as closed circles. Nozzle length R is normalized with the characteristic length of blast wave $R^* \equiv (E_i / p_a)^{1/3}$.

Experimental Result

Figure 5(a) shows the impulse measured in the case of $Z_f = 0$. The theoretical curves deduced from Eqs. (7) and (8) are also shown in the figure. The theoretical curve of $\eta_{bw} = 0.4$ agrees well with the measured point. This result is consistent with the measured result of blast wave energy conversion efficiency using the shadowgraph.

The relation between the impulse and Z_f is shown in Fig. 5(b). The impulse stayed almost constant when $Z_f / R_N < 0.3$, and began to decrease with Z_f when $Z_f /$

$R_N > 0.3$. It is natural that the impulse stays almost constant because the blast wave can fill the cone when $Z_f / R_N < 0.3$ before the shock compressed air is exhausted from the exit plane. On the other hand, since the shock-compressed air is exhausted before it fills the cone when $Z_f / R_N > 0.3$, the impulse decreased.

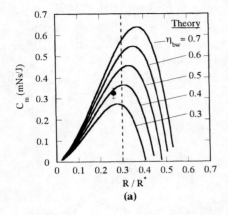 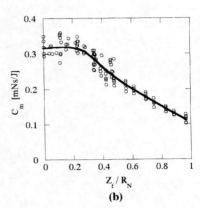

(a)   **(b)**

FIGURE 5. Experimentally deduced impulse a) in the case of $Z_f = 0$, and b) $0 \le Z_f / R_N < 1$. The theoretical curve is also potted in (a).

SUMMARY

The blast wave energy conversion efficiency was measured. It stands for the energy conversion process from laser to blast wave. As a result, the efficiency was found 0.48 ± 0.05 in an atmospheric air. Simple model was proposed to estimate the impulse imparted to conical nozzle.

REFRENCES

1. Y. P. Raizer, *Laser-Induced Discharge Phenomena*, Studies in Soviet Science Consultants Bureau, New York, 1977, p.199.
2. V. P. Ageev, A. I. Barchukov, F. V. Bunkin, V. I. Konov, V. P. Korobeinikov, B. V. Putjatin, and V. M. Hudjakov, Acta Astronautica **7**, 79 (1980) .
3. L. M. Myrabo, D. G. Messitt, and F. B. Mead, Jr., AIAA Paper, 98 –1001, 1998.
4. L. M. Myrabo, D. G. Messitt, and F. B. Mead, Jr., AIAA Paper, 2002–3783, 2002.
5. W. O. Schall, *Proceedings of High-Power Laser Ablation III, Santa Fe, NM*, 2000, SPIE, Bellingham, WA, 2000, p. 472.
6. K. Mori, K. Komurasaki, and Y. Arakawa, J. Appl. Phys, **92**, 10 (2002).
7. Kompaneets, A. S., Soviet Phys. Dokl. **5**, 46 (1960).

8. Sedov, L.I., Similarity and Dimension Methods in Mechanics (Academic Press, New York, 1959).
9. Y. B. Zel'dovich and Y. P. Raizer, *Physics of Shock waves and High-temperature Hydrodynamics Phenomena*, Academic Press, New York, 1967.
10. Brode, H, J. Appl. Phys. **26**, 766 (1955).

APPENDIX

Theory of Blast Wave Energy

According to the point blast explosion theory proposed by Sedov [8], E_{bw} is committed with the volume of blast wave $V_{bw}(t)$ and the Mach number of the expanding shock wave $M_s(t)$ as

$$\frac{16\pi\xi_0^5}{75\gamma}V_{bw}(t)M_s(t)^2 = \frac{E_{bw}}{P_a} \qquad \text{(A-1)}$$

where ξ_0 is a constant value, which is 1.03 for air ($\gamma = 1.4$). Although Sedov's theory assumes a spherical expansion of shock wave, Eq. (A-1) can be applied even to the elliptic blast wave expansion [7].

Unfortunately, Eq. (A-1) assumes an ultimately strong shock wave ($M_s \to \infty$), and hence it cannot be applied to general blast waves whose expanding speed has finite value. For finite-strength cases, Eq. (A-1) is corrected as

$$V_{bw}(t)f[M_s(t)] = \frac{E_{bw}}{P_a} \qquad \text{(A-2)}$$

where $f[M_s]$ is obtained from Brode's numerical computations[10]. It is expressed approximately as

$$f[M_s] = AM_s^{\alpha}$$
$$A = 6.6983 \qquad \text{(A-3)}$$
$$\alpha = 2.0433$$

as long as $2.0 < M_s < 48.0$.

The laser-induced blast wave shaped elliptically, and hence the longer and shorter radii were different from each other. However, the Mach number was unique during the adiabatic expansion of shock wave. Hence, the radii at the adiabatic expansion can be expressed as

$$r_A = r_{A,0} + r(t)$$
$$r_L = r_{L,0} + r(t) \qquad \text{(A-4)}$$

where r_A and r_L is the longer and shorter radius, respectively. At $t > t_0$, expansion speed is homogeneous.

From Eqs. (A-2)-(A-4), the following ordinary-differential equation represents the expanding motion of the shock wave.

$$r_A^{\frac{1}{\alpha}} r_L^{\frac{2}{\alpha}} \frac{dr(t)}{dt} = c_a \left(\frac{E_{bw}}{Ap_a} \right)^{\frac{1}{\alpha}} \qquad \text{(A-5).}$$

Here, c_a is the sound speed of the ambient air. Integrating Eq. (A-5), a linear expression

$$g(r_A, r_L) = c_a \left(\frac{E_{bw}}{Ap_a} \right)^{\frac{1}{\alpha}} (t - t_0)$$

$$g(r_A, r_L) \equiv \frac{1}{1 + 2/\alpha} r_A^{\frac{1}{\alpha}} r_L^{1 + \frac{2}{\alpha}} \left\{ 1 + \left(\frac{r_L}{r_L - r_A} \right) \right\}^{-\frac{1}{\alpha}} {}_2F_1 \left[1 + \frac{2}{\alpha}, -\frac{1}{\alpha}, 2 + \frac{2}{\alpha}; -\frac{r_L}{r_L - r_A} \right]$$

$$\text{(A-6)}$$

is obtained. Here, $_2F_1$ is a hyper-geometric function. At $t = 0$ the energy input is assumed to begin, and $t > t_0$, the adiabatic blast wave expansion with a homogeneous speed is assumed to begin.

The value of function $g(r_A, r_L)$ was calculated from the radii data obtained in the shadowgraph images at each t, and plotted on g-t plane. The blast wave energy was obtained from the inclination of line, which was least square fitted to the plots.

Vertical Launch Performance of Laser-driven In-Tube Accelerator

Naohide Urabe*, Sukyum Kim*,**, Akihiro Sasoh*,
and In-Seuck Jeung**

*Shock Wave Research Center, Institute of Fluid Science, Tohoku University, Sendai, Japan
**Department of Aerospace Engineering, Seoul National University, Seoul, Korea

Abstract. We studied the vertical launch performance of the Laser-driven In-Tube Accelerator (LITA). This device is primarily characterized by accelerating a projectile in a tube. Owing to the confinement effect, the thrust performance is enhanced. The driver gas can be specified and its pressure be turned so that the impulse performance is optimized. In the experiments, a 3.0-gram projectile was vertically launched. The effects of the projectile exit condition, the laser beam incident direction and the driver gas species were experimentally studied.

INTRODUCTION

Usually, in laser propulsion an object is accelerated in an open space. The laser beam is supplied from the ground. Focusing the laser beam, the local laser beam intensity exceeds a threshold value. Then, free electrons get generated, thereby further enhancing the laser beam absorption. The energy that is obtained by the electrons, is transferred to heavy particles through collisions. In this way, a core plasma is created around the focus. The expansion of the plasma core drives a blast wave. The blast wave generates a high-pressure region behind the object. A propulsive impulse is generated from the force balance between the frontal and rear surface. Kantrowitz proposed the idea of this principle in 1972 [1], demonstrations of lifting-up a light object with high power CO_2 lasers have been recently conducted [2] - [6].

Against practical applications, such conventional laser propulsion has the following drawbacks: (1) the propulsion performance is not satisfactory, (2) the tracking device and the posture control to the object are necessary, (3) danger due to the laser beam hazard, operation noise, vibration and air pollution, those are not favorable to the environment, cannot easily be eliminated.

The Laser-driven In-Tube Accelerator, which hereafter be referred to as LITA, has been developed at Shock Wave Research Center, Institute of Fluid Science, Tohoku University [7], [8]. This device accelerates an object (hereafter will be referred to as 'projectile') in a tube. The driver gas species can be changed and the operation pressure can be turned for favorable impulse performance. The tracking device and posture control to the projectile are not necessary. The danger against laser beam, noise, vibration and air pollution are vastly alleviated. Moreover, LITA can have high performance even in the high power operation by introducing the laser beam from the

CP664, *Beamed Energy Propulsion: First International Symposium on Beamed Energy Propulsion*,
edited by A. V. Pakhomov
© 2003 American Institute of Physics 0-7354-0126-8/03/$20.00

muzzle side. In this paper, operation characteristics of LITA under various conditions have been investigated.

OPERATION PRINCIPLE

The schematic illustration of LITA operation principle is shown in Fig. 1. Figure 1a shows the case of frontal incidence of the laser beam. A projectile is placed in an acceleration tube. The tube is filled with inert driver gas. The laser beam is supplied from the front side of the projectile, is focused behind it. The breakdown of the driver gas occurs near the focal point, and then a laser-driven blast wave is generated. With the laser beam being repetitively irradiated, repetitive impulses are produced. Figure 1b shows the case of rear incident operation. Except for the side of the laser beam supply, the operation principle in the latter is the same as that of the former.

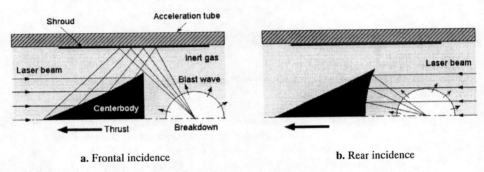

a. Frontal incidence   b. Rear incidence

FIGURE 1. LITA operation principle and methods of laser beam incidence

APPARATUS

Projectile

The design of the projectile for the frontal incident operation is shown in Fig. 2a. It comprises a centerbody, three struts and a shroud. They are all made of aluminum alloy, A7075-T6, and are glued together. The material has a high reflection rate for infrared light. Its total mass equals 2.1 gram. The fore shape of the centerbody is numerically designed [7] so that the focal point is placed at 7.9mm on the center axis behind the base of the centerbody. The diameter of the base is 16.2mm.

The design for the rear incident operation is shown in Fig. 2b. In order to compare the performance, the length from the focal point to the base of the centerbody and the diameter of the base are equated to those for the frontal incident operation. The projectile mass equals 3.0 gram.

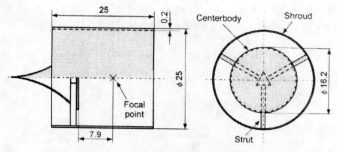

a. For frontal incidence

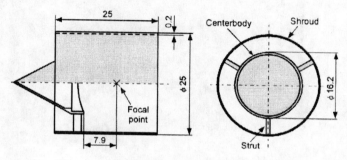

b. For rear incidence

FIGURE 2. Projectiles, lengths are in mm

Acceleration Tube Assembly

Vertical launch experiments and overpressure measurements have been conducted using a highly-repetitive-pulse CO_2 TEA (Transversely Excited Atmospheric) laser. It produces laser beams that have a maximum energy of 5J/pulse and a repetition frequency of up to 100Hz. Its wavelength is 10.6μ m. The effective input laser energy is measured with the energy meter (Gentic ED-500LIR). In the present study, the measured energy ranged from 2.4 J/pulse to 2.9 J/pulse.

The schematic of the acceleration tube assembly is shown in Fig. 3. The beam is horizontally supplied into the test room, and then vertically introduced to the acceleration tube with molybdenum plane mirrors. The acceleration tube is made of acrylic. It has an inner diameter of 25.2mm to 25.4mm and a length of either 0.5m or 1.0m. A tube end through which the laser beam is introduced is plugged with a NaCl window (thickness;5mm), the other end with a brass flange. The pressure transducer for the overpressure measurement is installed below the initial location of the projectile.

The detailed arrangement in the pressure measurement assembly is shown in Fig. 4. The projectile is initially placed above the pressure transducer. The pressure transducer is mounted in an aluminum tube. The inner diameter of the tube (24.6mm) is slightly smaller than the outer diameter of the projectile shroud (25.0mm). The pressure transducer is installed at 15.7mm below the focal point of laser beam. An overpressure history is measured in a single shot only in the rear incident operation.

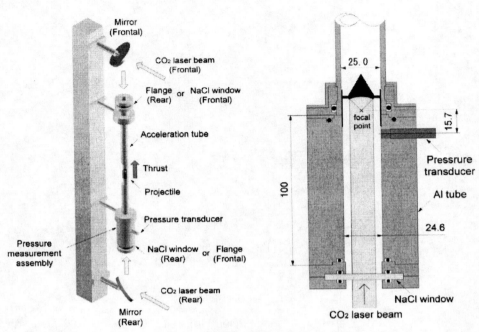

FIGURE 3. Acceleration tube assembly for frontal or rear incident operation

FIGURE 4. Pressure measurement assembly, lengths are in mm

RESURTS AND DISCUSSIONS

Effect of projectile exit condition

In this study, the momentum coupling coefficient: C_m, is defined as the ratio of a thrust to a laser power which is effectively incident on the projectile. Here, the thrust is determined from a projectile hovering condition. The laser repetition frequency is turned so that the gravitational force is balanced with the thrust.

$$C_m = \frac{mg}{(fE)_{hover}} \tag{1}$$

where m, g, f and E designate a projectile mass, the gravitational acceleration, a laser repetition frequency and the effective value of an input laser energy.

In the first series of the vertical launch experiments, the projectile was initially placed on a disk or a thin cylinder in the acceleration tube. In the former case, the acceleration tube was completely plugged (plugged condition); in the latter case, the cross-sectional area of the acceleration tube was decreased by 75% (unplugged condition). Under two different projectile exit conditions, plugged or unplugged, C_m was measured. In the rear incident operation, the plugged condition cannot be set up. Hence, the experiments for this purpose were performed only with the frontal incident configuration.

As shown in Ref. [9], C_m was measured under various fill pressures. C_m under the plugged condition always becomes larger than that under the unplugged at any fill pressure. In the former, the thrust was enhanced with reflected shock waves from the disk surface.

In practical operations, after lifting off, the unplugged condition should be applied. The plugged condition can be applicable only in the initial condition. Therefore, the performance under the unplugged condition is employed in all the following experiments. For the details, the reader should refer to [9].

Effect of beam incidence direction

The second subject is to examine the effect of the beam incident direction. Here, C_m was measured both in the frontal and in the rear incident operation. As shown in Ref. [9], C_m on the rear incidence always becomes higher than that on the frontal incidence at any fill pressure. A possible reason is the larger energy loss through two reflections in the case of frontal beam incidence. In the present study, the repetition frequency ranges up to 60Hz. In the numerical simulation of Ref. [10], the characteristic time for impulse production after a single laser irradiance is of the order of 1 ms. Therefore, under the present low frequency operation the flow-laser beam interference does not degrade the thrust performance. For further details, refer to Ref. [9]. Since higher performance is obtained, the rear incidence is employed in the following experiments.

Effect of atomic mass

Vertical launch

In order to study the effects of driver gas species on the impulse generation performance, vertical launch experiments have been carried out. In the experiments, argon, krypton or xenon was used as the driver gas. Figure 5 shows the relationship between C_m measured at 100kPa and a on double-logarithmic coordinates. Here, a designates the speed of sound of the driver gas. The broken line represents a relation of inverse proportion. As is seen in this figure, C_m is in almost inverse proportion to a.

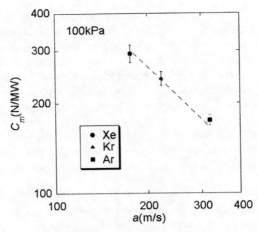

FIGURE 5. C_m vs. a, broken line; $C_m \sim 1/a$

In Ref. [11], the reason for this relation is discussed. The overpressure level does not depend on the atomic mass, whereas the duration time of the impulse force action is in inverse proportion to the speed of sound. In order to confirm these relations, the overpressure history on the tube wall was measured.

Overpressure measurement

We measured an overpressure below the projectile, see Figs. 3 and 4. The acceleration tube was filled with argon, krypton or xenon. In the present study, the fill pressure was set to 100kPa. The laser beam was supplied from the rear side of the projectile. Figure 6 shows the pressure history. ΔP and t designate the overpressure and time respectively. As is seen in Fig. 6, the overpressure history is extremely complicated since the blast wave interacts with the projectile and the acceleration tube wall. Nevertheless, as is seen in Fig. 7, the waveforms are similar once the time axis is multiplied by the speed of sound.

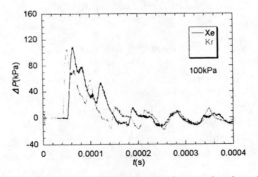

FIGURE 6. Overpressure history, measured on acceleration tube wall

110

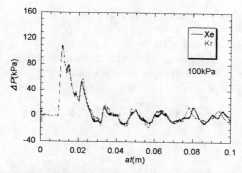

FIGURE 7. Overpressure history on the scaled time coordinate

Figure 8 shows the relationship between the duration time: τ and a on double-logarithmic coordinates. Here, τ is defined as the duration time during which a positive overpressure is kept in the first pressure event. The broken line represents a relation of inverse proportion. The plotted values are averaged one. As is seen in this figure, the exponent index approximately equals -1.

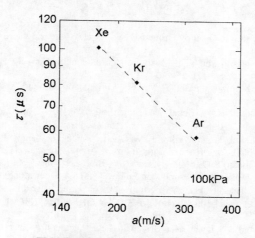

FIGURE 8. τ vs. a, broken line; $\tau \sim 1/a$

It follows from these results that the overpressure value is independent of the species, whereas the duration time of the overpressure is proportional to the reciprocal of speed of sound. Therefore, the impulse is inversely proportional to the speed of sound.

CONCLUSION

The LITA impulse performance with different projectile exit conditions, laser beam incidence directions and driver gas species, are experimentally analyzed. C_m is almost in inverse proportion to a. In the overpressure measurements, it is confirmed that the overpressure value is independent of a, while the duration time of force action is inversely proportional to a.

ACKNOWLEDGMENTS

We would like to thank H. Ojima and T. Ogawa with Shock Wave Research Center, and M. Kato, K. Asano and K. Takahashi with the machine shop of Institute of Fluid Science for the valuable helps in the present experiments.

This project was supported by Japan Society for the Promotion of Science as Grant-in-Aid for Scientific Research (S) # 13852014. The second author's stay in Tohoku University was supported by BK21 Program sponsored by Seoul National University.

REFERENCES

1. A. Kantrowitz, "Propulsion to Orbit by Ground-Based Lasers", Astronautics and Aeronautics: Vol.10, pp.74-76, 1972.
2. L.N. Myrabo, D.G. Messitt, and F.B. Mead Jr., "Ground and Flight Tests of a Laser Propelled Vehicle", AIAA paper: pp.98-1001, 1998.
3. F.B. Mead Jr., L.N. Myrabo, and D.G. Messitt, "Flight and Ground Tests of a Laser-Boosted Vehicle", AIAA paper: pp.98-3735, 1998.
4. W.O. Schall, W.L. Bohn, H.-A. Eckel, W. Mayerhofer, W. Riede, and E. Zeyfang, "Lightcraft experiments in Germany", High-Power Laser Ablation III, Proc. SPIE: Vol.4065, pp.472-481, 2000.
5. W.O. Schall, H.-A. Eckel, W. Mayerhofer, W. Riede and E. Zeyfang, "Pulsed power for space propulsion", International Conference on Pulsed Power Applications: No.H.05, 2001.
6. L.N. Myrabo, M.A. Libeau, E.D. Meloney, and R.L. Bracken, "Survivability of Thin Metal Mirror Coatings on Graphite Pulsed Laser Propulsion Engines", AIAA paper: pp.2002-3783, 2002
7. A. Sasoh, "Laser-propelled ram accelerator", Journal de physique IV: Pr11-41, 2000.
8. A. Sasoh, "Laser-driven in tube accelerator", REVIEW OF SCIENTIFIC INSTRIMENTS: Vol. 72, No.3, 1893-1898, 2001.
9. A. Sasoh, N. Urabe, and S. Kim, "Laser-Driven In-Tube Accelerator Operation Using Monoatomic Gases", AIAA paper: pp.2002-2201, 2002.
10. A. Sasoh, J.-Y. Choi, I.-S. Jeung, N. Urabe, H. Kleine, and K. Takayama, "IMPULSE ENHANCEMENT OF LASER PROPULSION IN TUBE", POLISH ASTRONAUTICAL SOCIETY:27, pp.40-50, 2001.
11. A. Sasoh, N. Urabe, S. Kim, and I.-S. Jeung, "Impulse scaling in laser-driven in-tube accelerator", Applied Physics A: submitted.

Laser Sustained Plasma Free Jet as a Tool for Propulsion

A. Lebéhot, M. Dupuy, V. Lago, and M. Dudeck

Laboratoire d'Aérothermique, CNRS
1C Avenue de la Recherche Scientifique, 45071 Orléans Cedex, France

Abstract. Although not really optimized for thruster use, the laser sustained plasma free jet issued from a sonic nozzle appears as an interesting step in the study of a thruster device. Even without any measurement of the propellant properties, the velocity distributions obtained at the end of the expansion give an estimation of thrust as well as specific impulse, thanks to the isentropic behavior of the free jet, and provided that some assumptions are made on the distributions of density, velocity, and temperature along the nozzle throat radius. It is expected that such assumptions could be got over by a complete calculation of the plasma upstream of the nozzle

INTRODUCTION

Different schemes including a laser beam can induce a propulsive effect.[1] According to the physical process used (surface plasma,[2] air breathing plasma[3] or plasma in a gas reservoir,[4] ablation,[5] evaporation,[6] radiation pressure[7]...), to the type of the operated laser (pulsed or cw, high power or micro-laser), the thruster can be used for launching a space-craft directly from the ground, for orbit or attitude control of a satellite, for inter-stellar missions. In any case, the main interest of these different types of thrusters lies in the fact that the laser beam can propagate over very large distances in space without significant attenuation, so that the energy source can be placed remotely on ground or on a platform in space. Then, the thruster itself may be very simple and its contribution to the total weight of the space-craft almost negligible.

In the present work, the thruster is just a nozzle where the usual stagnation cold gas is replaced by a laser sustained plasma. It is expected that the high temperature achieved in the plasma should increase the velocities of the flow at the exit of the nozzle, and consequently the thrust. Nevertheless, some other phenomena may have negative effects (for instance decrease of mass flow rate) which can reduce, or even cancel, the gains of the method. The present study tries to point out the actual effects of the laser sustained plasma when compared with the cold gas thruster, and to give some ways for the optimization of a laser thruster. It should be noticed that the experimental device has not been especially designed for propulsion studies, and that the thruster properties cannot be directly measured in the present state of the work.

CP664, *Beamed Energy Propulsion: First International Symposium on Beamed Energy Propulsion,*
edited by A. V. Pakhomov

EXPERIMENT

Plasma jets are obtained by focusing the TEM_{00} beam of a continuous infra-red laser just ahead of a sonic orifice as a nozzle, with diameter $D^* = 0.5$ mm (Fig. 1).[8] The waist-nozzle distance is about 1.3 mm. The laser beam energy is absorbed by the gas medium thanks to the inverse bremsstrahlung process which is a non-linear phenomenon in terms of the square of the gas density. The CO_2 laser, with maximum power 1500 W, allows a plasma to be sustained in argon or in a mixture including argon, after initiation by a pulsed auxiliary laser. The gas stagnation pressure is maintained below 10 bars (limitation due to the pumping speed in the expansion chamber and by the strength of the entrance window of the laser beam). A free jet is generated, which expands into a low pressure chamber (typical pressure 0.05 to 0.1 mbar). In such conditions, the free jet is surrounded by a shock structure (shock barrel) which protects it from any interaction with the background gas. Then, the expansion follows closely the ideal expansion laws, and even the isentropic behavior, as it would occur in perfect vacuum. The study in laboratory is then valid for an application in space conditions.

A set-up, combining a chopper located in a second chamber and a quadrupole mass selector, is used for the velocity analysis of the neutral species by a time-of-flight technique, after extraction of an atom beam by a skimmer.

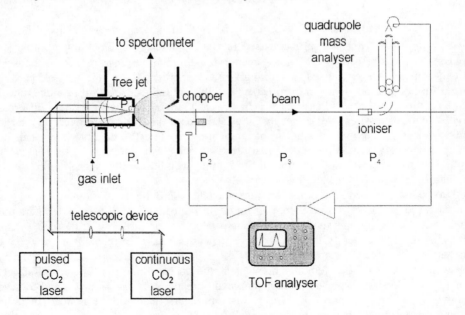

FIGURE 1. Experimental set-up

PRINCIPLE OF THE MEASUREMENTS AND CALCULATIONS

As the nozzle is a simple orifice in a plane wall, the thrust is calculated by using the sonic characteristics of the free jet. The number density flow rate through the nozzle is given by:

$$\dot{n} = n^* A^* a^* \qquad (1)$$

where n^* and a^* are respectively the number density and the flow velocity at the nozzle throat (equal here to the local sound velocity), and A^* is the area of the nozzle throat. Assuming an exact isentropic expansion, these parameters can be simply expressed as a function of the stagnation parameters (subscript 0) and of the isentropic exponent γ (assumed constant in the region of interest; see below):

$$n^* = n_0 \left(\frac{2}{\gamma+1} \right)^{\frac{1}{\gamma-1}} \qquad (2)$$

$$a^* = \sqrt{\frac{2\gamma}{\gamma+1} \frac{kT_0}{m}} \qquad (3)$$

using :

$$T^* = T_0 \frac{2}{\gamma+1} \qquad (4)$$

In usual stagnation conditions (gas in equilibrium at constant pressure and temperature), T_0 is the equilibrium temperature of the gas medium inside the nozzle reservoir, taken far upstream of the nozzle. At distances shorter than about 1 D^* from the nozzle throat, this temperature begins to decrease down to T^* at the nozzle throat, following the law of the adiabatic flow. When a plasma is generated close to the waist of a laser beam, the kinetic behavior of the plasma expansion, at least along the axis, is exactly the same as that of a neutral gas at a uniform temperature T_0, temperature of some intermediate point between the plasma core and the nozzle. Then, the thrust is given by:

$$F = \dot{m}a^* = m\,n^* A^* a^{*2} \qquad (5)$$

and, finally:

$$F = P_0\,\gamma\,A^* \left(\frac{2}{\gamma+1} \right)^{\frac{\gamma}{\gamma-1}} \qquad (6)$$

At constant pressure, the thrust does not depend on the stagnation temperature and, to this respect, heating the gas is of no use, as the increase of the velocity is counter-balanced by the decrease of mass flow rate. Then, the result is not affected by the fact that the flow tubes issued from the whole area of the nozzle are concerned with different values of T_0, due to the strong temperature gradients which occur around the plasma core.[8] The remaining strong hypothesis is the isentropic assumption, which is known to be valid for the expansion, but is highly questionable upstream of the nozzle in the present conditions. In fact, it appears that, within a distance of 1 D* upstream of the nozzle throat (where the effect of the nozzle begins to act upon the properties of the gas medium), the effects on γ of the changes of pressure and temperature nearly compensate.[9]

With the same hypothesis, the specific impulse is given by:

$$I_{sp} = \frac{a^*}{g} \qquad (7)$$

where g is the standard acceleration of gravity. So, with the help of Eq. 3:

$$I_{sp} = \frac{1}{g}\sqrt{\frac{2\gamma}{\gamma+1}\frac{kT_0}{m}} \qquad (8)$$

The specific impulse, to the contrary of the thrust, increases as the square root of the stagnation temperature. Also, the efficiency of the thruster can be defined as the ratio of the output kinetic power to the total input power :

$$E = \frac{\frac{1}{2}\dot{m}a^{*2}}{\dot{m}(C_P T_0)_{300} + P_L} \qquad (9)$$

The subscript "300" means "without laser power" or "at 300 K". It may be noted that this efficiency would equal nearly 1 without laser power, if the output kinetic power were taken at the end of the expansion, but the end of the expansion is not relevant here.

RESULTS AND DISCUSSION

Measurement of the Temperature T_0

From the preceding section, it appears that, for a given nozzle geometry, the thrust depends only on the measured stagnation pressure P_0 (Eq. 6), and the specific impulse on the equivalent stagnation temperature T_0 (Eq. 8). The value of this temperature is not directly measured, but deduced from the velocity distribution achieved in the final stage of the free jet. This distribution is analyzed on the atom beam extracted from the free jet thanks to a skimmer (Fig. 1). A time-of-flight device allows this distribution to be measured and then fitted by a Maxwellian distribution (Fig. 2). This best fit leads to the values of the final mean flow velocity U, the final translational temperature T, and the temperature actually experienced by the expansion, at least for its axial part, T_0.[8]

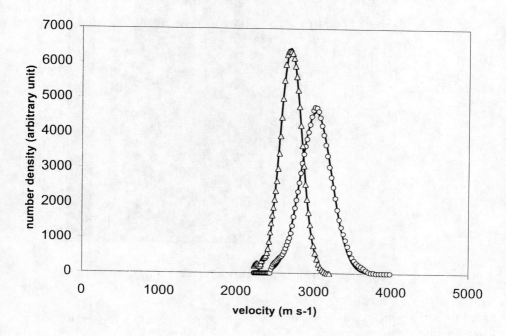

FIGURE 2. Velocity distributions of pure neutral argon beam obtained with plasma conditions: laser power 165 W and pressure P_0 = 7 bars (Δ), and 300 W with P_0 = 4 bars (o); the full lines are the corresponding best fit Maxwellian distributions.

The values obtained by following this procedure are reported in Fig. 3 in a frame laser power-gas pressure P_0. It is noteworthy that the temperature decreases as the gas pressure increases. This could appear surprising, as the inverse bremsstrahlung process is more efficient when the gas pressure increases, but then, the plasma can be sustained with lower laser intensity, and consequently the plasma core takes place farther upstream from the nozzle throat. Finally, the expansion is generated with lower temperature T_0.

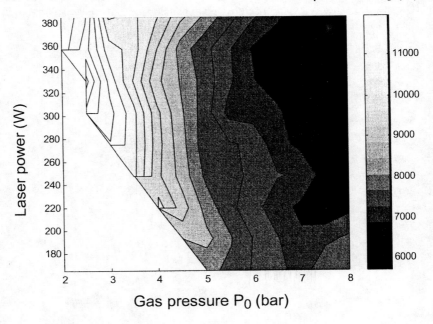

FIGURE 3. Equivalent stagnation temperatures T_0 as a function of laser power and gas pressure, for pure argon. Below the line connecting 320 W-2 bars and 170 W-5 bars, the plasma is not stable.

Thruster Properties

Pure Argon as a Propellant

In Figure 4 is reported the specific impulse, calculated as indicated above (Eq. 8), with the measured temperatures T_0, for constant pressure P_0. The maximum value of I_{sp} (180 s) is reached for P_0 = 2 bars with laser power = 400 W. The lowest point to the left in Fig. 4 corresponds to T_0 = 300 K, i.e. the situation without laser. With respect to these "cold" conditions, the laser heating increases the specific impulse by a factor 4 to 5, while the thrust remains constant in these conditions. Nevertheless, the operation at constant pressure P_0 does not qualify properly the thruster, because the mass flow rate through the nozzle changes with temperature. The mass flow rate is a more convenient parameter, as it characterizes the mass consumption of the thruster. Some values of the thrust with respect to T_0 are reported on Figure 5, for three different mass flow rates (Eq. 5). Then appears, for a given flow rate, the increase of the thrust as the square root of T_0, due to the corresponding increase of the sonic velocity (Eq. 3).

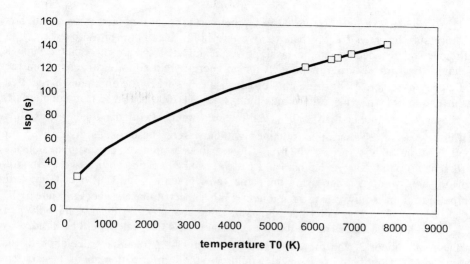

FIGURE 4. Specific impulse for pure argon as a function of T_0 (Eq. 8). Typical experimental conditions are: pressure $P_0 = 7$ bars, laser power between 170 and 400 W, or without laser, at 300 K (□); the corresponding thrust is 0.111 N.

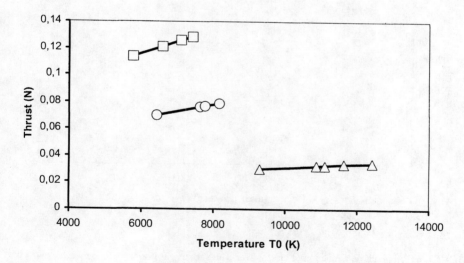

FIGURE 5. Thrust as a function of temperature T_0 (Eq. 5 with Eq. 3), for three different mass flow rates through the nozzle: 2.5×10^{-5} kg s^{-1} (Δ); 7×10^{-5} kg s^{-1} (o); 1.2×10^{-4} kg s^{-1} (□)

The efficiency (Eq. 9) is reported in Figure 6 as a function of laser power, with any gas pressure between 2 and 7 bars (above threshold). It is noteworthy that very high efficiency is reached when the laser power is lowered towards threshold. Such a behavior had already been mentioned for an energy yield slightly different from the efficiency used here.[8] The same kind of curve is found when the efficiency is reported as a function of T_0, with a sharp maximum at about 7000 K (Fig. 7). In counterpart, for laser power higher than 300 W and/or temperature T_0 higher than 9000 K, the efficiency becomes lower than obtained without laser. These results give a way for optimization of a thruster. The very high values obtained at maximum for the efficiency (up to 0.83 with 6.5 bars and 170 W) can be explained by an overestimation of the sonic velocity when using the same value given by Eq. 3 all over the nozzle throat. From mass flow rate measurements,[8] it is expected that the average value of T^* could be significantly smaller than the axial value given by Eq. 4. Then, the values of I_{sp}, as well as the values of the thrust at constant mass flow rate, could be divided by a factor of about 2, and the efficiency by a factor 4. A detailed calculation of the plasma in the flow field of the nozzle could give more information about these orders of magnitude.

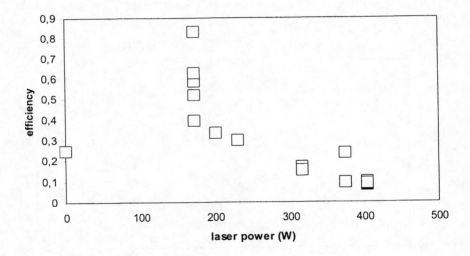

FIGURE 6. Efficiency as a function of laser power with various gas pressures P_0 between 2 and 7 bars.

As a conclusion, for a given gas, laser heating increases the specific impulse in terms of $\sqrt{T_0}$, whatever be the other conditions of the expansion. The thrust increases in the same way, provided that the flow rate is maintained constant. It may be recalled that

T_0 is not an independent parameter, but is given by a set of conditions: laser power, gas pressure, waist-nozzle distance, at given nozzle diameter and for a given gas.

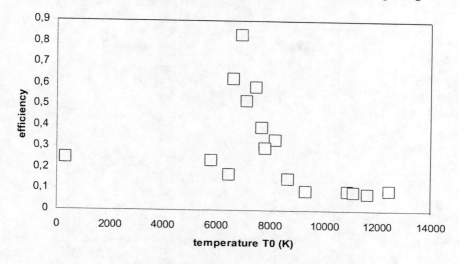

FIGURE 7. Efficiency as a function of temperature T_0.

Gas Mixture as a Propellant

In the simplified model, the effect of the atomic weight of the gas medium also vanishes for the thrust (Eq. 6), except if it is considered at constant mass flow rate (Eqs. 5 and 3). Then, the thrust is proportional to $(T_0/m)^{1/2}$. Also, the lighter the atomic weight of the gas, the higher the specific impulse of the thruster. As T_0 is not an independent parameter, the effect does not follow the simple law $1/\sqrt{m}$, even at constant pressure and laser power because, at given pressure and laser power, the value of T_0 depends on the nature of the gas (through the absorption parameter).

In the present experiments, a gas mixture is used by adding helium into argon at different concentrations. Then, each component behaves during the expansion nearly as if it had the average atomic weight of the mixture (within a slight "slip effect").[10] For instance, when adding helium into argon, T_0 increases (Fig. 8), and the effect amplifies the effect of mass in terms of $(T_0/m)^{1/2}$ (Fig. 9). For the efficiency, it is difficult to make a comparison with the pure gas case, because measurements are here available only for one single laser power of 340 W and two values of P_0: 6 and 7 bars (Fig. 10).

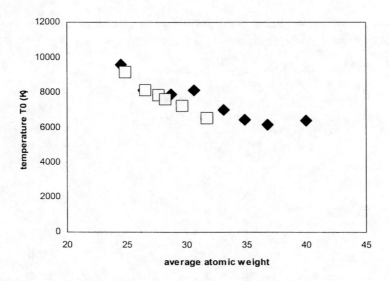

FIGURE 8. Measured values of the temperature T_0 as a function of the average atomic weight of the helium-argon mixture, for laser power 340 W and gas pressure $P_0 = 6$ bars (full lozenges) and 7 bars (open squares)

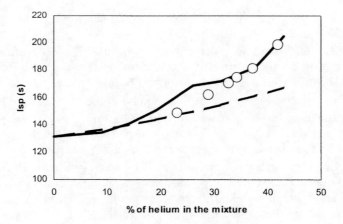

FIGURE 9. Specific impulse for different mixtures argon-helium as a function of helium percentage at pressure $P_0 = 6$ bars (full line) and 7 bars (o). For comparison, is also reported the simple effect of mass, as if the temperature T_0 would not depend on the nature of the mixture (dashed line).

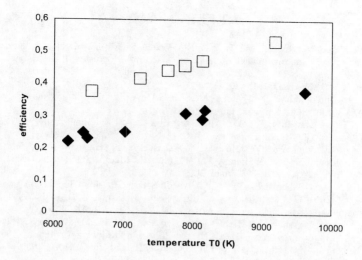

FIGURE 10. Efficiency as a function of temperature T_0 with laser power 340 W and P_0 = 6 bars (full lozenges) and 7 bars (open squares).

CONCLUSION

The experimental set-up described above has not been designed for thruster studies, and the results given here are calculated from experiments which have not been optimized for this purpose. In particular, it is not expected that the sonic nozzle could be a good candidate as a thruster geometry for at least two reasons: i) In space, there is no limitation to the free jet which could then impinge on the vehicle surfaces, and ii) The expansion is not here used for the thruster: only the subsonic to sonic part contributes to the thruster work. The only advantage of this device lies in the ease of calculation of all properties, if they can be considered as adiabatic and isentropic, and in the possibility of extracting the temperature T_0 from the velocity distributions of the extracted atom beam. Then, the results should be considered for their general trends which can give ways for optimization, and not for their absolute values. For instance, our thrust values are five times lower than those actually measured at the University of Tokyo.[1] Nevertheless, this is obtained with laser powers between 170 and 400 W instead of 500 to 1000 W in Tokyo, and with mass flow rates between 0.02 and 0.1 g s^{-1} instead of 0.5 to 1.5 in Tokyo.

REFERENCES

1. Lebéhot, A., Lago, V., Dudeck, M., Toyoda, K., Hosoda, S., Molina-Morales, P., Komurasaki, K., and Arakawa, Y., 3rd International Conference on Spacecraft Propulsion, Proceedings ESA SP-465, Cannes, 2000, pp. 235-241 ; Lebéhot, A., Lago, V., and Dudeck, M., *Revue Scientifique et Technique de la Défense* **53**, 39-43 (2001).
2. Horisawa, H. and Kimura, I., 37th AIAA/ASME/SAE/ASEE Joint Propulsion Conference and Exhibit, AIAA 2001-3662, Salt Lake City, 2001.
3. Mead,Jr., F.B. and Larson, C.W., 37th AIAA/ASME/SAE/ASEE Joint Propulsion Conference and Exhibit, AIAA 2001-3661, Salt Lake City, 2001.
4. Komurasaki, K., Arakawa, Y., Hosoda, S., Katsurayama, H., and Mori, K., 33rd AIAA Plasmadynamics and Lasers Conference, AIAA 2002-2200, Maui, 2002 ; Lebéhot, A., Lago, V., and Dudeck, M., ibid, AIAA 2002-2198, Maui, 2002..
5. Gonzales, D.A. and Baker, R.P., 37th AIAA/ASME/SAE/ASEE Joint Propulsion Conference and Exhibit, AIAA 2001-3789, Salt Lake City, 2001 ; Nehls, M., Edwards, D., and Gray, P., 33rd AIAA Plasmadynamics and Lasers Conference, AIAA 2002-2154, Maui, 2002.
6. Yabe, T., Phipps, C., Yamaguchi, M., Nakagawa, R., Aoki, K., Mine, H., Ogata, Y., Baasandash, C., Nakagawa, M., Fujiwara, E., Yoshida, K., Nishiguchi, A., and Kajiwara, I., preprint (2002).
7. Myrabo, L.N., Knowles, T.R., Bagford, J.O., Siebert, D.B., and Harris, H.M., 36th Joint Propulsion Conference and Exhibit, AIAA 2000-3336, Huntsville, 2000 ; Gray, P.A., Edwards, D.L., and Carruth, Jr., M.R., 33rd AIAA Plasmadynamics and Lasers Conference, AIAA 2002-2200, Maui, 2002, AIAA 2002-2178.
8. Girard, J.M., Lebéhot, A., and Campargue, R., *J. Phys. D : Appl. Phys.* **26**, 1382-1393 (1993).
9. Drellishak, K.S., Knopp, C.F., and Cambel, A.B., *Phys. Fluids* **6**, 1280-1288 (1963).
10. Lebéhot, A., Kurzyna, J., Lago, V., Dudeck, M., and Campargue, R., "Laser Sustained Plasma Free Jet and Energetic Atom Beam of Pure Argon or Oxygen Seeded Argon Mixture", in *Atomic and Molecular Beams, The State of the Art 2000*, edited by R. Campargue, Berlin, Springer, 2000, pp. 237-251.

Combined Theoretical and Experimental Flight Dynamics Investigation of a Laser-Propelled Vehicle

M. A. Libeau, L.N. Myrabo, M. Filippelli, and J. McInerney

Department of Mechanical, Aerospace, and Nuclear Engineering Rensselaer Polytechnic InstituteTroy, New York 12180-3590

Abstract: A six-degree of freedom (6-DOF) mathematical model of the laser lightcraft is developed to provide insight into the flight performance of the Model 200 lightcraft. Model validation is achieved by comparing simulation results to actual flight data obtained through video analysis of existing flights. The mathematical model is driven by forces and moments applied to the lightcraft by both the laser engine and vehicle aerodynamics. Because the forces and moments induced by the pulsed laser engine are time variant and act over very small time scales, they are examined with specially made test stands that measure linear and angular impulses. The aerodynamic forces and moments are found using a lightcraft model mounted in Rensselaer Polytechnic Institute's (RPI) 4' x 6' closed circuit wind tunnel. This study is similar to analyses performed on spinning objects such as artillery shells.

INTRODUCTION

The laser lightcraft is an ultralightweight vehicle propelled by repetitively-pulsed detonations induced by a ground-based laser. The concept of laser propulsion to orbit was first promoted by Arthur Kantrowitz [1-2]. The specific lightcraft geometry addressed herein was invented by the second author under contract to LLNL and the SDIO (Ref. 3); he later refined it under NASA and USAF support (Refs. 4-7) over the period of 1996-1999. The rear of the vehicle is a parabolic optic that focuses laser energy into an annular shroud, explosively heating the air and producing thrust. Once the hot gasses escape the engine, fresh air is automatically inducted into the engine.

A number of lightcraft designs have flown with varying success, but one design, the spin-stabilized Model 200, shows superior performance by attaining altitudes in excess of 230 feet on 2 October 2000 [Ref. 8]. This design was featured in a 1998 TV show produced by the BBC [Ref. 9]. The Model 200 design exhibits very good beam riding behavior because it is able to recenter itself in the beam after being perturbed. For these demonstration flights, the vehicle is driven by the 10 kW, Pulsed Laser Vulnerability Test System (PLVTS), a CO_2 electric discharge laser located at the High Energy Laser Systems Test Facility (HELSTF), White Sands Missile Range

CP664, *Beamed Energy Propulsion: First International Symposium on Beamed Energy Propulsion*,
edited by A. V. Pakhomov
© 2003 American Institute of Physics 0-7354-0126-8/03/$20.00

(WSMR). The laser produces a hollow square beam profile whose dimensions are varied with 'telescope' optics to match the size of lightcraft. Additionally, the laser can be configured to operate in a single pulse or repetitively pulsed mode.

The 6-DOF model of the lightcraft requires experimental data from four areas. First, the lightcraft's mass properties such as total mass and moments of inertia are measured. Secondly, the laser-induced reaction applied to the lightcraft engine is quantified for common operating conditions. Thirdly, an aerodynamic database is obtained from wind tunnel testing and Computational Fluid Dynamics (CFD) modeling. Finally, actual flight data obtained from video analysis is needed to validate the 6-DOF.

MASS PROPERTIES OF THE LIGHTCRAFT

The Model 200 lightcraft has historically been assembled from three parts: nose, shroud, and rear optic. The vehicles are thin shells, roughly 0.01-inch thick, machined from 6061-T4 aluminum. The masses of assembled lightcraft are typically measured just prior to flight, using a digital scale. However, lightcraft mass can also be calculated using available CAD software (SolidWorks, ProEngineer, etc.).

Because the lightcraft is a body of revolution, only two moments of inertia must be measured and they are measured using an Inertial Dynamics, Inc. MOI-005 device. The device uses a torsion pendulum to oscillate the test object about one axis; it then uses the period of oscillation to determine the moment of inertia of the object about the axis of revolution. To measure the axial MOI, the lightcraft is attached to a vertical shaft mounted to the MOI-005 device. Since the lateral MOI must be taken with respect to the lightcraft's center of mass, a special mount is used that allows the lightcraft to be positioned so that it's center of mass is centered directly over the MOI-005's axis of revolution. Two Model 200 lightcraft were tested having scales of ¾ and 5/6 (11-cm and 12.2-cm diameter, respectively). In addition to experimental measurements, drawings of the lightcraft are produced in Solid Works and the software calculates the inertial properties of the vehicles. The software determines the MOI's of a scale 11/10 Model 200 lightcraft (16.1-cm diameter), the vehicle examined by this modeling effort.

PLVTS AND THE LASER-INDUCED IMPULSE

The lightcraft engine receives a complex propulsive reaction from the laser-induced air detonation within its shroud resulting from a laser pulse; this reaction consists of three linear impulses and three angular impulses. Because the propulsive reaction is highly time-variant and occurs in less than a millisecond, the time-variant forces and moments are treated as six impulses. Although the lightcraft receives impulses about three axes, the impulses applied from a thrust force, side force, and a pitching moment dominate the reaction and are illustrated in Fig. 1.

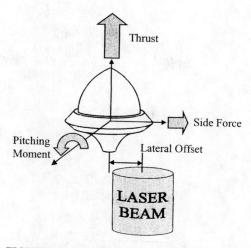

FIGURE 1. Dominant laser-induced engine reaction

The thrust impulse propels the vehicle skyward while the side impulse and pitching impulse determine the vehicle's beam riding performance. The restoring side force is perpendicular to axial thrust vector, and is defined to act in the direction of vehicle's lateral offset in the beam; the pitching moment is perpendicular to both the axial thrust and side force (Fig. 1). The side force attempts to center the vehicle in the beam and counteracts any tendency of the vehicle to drift from the center of the beam. Asymmetries in the laser beam's energy distribution received on the rear optic of the vehicle - due to beam misalignment - create the side force and this misalignment is termed lateral offset. If the restorative side force is too weak or acts in the wrong direction, the vehicle will depart from the beam and quickly loose thrust.

A pitching moment is also induced on the engine by asymmetric energy distribution due to beam misalignment. Such a moment is detrimental to engine performance because it tips the vehicle and thereby destroys the precise focus of the parabolic reflector. Tipping produces a loss of thrust if the tip angle becomes excessive.

The engine's performance and propulsive reaction are altered by the asymmetric distribution of laser energy around the engine. As the beam propagates away from the PLVTS laser, its cross-section changes shape with altitude and this in turn has an impact on the propulsive reaction produced by the engine. To understand the engine's performance with laser transmission distance, an understanding of the laser beam's altitude variation is needed.

As with any laser transmission path, the PLVTS beam starts out on the near-field, then moves through a long focal 'waist', and finally progresses into the far-field, as depicted in Fig. 2. In the near field, the original hollow square shape morphs into a roughly solid circular shape where the beam's hollow cross-section is completely filled-in. After emerging from the focus, the beam enters the far-field region where it takes on a distinctive cruciform structure marked by a solid central circular spot surrounded by four arms (the first diffraction 'ring', each composed of four smaller

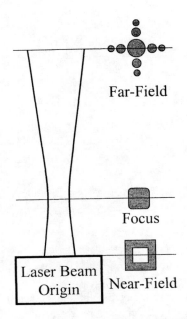

FIGURE 2. Beam propagation characteristics of the PLVTS CO_2 Laser

lobes); subsequent diffraction 'rings' appear at larger radii, with continually decreasing energy. The cruciform pattern increases in area as the beam moves farther away from PLVTS. Specialized telescope optics placed near the PLVTS trailer are used to adjust the size of the beam's cross-section and also the focal range.

For the purposes of this paper only the near-field character of the beam is examined. The near-field of the beam is chosen because all lightcraft flights begin in the near-field and many complete their entire flights without leaving this region. In fact to date, the only lightcraft flights that have flown through the focus and into the far field are the 2 Oct. 2000 altitude records set by Lightcraft Technologies, Inc. [Ref. 8]. Engine testing reveals how the lightcraft's engine performance varies with lateral offset of the vehicle in either the near-field or far-field beam. Note that lightcraft tip angle also effects the propulsive reaction but has not yet been comprehensively tested.

THRUST IMPULSE MEASUREMENTS

Under NASA/USAF sponsorship, the airbreathing pulsejet performance of the Model 200-11/10 lightcraft (16.1-cm diam.) was extensively studied in two separate campaigns at White Sands Missile Range using a ballistic pendulum apparatus. The first campaign on 8-9 Feb. 1999 examined the effects of lateral beam offset and lightcraft tip angle – for a constant pulse energy of ~450 joules. The second campaign on 23 May 1999, studied the effect of pulse energy on the delivered impulse and coupling coefficient – for a precisely aligned PLVTS beam. Both campaigns employed the PLVTS near field, 10-cm square beam with the 2X resonator optics.

For the May '99 tests, the peak impulsive performance was roughly 0.06 N-s at just over 500 joules of laser pulse energy. In the Feb. '99 campaign, the axial thrust/impulse component was generally found to decrease with increasing lateral offset (i.e., of the laser beam, from the engine's axis of symmetry). For modeling purposes, the performance can be crudely modeled by a straight line with the peak centerline value set at 0.06 N-s (i.e., zero offset), and the magnitude for 6 cm offset set at 0.015 N-s.

SIDE IMPULSE AND PITCHING ANGULAR IMPULSE MEASUREMENTS

The impulses from time-variant restoring forces and pitching moments acting on a laser-powered vehicle are measured as a function of the vehicle's lateral offset with the Angular Impulse Measuring Device (AIMD) [REF 11]. The device is used to measure both the impulse applied to the engine by a restoring force and the angular impulse applied to engine by a pitching moment; the device appears in Fig. 3.

A Model 200-11/10 lightcraft (16.1-cm diam.) was tested in the PLVTS laser beam at WSMR using the AIMD. The PLVTS beam did not pass through any telescoping optics and had outer dimensions of 10 cm x 10 cm upon entering the lightcraft; the 2X resonator optics were used and the laser pulse energy was about 460 joules. The test setup represented conditions that all lightcraft experience during the first few seconds of flight. The AIMD measured the side impulse and pitching impulses over a range of lateral offsets ranging from –3 cm to +10 cm. Ten tests were performed at each offset location and resulted in five side impulse and five pitching impulse measurements. The five values were then averaged together to produce a final value and the results appear in Fig. 3 as the experimental data.

(a)

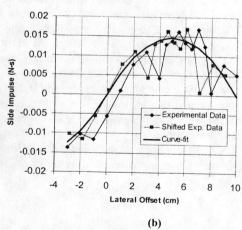

(b)

FIGURE 3. (a) AIMD and lightcraft during testing and (b) test results with curve-fit

The PLVTS beam propagated about 40 ft before entering lightcraft engine. The lightcraft was transversed through the beam in a direction parallel to the top and bottom sides of the square beam. The measured side impulse profile is unique to transversing the beam in this direction and will be different if the lightcraft transversed the beam with a nonzero angle - taken relative to one of beam's square sides. However time constraints limited the testing to one transverse profile.

For the purpose of modeling, the side impulse curve is shifted to left so that there is no side impulse for an offset of zero. Next a parabolic curve fit is assigned to the shifted side impulse data, the shifted data and curve-fit also appear in Fig. 3. At a lateral offset of zero, the curve fit has a slope of 0.59 N-s / m which is slightly less than the slope measured experimentally.

Like the side impulse, the pitching angular impulse also varies with lateral offset. For the Model 200 lightcraft, the pitching impulse is small over most of the range of offsets. The pitching impulse becomes positive at large offsets because only a section of the engine is illuminated by the laser beam and is experiencing laser-supported-detonation.

WIND TUNNEL TESTS

The experimental aerodynamic data on the lightcraft is collected through wind tunnel tests at Rensselaer Polytechnic Institute's Aerospace Laboratory. The testing environment is the 4' x 6' low-speed wind tunnel. The tunnel is a closed circuit, continuous flow model. The test section is four feet tall by six feet wide and was operated in closed configuration. The forces acting on the body in the test section are recorded with a 6-element load cell balance, located on the outside of the tunnel. The model is attached through the tunnel floor to a yoke surrounding the outside of the test section. The yoke is supported by 6 Lebow model 3167 load cells in such an orientation that six degrees of movement can be measured.

The model itself is a scaled version of the Model 200 Lightcraft having a 9" diameter. Its alterations from flight shape include an optic cut short to accommodate the sting. Prior to its installation, the model is carefully balanced on a static magnetic balance to eliminate vibrations during testing. The model contains an electric motor that spins it relative to its sting mount.

Seeking to characterize this body as best as possible, a broad test matrix is established that varies the three parameters of angle of attack, spin rate, and airspeed. Using these variables, the following aerodynamic characteristics are sought: drag, lift, Magnus force, roll damping, pitch moment and Magnus moment. A wind-based coordinate system is established to define these reactions, accommodate for the test fixture, and transfer the information from the tunnel to the to the 6-DOF program.

CFD ANALYSIS

The low speed aerodynamic properties of the Model 200 Lightcraft are explored using Fluent® CFD. The goal of these tests too is the prediction forces and moments acting upon the body under a variety of flight conditions. Tests are run at a variety of angles of attack to characterize the behavior and compare it with results from wind tunnel tests and actual flights.

The Fluent® CFD simulations are performed with a 9-inch diameter, non-spinning Model 200 lightcraft. For the simulations, a 500 000 cell 3D spherical grid with a radius 40 times the diameter of the Lightcraft is used. The standard kε turbulence model is used. The tests appear converged by 1000 iterations, but they are allowed to run beyond 2500 iterations to be certain of convergence. Analysis yields drag and lift data (no spin resulted in no Magnus force) in addition to three moments about the center of mass of the craft.

VIDEO ANALYSIS

Numerous lightcraft flights have been recorded on video for archival purposes, and many have been disclosed in 14 TV documentaries and 60 print media articles. Flights of the model 200 lightcraft that demonstrate the vehicle's beam riding behavior can be converted to digital files to be read into Sensor Applications Inc.'s *Image Express* video analysis software. The software tracks the lightcraft's position in the camera's viewing area and numerical algorithms are then applied to the screen position data to yield the lightcraft's real position. Complete trajectory data is obtained by combining data taken from two different camera angles of the lightcraft flight. One angle captures the motion of the lightcraft in the vertical plane; the camera is typically positioned 125 to 400 feet away from the launch site. Altitude data is gained from this side-looking camera. The second camera is positioned to peer up the center (or just outside) of the laser beam and into the lightcraft's optic; this angle captures the lightcraft motion in the horizontal plane and reveals the oscillations of the vehicle about the center of the beam. Two flights of the model 200 lightcraft have been thoroughly analyzed. These two flights are selected because the laser beam is most similar to the beam provided to the lightcraft for AIMD testing; the beam is passed from the laser directly into the lightcraft without being altered by mirrors, diffraction or atmospheric attenuation.

6-DOF MODEL

A 6-DOF model of the laser lightcraft is developed and implemented in FORTRAN 95. At the core of the model are Euler's Equations given as

$$F_x = m(\dot{u} + qw - rv)$$
$$F_y = m(\dot{v} + ru - pw) \qquad (1,2,3)$$
$$F_z = m(\dot{w} + pv - qu)$$
$$M_x = I_{xx}\dot{p} + (I_{zz} - I_{yy})qr$$
$$M_y = I_{yy}\dot{q} + (I_{xx} - I_{zz})pr \qquad (4,5,6)$$
$$M_z = I_{zz}\dot{r} + (I_{yy} - I_{xx})pq$$

Where u, v, and w represent the components of the lightcraft inertial velocity expressed in a body-fixed frame. Variables p, q, and r represent the components of the lightcraft inertial angular velocity expressed in the body-fixed axis. The variables F_x, F_y, and F_z denote forces acting along a body fixed axes; M_x, M_y, and M_z denote moments taken about the lightcraft center of mass acting about the body fixed axes. I_{xx} is the axial moment of inertia and the lateral moments of inertia I_{yy} and I_{zz}. Most of above variables are illustrated in Fig. 4. The equations of motion are augmented by four additional equations that describe Euler parameters necessary for the determination of vehicle angular orientation. Additionally three equations are used to determine the lightcraft's position relative to a reference frame fixed to the earth. A fourth-order Runge-Kutta numerical integration routine employing a fixed time step is used to solve the equations of motion between laser pulses.

The laser-induced forces and moments are considered impulses and angular impulses respectively and are incorporated into the mathematical model as velocity discontinuities according to the following equations

$$\int LF_x(t)dt = m(u_2 - u_1)$$
$$\int LF_y(t)dt = m(v_2 - v_1) \qquad (7,8,9)$$
$$\int LF_z(t)dt = m(w_2 - w_1)$$
$$\int LM_x(t)dt = I_{xx}(p_2 - p_1)$$
$$\int LM_y(t)dt = I_{yy}(q_2 - q_1) \qquad (10,11,12)$$
$$\int LM_z(t)dt = I_{zz}(r_2 - r_1)$$

Where LF's and MF's are laser-induced, time-variant forces and moments expressed in the body frame. And the subscripts 1 and 2 denote the velocities and angular velocities before and after the laser pulse, respectively. The equations of motion are integrated until a laser pulse occurs, after which, the simulation is stopped and new velocity initial conditions are determined based on the applied impulses. The equations of motion are then integrated until the next laser pulse occurs.

The mathematical model employs some assumptions. The lightcraft is assumed to be a rigid body with time-invariant mass properties. In the air-breathing mode addressed herein, laser pulse ablates some aluminum off the lightcraft shroud during flight but the amount of material removed is insignificant. Secondly, the 6-

DOF model uses flat Earth equations of motion since the lightcraft flights examined are all below several hundred feet altitude and the effects of a round, rotating Earth are negligible. Gravitational acceleration is constant at 32.17 ft/sec2. Density varies with a simple standard atmosphere model.

The 6-DOF model employs a database describing both laser impulse reactions (air-breathing engine mode) and vehicle aerodynamics. A table describing the variation of the three dominant engine reactions with lateral offset is included but the database contains no data for engine performance variation with range from the PLVTS laser source. The aerodynamic database is comprised of the three forces and three moments that have been described previously. The total angle of attack and spin rate are used to determine the appropriate aerodynamic loads on the vehicle.

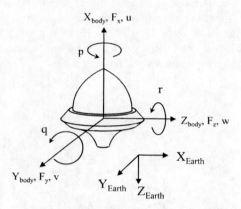

FIGURE 4. Axis and variable definition for 6-DOF model

6-DOF SIMULATION RESULTS

As of the writing of this paper, the wind tunnel aerodynamic database is not yet completed and installed fully in the 6-DOF simulation; however, the simulation does include the laser impulse reaction model discussed earlier in the paper. The laser model is slightly simplified so that the thrust impulse is constant with offset. Aerodynamic drag is the only aerodynamic force incorporated in the model at the writing of this report and a C_D value of 0.52 derived from wind tunnel tests is used. The simulation does correctly predict some important characteristics of the vehicle flight dynamics.

The pulsed laser generates a characteristic saw-tooth velocity profile as shown in Fig. 5. Here a laser-pulse applies a propulsive impulse instantaneously increasing velocity, after which aerodynamic drag acts to slow the lightcraft down prior to arrival of the subsequent laser pulse. When the pulse energy level is maintained, increasing the pulse rate increases the time average thrust given as

$$F_T = I_T \cdot PRF \qquad\qquad (13)$$

where F_T is time averaged thrust, I_T is the thrust impulse resulting from one impulse, and PRF is the laser pulse repetition frequency. Fig. 5 also shows the effect of increasing the PRF on lightcraft velocity.

The simulation also predicts oscillatory motion of the lightcraft about the laser beam's centerline after the lightcraft recieves a perturbation. The frequency of oscillation depends on the magnitude of lateral offset since the side impulse does not vary linearly with lateral offset at large offsets. Fig. 5 shows a frequency of about 2.5 Hz for lateral offsets below 0.05 ft. The lightcraft's velocity also affects the damping as seen in Fig 5. The plot shows vehicle lateral offset as a function of time for two flights; the first accelerates from 2 to 40 ft/sec and the second flight maintains 40 ft/sec velocity. The latter flight experiences larger initial damping because of its higher velocity. The slope of the side impulse curve around zero offset is similar to the spring constant in a second order oscillatory system in that it influences the frequency of oscillation of the lightcraft in the laser beam. Additionally, increasing PRF increases the "spring constant" of the system.

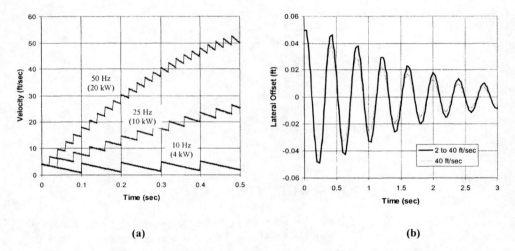

(a)   (b)

FIGURE 5. Simulation results showing (a) saw-tooth vertical velocity profile for various PRF's and (b) lightcraft lateral offset oscillatory history

When given an initial lateral offset combined with a suitable initial horizontal velocity, the lightcraft will enter orbital motion around the laser beam. The simulation models this behavior and also shows that the orbital lightcraft motion is highly dependent on PRF as evidenced in Fig. 6's look-down view of the simulated flight. Here the lightcraft is given an initial offset of 0.05 ft and an initial horizontal velocity of 0.6 ft/s. The horizontal trajectory of the lightcraft is segmented at a low PRF as the vehicle travels a visible distance between laser pulses. As the PRF increases the distance traveled between pulses decreases and the motion appears continuous. The simulation predicts that the lightcraft will depart from the laser beam if the PRF is too small.

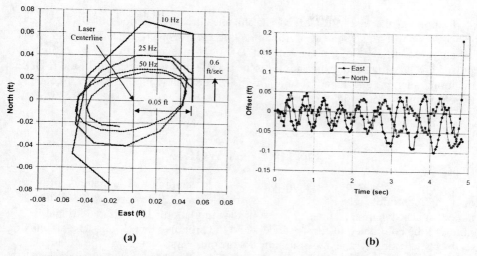

(a)   (b)

FIGURE 6. (a) Orbital motion for various PRF's for a flight with an initial lateral offset and horizontal velocity and (b) lateral offset oscillations obtained from video analysis

VIDEO ANALYSIS RESULTS

One particular flight video of a Model 200 lightcraft reveals a wealth of data about the lightcraft's flight dynamics. The vehicle climbs for nearly 5 seconds attaining a 37 ft altitude and 13 ft/sec maximum velocity; during the flight it completes about 8 ½ orbits around the laser beam. The flight begins with five counter clockwise rotations having an average radius of 0.39 inches and a frequency of 2.14 Hz. After about out 3 seconds of flight, the lightcraft begins to orbit clockwise for an additional three orbits having an average radius of 0.75 inches and a frequency of 1.51 Hz before exiting the beam. Fig. 6 shows the horizontal motion of the lightcraft as a function of time.

By assuming that the orbits are circular, the magnitude of a side force to maintain the orbit is found from to be

$$F_S = 4mr\left(\frac{\pi}{T}\right)^2$$

(14)

Where m is the mass of the ~50 g lightcraft, r is the radius of oscillation, and T is the period of oscillation. Dividing the force by the Laser's PRF yields a side impulse given by

$$I_s = \frac{4mr}{PRF}\left(\frac{\pi}{T}\right)^2 \qquad (15)$$

Applying the laser PRF of 28 Hz and the measured periods of 0.47 sec and 0.66 sec for the first and second oscillations with their respective lateral offsets – one calculates side impulses of 0.0035 N-s and 0.0034 N-s, respectively. These estimated impulses are less than those measured by the AIMD and implemented in the 6-DOF simulation; such a difference results in the discrepancy in oscillation frequency between simulation and flight test data.

CONTINUING WORK

Ongoing research will contribute to understanding the dynamics of laser propelled lightcraft flights. Wind tunnel testing and CFD work will continue to produce an aerodynamic database for use in the simulation. Likewise additional laser testing will be completed to further understand the propulsive reaction. The effects of laser beam asymmetry and transmission distortions verse range will be studied and incorporated in the model. Laser reactions besides the dominant three discussed in this report will be measured. Finally more video data will be analyzed and the results compared to simulation results to further improve the lightcraft model.

In addition to modeling flight data, the computer simulations will be used to examine the effects of tailoring physical parameters to improve vehicle performance. The side force and moment profiles as functions of lateral offset will be altered in the simulation, and the alterations' effects on vehicle flight characteristics will be discussed. Such an analysis could lead to vehicles that ride the laser beam more efficiently. The simulation will also be used to examine the effects of laser pulse repetition frequency on beam riding stability. The simulation will determine vehicle spin rates needed to provide adequate gyroscopic stability. Finally, control strategies for the lightcraft in flight will be an area of future study.

CONCLUSIONS

A 6-DOF model of the laser lightcraft has been created but awaits the complement of complete aerodynamic and propulsive databases. Mass properties of the lightcraft vehicles have been measured experimentally with electronic balances and an Inertial Dynamics MOI-005 device. The same mass properties have been estimated using CAD software. Engine performance is a function of lateral offset and the three dominant engine reactions of a thrust force, side force, and pitching moment have been measured as a function of lateral offset. Engine performance measurements have been taken with a ballistic pendulum and the AIMD with the PLVTS laser configured in a manner identical to flights. Finally, video analysis of existing lightcraft flights has been performed and the results have been compared to simulation results. The computer model and video analysis data both show that the model 200 lightcraft is definitely a beam riding design.

ACKNOWLEDGEMENTS

Archival videos of LTI lightcraft flights were generously provided by Lightcraft Technologies, Inc. - for motion analysis studies at RPI, using the *Image Express* software (produced by Sensor Applications Inc.). The authors would also like to thank the RPI Central Machine shop for their superior technical assistance and machining skills. The U.S. Army's PLVTS laser team (Stephen Squires, Chris Bearisto, and Mike Thurston) deserve a tremendous thanks for providing a laser beam for experiments and for excellent assistance during testing. Finally, Gabriel Williams must be acknowledged for his enthusiastic motion analysis efforts, and Jeff Alvarez deserves thanks for his work modeling the lightcraft in CAD software.

REFERENCES

1. Kantrowitz, A.R., "The Relevance of Space," Astronautics and Aeronautics, Vol. 9, No. 3, March 1971, pp. 34-35.
2. Kantrowitz, A.R., "Propulsion to Orbit by Ground-Based Lasers," Astronautics and Aeronautics, Vol. 10, No. 5, May 1972.
3. L.N. Myrabo, et.al., "Lightcraft Technology Demonstrator," Final Technical Report, prepared under Contract No. 2073803 for Lawrence Livermore National Laboratory and the SDIO Laser Propulsion Program, dated 30 June 1989.
4. L.N. Myrabo, D.G. Messitt, and F.B. Mead Jr., "Ground and Flight Tests of a Laser Propelled Vehicle," AIAA Paper No. 98-1001, 36th AIAA Aerospace Sciences Meeting & Exhibit, Reno, NV, 12-15 Jan. 1998.
5. F.B. Mead Jr., L.N. Myrabo and D.G. Messitt, "Flight Experiments and Evolutionary Development of a Laser Propelled, Transatmospheric Vehicle," Space technology & Applications International Forum (STAIF-98), Albuquerque, NM, 25-29 Jan. 1998.
6. F.B. Mead Jr., L.N. Myrabo and D.G. Messitt, "Flight and Ground Tests of a Laser-Boosted Vehicle," AIAA Paper No. 98-3735, 34th AIAA Joint Propulsion Conference & Exhibit, Cleveland, OH, 13-15 July 1998.
7. D.G. Messitt, L.N. Myrabo, and F.B. Mead Jr., "Laser Initiated Blast Wave for Launch Vehicle Propulsion," AIAA 2000-3848, 36th AIAA Joint Propulsion Conference, 16-19 July 2000, Huntsville, AL.
8. Myrabo, L.N., "World Record Flights of Beam-Riding Rocket Lightcraft: Demonstration of 'Disruptive' Propulsion Technology," AIAA Paper N. 2001-3798, 37th AIAA/ASME/SAE /ASEE Joint Propulsion Conference, 8-11 July 2001, Salt Lake City, UT.
9. "Tomorrow's World," produced by B. Ley for the British Broadcasting Channel TV, first aired 18 June 1998 in the UK.
10. F.B. Mead, Jr., S. Squires, C. Beairsto, and M. Thurston, "Flights of a Laser-Powered Lightcraft During Laser Beam Hand-off Experiments," AIAA Paper 2000-3484, 36th AIAA/ASME/ SAE/ ASEE Joint Propulsion Conference, 16-19 July 2000, Huntsville, AL.
11. Libeau, M.A., "Experimental Measurements of the Laser-Induced Reaction on a Lightcraft Engine," AIAA Student Paper 2002
12. A.D. Panetta, H.T. Nagamatsu, L.N. Myrabo, M.A.S. Minucci, and F.B. Mead, Jr., "Low Speed Wind Tunnel Testing of a Laser Propelled Vehicle," Paper #1999-01-5577, 1999 World Aviation Conference 19-21 October 1999, San Francisco, CA.

Numerical Modeling of Laser Supported Propulsion with an Aluminum Surface Breakdown Model

Yen-Sen Chen*, Jiwen Liu* and Ten-See Wang[†]

*Engineering Sciences Incorporated
Huntsville, AL 35802, USA

[†]NASA/Marshall Space Flight Center
Marshall Space Flight Center, AL 35812, USA

Abstract. In this paper, a multidimensional unstructured-grid computational plasma aerodynamics methodology is developed for pulsed wave devices, with emphases on nonequilibrium effects and laser-induced physics. The present numerical model couples the Navier-Stokes equations, three-temperature energy equations, finite-rate chemistry model and ray-tracing model with radiative energy absorption using realistic absorption coefficient formulation. With the model, the focused laser beam creates high energy concentration at focal point on aluminum surface and causes the ionization of aluminum, which starts the air plasma. Laser lightcraft designs of Myrabo, using pulsed laser as the energy source, are selected for numerical simulation and data comparisons.

INTRODUCTION

Currently, NASA's aim of operating low cost launch and space vehicles requires the research and development of advanced propulsion technologies and concepts. One plausible advanced concept is the utilization of off-board pulsed laser power source to propel small payload (e.g. 100kg) into earth orbit. The merit of the laser-propelled vehicles is in its high efficiency (do not need to carry fuel) and high specific impulse. Previous SDIO research led to the invention of the one of the laser powered launch vehicle concept – the Laser Lightcraft concept, currently being tested at the High Energy Laser Test System Facility, White Sands Missile Range, New Mexico. Although the spin-stabilized small scale Lightcraft model (invented by Myrabo) has been flown successfully up to an altitude of 30 meters using a 10 kW pulsed-laser at 10 Hz, many technical issues need to be addressed before an optimized design of the vehicle and its operation can be achieved.

The purpose of this study is to establish the technical ground for modeling the physics of laser powered pulse detonation phenomenon. The principle of the laser power propulsion is that when high-powered laser is focused at a small area near the surface of a thruster, the intense energy causes the electrical breakdown of the working

CP664, *Beamed Energy Propulsion: First International Symposium on Beamed Energy Propulsion*,
edited by A. V. Pakhomov
© 2003 American Institute of Physics 0-7354-0126-8/03/$20.00

fluid (e.g. air) and forming high speed plasma (known as the inverse Bremsstrahlung, IB, effect). The intense heat and high pressure created in the plasma consequently causes the surrounding to heat up and expand until the thrust producing shock waves are formed. This complex process of gas ionization, increase in radiation absorption and the forming of plasma and shock waves will be investigated in the development of the present numerical model. In the first phase of this study, laser light focusing, radiative absorption and shock wave propagation over the entire pulsed cycle are modeled. The model geometry and test conditions of known benchmark experiments such as those in Myrabo's experiment will be employed in the numerical model validation simulations. The calculated performance data (e.g. coupling coefficients) will be compared to the test data. Plans for the numerical modeling of the detailed IB effect will also be described in the proposed investigation. The final goal will be the design analysis of the full-scale laser propelled flight vehicle using the present numerical model.

In previous study, Engineering Sciences, Inc. has developed a laser powered launch vehicle performance analysis tool based on its in-house flow and radiation codes. UNIC-UNS unstructured-grid flow code and GRADP-UNS unstructured-grid radiation code are two advanced numerical models. Many complex engineering design problems related to fluid dynamics and radiative heat transfer have been solved using these two codes. High-temperature thermodynamics and plasma dynamics models have been developed with benchmark data validations presented for laser powered launch vehicles. The development work has included transient shock capturing algorithm using unstructured-grid method with dynamic local refinement and coarsening adaptive grid strategy. High temperature thermodynamics and plasma gas dynamics physics are modeled and validated. Non-equilibrium radiation model with the effects of gas breakdown and laser energy absorption has also been addressed and modeled. These advanced thermodynamics and radiation models serve as the fundamental building blocks for the present model development and for the ultimate utilization of the laser powered launch vehicle performance analysis tool in real designs.

In the present research, the development work for modeling the mechanisms for initializing the aluminum surface ablation and ionization due to focused laser energy is completed and tested for a laser Lightcraft model-200. At the start of the laser pulse, the beams are focused at a small area on the aluminum surface of the test vehicle. Part of the energy is absorbed by the material and causes to overcome the bounding energy of the molecules. According to Harada [1-4], this process is accounted for through stopping power that is caused by collisions between ions and atoms of the material. Since the energy source is different in the present application that laser beam instead of ion beam is the energy source, the energy source modeling would consider the absorption of radiative energy at the wall surface. In the present case, the energy absorbed, as output from the radiation model, is used to calculate the heat conduction and energy balance at the wall surface through which the ablation rates of the aluminum material from the surface is calculated. The ionization of the ablated material is then initiated by using the Saha-Eggert equation under the thermal and chemical equilibrium assumption [4]. The developed model is described below.

COMPUTATIONAL FLUID DYNAMICS MODEL

Governing Equations

For The Continuity, Navier-Stokes and Energy (Total Enthalpy) Equations, can be written in a Cartesian tensor form:

$$\frac{\partial \rho}{\partial t} + \frac{\partial}{\partial x_j}\left(\rho u_j\right) = 0 \tag{1}$$

$$\frac{\partial \rho u_i}{\partial t} + \frac{\partial}{\partial x_j}\left(\rho u_j u_i\right) = -\frac{\partial p}{\partial x_i} + \frac{\partial \tau_{ij}}{\partial x_j} \tag{2}$$

$$\frac{\partial \rho H}{\partial t} + \frac{\partial}{\partial x_j}\left(\rho u_j H\right) = \frac{\partial p}{\partial t} + Q_{ec} + Q_v + \frac{\partial}{\partial x_j}\left(\frac{\mu}{P_r}\nabla H\right) + \frac{\partial}{\partial x_j}\left(\left(1 - \frac{\mu}{P_r}\right)\nabla\left(V^2/2\right)\right) \tag{3}$$

$$\frac{\partial}{\partial t}\left(\frac{2}{3}k_b n_e T_e\right) + \frac{\partial}{\partial x_j}\left(\frac{2}{3}k_b n_e T_e u_j\right) = \frac{\partial}{\partial x_j}\left(\lambda_e \frac{\partial T_e}{\partial x_j}\right) + Q_r - Q_{ec} \tag{4}$$

where ρ is the fluid density, u_i is the ith Cartesian component of the velocity, p is the static pressure, μ is the fluid viscosity, P_r is the Prandtl number, H is the gas total enthalpy and V stands for the sum of velocity squared. In Eq. (4), k_b, n_e, T_e, λ_e, Q_r, Q_v and Q_{ec} are the Boltzmann's constant, electron number density, electron temperature, electron thermal conductivity, radiative heat source from laser absorption and radiative transfer, vibrational-translation energy transfer source term, and the energy transfer due to eletron/particle elastic collisions, respectively. The shear stress τ_{ij} can be expressed as:

$$\tau_{ij} = \left(\mu + \mu_t\right)\left(\frac{\partial u_i}{\partial x_j} + \frac{\partial u_j}{\partial x_i} - \frac{2}{3}\frac{\partial u_k}{\partial x_k}\delta_{ij}\right) - \frac{2}{3}\rho k \delta_{ij} \tag{5}$$

The species conservation equation is expressed as:

$$\frac{\partial \rho Y_i}{\partial t} + \frac{\partial}{\partial x_j}\left(\rho u_j Y_i\right) = \frac{\partial}{\partial x_j}\left[\left(\rho D + \frac{\mu_t}{\sigma_Y}\right)\frac{\partial Y_i}{\partial x_j}\right] + \dot{\omega}_i \tag{6}$$

where Y_i is the ith species mass fraction, D is the mass diffusivity, σ_Y is the turbulent Schmidt number, and $\dot{\omega}_i$ is the chemical reaction rate for species i respectively.

Vibrational Energy Equation

For high temperature gas flows, thermal non-equilibrium state may be important. In Landau and Teller's derivation, a master equation is employed to describe the evolution of the population of quantum level N_i. This master equation is written as:

$$\frac{dN_i}{dt} = N\sum_{j=0}^{l_{max}} K_{j\Rightarrow i} N_j - N\sum_{j=0}^{l_{max}} K_{i\Rightarrow j} N_i; \quad i = 0,1,2,...,l_{max} \tag{7}$$

Results from the quantum mechanical solution of the harmonic oscillator are used to relate the various quantum transition rates to one another, and then the master equation may be summed over all quantum states to arrive at the Landau-Teller equation:

$$\frac{D\rho e_v}{Dt} = \frac{\partial}{\partial x_i}\left(k_v \frac{\partial T_v}{\partial x_i}\right) + \rho\frac{e_v^{eq}(T_t) - e_v}{\tau_{LT}} \approx \frac{\partial}{\partial x_i}\left(k_v \frac{\partial T_v}{\partial x_i}\right) + \rho\frac{C_{v,v}(T - T_v)}{\tau_{LT}} \tag{8}$$

where ρ, e_v, e_v^{eq} and τ_{LT} represent the gas density, vibrational energy, effective (equilibrium) vibrational energy and the vibrational-translational relaxation time scale respectively. An empirical expression (to be discussed in the next section) is used to model the Landau-Teller relaxation time scale.

As discussed by Gnoffo (1989), the vibrational-translation energy relaxation time scale can be evaluated.

Three-Temperature Energy Equations Point Implicit Coupling

For solution accuracy and stability, a point-implicit procedure is employed for solving the coupled three-temperature energy equations. The three-temperature energy equations in time domain can be written as:

$$\rho\frac{\partial h_g}{\partial t} = -C_{coll}(T_g - T_e) - \sum_i \frac{\rho_s}{\tau_s}(e_{v,s}^* - e_{v,s})$$

$$\rho\frac{\partial e_v}{\partial t} = \sum_i \frac{\rho_s}{\tau_s}(e_{v,s}^* - e_{v,s})$$

$$\rho\frac{\partial e_e}{\partial t} = C_{coll}(T_g - T_e) + Q_r$$

where C_{coll} represents the elastic-inelastic collision energy transfer coefficient. The left-hand-side terms of the above equations can be discretized as:

$$\rho\frac{\partial h}{\partial t} = \rho\left(\frac{\partial h}{\partial T}\right)\frac{\partial T}{\partial t} = \rho C_p\frac{\partial T}{\partial t} \approx \rho C_p\frac{(T^{n+1} - T^n)}{\Delta t}$$

$$\rho\frac{\partial e}{\partial t} = \rho\left(\frac{\partial e}{\partial T}\right)\frac{\partial T}{\partial t} = \rho C_v\frac{\partial T}{\partial t} \approx \rho C_v\frac{(T^{n+1} - T^n)}{\Delta t}$$

The last term in the above equations represent finite difference discretization. Also, applying the simplifying assumption for the vibrational energy relaxation term: (Gnoffo, 1989)

$$\sum_i \frac{\rho_s}{\tau_s}(e_{v,s}^* - e_{v,s}) \approx \rho\frac{C_{v,v}}{\tau}(T_g - T_v)$$

The linearized form of the equations is used in the computation.

Aluminum Wall Energy Equation

The energy balance of the aluminum wall material is described by a 1-D heat conduction equation with energy input at one end with material ablation rates. That is,

$$\rho_w\frac{d(CpT)_w}{dt} = \frac{d}{dx}\left(k\frac{dT_w}{dx}\right) + A\left\{Q_r - \dot{m}\Delta H\right\} \qquad A = 1 \text{ at surface, } 0 \text{ otherwise}$$

141

where the subscript $_w$ stands for wall material, Q_r is the energy input due to absorption of radiative energy, $\dot{m}$ denotes the ablation rates and ΔH is the ablation energy that includes the sum of latent heats of melting, ΔH_{sl}, and evaporation, ΔH_{lg}, (i.e. 38,738 J/kg and 491,675 J/kg at 1 ATM respectively) [11]. The melting and boiling points for pure aluminum are 933.47K and 2792K respectively. The heat capacity, Cp, and thermal conductivity, k, of aluminum is 895.7056 J/Kg-K and 249.1023 watts/m-K respectively. An enthalpy form of the above equation is solved such that the latent heat effects can be included implicitly. That is,

$$\rho_w \frac{dH_w}{dt} = \frac{d}{dx}\left(k\frac{dT_w}{dx}\right) + AQ_r$$

When the saturated gas enthalpy point is reached from the above equation, the ablation mass flow rate is calculated as the excess enthalpy at wall divided by the latent heat of evaporation. That is, $\dot{m} = \rho_w(H_w - H_{sg})/(\Delta H_{lg}\,\Delta t)$. The wall heat conduction equation and the ablation mass flow rate equation are solved through an iterative procedure until convergence for every time step.

In the present model, the surface energy absorption of the laser energy is calculated through the ray-tracing module. All near wall cells that are active due to energy absorption are then analyzed for the thermal conditions based on the above heat conduction equation. The density of the ablated aluminum atoms in the near wall cell is than calculated based on the following equation.

$$\frac{D\rho_{AL}}{Dt} = \dot{m}$$

Wall cells (elements) subjected to laser absorption are registered and counted after the laser ray tracing process. The 1-D heat conduction models for these wall points are solved for the wall surface temperature evolution. The outside wall is assumed to be always isothermal at 300 K. For simplicity, the aluminum atom generation is ignored in the present model. It is found that only the calculation of electron concentration production rate is important for initiating the air breakdown process. The electron species source term is used in the finite-rate chemistry model for air plasma. Aluminum gas species, which is only involved in equilibrium ionization calculation at wall surface, is ignored in the finite-rate chemistry calculation in the flowfield.

Ionization Modeling

The resultant surface temperature and ablation rates are then used for material ionization calculation through the Saha-Eggert equation. That is,

$$S_r(T) = \frac{n_{r+1}n_e}{n_r} = \frac{2U_{r+1}(T)}{U_r(T)}\frac{(2\pi m_e kT)^{3/2}}{h^3}e^{-\frac{E_r-\Delta E_r}{kT}}$$

where n stands for number density ($1/cm^3$), r denotes the r-fold ionization level, e denotes electronic property, U is the partition function of ionization potential, me is the rest mass of electron (= 9.108E-28 g), k is the Boltzmann's constant (=1.381E-16 erg/deg-K) and h is the Plank's constant (= 6.626E-27 erg-sec). For first-fold ionization (r=0, $AL \rightarrow AL^+ + e^-$ and $n_{AL+} = n_e$), the above equation can be written as (after plugging in all constants):

$$\frac{n_e^2}{n_{AL+}} = 2.415130256 \times 10^{15} \frac{2U_1}{U_0} T^{3/2} e^{-\frac{E_0 - \Delta E_0}{kT}}$$

where the Unsold's formula for the lowering of ionization energy is used here. That is, $\Delta E_0 = 6.96 \times 10^{-7} (n_e)^{1/3} Z_{eff}^{2/3}$ (in eV), and $Z_{eff} = (\sum_{i=1}^{r+2} Z_i^2 n_i)/n_e = 1$ for $r = 0$. Also, $E_0 = 5.984$ eV for aluminum (1 eV $= 1.602 \times 10^{-12}$ erg). Tabulate values for U1 from [5] is generated for the present model.

A module for the aluminum surface ablation has been developed and tested. To apply the present model to the laser lighcraft aluminum-surface-breakdown process, the aluminum surface emissivity is needed to start the calculation. Some emissivity data for aluminum is given in Ref. 6. For the present application, aluminum emissivity of 0.1 is used in the simulation (Note: This is a just best guess value of the authors, which is close to the value for commercial aluminum sheet). This means that 90% of the laser energy is absorbed at the focal region, which gives very rapid temperature rise and creates high values of aluminum ablation rates for subsequent ionization. High values of electron number density are generated as a result of the ionization process.

MODEL-200 PERFORMANCE ALANYSIS

The present model is tested for the laser Lightcraft Model-200-3/4 and compared with experimentally measured data for a range of laser power levels. Two models are used for comparison purpose. The spark ignition model is used to provide the baseline of the predictions. Then, the same cases are computed with the present aluminum-surface breakdown/ionization model. The same laser absorption model based on a single ionization formula [7] modified by a tabulated augmentation factor described by Raizer [8] used to simulate multiple ionization effects.

The impulse thrust of the laser Lightcraft Model-200-3/4 was tested using a 10 kW CO_2 pulse laser, which can provide up to 800J for each single pulse, of the Pulsed Laser Vulnerability Test System at the High Energy Laser System Test Facility, White Sands Missle Range, NM. The laser pulse energy was measured with a calorimeter with an estimated uncertainty of $\pm$ 10 J in total laser energy delivered. Several variations of the basic Laser Lightcraft design (Model-200 series) similar to those described in Ref. 9 and 10 were tested experimentally. The test results of the 6061-T6 all aluminum Model #200-3/4 vehicle with 18 μs pulse width [10] are chosen for this study. The impulse measurements were conducted with a pendulum apparatus, which has an estimated impulse measurement uncertainty of 1% or better [9,10].

A hybrid unstructured mesh system, shown in figure 1, was created for the present computation. Also shown, as yellow spots in figure 1, are the laser traces from the downstream (right) side of the vehicle. The figure on the right shows a close-up view in the cowl region that reveals the hybrid grid arrangement. The element and node numbers for this grid are 26,142 and 14,441 respectively. For spark ignition model,

25% of laser power is assumed as the initial heat source near the focal region. The spark heat source region is bounded by the laser traces and a radius of 1.355 mm from the focal point of the parabolic optical surface of the vehicle. The spark heat source is turned off when the air plasma reaches the self-sustained state (specified as when laser energy absorption efficiency reaches 15%).

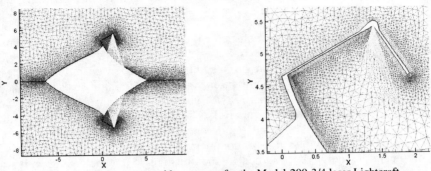

Figure 1. Mesh system and laser traces for the Model-200-3/4 laser Lightcraft.

In the present model, the spark ignition model is totally eliminated. The aluminum surface breakdown/ionization model replaces the air plasma initiation process. Assuming that the aluminum surface emissivity of 0.1, 90 percent of the laser energy is therefore used for the heat-up and breakdown of the aluminum near the focal region. One-dimensional heat conduction equation for the aluminum wall segments, where laser energy absorption is registered, is solved for the time history of the aluminum surface conditions. Surface breakdown and ionization process is then calculated based on the aluminum surface conditions. This provides the electron species production rate in the near wall region, which is responsible for initiating the air plasma. As a result of this model, the surface heat-up time and ionization time scale is directly proportional to the input laser power and the focus condition of laser beam (i.e. better focus produces faster heat-up and plasma iginition).

In the numerical computations, the initial time step size is specified at 0.05 microseconds and linearly increased to 1 microsecond between 50 to 500 microseconds. This arrangement allows better time resolution for accuracy at the start-up due to expected fast temperature and pressure rise near the focal region and allow better computing time as the blast wave expands and its strength weakened. During every time marching step, the point implicit chemistry solution procedure and the three-temperature coupled equations are solved with reduced time step size based on the chemistry time scale. This multiple time stepping treatment for chemistry is realized to provide the time accuracy for the species equations and has shown great improvement in the smoothness of the flowfield.

During the course of performing the computations, it was found that the pressure and velocity fields near the axis had developed spurious oscillations as the solution proceeded beyond 100 microseconds. The source of this anomaly was traced to be stemming from the cumulative round-off errors in computational cell volume and area calculations. This error becomes obvious for transient flows with quiescent freestream conditions, which is the case for the present application. A better-structured pressure-smoothing scheme is therefore introduced to damp out these background oscillations. Further test of the new scheme has shown good results.

The flowfield solution plots of density and heavy gas temperature, for the 400J case at 10 microseconds time level are shown in Figure 2. The laser traces are also shown in the plots. Figure 3 shows the solution plots of Mach number and heavy gas temperature, for the same case at 25 microseconds time level. These plots show the forming and evolution of the blast wave and the high temperature regions. The heavy gas temperature at this time is closely follow the distributions of the electron temperature, which is indicative of the very high energy-transfer rate between gas species and the electrons of the air plasma. The solution also indicates that the vibrational temperature follows the blast wave, which is due to shock heating and departure from heavy gas temperature in high temperature regions due to increased relaxation time scale.

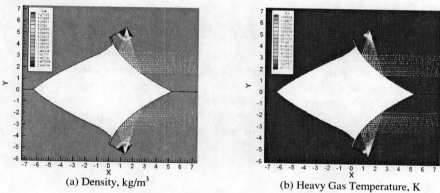

(a) Density, kg/m^3 (b) Heavy Gas Temperature, K

Figure 2. Flowfield solution for 400 J laser energy case at 10 μsec.

At the beginning, as the shock wave develops, negative thrust is produced due to the shape of cowl inner surface. Later, the thrust becomes positive after the shock wave bounce back from the upstream corner of the cavity. As indicated in the solution, the pressure level is increased to about 41 ATM due to shock reflection off the upstream corner of the cavity. The high-pressure region is attached to the parabolic optical surface. During this time period for the high-pressure wave to travel along the optical surface, the positive thrust is produced and can be referred to as thrust-producing period. As the shock wave expands to wrap around the cowl outer surface and leaves the tip of the optical surface, negative contribution to the thrust will resume until the thrust curve levels off. The time-integrated thrust curves are calculated for the present model using aluminum surface breakdown model to initiate the air plasma. The final vehicle thrust coefficients (expressed as coupling coefficient) are then recorded based on the level-off thrust curves and used for data comparisons.

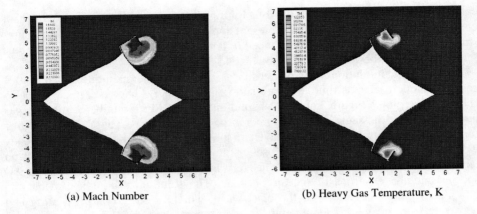

(a) Mach Number

(b) Heavy Gas Temperature, K

Figure 3. Flowfield solution for 400 J laser energy case at 25 μsec.

The predicted coupling coefficients (in dyne-s/J) for the 75J, 100J, 150J, 200J, 300J and 400J cases are 4.61, 6.90, 11.27, 13.72, 14.65 and 14.85 respectively with the present model. These results are slightly lower than the predicted values of the previous model using the spark ignition mechanism. Both are in good agreement with the experimental data. This is understandable since the laser absorption coefficient was tuned for the spark ignition model. Further tuning of the absorption coefficient is needed to fit the present model predictions to the measured data. Figure 4 shows the data comparisons between the predictions of the present and spark ignition models and the experimentally measured data with error bends. The overall trend of the model predictions shows very good correlation with the measured data.

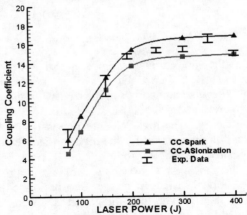

Figure 4. Comparisons of computed and measured coupling coefficients for different laser power levels. CC-Spark stands for the previous model with empirical spark ignition and CC-ASIonization represents the current model with aluminum surface breakdown ionization process.

The impulse thrust of the lightcraft vehicle can be correlated with the laser energy absorption efficiency. This also explains the differences between the predictions of the spark ignition model and the present aluminum-surface breakdown model. Table 1 summaries the energy absorption efficiency of the models and the heat-up time, T, for the present model. Es in Tables is the total energy absorbed by the air plasma. Espark represents the total spark energy required to reach the 15 percent laser absorption efficiency threshold for sustaining the plasma. T (start) indicates when the laser absorption starts and T (0.001J) signifies the time for the absorbed laser energy reaches 0.001J. It is seen clearly from Table 1 that the heat-up and plasma initiation time is directly proportional to the laser power and the low thrust produced for the present model can be attributed to the lower laser absorption efficiency, Eff, indicated.

Table 1. Energy absorption efficiency and heat-up time scale summary

Laser Power	Spark Ignition Model			AL Surface Breakdown/Ionization Model			
	Es (J)	Eff (%)	Espark (J)	Es (J)	Eff (%)	T (start) (sec)	T (0.001J) (sec)
75J	24.12	32.2	2.893	16.49	22.0	1.8×10^{-06}	3.5×10^{-6}
200J	152.04	76.0	4.486	131.53	65.8	7.5×10^{-7}	1.3×10^{-6}
400J	339.92	85.0	5.964	316.34	79.1	4.0×10^{-7}	7.0×10^{-7}

The present calculation also reveals the amount of aluminum material that is ablated off the vehicle surface for each pulse in the area where the laser beam is focused. This mass flow rate is then used to estimate the specific impulse (ISP) of the vehicle. The calculated ablation mass flow rates and vehicle ISP are summarized in Table 2. It is clear that very high ISP values are predicted for this type of propulsion system as expected. This indicates that it takes only small amounts of surface material to initiate the air breakdown and subsequent plasma flow.

Table 2. Summary of vehicle thrust and ISP

Laser Power	Thrust (N)	Ablation Mass Flow Rate (kg/sec)	ISP (sec)
75J	0.00346	4.9599×10^{-10}	7.1183×10^{5}
100J	0.00690	5.3506×10^{-10}	1.3159×10^{6}
150J	0.01690	5.6469×10^{-10}	3.0537×10^{6}
200J	0.02744	5.9278×10^{-10}	4.7235×10^{6}
300J	0.04395	6.9750×10^{-10}	6.4297×10^{6}
400J	0.05943	7.4143×10^{-10}	8.1792×10^{6}

CONCLUSIONS

In the present study, an aluminum surface breakdown/ionization model has been developed and tested for a laser Lightcraft launch vehicle design. The aluminum surface temperature rise due to laser energy input has been modeled using a one-dimensional heat conduction equation with laser radiation heat source applied on the wall surface. An averaged emissivity of 0.1 is assumed for the aluminum surface. The subsequent ionization process near the aluminum surface is calculated based on the wall surface conditions obtained for every time step. This provides the needed electron generation source term near the laser focal region to initiate the air plasma and the associated blast waves. The present model has been demonstrated to replace the need for an artificial spark ignition mechanism as used in the previous model. Hence, this research has provided one step further in making the present analytical model as a true predictive tool for laser supported propulsion system with reduced empiricism.

Benchmark testing of the laser Lightcraft Model-200-3/4 has demonstrated the effectiveness of the present model in predicting the integrated coupling coefficient of the vehicle. The predicted trends are in good agreements with the previous model and the measured data. Further tuning of the modeling constants in estimating the laser absorption coefficient is needed to better fit the present predictions with the measured data. The current development work has laid a concrete foundation for future research in advanced propulsion concepts that may involve short pulses of high intensity energy sources such as microwave, electromagnetism, etc.

ACKNOWLEDGEMENTS

This work was supported by the National Aeronautics and Space Administration, Marshall Space Flight Center, under the Contract number H-33325D.

REFERENCES

1. Harada, Nob., Kagihiro, M., Shinkai, H., Jiang, W. and Yatsui, K., "Flyer Acceleration by Ablation Plasma Using an Intense Pulsed Ion Beam," AIAA 99-3485, 30th Plasmadynamics and Lasers Conference, 28 June – 1 July, 1999, Norfolk, VA.
2. Harada, Nob., Yazawa, M., Kashine, K., Jiang, W. and Yatsui, K., "Numerical Simulation of Foil Acceleration by Intense Pulsed Ion Beam," AIAA 2000-2272, 31st AIAA Plasmadynamics and Lasers Conference, 19-22 June, 2000, Denver, CO.
3. Harada, Nob., "Acceleration of Multi-Layer Foil by Intense Pulsed Ion Beam," AIAA 2001-3005, 32nd AIAA Plasmadynamics and Lasers Conference, 11-14 June, 2001, Anaheim, CA.
4. Harada, Nob., (Private Communication), September, 2001.
5. Drawin, Hans-Werner and Felenbok, Paul, Data for Plasmas in Local Thermodynamic Equilibrium, Gauthier-Villars, Editeur, Paris, 1965.
6. Reynolds, William C., and Perkins, Henry C., Engineering Thermodynamics, McGraw-Hill Book Company, New York, NY, 1970.
7. Raizer, Y.P., and Tybulewicz, A., "Laser-Induced Discharge Phenomena", Studies in Soviet Science, Edited by Vlases, G.C., and Pietrzyk, Z.A., Consultants Bureau, New York, 1977.
8. Zel'dovich, Y.B., and Raizer, Y.P., "Physics of Shock Waves and High Temperature Hydrodynamic Phenomena", Vol. 1, Edited by Hayes, W.D., and Probstein, R.F., Academic Press, New York and London, 1966.
9. Myrabo, L.N., Messitt, D.G., and Mead, F.B., Jr., "Ground and Flight Tests of a Laser Propelled Vehicle," AIAA Paper 98-1001, Jan., 1998.
10. Mead, F.B., Jr., and Myrabo, L.N., Messitt, D.G., "Flight and Ground Tests of a Laser-Boosted Vehicle," AIAA Paper 98-3735, July, 1998.
11. J. D. Cox, D. D. Wagman, S. and V. A. Medvedev, CODATA Key Values for Thermodynamics, Hemisphere Publishing Corp., New York, USA, 1989.

Numerical Analysis of Gasdynamic Aspects of Laser Propulsion

Yu.P.Golovachov[*], Yu.A.Kurakin[*], Yu.A.Rezunkov[♣], A.A.Schmidt[*], V.V.Stepanov[♣]

[*] - Ioffe Institute of Russian Academy of Sciences,
Politekhnicheskaya 26, Saint Petersburg, 194021 Russia,
[♣] - Research Institute for Complex Testing of Optic-electronic Devices
Sosnovy Bor, 188540 Russia.

Abstract. The paper is focused on numerical investigation of the gasdynamic processes accompanying laser beam energy deposition in nozzle of the laser propulsion engine (LPE). Various gasdynamic models were implemented in order to compare their applicability for simulation of the phenomena: (a) perfect ideal gas; (b) equilibrium plasma; (c) non-equilibrium multi-temperature plasma, and (d) two-phase mixture.
Numerical method of the simulation is based on high-resolution Godunov-type scheme and unstructured adaptive grid technique. Two types of LPE nozzles were considered: parabolic and "toroidal" parabolic. Non-stationary gasdynamic function distribution and the thrust of the LPE were obtained. Time of gasdynamic parameter relaxation in the nozzle was estimated.

INTRODUCTION

Since the beginning of the seventies the idea of laser propulsion has been very attractive [1, 2]. Beamed energy propulsion reveals many advantages resulting from separation of the energy source and the vehicle, attainable on board high density of energy, and therefore, high specific impulse of the laser propulsion engine.

Complexity of physical phenomena occurring in course of LPE operation makes obvious necessity of fundamental theoretical investigations of the problem, in particular, a detailed numerical simulation is required of gasdynamic processes: shock wave origination, propagation, and interaction with each other and with elements of construction of the LPE, initiation of jets, etc. A number of attempts have been made in this direction starting from rather simple analytical approach to 1D flow at constant pressure [3], in [4] a model of isoentropic flow at the adiabatic exponent $\gamma = c_p / c_v = \text{const}$ in a nozzle of the LPE was proposed and a selfsimilar solution based on the Sedov's approach was obtained, recent investigations [5] were based on non-equilibrium physical gasdynamic model. Nevertheless, the problem still remains of determination of adequate but simplest description of LPE gasdynamics.

The goal of the present paper is to simulate gasdynamic processes resulting from deposition of laser pulse energy within the working medium of the LPE, to compare results provided by various models (perfect ideal gas, equilibrium plasma, non-equilibrium multi-temperature plasma, and two-phase mixture), and on the basis of the

CP664, *Beamed Energy Propulsion: First International Symposium on Beamed Energy Propulsion*,
edited by A. V. Pakhomov

obtained results to develop efficient tool for numerical investigation of gasdynamic aspects of LPE operation.

MATHEMATICAL MODELS

Three models of the gas-phase working medium of the LPE were considered. All the models were based on general conservation laws for mass, momentum, and total energy, which can be written in the following form:

$$\frac{\partial \rho}{\partial t} + \nabla(\rho \vec{V}) = 0,$$

$$\frac{\partial \rho \vec{V}}{\partial t} + \nabla(\rho \vec{V}\vec{V} + p) = 0, \tag{1}$$

$$\frac{\partial E}{\partial t} + \nabla[(E + p)\vec{V}] = 0,$$

where ρ, p, E, V are the density, the pressure, the total energy, and the velocity of the working medium, respectively. The above set of the governing equations is closed by the constitutive equations:

$$p = p(\rho, T), \qquad E = e(\rho, p) + \frac{\rho V^2}{2}, \tag{2}$$

where $e(\rho, p)$ is the internal energy.

Ideal perfect gas model: In the framework of this model the closing constitutive equations are written in the form:

$$p = \rho R T, \qquad e = \frac{p}{\gamma - 1}, \qquad \gamma = \frac{c_p}{c_v} = \text{const}. \tag{3}$$

Equilibrium gas model: In fact equations (3) are not valid within the high temperature region initiated by the laser breakdown and behind strong shock waves in the induced flow. Under these conditions it is necessary to consider variation of the specific heats due to physical-chemical processes occur. Here it is more suitable to introduce the model of equilibrium gas [6]. In the framework of this model special curve fits for the thermodynamic properties of the working medium are used [7, 8]:.

$$p = p(\rho, T), \qquad e = e(\rho, p). \tag{4}$$

Non-equilibrium gas model: This model considers non-equilibrium physics in three-temperature plasma. Processes of excitation of inner degrees of freedom and chemical reaction are considered. Set of governing equations (1) is completed in this case by the continuity equations for plasma components and conservation equations for electron and vibrational energies. In form proposed by C. Park [9, 10] these additional equations look as follows:

$$\frac{\partial \rho_s}{\partial t} + \nabla(\rho_s \vec{V}) = \dot{\omega}_s,$$

$$\frac{\partial \rho e_v}{\partial t} + \nabla(\rho e_v \vec{V} + p) = \sum_{s=mol} \rho_s \frac{e_{v,s}^* - e_{v,s}}{\langle \tau_s \rangle} + \sum_{s=mol} \dot{\omega}_s \widetilde{D}_s, \qquad (5)$$

$$\frac{\partial \rho e_e}{\partial t} + \nabla(\rho e_e \vec{V} + p_e) = \vec{V} \cdot \nabla p_e + 2\rho_e \frac{3}{2} R(T - T_e) \sum_s \frac{\nu_{es}}{M_s} - \sum_s \dot{n}_{e,s} \widetilde{I}_s$$

where $\rho = \sum_s \rho_s$, $p = \sum_s p_s$; $p_s = \rho_s RT / M_s$, ρ_s are the partial pressure and the density of s-component of the plasma, $p_e = \rho_e RT_e / M_e$, T_e are the electron pressure and temperature, T is translational-rotational temperature, $e = e_{tr} + e_v + e_e + \sum_s \rho_s h_s^0$, e_{tr} is the total translational-rotational energy, e_v is the total vibrational energy of the plasma, e_e is the total energy of the electrons and electronic excitation, e_v^* is the equilibrium vibrational energy, h_s^0 is the specific enthalpy of formation of s-component, $\widetilde{D}_s$ is the vibrational energy produced in dissociation of s-component, ν_{es} is the electron-heavy particle collision frequency, $\widetilde{I}_s$ is the first ionization energy, $\dot{n}_{e,s}$ is the electron production rate due to ionization of s-component, $\dot{\omega}_s$ is the s-component production rate, M_s is the molecular mass of s-component, $\langle \tau_s \rangle$ is the vibrational relaxation time.

Sets of equations (1) and (5) is closed by descriptions of chemical kinetics and thermodynamic relations.

The below simulations of non-equilibrium flows were carried for the nitrogen as the working medium of the LPE. It was supposed that the plasma consists of five components: N_2, N, N^+, N_2^+, and electrons e. The following chemical reactions occur in such a mixture [9]:

Dissociation:	$N_2 + M \leftrightarrow N + N + M$, (M=N, N_2, N^+, N_2^+, e);
Associative ionization:	$N + N \leftrightarrow N_2^+ + e$;
Charge exchange:	$N^+ + N_2 \leftrightarrow N_2^+ + N$;
Ionization by electron impact:	$N + e \leftrightarrow N^+ + e + e$.

The reaction rate constants necessary for calculation of the source term $\dot{\omega}_s$ were taken from [9]. Curve fits proposed in [11] were used for specific heats in the thermodynamic relations.

Two-phase mixture model: It is well known that the laser breakdown threshold decreases significantly in presence of a dispersed phase in the medium, on the other hand the two-phase working medium can increase the efficiency of the laser energy transform. In this study the gasdynamic processes induced by laser energy deposition in a two-phase droplet mixture is considered. Necessary concentration of the dispersed phase is determined by condition that at least one of the additives should be in the laser focus. It was supposed that the additives are the spherical monodispersed drops, with initial diameter is 10^{-6}m. Interphase mass transfer was taken into account.

151

Estimates have shown that at conditions under study the Stokes number characterizing interphase relaxation process,

$$Sk = \tau_{12}/\tau_1 \approx 10^{-2} \ll 1, \tag{6}$$

where $\tau_{12} = d_2{}^2\rho_2/18\mu_1 F(\mathrm{Re}_{12}, \mathrm{M}_{12}, ...)$ is the interphase relaxation time, $\tau_1 = L/V_1$ is the characteristic time of the problem, d_2 is the droplet diameter, ρ_2 is the dispersed phase density, μ_1 is the gas phase viscosity, $F(\mathrm{Re}_{12}, \mathrm{M}_{12}, ...)$ is a function accounting for specific features of gas flow about droplets, $\mathrm{Re}_{12}, \mathrm{M}_{12}$ are the Reynolds and the Mach numbers of relative motion of the phases, L is characteristic size of the LPE nozzle, indices 1 and 2 corresponds to the carrier and dispersed phases, respectively.

On the basis of estimation (6) it is possible to introduce a model of equilibrium two-phase mixture, so-called "equivalent gas" model, which implies

$$V_1 = V_2 \equiv V, \quad T_1 = T_2 \equiv T. \tag{7}$$

In this case the governing equations may be written in form:

$$\frac{\partial \rho}{\partial t} + \nabla(\rho \vec{V}) = 0,$$

$$\frac{\partial n_2}{\partial t} + \nabla(n_2 \vec{V}) = 0,$$

$$\frac{\partial \alpha_2}{\partial t} + \nabla(\alpha_2 \vec{V}) = \dot{\alpha}_2, \tag{8}$$

$$\frac{\partial \rho \vec{V}}{\partial t} + \nabla(\rho \vec{V}\vec{V} + p) = 0,$$

$$\frac{\partial E}{\partial t} + \nabla[(E + p)\vec{V}] = 0,$$

here $\rho = (1 - \alpha_2)\rho_1 + \alpha_2\rho_2$, p are the density and pressure of the "equivalent gas", $\alpha_2 = n_2\pi d_2^3/6$ is the volume fraction of the dispersed phase, l is the latent evaporation heat, $E = (1 - \alpha_2)p/(\gamma - 1) + \alpha_2\rho_2(c_p^0 T - l) + \rho V^2/2$ is the total energy of the "equivalent gas", ρ_1, n_2 are the carrier phase density and the numerical density of the dispersed phase, c_p^0. is the carrier phase specific heat at constant pressure.

Interphase mass exchange due to evaporation/condensation processes results in variation of diameters of the dispersed additives. In the framework of the Hertz-Knudsen model the rate of the diameter variation may be written in the following form [12]:

$$\dot{d}_2 = \frac{\beta p}{\rho_2\sqrt{2\pi RT}}\left[1 - \frac{p_s}{p}\right], \tag{9}$$

here β is condensation coefficient (for water 0.04 is recommended [13]), $p_s(T)$ is the saturation pressure. Relation for α_2 and the Hertz-Knudsen formula enable one to obtain the rate of variation of the dispersed phase volume fraction $\dot{\alpha}_2$.

NUMERICAL METHOD

The method for solving gasdynamic equations is based on the explicit Godunov-type high-resolution finite-volume scheme. In vector form the system of governing equations looks as follows:

$$\frac{\partial U}{\partial t} + \nabla \vec{F} = R$$

where U is the vector of conservative gas variables, $\vec{F}$ is the gasdynamics convective flow tensor and R is the vector of the right parts, arising, particularly, in the model of non-equilibrium plasma or in the model of two-phase flow. For the set of equations (1) vectors U and $\vec{F}$ have a form:

$$U = \begin{Bmatrix} \rho \\ \rho\vec{V} \\ E \end{Bmatrix} \qquad \vec{F} = \begin{Bmatrix} \rho\vec{V} \\ \rho\vec{V}\cdot\vec{V} \\ (E+p)\vec{V} \end{Bmatrix}.$$

Computational domain is divided by grid into set of cells. Assuming gas dynamic variables to be constant within a computational cell and making use of the integral formulae of vector analysis, one can write out finite-volume notation of the above equation for a computational cell:

$$\frac{dU}{dt} = -\frac{1}{V}\int_{\sigma} \vec{F}\cdot\vec{n} \, d\sigma + R \qquad (10)$$

where V is cell volume, σ is cell surface and $\vec{n}$ is unit vector normal to σ. Convective flows in integral of Eq.(10) can be calculated using the solution of Riemann problem between the values of gasdynamics variables on the cell surface from inside and outside of the cell. Gradient components implemented in calculation of these values are evaluated using the special TVD technique through limitation procedure. Particularly, minmod and Van Leer limiters were used.

With known fluxes (integral of Eq.(10)) evaluated from the previous time level data, one comes to the set of ordinary differential equations which is solved using the simplest two-layer implicit scheme.

The above computational procedure is second-order accurate with respect to spatial coordinates in the flow regions with smooth function behavior.

The admissible time step is limited by the CFL stability condition. More detailed description of the numerical method can be found in [14].

RESULTS

Relaxation of the laser spark in nitrogen at the normal conditions was simulated on the basis of the above gas-phase models to compare with experimental data and to elucidate effects of accounted physical-chemical processes on adequacy of the models. Computations were carried out of one dimensional flow and the results were compared with experimental data obtained with Nd-laser. Energy of the laser pulse was 13.3J, pulse duration was 10ns. In the experiments two pressure gages were installed at the distances of 6cm and 10.2cm from the laser focus. One of the main problems of the simulation is determination of the laser beam energy absorbed by the working medium. Characteristics of the breakdown depend on the radiation wavelength, the pulse energy, physical properties of the working medium, and the way of laser beam focusing. Experiments [15] showed that about 30% of Nd-laser pulse energy are absorbed by the plasma. Another problem is the plasma composition just after the laser pulse, which is used as initial data for computations. This composition as well as the plasma absorption coefficient were calculated utilizing the approach proposed in [16].

Figures 1 (a, b, c) present $(x-t)$-diagrams of the density for all the above models:

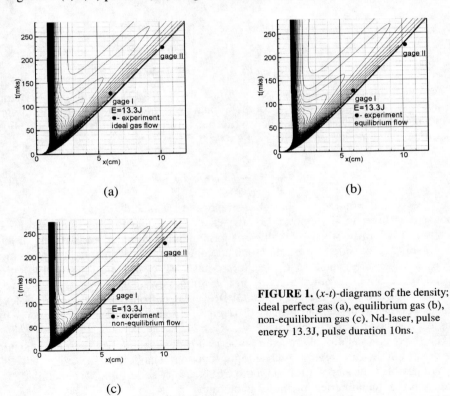

(a)

(b)

(c)

FIGURE 1. $(x-t)$-diagrams of the density; ideal perfect gas (a), equilibrium gas (b), non-equilibrium gas (c). Nd-laser, pulse energy 13.3J, pulse duration 10ns.

ideal perfect gas (a), equilibrium gas (b), and non-equilibrium gas (c). Figures 2 (a, b)

154

demonstrate profiles of the pressure correspond to 129µs and 228µs, the experimental time moments of the shock wave arrival on 6cm and 10.2cm gages, respectively. It is seen that all the models provide results, which are close to experimental data.

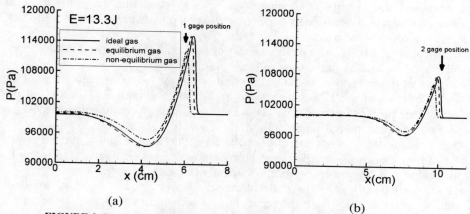

(a)

(b)

FIGURE 2. Pressure profiles correspond to 129µs (a) and 228µs (b) time moments. Arrows indicate positions of the shock wave in the experiment.

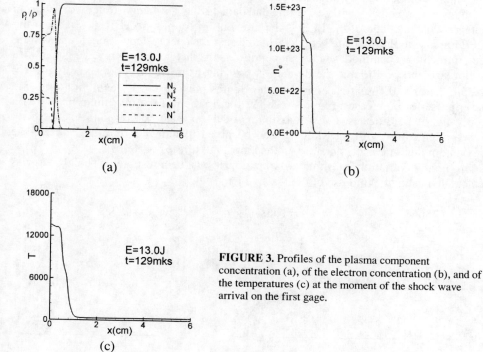

(a)

(b)

(c)

FIGURE 3. Profiles of the plasma component concentration (a), of the electron concentration (b), and of the temperatures (c) at the moment of the shock wave arrival on the first gage.

Figs. 3 (a, b, c) present the temperature, the electron number density, and the plasma component concentration profiles, respectively. It is worth noting that at conditions under study the characteristic times of relaxation processes is significantly shorter than the time of the flow. Practically the flow is appeared to be in equilibrium, the vibrational and electron temperatures are equal to translational one, for distances greater than 1cm it is possible to neglect by all components of the working medium excepting molecular nitrogen.

Thus the results demonstrate minor effect of the non-equilibrium processes on the shock wave propagation in the laser-induced plasma.

In view of this it is possible to use the equilibrium gas model in simulations of the flows in nozzles of the LPE which saves significantly computer resources, in our computations equilibrium model CPU time was 11 times greater than that of ideal gas model and the same ratio for non-equilibrium model was 280.

Some results of the computations of operation of the LPE are shown in the figures 4 and 5. Two types of the nozzles were considered: a parabolic nozzle (Fig. 4) and a "toroidal" nozzle formed by rotation of a parabola about an external axis parallel to the parabola axis (Fig. 5). The outlet diameter of the nozzles was equal to 10cm. Distances of the focuses from the parabola tops were 1cm and 0.5cm, respectively. The working medium was air. The ambient pressure was supposed to be equal to 10^5Pa. Laser energies released in the focuses of the nozzles were equal to 100J. Determination of domain of the initial energy deposition is the important problem of the simulation but in [17], where different ways of determination of this domain were used, good agreement with experimental data [18] was demonstrated for such an integral characteristic of the LPE as the momentum coupling coefficient (ratio of the LPE impulse to the deposited energy). In this case the energy deposition domains were determined to accommodate at given conditions the laser pulse energy. To get an idea about size and position of the domain is possible from figures corresponding to t = 0.25μs and t = 0.13μs, respectively. The figures present density evolution after the laser pulse for both cases. Figures 6 shows distribution of the maximum pressure on the walls of the nozzles. Local maximums of the pressure correspond to the shock wave interactions with the walls. It can be noted that in the "toroidal" nozzle the maximum pressure occurs mainly on the inner wall of the nozzle.

In Fig. 7 variation of thrust of the nozzles is demonstrated. The thrust was calculated using the impulse of the reactive jet.

t = 0.25μs   t = 10μs   t = 50μs

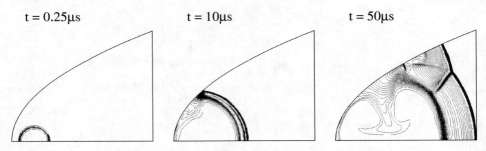

FIGURE 4. Density contours in the parabolic nozzle.

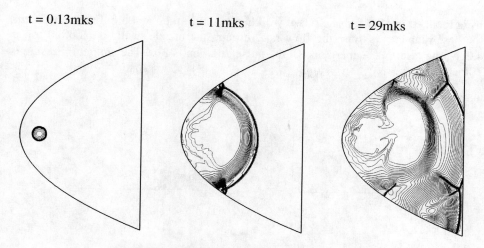

t = 0.13mks   t = 11mks   t = 29mks

FIGURE 5. Density contours in the "toroidal" nozzle.

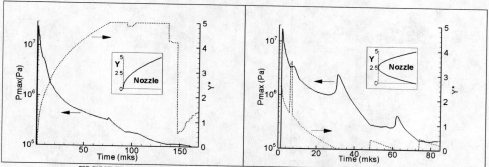

FIGURE 6. Evolution of the maximum pressure on walls of the nozzles.

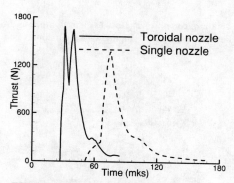

FIGURE 7. Jet impulses as functions of time.

Some results of the numerical simulation of the LPE operation with the two-phase working medium are presented below. The initial parameters are the same ones that were realized in [17] for one-phase working medium: the paraboloid height is 10cm, the base diameter is 7cm, the flow rate through the nozzle in the paraboloid top is 4.2g/s, the laser pulse energy is 8J, the pulse duration is $3\mu s$. The ambient pressure is 10Pa.

t=5μs

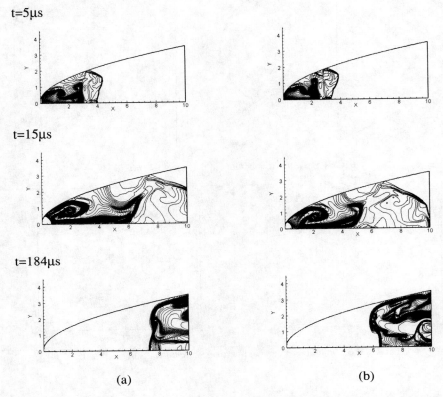

FIGURE 8. Temperature contours for nitrogen flow (a) and two-phase droplet flow (b). Initial mass loading ratio $m_2 = 0.4$, droplet diameter $d_2 = 10^{-6}$m, the working medium flow rate 4.2g/s.

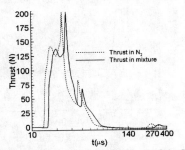

FIGURE 8 (c). Time variation of the thrust of the LPE.

In figures 8 (a, b) evolution of the temperature after the end of the laser pulse is shown for two cases: pure nitrogen (a) and droplet medium (b). More high density of the two-phase mixture and energy losses on the evaporation process results in lower velocity of the working medium in this case. In figure 8 (c) variation in time of the LPE thrust for the above cases is presented. At the given parameters the presence of the dispersed phase demonstrates itself mainly in decrease of the breakdown threshold. Small time shift of the thrust in case of two-phase medium is due to lower velocity of the medium.

CONCLUSIONS

- The numerical simulation of laser spark relaxation has been carried our in the framework of three models. Comparison with the experimental data confirms the opportunity to use rather simple models for investigation of operation of the LPE at conditions under study.
- 2D simulations of the LPE operation were carried out for various shapes of the nozzles.
- Effect of droplets on formation of the LPE thrust was investigated using the "equivalent gas" model.

ACKNOWLEDGMENTS

Support of this study by ISCT grant # 1801 is gratefully acknowledged.

REFERENCES

1. Kantrovitz A., *Astronautics and Aeronautics*, **10**, 74-76 (1972).
2. Bunkin F. V., Prokhorov A. M., *Uspekhi Fizicheskikh Nauk*, **119**, 425-446, (1976) (in Russian).
3. Raizer Y.P., *Soviet Physics JETP*, **31**, 1148-1154, (1970)
4. Simons G. A., Pirri A. N., *AIAA Journa,l* **15**, 835-842, (1977)
5. Wang, T.S., Cheng Y.S., Lui J, Myrabo L.N., Mead Jr. F.B., "Performance and Modeling of an Experimental Laser Propelled Lightcraft", *AIAA Paper*, 2000-2347.
6. Tarnavsky G.L., Shpak S.I., *Thermo-physics and air-mechanics* **8**, 41-58, (2001), (in Russian).
7. Kraiko A.N., *Engineering Journal*, **4**, 548-550, (1964), (in Russian).
8. Tannehill J.C., Mugger P.H., "Improved curve fits for the thermodynamicc properties of equilibrium air suitable for numerical computation using time dependent or shock-capturing methods", *NASA CR-2470*, (1974).
9. Park C. *Nonequilibrium Hypersonic Aerothermodynamics*. NY, J.Wiley & Sons, 1990.
10. Gnoffo P.A., Gupta R.N., Shinn J.L., "Conservation equations and physical models for hypersonic air flows in thermal and chemical non-equilibrium", *NASA Technical Paper 2867*, (1989).
11. Balakrishnan A., "Correlations for specific heats of air species to 50000K", *AIAA-86-1277*, (1989).
12. Nigmatulin R.I. *Foundations of heterogeneous medium mechanics*, M., Nauka, 1978 (In Russian).
13. Saltanov G.A. *Non-equilibrium and non-stationary processes in gasdynamics*, M., Nauka, 1979 (In Russian).
14. Golovachov Yu.P., Kurakin Yu.A., Schmidt A.A., "Numerical investigation of non-equilibrium MGD flows in supersonic intakes," *AIAA 2001-289*, (2001).
15. Ageichik A.A., Rezunkov Yu.A., Safronov A.L., et al., *Journal of Optical Technology* (to be published).
16. Danilov M.F., *Journal of Technical Physics*, **70**, 21-26, (2000) .(In Russian).
17. Rezunkov Yu.A., Ageichik A.A., Golovachov Yu.P., Kurakin Yu.A., Stepanov V.V., Schmidt A.A., *The Review of Laser Engineering*, **29**, 268-273, (2001)
18. Nebolsine P.E., Pirri A.N., Simons G.A., *AIAA journal*, **19**, 127-128, (1981).

Numerical Simulation of Flow Characteristics of Supersonic Airbreathing Laser Propulsion Vehicle

Sung-Don Kim[†] , Jun-Sik Pang[*], In-Seuck Jeung[†] , and Jeong-Yeol Choi[‡]

[†] Department of Aerospace Engineering, Seoul National University, Seoul 151-742, Korea
[‡] Department of Aerospace Engineering, Pusan National University, Pusan 609-735, Korea
[*] Korea Aerospace Industries, Ltd., Sachon, 664-942, Korea

Abstract. LITA(Laser-Driven In-Tube Accelerator) is a new device for the propulsion of projectile under high velocity condition. LITA is a little different from other accelerators in that it needs continuous laser source energy for acceleration process. One of the issues for LITA is the optical design of the projectile, because the focusing point of laser behind projectile decides its performance. Laser-supported detonation wave is a main energy source mechanism. Present study shows the performance analysis of LITA using computational fluid dynamics (CFD). Laser power, laser energy, laser frequency, laser focusing point, and projectile base geometry play important roles in LITA's performance. In this research, blast wave produced by explicitly energy input is used.

INTRODUCTION

During the last decade, laser propulsion concept was studied by many investigator.[1-9]. Myrabo and Messitt succeeded to lift off a small body with a pulse type laser propulsion.[7] Recently, Sasoh suggested a laser-driven in-tube accelerator (LITA)[8], a concept of laser propulsion of a projectile in a barrel. Figure 1 shows a principle of LITA at various velocity regimes.

Laser propulsion consist of using energy from a remotely located laser to heat propellant to extremely high temperature and pressure and then expand the gas or push projectile repetitively. Chemical rocket provides high thrust level and low overall weight but, specific impulse tends to be relatively low. Conversely, electric propulsion produces high specific impulse but low thrust level. Even nuclear engine has high thrust and high specific impulse, a massive nuclear reactor must be carried onboard. Instead of using energy from nuclear or chemical reactions to heat a propellant, a power beam of laser energy is employed. This laser propulsion type device relies on high temperature pulse of repetitively breakdowned propellant to offer high specific impulse and high thrust. These can be grouped into two district categories depending on the type of laser used. One concept is to operate the laser in a steady-state mode known as continuous wave (CW) operation. The other is to use a repetitively pulsed (RP) laser which operates by producing high-frequency pulses of intense laser radiation.[6-7]

CP664, Beamed Energy Propulsion: First International Symposium on Beamed Energy Propulsion,
edited by A. V. Pakhomov
© 2003 American Institute of Physics 0-7354-0126-8/03/$20.00

This investigation treats LITA as one of the RP type laser-propulsion. In this concept, an incident laser light from a muzzle reflects at optically contoured projectile front area. At the focus of laser beams, propellant gas breaks down into plasma flow having core temperature of higher than 10,000K. Expansion of hot plasma flow gives an impact on the rear side of projectile remains at sufficiently high pressure for acceleration. Figure 1 shows operation of the LITA.

The purpose of this article is to examine the feasibility of LITA. This research treat about laser focusing point location, variable power source, frequency, and projectile base geometry of the LITA for performance analysis assuming simplified energy source model.

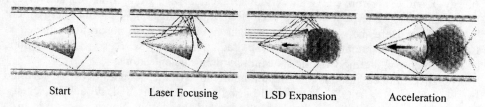

| Start | Laser Focusing | LSD Expansion | Acceleration |

FIGURE 1. Schematics of LITA operation

COMPUTATIONAL MODEL

The concept of LITA is suggested by Sasoh, i.e., air breathing ram accelerator propulsion in tube[8], energy supplied by laser, and projectile surface contoured for laser focusing. Figure 1 shows four step operations from starting initial velocity to acceleration continuously. After laser beam from front projectile surface is emitted, it reaches at a focusing point. And then a hot spot at laser focusing point causes plasma expansion. The expansion phenomenon is so called LSD (Laser-Supported Detonation). The plasma propagates toward projectile rear surface at supersonic velocity, which represents blast wave phenomenon.

This is caused by the plasma heating the fluid in front of it, which in turn begins absorbing and heating up. The main laser absorption process is Inverse-Bremsstahlung (IB) absorption. It occurs when a photon is absorbed by a free electron during a collision with either a neutral atom or ion. The excited electron eventually transfers its energy to the surrounding gas through collisions, raising the bulk temperature of gas.

This investigation is based on the fact that the hot spot plasma expansion is like supersonic blast wave. The detailed process of molecular breakdown should be modeled for an accurate numerical study of the plasma core formation. However, at present status of preliminary study on LITA, the process of periodic laser focusing is simply modeled as a periodic input of radiation energy at a focus of small diameter. Also, the dissociation process of propellant gas is neglected because the energy loss due to chemical dissociation would be relatively small compared with radiation energy input for laser focusing. More delicate modeling of physical process would be included in the continuing studies.

In this case, fluid dynamics assumption is accomplished with neglecting diffusion processes; mass diffusion, momentum diffusion (of viscosity and turbulence), and thermal diffusion. As a result, Euler axisymmetric equation is adoptted as governing equations in this high velocity research region. The equation is expressed as (1)

$$\frac{\partial Q}{\partial t} + \frac{\partial F}{\partial x} + \frac{\partial G}{\partial y} + \alpha H = W$$

$$Q = [\rho,\ \rho u,\ \rho v,\ e,\ 0,\ 0]^T \tag{1}$$

$$W = [0,\ 0,\ 0,\ q_{IB} - q_{B,}\ 0,\ 0]^T$$

Q is a vector of conservation variables consists of density, momentums and total energy. F and G are convective and viscous fluxes in x and y direction, respectively. H is an axisymmetric term and W is a vector of source terms. q_{IB} $-q_B$ is net energy input by radiation. Here q_{IB} is Inverse-Bremsstahlung for radiation absorption and q_B is Bremsstahlung for radiation emission. The radiation energy sources are the functions of laser intensity, thermodynamic state and particle concentrations.

The governing equation is integrated numerically with a second order time accurate implicit scheme (LU-SGS). An upwind-biased third-order MUSCL type TVD scheme is used for the spatial discretization of the invicid flux terms. Source energy term is calculated with explicit treatment. More detailed computational algorithm is addressed in the reference.[10]

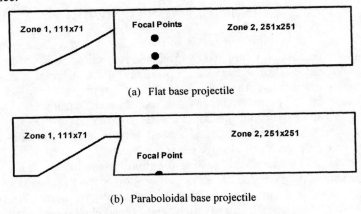

(a) Flat base projectile

(b) Paraboloidal base projectile

FIGURE 2. LITA configurations used for numerical simulation.

The projectile surface of LITA is optically designed by Prof. Sasoh[9], so that laser beam can focus at a point (2.0,0.5) in computational domain or at the different various locations. The typical LITA configuration considered are plotted in Figure 2. Computational domain Zone 1 is covered by (111 X 71) grid point and Zone 2 (251 X 251) grid points. This is two-block grid which boundary condition is exchanged at block interfaces with the second order space accuracy. The resolution of the grid is sufficient for capturing shock flow physics. Flight initial condition of Mach number 4 is considered because LITA performance is demonstrated in this flow condition or higher Mach number conditions. The projectile radius is 1.64cm for the case of (a) or 1.62cm for the case of (b), respectively. Barrel radius is 1.25cm based on the Sasoh's data. Laser focusing point diameter and projectile mass are assumed to be 0.4mm and 3g alloy, respectively.

The total amount of laser energy was numerically inputted as the form of torus geometry at laser focusing location. The conversion value of total input energy amount explicitly in cross section is used for computational process.

For numerical calculation, the inflow boundary condition is fixed, solid walls (projectile, tube) is slip line condition and outflow zeroth-order extrapolation.

RESULT

Flat Base Projectile

To justify computational modeling, the property of supersonic blast wave must first be demonstrated. Sasoh's description shows shock propagation when single focus is accomplished[8]. For validity of numerical approach, the properties of blast wave propagation are measured at the same condition which Sasoh performed. Figure 3 expresses shock propagation.

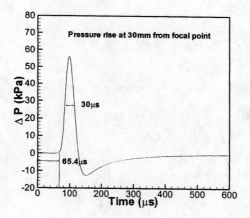

FIGURE 3. The calculation of the pressure rise at 30mm from the focal point

Sasoh presented the relation of the difference between incident laser energy and transmitted energy. Present result considered Sasoh's relation of energy conversion[8]. Therefore laser energy of 1.2J is used at focus point. Figure 3 shows good agreements between the two results with respect to peak pressure rise, shock arrival time.

Initial cold flow condition is described in Figure 4. Since the projectile shape is designed only with optical consideration, incident oblique shock wave induces regular reflection at barrel surface. Expansion wave is observed at projectile edge corner. In this region, shock reflection and expansion wave is overlapped each other.

Second reflection shock is observed at the rear of projectile. Streamline pattern shows recirculation zone and shock structures in very high velocity range

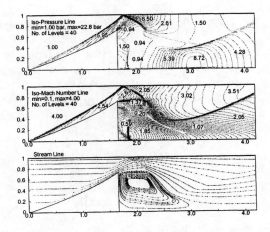

FIGURE 4. Cold flow condition (initial condition) Pressure, mach number, stream line contour

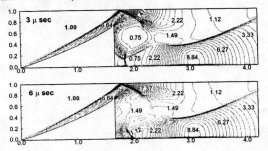

FIGURE 5. Time sequence of contour plots after laser focusing

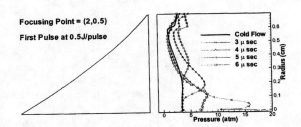

FIGURE 6. Pressure distribution at projectile rear surface after first impulse

Second reflection shock is observed at the rear of projectile. Streamline pattern shows recirculation zone and shock structures in very high velocity range

Figure 5 presents blast wave expansion according to time under a single pulsed laser focusing. Strong circular shock wave expands outward, but pressure inside the shock wave is not so much high due to the low initial pressure. Circular shock shape starts to fade away after 6 μsec but shock wave propagation still affects the rear projectile surface. Figure 6 shows time histories of rear surface pressure distribution.

Pressure rises as time passes and peak pressure is observed at 5 μsec. But rear surface pressure distribution begins to decay down at 6 μsec. Laser hot spot at (2.0,0.5) propagates to the projectile surface and then high pressure distribution moves near the center line because of the flow condition.

The thrust of LITA is produced by this blast wave striking process. If this striking process occurs successively, the projectile would obtain more stable thrust. It is important to consider repeated pulse type propulsion for the practical operation of LITA. It is necessary to produce repeated impulse for continuous and stable thrust. Figure 7 shows the operation of the LITA at 0.5J/pulse and 1 kHz case. High oscillation line means the change of the net thrust because of the pulse type periodic laser focusing. Solid line (Time Average Thrust History, TATH) is calculated from net thrust and time matching. The amount of focusing energy, the frequency of the laser beam, and the significance of projectile geometry related to focusing point are dealt with. Tables 1 and 2 show the performances indexes, ballistic efficiency and time average thrust, of variable cases as defined in Eqs. (2) and (3). In Table 1, frequency and energy of laser beam are varied, with focusing point fixed to (2.0, 0.5). However, in Table 2, focusing point and frequency of laser beam are changed, with its energy set up at 0.5J/pulse. The net thrust imposed on projectile is calculated by integrating the difference between front surface and rear surface pressure.

$$Time \ \ Average \ \ Thrust = \frac{\sum \Delta Time * Net \ Thrust}{Total \ \ Time \ \ Elapsed} \qquad (2)$$

$$\eta_b = \frac{Time \ \ Average \ \ Thrust * Velocity \ \ Increase}{Input \ \ Power} \qquad (3)$$

Ballistic efficiency (η_b) is used to quantify LITA's performance and time average thrust is introduced to evaluate the amount of thrust stably accelerating projectile. And, 40msec is thought to be suitable for integration time because stable thrust is achieved at this time.

Time average thrust histories under numerical conditions of Table 1 are graphically expressed in Figure 8. And Figure 9 shows ballistic efficiencies with respect to power in Tables 1 and 2. The power is a factor capable of considering overall performance of frequency and energy per pulse of laser beam.

Figure 8 exhibits that TATH is augmented, as laser frequency or focusing energy per pulse increase. Also, it is noted in Figure 8 that under the condition of the same power, TATH is almost identical regardless of frequency and energy per pulse. Six results of Fig. 8 all have similar increasing trends of THAT as time. And, steady thrust is observed after 20msec in Figures 7 and 8.

Ballistic efficiency is maximized when focusing point is (2.0, 0.2) and frequency is 2 kHz. Generally, ballistic efficiency is higher in focusing case of (2.0, 0.2) than that of (2.0, 0.5), when other conditions are the same. It is because heating energy per torus volume becomes higher as laser beam focuses on near the centerline. Therefore, the higher laser intensity per torus volume produces the stronger blast wave, which provides projectile for more thrust.

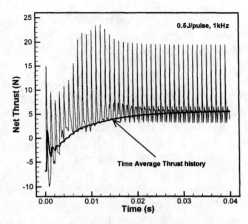

FIGURE 7. Time vs. Net thrust and TATH contour at 0.5J/pulse, 1kHz

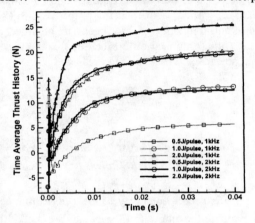

FIGURE 8. Time vs. TATH contour at variable laser performance cases.

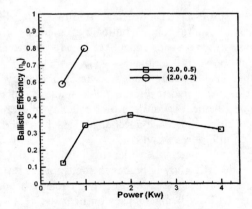

FIGURE 9. Power vs. Ballistic efficiency at focusing point (2.0, 0.5) and (2.0, 0.2)

166

TABLE 1. LITA performance at focusing point (2.0, 0.5)

Energy (J/pulse)	Freq. (kHz)	Power (Kw)	Time Average Thrust (N)	η_b
0.5	1	0.5	5.57	0.12
	2	1.0	12.52	0.31
1.0	1	1.0	13.16	0.35
	2	2.0	19.58	0.38
2.0	1	2.0	20.15	0.40
	2	4.0	25.39	0.32

TABLE 2. LITA performance at 0.5J/pulse energy comparing different focusing point

Freq. (kHz)	Focusing Point	Power (Kw)	Time Average Thrust (N)	η_b
1	(2.0,0.5)	0.5	5.57	0.12
	(2.0,0.2)	0.5	12.21	0.59
2	(2.0,0.5)	1.0	12.52	0.31
	(2.0,0.2)	1.0	20.0	0.80

Paraboloidal base projectile

In paraboloidal base projectile, incident laser beam is supplied from the rear side of the projectile and reflected on the parabolic shaped base plate of the centerbody. Figure 10 presents blast wave expansion according to time under a single pulsed laser focusing and Figure 11 shows the time history of impulse

t = 2.000 [μsec.]

t = 4.000 [μsec.]

(a) Input laser power = 1J

t = 2.000 [μsec.]

t = 4.000 [μsec.]

(b) Input laser power = 4J

FIGURE 10. Time sequence of contour plots after laser focusing

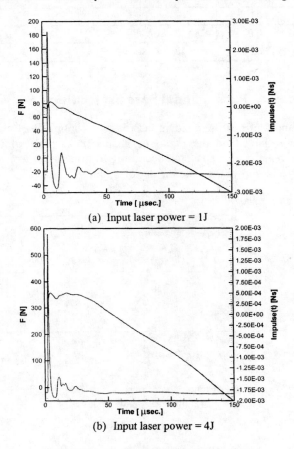

(a) Input laser power = 1J

(b) Input laser power = 4J

FIGURE 11. Time history of impulse

SUMMARY

In the present calculation, laser-supported detonation (LSD) is simplified by energy source model because LSD has similar characteristics to supersonic blast wave. It is believed that energy source model is eligible for simulating LITA. Sasoh's pressure rise data coincides with explicit energy source model in reasonable degree. In this study, amount of the energy source inputted is not incident laser energy but energy transmitted to focusing point, and that molecular breakdown process, light emission, absorption processes, and flux diffusion process are all ignored. Results of one pulse laser show the tendency for blast wave to expand. To obtain thrust of LITA laser energy should successively be provided and it is observed that stabilized operation begins after roughly 20msec in all cases of present study. In conclusion, the performance of LITA is determined by four parameters: laser energy, laser frequency, laser focusing point, and projectile base geometry.

The consideration of radiation, molecular breakdown and diffusion is required to obtain more delicate results.

More extensive studies are expected on the operating conditions of LITA such as other flight Mach number and initial pressure condition.

ACKNOWLEDGEMENTS

Authors thank to Prof. A. Sasoh of Shock Wave Research Center, Institute of Fluid Science in Tohoku University for providing information and experimental results. This study was partly supported by BK21 Program sponsored by Seoul National University.

REFERENCES

1. Kantrowitz, A., "Propulsion to Orbit by Ground-Based Lasers," *Astronautics & Aeronautics*, Vol 10, May, 1972, pp. 74-76.
2. Weiss, R. F., Pirri, A. N., and Kemp, N.H.,"Laser Propulsion," *Astronautics & Aeronautics*, Vol 17, March, 1979, pp. 50-58.
3. Jones, L. W., "Laser-Propulsion," AIAA Paper 80-1264, June, 1980.
4. Legner, H. H. and Douglas-Hamilton, D. H., "CW Laser Propulsion," *Journal of Energy*, Vol. 2, March-April, 1978, pp. 85-94.
5. Glumb, R. J., and Krier, H.,"Concepts and Status of Laser-Supported Rocket Propulsion," *J. Spacecraft*, Vol. 21, No. 1, 1984, pp. 70-79.
6. Birkan M. A., "Laser Propulsion: Status and Needs," *Journal of Propulsion and Power*, Vol 8, No. 2, 1992, pp. 354-360.
7. Myrabo, L. N. and Messitt, D. G., "Ground and Flight Test of a Laser Propelled Vehicle," AIAA 98-1001.
8. Sasoh, A., "Laser-driven in-Tube Accelerator (LITA)," *Rev. Sci. Instrum.*, Vol.72, No.3, 2000.
9. Sasoh, A., Choi, J.-Y., Jeung, I.-S., Urabe. N., Kleine. H., and Takayama.K., "Impulse Enhancement of Laser Propulsion in Tube," *Postepy Astronautyki*. Vol.27, 2001, pp.40-51
10. Choi, J.Y., Jeung, I,-S., Yoon, Y., "Computational Fluid Dynamics Algorithms for Unsteady Shock-Induced Combustion, Part I ; Validation." *AIAA Journal*, Vol, 38, No.7, 2000, pp.1179-1187

Energy conversion in laser propulsion III

C. William Larson, Franklin B. Mead, Jr., and Wayne M. Kalliomaa

Propulsion Directorate, Air Force Research Laboratory, Edwards AFB, CA 93524-7680

Abstract. Conversion of pulses of CO_2 laser energy (18 microsecond pulses) to propellant kinetic energy was studied in a Myrabo Laser Lightcraft (MLL) operating with laser heated STP air and laser ablated delrin propellants. The MLL incorporates an inverted parabolic reflector that focuses laser energy into a toroidal volume where it is absorbed by a unit of propellant mass that subsequently expands in the geometry of the plug nozzle aerospike. With Delrin propellant, measurements of the coupling coefficients and the ablated mass as a function of laser pulse energy showed that the efficiency of conversion of laser energy to propellant kinetic energy was ~ 54%. With STP air, direct experimental measurement of efficiency was not possible because the propellant mass associated with measured coupling coefficients was not known. Thermodynamics predicted that the upper limit of the efficiency of conversion of the internal energy of laser heated air to jet kinetic energy, α, is ~ 0.30 for EQUILIBRIUM expansion to 1 bar pressure. For FROZEN expansion α ~ 0.27. These upper limit efficiencies are nearly independent of the initial specific energy from 1 to 110 MJ/kg. With heating of air at its Mach 5 stagnation density (5.9 kg/m^3 as compared to STP air density of 1.18 kg/m^3) these efficiencies increase to about 0.55 (equilibrium) and 0.45 (frozen). Optimum blowdown from 1.18 kg/m^3 to 1 bar occurs with expansion ratios ~ 1.5 to 4 as internal energy increases from 1 to 100 MJ/kg. Optimum expansion from the higher density state requires larger expansion ratios, 8 to 32. Expansion of laser ablated Delrin propellant appears to convert the absorbed laser energy more efficiently to jet kinetic energy because the effective density of the ablated gaseous Delrin is significantly greater than that of STP air.

INTRODUCTION

Laser propulsion is limited by laser power, so optimization of the laser propulsion mission may be factored into optimization of four energy conversion efficiencies, which, in a first approximation, are independent of each other. In this idealization the kinetic energy of the propelled vehicle at the end of the mission may be expressed simply:

$$E_f = \tfrac{1}{2}m_f v_f^2 = \eta\alpha\beta\gamma E_L. \qquad (1)$$

The "propulsion efficiency", η, is the efficiency with which jet kinetic energy is converted into vehicle kinetic energy. Sutton[1] pointed out, more than 50 years ago, that the instantaneous propulsion efficiency varies during a rocket mission and that it is unity only when the vehicle velocity in the inertial frame is equal to the jet velocity in the rocket frame. Unit propulsion efficiency is achieved when the jet is deposited as a stationary mass relative to an observer in the inertial frame of reference.

CP664, *Beamed Energy Propulsion: First International Symposium on Beamed Energy Propulsion,*
edited by A. V. Pakhomov
© 2003 American Institute of Physics 0-7354-0126-8/03/$20.00

Then, 25-years ago, Moeckel[2] and Lo[3] independently and nearly simultaneously published analyses of the optimization of laser rocket propulsion by maximizing η, and most recently, Phipps, Reilly and Campbell (2000, 2001)[4] cited Moeckel's paper in their comprehensive analysis of the single stage, constant I_{sp} Earth to LEO rocket mission. They reiterated the fundamental limit that Newton's second law imposes: for rocket missions that start at zero initial velocity, the maximum η is 0.648, which is achieved when f = 0.203 and v_f/v_e = 1.595. For the Earth to LEO mission the effective "delta v" (v_f) is about 10 km/s, so the optimum single stage to orbit jet velocity is ~ 6.27 km/s, or specific impulse ~ 640 s.

In this paper we report a continuation of our previous work[5] and analyze measurements of the overall efficiency of conversion of laser energy to propellant kinetic energy, αβ, based on various ballistic pendulum and flight experiments with Myrabo Laser Lightcraft, MLL [Messitt, Myrabo, and Mead (2000)[6], Mead, Squires, Beairsto, and Thurston (2000)[7]]. The Phipps, et al.[4] study defined an "ablation efficiency " and analyzed the Earth to LEO mission with unit ablation efficiency. Their ablation efficiency is equivalent to the product of α and β. Comparison of experimental results with thermodynamic analysis enables confining the range of permissible β that operates during the heating process. Wang's[8] CFD plasma model of laser energy absorption by air have predicted plasma temperatures up to 30000 K with attendant low β values ~ 0.3 to 0.4. It has been pointed out that β approaches zero as the plasma temperature approaches ~ 40,000 K, where the plasma frequency approaches the laser frequency.[8, 9]

The coupling coefficient

Newton's second law expresses the thrust that results from expulsion of matter from a vehicle of mass m
at velocity v_e as

$$F = -\frac{d(mv_e)}{dt},$$
(2)

where mv_e is the momentum of the jet exhaust in the vehicle frame of reference, [Corliss, (1960)][10]. For the case where v_e is constant,

$$F = -v_e \frac{dm}{dt}.$$
(3)

Equation (2) may be used to define an average exit velocity for rockets where v_e is not constant, such as blowdown of a specified mass of hot propellant from a fixed volume, e.g., as in laser rockets and pulse detonation rockets:

$$\langle v_e \rangle = -\frac{\int_0^t \mathbf{F}\,dt}{\int_{m_i}^{m_f} dm} = \frac{\int_{m_i}^{m_f} d(m\,v_e)}{\int_{m_i}^{m_f} dm} = \frac{\int_{\rho_i}^{\rho_f} d(\rho\,v_e)}{\int_{\rho_i}^{\rho_f} d\rho}. \tag{4}$$

so that $\langle v_e \rangle$ is the mass weighted average exit velocity and $\mathbf{F} = -\langle v_e \rangle dm/dt$. Chemical thermodynamics may be used to rigorously establish an upper limit to $\langle v_e \rangle$ when the propellant equation of state is known and the initial and final states of the propellant expansion are specified.

The efficiency of conversion of laser energy to propellant kinetic energy, $\alpha\beta$, may be defined by energy conservation in terms of $\langle v_e^2 \rangle$ for the general case of variable $\mathbf{v_e}$,

$$E_p = \tfrac{1}{2} m_p \langle v_e^2 \rangle = \alpha\beta\, E_L, \tag{5}$$

where the mass weighted average of the square of the propellant exit velocity is

$$\langle v_e^2 \rangle = \frac{\int_{\rho_c}^{\rho_f} d(\rho\, v_e^2)}{\int_{\rho_c}^{\rho_f} d\rho}. \tag{6}$$

The impulse, $\mathbf{I} = \mathbf{F}dt$, imparted to a test article by expansion of its propellant has been accurately measured with a ballistic pendulum in the past[5]. Momentum conservation requires equivalence between the measured impulse and the propellant impulse so that

$$\mathbf{I} = m_p \langle v_e \rangle. \tag{7}$$

Thus, when $\mathbf{I}$ and m_p are both measured an experimental $\langle v_e \rangle$ may be obtained.

The momentum coupling coefficient, also a measured quantity, is the impulse imparted to a test article per unit laser energy incident on the propellant,

$$\mathbf{C} = \frac{\mathbf{I}}{E_L}. \tag{8}$$

Using the definitions embodied in Equations (5) – (8), $\mathbf{C}$ may be expressed in terms of α, β, $\langle v_e \rangle$, and $\langle v_e^2 \rangle$:

$$\mathbf{C} = \frac{2\alpha\beta}{\langle v_e \rangle}\left[\frac{\langle v_e \rangle^2}{\langle v_e^2 \rangle}\right] = \frac{2\alpha\beta\Phi}{\langle v_e \rangle} \tag{9}$$

If $\mathbf{v_e}$ is constant, $\Phi = \langle v_e \rangle^2/\langle v_e^2 \rangle = 1$. Thermodynamics may be used to rigorously establish an upper limit to Φ and α for any specified free-expansion blowdown process when the propellant equation of state is known[5]. The Φ factor depends on the distribution of exit velocities, and is mathematically limited to $0.5 \le \Phi \le 1$. It has been shown[5b] that

Φ for optimum blowdown of laser heated air to 1 bar pressure increases from 0.95 at low specific energy (2 MJ/kg) to 0.98 at high specific energy (60 MJ/kg). The Φ factor arises in Equation (9) because the measured quantity, the jet impulse, is proportional to mass weighted average velocity whereas the jet kinetic energy is proportional to the mass weighted average of the squared velocity. Figure 1 shows the relationship between $C^* = C/\beta$, α, and $v_e^* = v_e/\Phi$. For a given α value, C^* <u>decreases</u> as $<v_e>^*$ increases.

Isentropic conversion of internal energy of propellant to kinetic energy

Perfect isentropic conversion of internal energy to propellant kinetic energy occurs with no losses so that

$$<v_e^2> = 2<u_c - u_e> = 2\alpha(u_c - u^o), \text{ where} \tag{10}$$

$$\alpha = <u_c - u_e>/(u_c - u^o) = <v_e^2>/2(u_c - u^o). \tag{11}$$

These definitions generate a second expression for C as a function of α that applies to isentropic energy conversion from initial states defined by $(u_c - u^o)$:

$$C = \beta \, [2\alpha\Phi/(u_c - u^o)]^{1/2}. \tag{12}$$

Figure 1 shows expansion isentropes in the C^*-α plane for various initial state values of $(u_c - u^o)^* = (u_c - u^o)/\Phi$ ranging from 1 MJ/kg to 110 MJ/kg. At constant α, C^* <u>decreases</u> as $(u_c - u^o)^*$ increases.

EXPERIMENTAL

Figure 2 shows a cross-section of the test article (MLL model 200-3/4) with a ring of Delrin installed in the shroud. The Delrin shown weighs ~ 10 g, occupies a volume of ~ 7 cm^3, and has a surface area of ~ 25 cm^2. The exit area of the idealized plug-nozzle[11] is ~ 350 cm^2. Previous measurements[5] of Delrin coupling coefficients as a function of laser pulse energy (18 μs pulses) showed that they rise to a plateau of ~ 350 Ns/MJ above $E_L \sim 250$ J. At $E_L \sim 350$ J, the measured mass of ablated Delrin was ~ 40 mg, which is the mass of a uniform thin-film layer ~ 11 micrometers thick. Thus, from Equation (7) $v_e \sim 3100$ m/s and from Equation (9) $\alpha\beta\Phi \sim 0.54$. If $\beta = 1$, $C = C^*$, and Figure 1 shows that the initial specific internal energy that produces $\alpha = 0.54/\Phi$ (with $\Phi = 0.98$) in an isentropic expansion is u_c-$u^o \sim 9$ MJ/kg. At the other extreme, if $\alpha\Phi = 1$, then $\beta = 0.54$, $C^* = 1.85$ C, and u_c-$u^o \sim 4.8$ MJ/kg. By any analysis, these results show that Delrin is a remarkably efficient propellant for laser ablation propulsion.

For the case of air, the propellant mass m_p is unknown. Figure 2 may be used to visualize a reasonable absorption volume for the case where Delrin is absent and air is the heated material. The notion of an energy absorption volume, V_{abs}, may be invoked, which contains a mass of air propellant $m_p = \rho_c V_{abs}$ into which an amount of energy βE_L

is deposited. In the limit where the time scale for energy absorption is much shorter than that for expansion, the propellant density within V_{abs} (the chamber) remains constant during energy absorption. This enables the initial specific internal energy of the propellant to be defined,

$$u_c - u^o = \beta E_L / \rho_c V_{abs}, \tag{13}$$

where $\rho_c = 1.18$ kg/m^3 and $u^o = -0.09 \times 10^6$ J/kg for air at STP. Table 1 provides a convenient list of values of the normalized absorption volume, $V_{abs}* = V_{abs}/\beta$, derived from Equation (13) for values of $u_c - u^o$ and E_L that lie within the conceivable parameter space explored in experiments.

Table 1 shows that an absorption volume for air, ~ 7 cm^3, would produce, with unit β and nominal E_L values between 100 and 400 J, heated air with internal energy between 10 and 40 MJ/kg. If the Delrin surface shown in the figure, about 25 cm^2, is a suitable representation of the sonic surface of expanding air, then, with an idealized plug-nozzle exit area[11] of ~ 350 cm^2, the expansion ratio in this test article may be as large as ~ 14. Previously reported experiments[5a] showed that air coupling coefficients with a "loosely" focused laser increased to a plateau value of C(loose focus) ~ 150 Ns/MJ above $E_L \sim 150$ J. With a "tightly" focused laser they increased to a plateau value of C(tight focus) ~ 100 Ns/MJ above $E_L \sim 300$ J. As shown below, the maximum value of $\alpha\Phi$ for air in a chemical equilibrium isentropic expansion is $\alpha\Phi \sim 0.25$ and it is almost independent of initial specific internal energy and initial temperature. Table 2 summarizes the minimum values of β, u_c-u^o, $C*$, and temperature for various reasonable values of V_{abs} and m_p. For $\beta = 1$, $m_p = 5.1$ mg (tight focus) or $m_p = 5.6$ mg (loose focus). These are the smallest values of m_p consistent with the measured C and E_L.

DISCUSSION

Figure 3 shows the chemical equilibrium Mollier diagram (u-s plane) for air up to 24,000 K. Figure 3 is based on the database maintained at NASA/Glenn [McBride and Gordon (1996)][12], which is certified accurate up to 20,000 K and which is based on extended 9-parameter fits to enthalpy, heat capacity, and entropy of neutral species and singly charged ions. Above 20,000 K doubly charged ions begin to contribute but these are not included in the database. This limitation leads to predictions of temperatures (at specified u and ρ) that are too high for plasmas above $\sim 20,000$ K.

Figure 4 shows a series of isentropes (vertical lines) on the Mollier diagram. These are representations of equilibrium isentropic expansions that originate from initial states located along the constant density line, $\rho = 1.18$ kg/m^3, and specific internal energies ranging from 1 to 100 MJ/kg. Table 3 summarizes other interesting thermodynamic properties under conditions of chemical equilibrium: T, P, h, s, M_m, c_p, v_a, c_p/c_v, and X(e$^-$). Table 4 provides properties of Mach 5 air at its stagnation density[13], 5.9 kg/m^3. Since the entropy of the initial and final states are equal, the thermodynamic state of the propellant in the exit surface is uniquely defined when only one additional property in the exit surface is specified, such as the exit pressure or the expansion ratio, which are also indicated in Figure 4. The expansion ratio, ε, is the ratio of the area of the exit surface to the area of the sonic surface or nozzle throat, and for isentropic expansions this may be represented in terms of thermodynamic properties in the nozzle throat and exit plane: $\varepsilon = A_e/A_t = \rho_t v_a / \rho_e \mathbf{v_e}$.

Table 1. Normalized absorption volume for air at 1.18 kg/m^3 as a function of internal energy, u, and laser energy, EL.

u MJ/ kg	V_{abs}/β normalized absorption volume, cm^3					
	$E_L=$ 50 J	$E_L=$ 100 J	$E_L=$ 150 J	$E_L=$ 200 J	$E_L=$ 300 J	$E_L=$ 400 J
1	42.3	84.7	127.1	169.4	254.2	338.9
2	21.1	42.3	63.5	84.7	127.1	169.4
3	14.1	28.2	42.3	56.5	84.7	112.9
4	10.5	21.1	31.7	42.3	63.5	84.7
5	8.47	16.9	25.4	33.9	50.8	67.8
6	7.06	14.1	21.1	28.2	42.3	56.5
7	6.05	12.1	18.1	24.2	36.3	48.4
8	5.30	10.5	15.8	21.1	31.7	42.3
9	4.71	9.42	14.1	18.8	28.2	37.6
10	4.24	8.47	12.7	16.9	25.4	33.9
15	2.82	5.65	8.47	11.3	16.9	22.6
20	2.12	4.24	6.36	8.47	12.7	16.9
30	1.41	2.82	4.24	5.65	8.47	11.3
40	1.06	2.12	3.18	4.24	6.36	8.47
50	0.85	1.69	2.54	3.39	5.08	6.78
60	0.71	1.41	2.12	2.82	4.24	5.65
70	0.61	1.21	1.82	2.42	3.63	4.84
80	0.53	1.06	1.59	2.12	3.18	4.24
90	0.47	0.94	1.41	1.88	2.82	3.77
100	0.42	0.85	1.27	1.69	2.54	3.39
110	0.39	0.77	1.16	1.54	2.31	3.08

Table 2. Measured and calculated quantities for expansion of laser heated STP air.

Quantity	tight focus	loose focus
Measured Quantities		
C (Ns/MJ)	100	150
E_L (J)	300	150
I (Ns)	0.030	0.022
Calculated Quantities with $(\alpha\Phi)_{max} = 0.25$		
V_{abs} (cm^3)	5.08	5.59
$m_p = \rho_c V_{abs}$ (mg)	6.00	6.60
$<v_e> = I/m_p$ (km/s)	5.00	3.33
$\beta_{min} = C<v_e>/2(\alpha\Phi)_{max}$	1.00	1.00
$C^*_{max} = C/\beta$ (Ns/MJ)	100	150
$(u_c-u^o)_{min} = \beta E_L/m_p$ (MJ/kg)	50	22.7
T_{min} see Table 3 (K)	14400	8700
V_{abs} (cm^3)	6.78	6.78
$m_p = \rho_c V_{abs}$ (mg)	8.00	8.00
$<v_e> = I/m_p$ (km/s)	3.75	2.81
$\beta_{min} = C<v_e>/2(\alpha\Phi)_{max}$	0.76	0.84
$C^*_{max} = C/\beta$ (Ns/MJ)	131	178
$(u_c-u^o)_{min} = \beta E_L/m_p$ (MJ/kg)	28.5	15.7
T_{min} see Table 3 (K)	9500	7500
V_{abs} (cm^3)	13.6	13.6
$m_p = \rho_c V_{abs}$ (mg)	16	16
$<v_e> = I/m_p$ (km/s)	1.8	1.4
$\beta_{min} = C<v_e>/2(\alpha\Phi)_{max}$	0.38	0.42
$C^*_{max} = C/\beta$ (Ns/MJ)	262	356
$(u_c-u^o)_{min} = \beta E_L/m_p$ (MJ/kg)	14.3	7.9
T_{min} see Table 3 (K)	7000	5400

Figure 5 shows transformations of the isentropes in the Mollier plane to the $\mathbf{C}^*$-α plane for equilibrium expansions from an initial density of STP air, 1.18 kg/m^3. Lines of constant ε and ρ_e are almost exactly coincident. Lines of constant exit pressure run nearly parallel to lines of constant ε and ρ_e, and all are nearly vertical, indicating that alpha is nearly independent of $\mathbf{v_e}$ and u_c-u^o. At five times higher density, the constant exit pressure lines are nearly coincident with the STP constant pressure lines but their exit

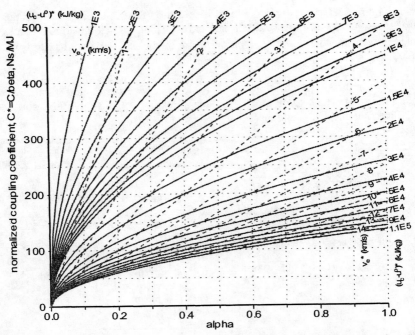

Figure 1. Defined relationships between six variables of interest: C*, α, β, Φ ,<v$_e$>*. and [u$_c$ – u°]*. The plots show C* = C/β as a function of α, with lines of constant v$_e$* = <v$_e$>/Φ and constant [u$_c$-u°]* = [u$_c$-u°]/Φ. The plots may alternatively be interpreted as a C vs α plots with lines of constant v$_e$* = <v$_e$>/βΦ and constant [u$_c$-u°]* = [u$_c$-u°]/ β²Φ.

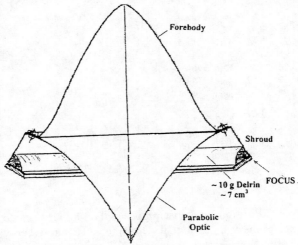

Figure 2. Cross-sectional view of Myrabo Laser Lightcraft, Model 200-3/4. The maximum diameter of the test article at the shroud is ~ 10 cm. The indicated ring of Delrin weighs ~ 10 g and has a volume of ~ 7 cm³ and a surface area ~ 25 cm². The idealized maximum plug nozzle exit area is ~ 350 cm².

Table 3. Thermodynamic properties of equilibrium air, $\rho = 1.18$ kg/m³.

u /MJ/kg	T/10³ K	P/bar	h/MJ/kg	s/KJ/kg/K	c_p/KJ/kg/K	M(kg/kmol)	X(e⁻)	$v_{a\,(km/s)}$	c_p/c_v
0.102	0.560	9.492	0.263	6.864	1.042	28.965	0	0.471	1.38
1	1.6	27.1	1.5	7.7	1.25	28.97	4e-13	0.77	1.30
2	2.6	43.2	2.7	8.2	1.45	28.95	6.E-11	0.96	1.25
3	3.3	56.5	4.0	8.6	1.85	28.73	2.E-08	1.08	1.21
4	3.9	67.7	5.1	8.9	2.33	28.19	3.E-07	1.17	1.20
5	4.4	78.2	6.3	9.1	2.65	27.46	2.E-06	1.26	1.20
6	4.8	88.9	7.5	9.3	2.71	26.69	6.E-06	1.35	1.22
7	5.3	100.3	8.7	9.5	2.61	25.96	2.E-05	1.45	1.23
8	5.8	112.4	9.9	9.7	2.55	25.32	4.E-05	1.53	1.23
9	6.3	124.5	11.1	9.9	2.69	24.79	8.E-05	1.61	1.22
10	6.7	135.8	12.3	10.0	3.04	24.32	1.E-04	1.67	1.21
15	8.2	182.0	18.1	10.7	5.49	22.19	6.E-04	1.91	1.18
20	9.2	222.3	23.8	11.2	7.36	20.32	1.E-03	2.11	1.18
30	10.8	304.9	35.2	12.2	8.05	17.41	3.E-03	2.49	1.20
40	12.7	404.9	46.9	13.1	5.52	15.45	1.E-02	2.92	1.24
50	15.6	534.8	59.1	13.8	4.28	14.33	3.E-02	3.39	1.27
60	18.4	667.9	71.3	14.4	5.20	13.54	8.E-02	3.78	1.26
70	20.8	794.6	83.5	14.9	6.32	12.81	1.E-01	4.13	1.27
80	22.8	919.9	95.6	15.4	7.26	12.14	2.E-01	4.45	1.27
90	24.6	1046	107.7	15.8	7.99	11.52	2E-01	4.76	1.28

Table 4. Thermodynamic properties of air at $\rho = 5.90$ kg/m³.

u /MJ/kg	T/10³ K	P/bar	h/MJ/kg	s/KJ/kg/K	c_p/KJ/kg/K	M(kg/kmol)	X(e⁻)	$v_{a\,(km/s)}$	c_p/c_v
-0.9	0.298	1.000	-.004	6.864	1.005	28.965	0	0.346	1.40
1	1.6	5.4	1.5	8.2	1.25	29.0	4E-10	0.77	1.30
2	2.5	8.6	2.7	8.7	1.51	28.9	3.E-09	0.95	1.24
3	3.2	11.1	3.9	9.0	2.16	28.6	3.E-08	1.06	1.20
4	3.7	13.1	5.1	9.3	2.83	27.8	3.E-07	1.15	1.19
5	4.1	15.0	6.3	9.6	3.15	26.9	2.E-06	1.23	1.19
6	4.5	16.9	7.4	9.8	3.04	26.1	5.E-06	1.32	1.21
7	4.9	19.1	8.6	10.0	2.69	25.3	2.E-05	1.41	1.23
8	5.4	21.5	9.8	10.2	2.56	24.7	4.E-05	1.50	1.23
9	5.9	23.9	11.0	10.4	2.86	24.2	8.E-05	1.57	1.21
10	6.3	26.0	12.2	10.6	3.43	23.8	1.E-04	1.62	1.19
15	7.5	34.1	17.9	11.3	6.70	21.7	5.E-04	1.84	1.17
20	8.3	41.3	23.5	11.9	8.93	19.8	9.E-04	2.02	1.17
30	9.7	56.2	34.8	13.0	9.09	16.9	3.E-03	2.38	1.19
40	11.5	75.4	46.4	14.0	5.13	15.0	1.E-02	2.81	1.24
50	14.4	101.0	58.5	14.8	4.81	14.0	4.E-02	3.26	1.25
60	16.6	124.0	70.5	15.4	6.62	13.2	1.E-01	3.60	1.24
70	18.4	145.0	82.3	16.0	8.25	12.4	1.E-01	3.91	1.24
80	19.9	167.0	94.1	16.5	9.51	11.7	2.E-01	4.20	1.24
90	21.3	189.0	106.0	17.0	10.40	11.1	2.E-01	4.48	1.25
100	22.6	211.0	118.0	17.4	10.90	10.5	3.E-01	4.76	1.26
110	23.9	235.0	130.0	17.9	11.10	10.0	3E-01	5.03	1.27

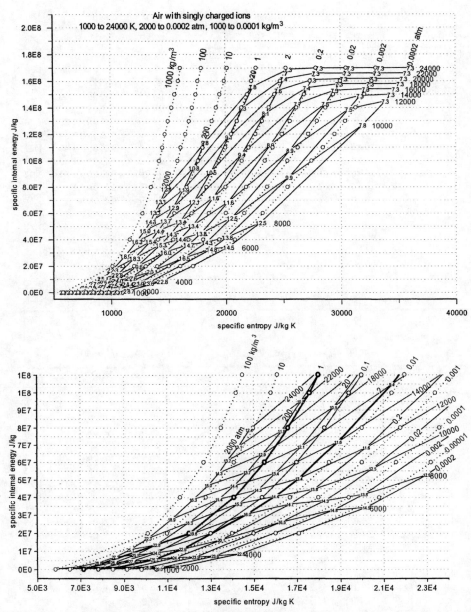

Figure 3. Mollier diagram for air including singly ionized species. Molecular weights are indicated at intersections of isobars and isotherms. The lower diagram shows a heavy constant density line, $\rho = 1.18$ kg/m³ above a heavy constant pressure line, P = 1 atm. The maximum energy initial states of laser heated STP air lie on the constant density line and the optimally expanded states lie vertically below on the constant pressure line.

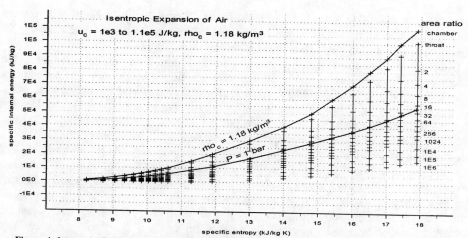

Figure 4. Isentropes for equilibrium expansions originating from the constant density line at 1.18 kg/m³ and terminating on the constant pressure line at 1 bar. Lines of constant area ratio are nearly coincident with lines of constant density.

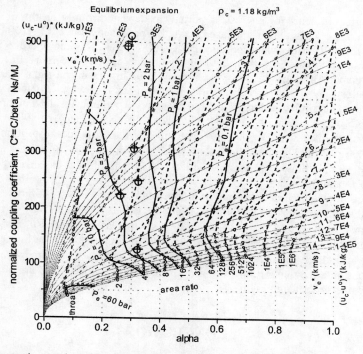

Figure 5. Isentropic expansions of laser heated STP air (1.18 kg/m³) from initial states of specific internal energy ranging from 1 to 110 MJ/kg. The circles and nearby crosses represent the blowdown quantities obtained from initial $[u_c-u^o]^*$ states of 2E3, and 1E4 J/kg for the frozen expansion and 2E3, 6E3, 1E4, and 4E4 kJ/kg for the equilibrium expansion.

pressures are about five times higher. Thus, in the C^*-α plane, the 0.2 atm exit pressure line at $\rho_c = 1.18$ kg/m^3 is almost coincident with a 1 atm exit pressure line at $\rho_c = 5.90$ kg/m^3. The effect of higher ρ_c is to increase α from ~ 0.4 to 0.6.

Coupling coefficients with Delrin installed in the Figure 2 test model were measured and reported in previous papers[5]. With increasing laser energy they rise to plateaus above about 300 J. At $E_L \sim 350$ J, C(Delrin) ~ 350 Ns/MJ and the ablated/vaporized mass was $m_p \sim 40$ mg. This means that $<v_e> \sim 3100$ m/s by Equations (7) and (8), and $\alpha\beta\Phi \sim 0.54$ by Equation (9). Thus, since $\beta\Phi < 1$, $\alpha > 0.54$, which is remarkably high. Air and Delrin will show very similar expansion behavior. As shown previously[5b], the dependence of α on the density of the heated air is quite strong. At $[u_c-u^o]^* = 10$ MJ/kg and expansion to 1 bar, the <u>instantaneous</u> α increases from about 0.43 to about 0.60 when the density increases from the STP value (1.18 kg/m^3) to its Mach 5 stagnation value (5.9 kg/m^3). These instantaneous α values decrease to about 0.32 and 0.50 respectively for the free-expansion blowdown process[5b]. An α value around 0.5 is reasonable when the density of the ablated and vaporized Delrin is as high as ~ 6 kg/m^3 and the blowdown expansion is near perfect with $\varepsilon \sim 16$. Most importantly, it appears that most of the inefficiency in the composite $\alpha\beta\Phi$ efficiency is carried by $\alpha\Phi$ and that β is very close to unity.

Coupling coefficients for air were reported[5] to depend strongly on the quality of the laser beam, as between a tightly focused beam that produced a lower $C \sim 100$ Ns/MJ than a loosely focused beam, which produced $C \sim 150$ Ns/MJ. This may be due to the tight beam heating a smaller mass of air to a higher energy than the more diffuse loosely focused beam. Although the exit velocity would be higher in the tight beam case, the total impulse may be lower because the heated mass is lower. Figure 2 shows the geometry and size relationship of a 7 cm^3 absorption volume inside the shroud, which contains ~ 8 mg of air. With C(air, loose focus) = 150 Ns/MJ at $E_L = 300$ J, and $m_p = 8$ mg we may deduce $<v_e> \sim 5600$ m/s, and $\alpha\beta\Phi \sim 0.42$. If the absorption volume is double, then $<v_e>$ and $\alpha\beta\Phi$ are halved. Figure 5 shows that STP air heated to 10 MJ/kg for example would blowdown to 1 bar with $\alpha = 0.32$ (equilibrium expansion) or $\alpha = 0.27$ (frozen expansion).

If we accept that a reasonable upper limit operational alpha is ~ 0.30 in our experiments $\alpha < 0.3$, then the measured $C \sim 150$ Ns/MJ and Equations (7-9) with $\beta\Phi = 1$ require $<v_e>$ < 4000 m/s, and $m_p > 11$ mg. Now if $\beta\Phi$ is ~ 0.3 as has been suggested by CFD modeling,[8] then the upper limit of $<v_e>$ decreases to 1200 m/s and the lower limit of m_p increases to 36 mg. It would seem apparent that the value of β is somewhat larger than 0.3 because both the upper limit $<v_e>$ and lower limit m_p with $\beta = 0.3$ are not reasonable for the geometry shown in Figure 2.

CONCLUSIONS

Experimental studies of the propulsion of a 200-3/4 model Myrabo Laser Lightcraft heated by 10.6 μ radiation from a CO_2 laser showed that the efficiency of conversion of laser energy to propellant kinetic energy was 54% for Delrin propellant.

Thermodynamic analysis of isentropic expansion of a unit mass of air after laser heating at constant volume was examined under conditions where either chemical equilibrium or frozen composition was maintained. The upper limit for the efficiency of conversion of internal to kinetic energy in optimum blowdown to 1 bar pressure is $\alpha \sim 0.25$, which is nearly independent of the initial energy. For STP air, blowdown to 1 bar is achieved with expansion ratios from $\varepsilon \sim 4$ at high energy to $\varepsilon \sim 8$ at low energy. With this small effective $\varepsilon \sim 4$, equilibrium expansion was only slightly more efficient than frozen expansion. Heating of propellant to higher energy states resulted in only slightly lower α but much higher exit velocity. The thermodynamic limitations were illustrated by process representations of blowdown in the Mollier plane.

REFERENCES

1. Sutton, G. P., *Rocket Propulsion Elements, An Introduction to the Engineering of Rockets*, John Wiley and Sons, Inc., New York, 1949, p. 17. See also: Sutton, G P., and Biblarz, O, *Rocket Propulsion Elements, Seventh Edition*, John Wiley and Sons, Inc., New York, 2001, p. 38.
2. Moeckel, W. E., *J. Spacecraft*, **12**, 700-711 (1975).
3. Lo, R. E., "Propulsion by laser energy transmission," IAF PAPER 76-165, October 1, 1976, 11 pages.
4. Phipps, C. R, Reilly, J. P., and Campbell, J. W., *Lasers and Particle Beams*, **18**, 1-35 (2001). See also: Phipps, C. R, Reilly, J. P., and Campbell, J. W., "Laser launching a 5-kg object into low earth orbit," in *Proceedings of SPIE* **4065**, 502 (2000).
5. (a) Larson, C. W., and Mead, F. B. Jr., "Energy Conversion in Laser Propulsion," in *39th AIAA Aerospace Sciences Meeting, 8-11 January 2001, Reno, NV, Paper No. 2001-0646*, (b) Larson, C. W., Mead, F. B. Jr., and Kalliomaa, W. M., "Energy Conversion in Laser Propulsion II" in *40th AIAA Aerospace Sciences Meeting, 14-17 January 2002, Reno, NV, Paper No. 2002-0632, and 33rd Plasmadynamics and Lasers Conference, 20-23 May 2002, Maui, Hawaii, Paper No. 2002-2205*.
6. Messitt, D. G., Myrabo, L. N. and Mead, F. B. Jr., "Laser initiated blast wave for launch vehicle propulsion," in *36th AIAA Joint Propulsion Conference, 16-19 July 2000, Huntsville, AL, paper 2000-3035*.
7. Mead, F. B. Jr., Squires, S., Beairsto, C., and Thurston, M., "Flights of a laser-powered Lightcraft during laser beam hand-off experiments," *36th AIAA Joint Propulsion Conference, 16-19 July 2000, Huntsville, AL, paper 2000-3484*.
8. (a) Wang, T.-S., Mead, F. B. Jr., and Larson, C. W., "Analysis of the Effect of Pulse Width on Laser Lightcraft Performance," *37th AIAA Joint Propulsion Conference, 8-11 July 2001, Salt Lake City, UT, Paper No. 2001-3664*. See also: (b) Liu, J., Chen, Y.-S., and Wang, T.-S., "Accurate prediction of radiative heat transfer in laser induced air plasmas," *34th AIAA Thermophysics Conference, 19-22 June 2000, Denver, CO, paper 2000-2370*. See also (c) Wang, T.-S., Chen, Y.-S., Liu, J., Myrabo, L. N., and Mead, F. B. Jr., "Advanced Performance Modeling of Experimental Laser Lightcrafts," *39th AIAA Aerospace Sciences Meeting and Exhibit, 8-11 January 2001, Reno, NV, Paper 2001-0648*.
9. Pakhomov, A. V., and Gregory, D. A., *AIAA Journal*, **38**, 725 - 735 (2000).
10. Corliss, William R., *Propulsion Systems for Space Flight*, McGraw-Hill Book Company, New York 1960, Chapter 2.
11. (a) Vinson, J., *Aerospace America*, February 1998, pp.30-33 (1998). See also: (b) Weegar, R., *Launchspace Magazine*, August 1996, p. 1719 (1996).
12. McBride, Bonnie J., and Gordon, Sanford, "Computer program for calculation of complex chemical equilibrium compositions and applications, II. Users manual and program description," NASA Reference Publication 1311, Lewis Research Center, Cleveland, OH 44135, June 1996.
13. Shapiro, Ascher H., *The Dynamics and Thermodynamics of Compressible Fluid Flow, Volume 1*," Ronald Press, New York, N. Y., 1953, p. 625.

ABLATIVE LASER PROPULSION
AND MICROTHRUSTERS

Simulation and Experiments on Laser Propulsion by Water Cannon Target

Takashi Yabe, Ryou Nakagawa , Masashi Yamaguchi, Tomomasa Ohkubo, Keiichi Aoki, Choijil Baasandash ,Hirokazu Oozono, Takehiro Oku, Kazumoto Taniguchi, Masamichi Nakagawa, Masashi Sakata, Youichi Ogata and Gen Inoue*

Tokyo Institute of Technology, Dept. of Mechanical Engineering and Science
2-12-1 O-okayama, Meguro-ku, Tokyo 152-8552, Japan
*National Institute for Environmental Studies, Center for Global Environmental Research,
Onogawa 16-2, Tsukuba 305-8506, Japan

Abstract. In previous papers, we reported the successful flight of paper-airplane about 5 cm. Application of such micro-airplane to CO_2 measurement and tornado observation is proposed. For practical application, repetitive water supply system and levitation system are proposed and examined by experiments. The latter can also be used for launching waste of nuclear reactor and structural materials for space station. Some future applications like stratospheric airplane and microship in human blood vessel are discussed.

1.INTRODUCTION

Kantrowitz proposed using laser ablation as an alternative to chemical propulsion for space vehicles in 1972[1]. Recent developments in high power lasers show that this idea is realistic [2,3]. In fact, the vertical launch of a 100-g rocket has already been demonstrated[3].

As an important near-future application of laser propulsion, we proposed a propulsion concept in which a laser drives a micro-airplane, and demonstrated it by experiments with a 590mJ/5ns YAG laser[4,5]. In these experiments, target is covered with overlay which is transparent to laser. The use of such concept for target acceleration appeared as an application to laser-driven fusion[6,7]. The efficiency enhancement is attributed to the tamped flow effect[6] or cannon-ball effect[7]. Winterberg used an enclosed configuration (like a real cannon) and Yabe employed a transparent overlay which was introduced to laser propulsion by Fabbro[8] and Phipps[9]. Although the enhancement is encouraging, the repetitive use of the structure is the key issue for realistic application. For this purpose, Yabe et al.[4] used water overlay and proposed repetitive water supply concept[5].

This paper provides various achievement and further development of this "water cannon" technology in variety of propulsion system.

CP664, Beamed Energy Propulsion: First International Symposium on Beamed Energy Propulsion,
edited by A. V. Pakhomov

2. MICRO-AIRPLANE

Coupling Efficiency

In order to investigate this coupling, we did a systematic survey with experiments and simulations. The numerical simulation code PARCIPHAL used here is based on the CIP-CUP(Constrained Interpolation Profile-Combined Unified Procedure) method[10-14]. Figure 2 shows the simulation and experimental results. The simulation results verify that the momentum coupling is greatly enhanced by this transparent overlay structure as shown by the solid line. The momentum coupling efficiency $C_m = m\Delta u / W$ is the ratio of the momentum $m\Delta u$ of target to the incident laser energy W.

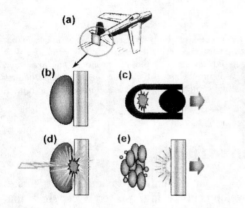

FIGURE1. The water cannon target. (a) Airplane with water cannon. (b) water droplet on metal, (c) real cannon, (d) laser is penetrate through water, (e) water and metal are propelled in opposite directions.

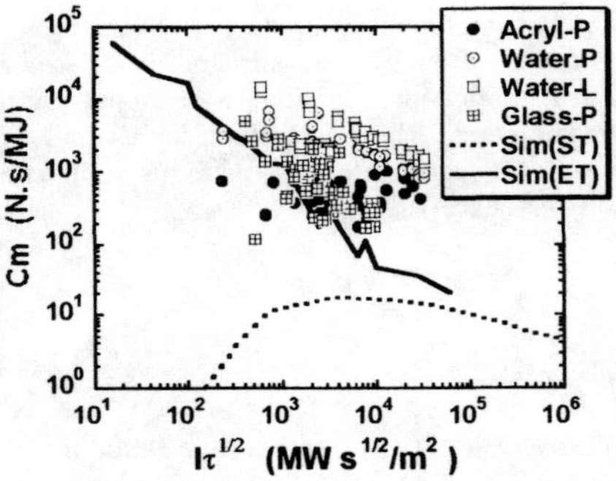

FIGURE2. Coupling coefficient vs laser intensity. τ is laser pulse width. Water-P means water cannon observed by pendulum and Water-L by load cell. Acrylic, glass, water overlay are compared.

In the simulation of the standard target (ST) (the single-layer target), C_m has the maximum as shown in Fig.2 because laser energy is too small to drive the aluminum target when I is less than the optimal intensity. In ST's, most of the energy is transferred to low density gas and only a small fraction of energy (normally less than a few percent) is used to drive a target. On the other hand, when I is larger, laser-produced plasma absorbs most of the laser energy by inverse Bremsstrahlung and shields the target surface.

In contrast to the ST, the C_m of the exotic target (ET) increases even with decreasing intensity I because the space between the two layers is filled with evaporated gas , providing a large amount of energy to drive the metal target, while the reduced C_m at high laser intensity is the same as the case of ST, i.e., it is dominated by plasma shielding.

The results of glass overlay was obtained by LHMEL Nd:glass laser of 70J /25-100ns at Wright-Patterson Air Force Base, Ohio[9], while those of water and acrylic overlay are obtained by 590mJ/5ns YAG laser[4]. The figure shows the systematic increase of C_m we observed for water overlay compared with acrylic and glass overlay. In the case of water, both pendulum and load cell (Model FPA-6/10 N made by Toyoda Machine Works, Ltd.) measurements are compared for the calibration of the load cell.

Flying Micro-airplane

Since the water overlay is sufficiently effective, we tried to fly paper airplane shown in Fig.3 whose size is 39mm x 56mm x 15mm. At the rear edge of the airplane, an aluminum foil of 3.5mm x 3.5mm x 0.1mm-thick is pasted and the total weight is 0.2g. In addition, water droplet of 3mm-diameter, 0.014g is attached to the aluminum foil as shown in Fig.3 and Figs.1(a) and (b).

At first, the airplane is placed on the platform with a guiding groove and is irradiated by one pulse of YAG laser. Figure 4 shows the flight path of this airplane. With similar airplane, we obtained Cm=237N.sec/MJ for $I\tau^{1/2}= 1.8\times10^4$ MW.sec$^{1/2}$ /m^2.

FIGURE3. The paper airplane with water overlay on aluminum foil.

FIGURE4. The flight trajectories of paper airplane with water overlay.

Application of Micro-airplane

Global warming owing to CO_2 is internationally recognized and Kyoto treaty on the restriction of CO_2 exhaust will be active forcing most of the countries to reduce the CO_2 exhaust by specific amount over 5 years starting from 2008. For this purpose, we need a measuring instrument for three-dimensional distribution of CO_2. In this paper, we shall propose to use a paper plane for this observation.

We can use number of small paper planes carrying corner cube. By choosing an observation laser that can be absorbed by CO_2, we can measure the distance and absorption rate by the reflected light from the airplanes. By using number of airplanes, we can observe the time-resolved three-dimensional structure of CO_2 in atmosphere. This will be an important measuring tool for the study of global warming on Earth.

As the first stage, we can use a balloon that contains a number of paper planes and lift them up to high altitude, where the balloon will be broken by irradiation of laser. Then a number of airplanes fly out and scatter away in many directions. Scanning laser may hit some of the airplanes by chance and then reflection gives us information of distance and absorption as shown in Fig.5. At the next stage, a group of airplanes will be guided to any direction by laser propulsion. In this case, we do not need to accurately shoot and guide each plane but statistically guide the group of planes in desired direction by randomly shooting laser as shown in Fig.6.

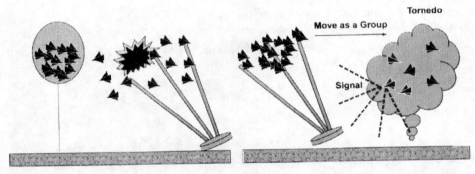

FIGURE5. The paper airplanes are used for CO_2 observation.

FIGURE6. The paper airplanes with signal emitter are led to tornado by laser..

Such a guiding technique can also be used to tracing tornado. If each airplane is equipped with a signal emitter, then we can lead some of the airplane towards tornado and once they are trapped by the tornado, dynamical behavior of tornado can be observed as shown in Fig.6.

3. PRACTICAL USE

Water Supply for Repetitive Propulsion

For practical application, the overlay structure must be repetitively constructed. Thus water overlay has several advantages : (1) higher efficiency, (2) easier arrangement, (3) automatic collection from air, (4) no environmental problem and so on. For water supply, we proposed a structure where water is contained in a narrow space between two plates with a small hole of 2mm in one side. We have demonstrated that the water did not escape from the hole due to surface tension and was quickly supplied after irradiation[5]. After several investigations, we found the following configuration is the best for stable supply and large volume of water containment.

In this system, water is contained in the cylindrical tube. The aluminum target is fixed to this cylinder as shown in Fig.7. The target has a quadrilateral shape that is elongated vertically and thus giving a path of water through the gaps in both sides from the large water reservoir on the left of side-view figure. After laser is irradiated through a hole shown in the front-view figure, small fraction of water between target and hole is ejected. Then water will be quickly fed through the gap.

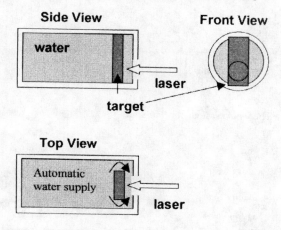

FIGURE7. Water supply system for repetitive propulsion.

Air Slider for Levitation

In practical situation, the objects should be flying in the sky. It would be very difficult to realize this situation because we need tracing system and large lasers. In order to simulate this situation, we try to realize a friction-free system and repetitive propulsion.

For this purpose, we built a new system called "air-slider" which suspends objects using air-flow from air-compressor (1.57×10^{-3} m^3/s). Figure 8 shows the whole system. We use the edge of the quadrilateral pillar to guide the object in one-dimensional direction so that the laser is easily irradiated. Along the surface of the quadrilateral pillar, many holes are drilled in order to supply air-flow. This system reduces the effect of friction by air-flow from the holes made in intervals of 1cm, and levitates the object so that the object can move very smoothly. The acrylic plate that is bent in the right angle and is fitted to the corner of the quadrilateral pillar as shown in Fig.9 is used to mount the object.

Such a levitation system can also be used for launching waste of nuclear reactor and structural materials for space station into space.

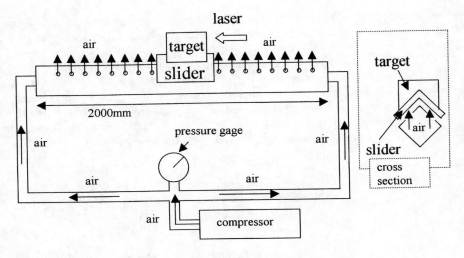

FIGURE 8 The whole system of "air-slider".

FIGURE 9. The multi-exposure photograph of 10.7g object with water supply moving along the air-slider.

Driving 300g Objects by 0.5J Laser

The right figure in Fig.9 shows multi-exposure photograph of accelerated objects which carry water supply system. The weight of the object is 10.7 g including water supply. Since the exposure time interval is equal, the acceleration of object is clearly shown after several shots of laser.

By combining the above air-slider and repetitive propulsion system, we are able to move a heavy object. Figure 10 shows the time evolution of 300 g object. Since the distance between the object and laser is largely changing, unfocused beam is used and laser radius is about 3mm. Water supply system is not fast enough and 2Hz repetition has the maximum efficiency. For higher repetition, efficiency was greatly reduced and did not profit from additional pulses.

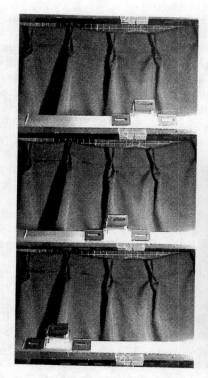

FIGURE 10. (Left) Acceleration of 300g object with water supply on the slider. (Right) whole view of air slider..

4. FUTURE APPLICATION

Stratospheric Airplane

Stratosphere is an atmospheric region at an altitude of about 11~50km. The density and the pressure of this region is less than 1/10 of surface of earth, and temperature is -60~-70 degrees centigrade. There is no cloud and therefore is always sunny. Using this feature, the idea to communicate by using flight vehicle in the stratosphere instead of the satellite is proposed.

The difficulty here is the realization of the engine. Due to low density, it's difficult to propel by the propeller. The efficiency of jet engine is low, thus a lot of fuel must be loaded. There exists more important and serious problem. That is the existence of the ozone layer. About 90% of the ozone of the entire globe exists in stratosphere. Using a reciprocal engine or a jet engine that burns fossil fuel, NO_x or CO of exhaust gas indirectly contributes to the destruction of ozone layer by catalysis.

While the influence of chloro-fluorocarbon appears after several decades, this influence is direct and thus a serious damage can be expected.

Therefore the application of laser propulsion by water cannon has big advantages such as no fossil fuel leading to harmful emission (emission of only water), easy handling and the possibility of getting water from atmosphere.

Towards this airplane, the result in Fig.2 is very encouraging. The maximum value $\sim 10^5$ N.sec/MJ obtained by simulations means that one pulse of 1MJ laser can give a velocity 10m/sec to an airplane of 10tons. Driving such a large airplane is no more dream.

Tracing the airplane is another difficult subject. It is not clever to focus the light at a large distance from the source to the airplane. Thus we propose to use the Fresnel lens placed at the airplane. Then the unfocused light propagated through the atmosphere is focused only at the point close to the target.

If we use the phase-conjugation mirror in lasers and array of prisms or corner cubes on the airplane, we can scan over the neighboring area of airplane by small signal laser and only signal that will return back from prisms is amplified by the amplifier with phase-conjugation mirror and is directed back to the airplane again.

Micro-ship in Inner Space

In 1948, a famous Japanese cartoonist, Osamu Tezuka proposed the idea of miniaturized human getting into human body in series of his famous cartoon called "Tetsuwan Atom (in Japanese)". In 1966, Hollywood adopted this idea to make the film"Fantastic Voyage". Such a dream might be coming into reality.   We can drive a micro-ship or micro-capsule of medicine by a X-ray whose wavelength is ~ 1 A whose penetration depth is around 10cm. Such X-ray will be able to shine the micro-ship from outside (see Fig.11). The ship can have a structure as shown in Fig.11(Left) and is equipped with focusing lens made by Fresnel zone plate. By this method, X-ray need not be focused onto the micro-ship and low intensity of unfocused X-ray may not cause serious damage to the human body. A microscopic Fresnel zone plate of 47μm-size has already been developed and 1.45A X-ray was successfully focused into 0.5μm spot. If we need the laser intensity of $I\tau^{1/2} =10$MW.sec$^{1/2}$/m^2 for ablation as in Fig.2, the laser power and energy delivered to this spot size are 28mW and 142pJ for 5ns pulse.

Actually, we tried to simulate the heating of water adjacent to the microship irradiated by X-ray pulse of 1nJ/5ns. Taking into account of the absorption depth of X-ray, we obtained the temperature rise of 70-90 degrees in water depending on material of microship.   In practical use, the temperature must be less than boiling temperature of water. Therefore, the process should not be the ablation but is merely imposing static pressure on microship or changing the surface tension and/or viscosity caused by temperature increase.

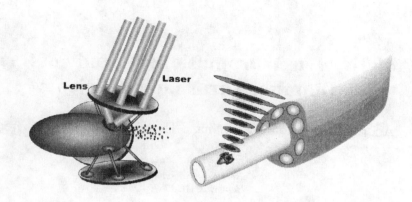

FIGURE 11. (Left) The microship is equipped with several Fresnel zone plates and then X-ray needs not be focused onto the ship. This is important to avoid un-necessary damage to the body.(Right) The microship can go along the blood or lymphatic vessel driven by "water-window" X-ray laser placed outside of human body.

REFERENCES

[1] A.Kantrowitz, ``Propulsion to Orbit by Ground-Based Laser," Astronaut. Aeronaut. 10, p.74 ,1972.
[2]C.R.Phipps,Jr. et.al., ``Impulse coupling to targets in vacuum by KrF,HF,and CO2 single-pulse lasers," J.Appl.Phys. 64, pp.1083-1096, 1988.
[3]L.N.Myrabo and F.B.Mead, Jr., ``Ground and Flight Tests of a Laser Propelled Vehicle ," AIAA98-1001, Aerospace Sciences Meeting & Exhibit,36th, Jan12-15,1998
[4]T.Yabe et.al., Micro-airplane Propelled by Laser-Driven Exotic Target   Appl.Phys.Lett. 80, 4318(2002).
[5] T.Yabe et.al., Laser-Driven Vehicles - from Inner-Space to Outer-Space Proc.SPIE 4760 High-Power Laser Ablation IV, 21-26 April 2002,Taos, NM, USA
[6]T.Yabe and K.Niu, Numerical Analysis on Implosion of Laser-Driven Target Plasma.   *J.Phys.Soc.Japan*, **40** , 863(1976).
[7]F.Winterberg, Recoil Free Implosion of Large-Aspect Ratio Thermonuclear Microexplosion, *Lettere al Nuovo Cimento* **16**, 216 (1976).
[8]F.J.Fabbro, P.Ballard, D.Devaux and J.Virmont, Physical Study of Laser-Produced Plasma in Confined Geometry, *J.Appl.Phys.* **68**, 775(1990).
[9] C.R.Phipps, D.B.Seibert II, R.Royse, G.King and J.W.Campbell, III International Symbosium on High Power Laser Ablation, Santa Fe, 2000, SPIE Vol.4065.
[10]H.Takewaki, A.Nishiguchi and T.Yabe, Cubic Interpolated Pseudoparticle Method (CIP) for Solving Hyperbolic Type Equations. *J.Comput.Phys.*, **61** , 261 (1985) .
[11]T.Yabe and T.Aoki, A Universal Solver for Hyperbolic Equations by Cubic-Polynomial Interpolation I. One-Dimensional Solver. *Comput.Phys.Commun.* **66**, 219 (1991)
[12]T.Yabe, F.Xiao and T.Utsumi, Constrained Interpolation Profile Method for Multiphase Analysis. *J. Comput. Phys.* 169,   556 (2001)
[13]T.Yabe and P.Y.Wang, Unified Numerical Procedure for Compressible and Incompressible Fluid.   *J.Phys. Soc. Japan*, **60**, 2105 (1991).
[14]F.Xiao, T.Yabe, N.Konma, A.Uchiyama, K.Akutsu and T.Ito, An Efficient Model for Driven Flow and Application to a Gas Circuit Breaker *Comput. Model. Simul. Eng.* **1**, 235 (1996).

Ablative Laser Propulsion: A Study of Specific Impulse, Thrust and Efficiency

Andrew V. Pakhomov, M. Shane Thompson and Don A. Gregory

Department of Physics
The University of Alabama in Huntsville,
Huntsville, AL 35899, USA

Abstract. This paper represents a review of studies devoted to Ablative Laser Propulsion and conducted by the Laser Propulsion Group (LPG) at the University of Alabama in Huntsville.

INTRODUCTION

The concept of *ablative laser propulsion* (ALP), formulated in Ref. 1, is essentially a slightly refined return to the original idea of *laser propulsion* introduced by Arthur R. Kantrowitz in his seminal paper from 1972 [2]. The word *ablative* was added to emphasize that the momentum transfer to the vehicle is indeed defined by direct laser ablation of matter. In this case the role of an *ablatant* should be played only by condensed matter, which is a crucial feature of ALP, as will be shown below. The direct ablation of matter in laser propulsion could be counterposed to the leading alternative momentum transfer mechanism utilizing shock waves that originate from the laser breakdown and propagate in gaseous or plasma media. It should be noted that such shockwave-driven techniques are the only methods to achieve successful flight demonstrations to date [3-6]. In all these cases, ambient air [3-5] or inert gases [6] were used as necessary shock wave media, and therefore served as *propellants*. However, an absence of such agents in space, on the moon, or even above the troposphere, poses a serious limitation to pressure-wave driven lightcrafts. The use of stored on-board hydrogen was proposed as a possible solution to the problem [3]; however, such an approach would have its own disadvantages. After all, the need to get rid of carrying on-board heavy containers with liquefied gases was one of the reasons for invoking the idea of laser propulsion in the first place. For this reason, ablative laser propulsion could serve as the more attractive alternative. Moreover, the development of in-space propelled micro- and nano-satellites will demand propulsion techniques with power systems reduced to an absolute minimum. In this case ALP could be *the only* working LP concept. Lightweight concentrating optics and a bit of solid ablative will be the only propelling components, which such systems could afford.

Important advantages of ablative laser propulsion include [1]:
- *Directionality*: plasma jets are normal to the *ablatant (or propellant)* surface, independent from the direction of the incident laser pulse, and hence the

CP664, Beamed Energy Propulsion: First International Symposium on Beamed Energy Propulsion,
edited by A. V. Pakhomov
© 2003 American Institute of Physics 0-7354-0126-8/03/$20.00

imparted net momentum is likewise directed. Therefore, one can control the direction of imparted momentum without any special confinements, such as chambers with nozzles, air pockets, and likewise, all of which are needed for pressure-wave driven lightcrafts.

- *Collimation*: although this property is directly related to directionality, it has its own merit. The angular distribution of ejected ablatant is not only a maximum normal to the surface, but it also can be quite *narrow*. In general, it follows the function [1]:

$$N(\theta) = N_0 Cos^\gamma(\theta) \qquad (1),$$

where θ is the angle of ejection with respect to the surface normal, $N(\theta)$ is the density of ejected ions, $N(0) = N_0$, and γ varies with ablation conditions from unity [7,8], to 4.0-8.0 [9]. Thus, in the condition of uniform velocity distribution and for $\gamma = 1$, the net vector momentum imparted to the vehicle will be 2/3 of a scalar sum of all individual momenta of ejected particles [8]. Apparently, an increase in γ will lead to an increase in this ratio, and, therefore, *efficiency*.

- *Low Breakdown Thresholds*: in comparison to gases, solids (especially metals) have essentially lower breakdown thresholds within the same irradiation conditions. The difference of up to a factor of 10^4 [10] would mean 100-fold gain in the distance, over which a remote laser could accelerate the vehicle. Since the distances are limited by both diffraction and available laser powers, this factor could be of extreme importance.

- *Low Breakdown Thresholds*: this is not a misprint. Higher payloads, especially when launched from earth's surface, would require higher fluences. At some point, the breakdown in air before the laser pulse reaches the lightcraft would become a bane (this problem was originally pointed out by Kantrowitz in Ref. 2). Using solid ablatants with much lower breakdown thresholds would obviate such a problem to a great extent.

- *Material Versatility*: a great variety of specially-crafted solid ablatants, which modern materials science could provide, offers many possibilities for experimentation. Any combinations of specific impulses (up to 2×10^4 s, as reported in Ref. 8) and coupling coefficients (up to 10^3 dyne/W, as reported in Ref. 11) would be possible as long as the energy conservation law is observed. The choice of ablatant, from delrin to lead, would be dictated by specific requirements or tasks of the mission. Additional options are available by ablation in a confined geometry (see Ref. 12 and references therein).

- *Efficient Energy Conversion:* regimes that convert 90% (or even more) of laser pulse energy into kinetic energy of ablated species are possible [13]. This leaves the surface of ablatant "cold" and, therefore, ready for absorption of the next laser pulse.

With all these advantages in mind, we will turn to the issue of the universal figures of merit for propulsion systems: specific impulse, coupling coefficient (or thrust) and efficiency for ablative laser propulsion. However, some definitions must be given first.

A possible ALP process can be described as follows: first, a short, high-power laser pulse from a remote source is absorbed by the surface layer of the propellant, located at the rear of the vehicle. Since the light delivers the energy, not momentum, the light incidence can be at practically any angle to the normal to the surface of propellant. The absorbed light initiates a laser-induced breakdown on the surface, predominantly by inverse bremsstrahlung, leading to transformation of the portion of solid propellant at the focal spot into a highly energetic plasma jet. The ionized jet normal to the propellant surface is then accelerated by ponderomotive forces. The momentum opposite in direction and equal in value to the net momentum of ejected matter propels the vehicle. Thus, as follows from the described scenario, *ablative* laser propulsion means that the momentum is delivered to the vehicle by direct mass removal from the surface of the ablatant. This follows from the conservation of momentum in the system vehicle – the ejected ablative is then essentially a rocket system. Such a definition of ALP naturally follows from the semantics of the term *laser ablation,* which can be defined in a broad sense as matter removal under laser irradiation. The scenario described above is just one possible chain of events, *i.e.* the physics leading to ALP surely could be different. Depending on irradiation intensity, the physical phenomena underlying matter removal could be covered by a broad range of processes, from thermal desorption to photonuclear fission. Between these extremes one can find thermal (vaporization, explosive boiling) and non-thermal (plasma generation and acceleration) mechanisms. As it will be shown below, some of these processes can occur concurrently. In spite of the differences in the physics of the above mentioned phenomena, as long as momentum is provided directly by ablative mass removal, it will be ALP. However, in the further discussion we stand by the scenario based on plasma generation and acceleration.

Any study of ablative laser propulsion must address the question on how well the subject fits to the given above definition of ALP, or, in other words, is this really ALP? In our case, this can be re-phrased as follows: how, actually, one can tell that the momentum transfer occurs via direct mass removal of the ablatant and not formation of pressure waves in the *laser-induced plasma* (LIP)? The early statement, given in Ref. 1 that ALP must be strictly material dependent is certainly true [8]. As demonstrated in Ref. 8, ion velocity and, therefore, specific impulse (I_{sp}), imparted to the target by the ejected ions, is inversely proportional to the square root of atomic mass of ablative. However, using this as a proof of ALP could be an oversight. Although some works report that momentum transfer via *laser-sustained detonation* waves appears much less ablatant-dependent [14], the dependence of specific impulse on molecular weight in case of pressure-wave momentum transfer was also predicted [15]. Thus, materials dependence of I_{sp} would be certainly supportive, but not an essential sign of ALP. For this reason, a slightly different approach to this issue was taken by Pakhomov *et al.* [8]. According to Pakhomov *et al.* [8], the laser pulse must be short so it will be delivered to the target surface prior to formation of a dense LIP, which will hinder further light passage to the target and serve as a medium for momentum transfer. A simple estimation shows that critical time (t_c), *i.e.* time over which plasma achieves critical density, is ~ 0.1 ns [8]. The first experimental effort to determine t_c for our irradiation conditions is presented in Ref. 16 (this volume). These experiments, based on probing the LIP with a second pulse incident with relative time delays in the range of ~ 0.1 – 6.5 ns, have shown that t_c is indeed ~ 0.1 ns. Thus, within the realm of described physical phenomena (plasma generation), only systems employing pulses of width 0.1 ns, or shorter, could provide real ALP regimes. For the same reason, ALP was practically out of scope in all the wealth of early LP research, developed since the early 70's. In these studies, relatively long-pulsed (microseconds) if not cw irradiation was used. Therefore, the majority of employed lasers were imparting momentum to the targets via plasma pressure waves, with minuscule fraction of thrust applied by an actual ablation. This conclusion, originally stated in Ref. 1, corroborates early work of Pirri [17], where the negligible role of direct ablation was indeed postulated. In terms of critical time, this condition would imply that LP based on laser pulses significantly exceeding t_c in width would be similar in efficiency to LP based on cw-irradiation. Such equivalency was indeed demonstrated in Ref. 15.

SPECIFIC IMPULSE

Keeping in mind that ALP is primarily defined by relation $t_c \geq \tau_p$, where the latter is laser pulse width, one can now recall the definition of specific impulse (I_{sp}) imparted to the target by a single pulse:

$$I_{SP} \equiv \frac{1}{W} \int_0^{\tau_a} F(t)dt \qquad (2),$$

where W is the weight of the propellant ablated per pulse and $F(t)$ is the thrust as a function of time t. The integral presents an impulse applied to the target, with the force integrated over the period τ_a, i.e. the time over which the *essential* mass ablation within laser-induced plasma occurs. In our experimental conditions [8], when $\tau_p = 0.1$ ns, $\tau_a \approx 10^4 \tau_p$. It is worth noting that even when the width of ablating pulses is 10^3 times shorter (0.1 ps), τ_a is still in the microsecond range [7,18]. This fact certainly serves as an indicator that the direct interaction of light with LIP is not a critical issue at such short pulse widths, as we have already discussed above.

The straightforward measurement of $F(t)$ would provide an ideal way of determining I_{sp} for ALP on the condition that the rise time of the force sensor is smaller than τ_a. In our first attempt to measure $F(t)$ directly with a piezo force sensor [18], we encountered this problem. As a result, an independent assessment of τ_a became a necessity.

In early studies of plasma pressure-driven momentum transfers, the thrusting time was assumed equal to the time over which hot gas exits the system nozzle [19]. Applying a similar approach to ALP, one may assume that τ_a can be approximated by the time of plasma ejection from the target surface. The duration of this process can be determined with nanosecond-level temporal resolution using ICCD-imaging [18]. Or, independently, τ_a can be estimated from the width of ionic time-of-flight (TOF) waveforms. In both cases, measurements yield τ_a in the range of 1.0 - 2.5 μs for different elements [18]. Therefore, for reliable determination of I_{sp} from Equation 2, one will need a force sensor with a rise time of ~ 0.1 μs or better. In the absence of such sensors (our case), the determination of I_{sp} turns into a combined task of several independent techniques: ICCD-imaging and TOF for assessment of τ_a, precise sample weighing, and, finally, actual force measurements [18]. Although this technique yielded overall satisfactory results, it appeared too laborious to be worth any further development (see Ref. 18 for details), unless a faster force sensor would be employed. Currently we are considering alternative ways of impulse measurements, such as a ballistic pendulum.

Another alternative to force measurements involves deriving specific impulse from ion velocity. Applying Newton's second law to Equation 2 yields:

$$I_{SP} = \frac{1}{W} \int_0^{\tau_a} \frac{dP(t)}{dt} dt = \frac{\overline{P}}{W} \qquad (3),$$

where $\overline{P}$ denotes the net imparted momentum per pulse. According to the scenario described in the *Introduction*, it would be reasonable to assume that the ionic component of the ejected mass comprises the main propelling body. In that case the rest of the ablated propellant mass, including neutrals and all matter removed via thermal processes, would increase the denominator of Equation (3) without any significant addition to the numerator, *i.e.* accounted as losses. Assuming that $\overline{P}$ is imparted by a flux of monoenergetic ions ablated with the mean velocity $\overline{v}$ (see the justification for such assumption in Ref. 13), I_{sp} can be directly deduced from a measurement of the ionic velocity. However, the assumption that ions are monoenergetic does not imply that they are all ejected in direction normal to the target surface. In our experimental condition the angular dependence of ion number density was in accordance with Equation (1) with $\gamma = 1$, while ion velocity appeared independent of the direction [8]. Vector integration over such an ion distribution led to the net imparted momentum:

$$\vec{P} = \frac{2}{3} m_i \overline{v} \hat{n}, \qquad (4)$$

where m_i is total ionic mass and $\hat{n}$ is unit vector normal to the target surface. Substitution of an absolute value of momentum $\vec{P}$ into Equation (3) will yield:

$$I_{SP} = \frac{2}{3} \xi \frac{\overline{v}}{g} \qquad (5)$$

Thus, I_{sp} can be determined from ion velocity with an accuracy limited to the factor of ξ, which represents the mass fraction of ionized matter ejected with average velocity $\overline{v}$. The geometrical factor (2/3 in our case) must be determined from the angular dependence of ion density and velocity. Since ξ cannot be derived from the ion velocity measurements, it places a serious limitation on determination of I_{sp} from this technique. In other words, it sets sort of "upper boundary" of I_{sp} assuming that ξ is unity. Of course, this would hardly answer the question how far below this boundary the actual I_{sp} is. At this point the determination of ξ is crucial. The ionization ratio in LIP can be determined from plasma emission spectra, and, if the mass-removal rate is known, ξ can be estimated from such data, though this still would not account for mass losses other than via LIP). However, such measurements show that ion fraction is a function of a distance from the target and time [20], and therefore, ξ in Equation (5) should actually be replaced with an integral of $\xi(t)dt$. This would hardly simplify the determination of I_{sp}.

A possible solution to this problem is to use a combination of the force and velocity measurements, conducted within the same experimental arrangement. To some extent, this effort was undertaken by LPG [8,18]. The results are presented in Figure 1.

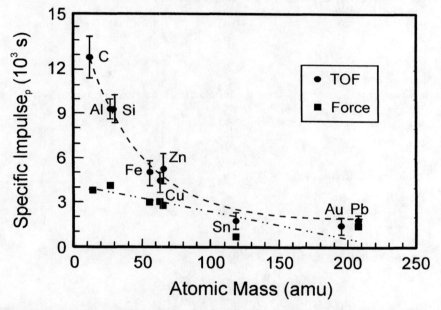

FIGURE 1. Specific impulse *vs.* atomic mass of propellant as derived from ion velocity (Ref. 8, dots) and force measurements (Ref. 18, squares). Continuous lines represent numerical fit.

As one can see from Figure 1, specific impulse derived from velocity measurements is presented in assumption of $\xi = 1$, which is known *a priori* as untrue. For this reason, and as expected, the I_{sp} derived from force measurements are reduced by a factor of ξ. Assuming that the force data represents the actual I_{sp}, and following Equation (5), an efficient ξ can be estimated as a ratio of I_{sp} measurements. The results are given in Table 1.

Thus, as it follows from Table 1, I_{sp} derived from ion velocity, has to be reduced 1.5 - 3 times in order to take into account the mass losses via ablation of neutrals. The values of ξ estimated via such a rather indirect technique seem reasonable, except the result for lead (0.93). As we have shown earlier, about a half of the ablative mass removal rate for lead is due to relatively low energetic phase explosions [22], which make such a high value of ξ rather unlikely. In general, the presented ξ must be less than the time-averaged ionization ratio in LIP, which can be deduced, for example, from plasma emission. However, emission spectroscopy does not account for any component removed via post-plasma ablation mechanisms, such as, for example, vaporization or phase explosion [22,23].

EFFICIENCY

The earliest version of the derivation of laser propulsion efficiency, known to the authors, was published by Pirri *et al.* [19], and another derivation was performed by Phipps and Michaelis [13]. We will perform an independent derivation of ALP efficiency, which, not surprisingly, yields similar results.

We define coupling coefficient C_m as:

$$C_m \equiv \frac{\overline{P}}{E}, \qquad (6)$$

where E is laser pulse energy. Then the product of Equations (3) and (6) will yield:

$$I_{sp}C_m = \frac{\overline{P}^2}{EW}, \qquad (7)$$

Combining Equations (4) and (7) will result in:

$$I_{sp}C_m = \frac{2\eta\xi}{g}, \qquad (8)$$

where η can be defined as:

$$\eta \equiv \frac{\overline{P}^2}{2m_i E}. \qquad (9)$$

The product $\eta\xi$ will represent the ALP efficiency of a given propellant, which we will denote further as χ. It is worth noting that this efficiency would characterize an ablatant, but not an ALP-vehicle. Consideration of a particular design would bring additional factors to χ, and this is out of scope of this discussion.

The form of Equation (9) suggests η is a measure of energy efficiency, and this definition was used in our original derivation [21]. However, it is worth noting that due to the directionality of net imparted momentum $\overline{P}$, the actual (or directional) energy efficiency η is not identical to total energy efficiency η^*, defined as a total kinetic energy of ejected ions divided by laser pulse energy:

$$\eta^* \equiv \frac{m_i \overline{v}^2}{2E}. \qquad (10)$$

200

The relationship between η^* and η can be found by taking a square of Equation (4), substituting it into Equation (9), and comparing it with Equation (10). This leads to:

$$\eta^* = \left(\frac{3}{2}\right)^2 \eta \qquad (11)$$

The values of η^* evaluated from TOF data together with η derived using Equation (11) are posted in Table 1. Omitting the extremes, the median ALP efficiency χ presented in the bottom row of Table 1 appears near 17 %.

TABLE 1. Efficiency Components.

Element	C	Al	Fe	Cu	Zn	Sn	Pb
η^*	0.53	0.88	0.66	0.59	0.92	0.22	0.39
η	0.24	0.39	0.29	0.26	0.41	0.10	0.17
ξ	0.30	0.45	0.60	0.66	0.53	0.38	0.93
χ	0.07	0.18	0.17	0.17	0.22	0.04	0.16

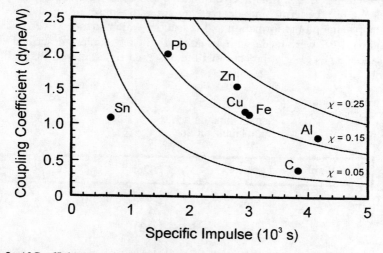

FIGURE 2. ALP efficiency curve based on the data from Refs. 8, 18 (see the text for details). Continuous lines represent $\chi = 0.05$; 0.15 and 0.25 respectively.

Using the data from Table 1 we have "combined" our experimental points from Figure 1 in a graph presented on Figure 2. The graph illustrates Equation (8) with C_m and I_{sp} derived primarily from TOF data [8] and corrected by ξ factor deduced from force measurements [18]. The hyperbolic iso-efficiency lines set a theoretical grid, indicating that presented points are well below the $\chi = 1.0$ margin. It is worth noting that except for tin and graphite, the studied elements closely follow the 0.17 iso-efficiency line, in spite of a quite broad variation of C_m and I_{sp}! This confirms the original expectation that elementary metal targets can provide a broad choice of C_m

and I_{sp} combinations, making ALP flexible to the needs of any particular mission or operation.

The elements with lower χ, such as C, Sn, and possibly Pb (see the comments on ξ in the previous section), presumably suffer from some energy sink mechanism. Such mechanism would lead to ablative removal of an essential fraction of mass with negligible momentum transfer either during or after the LIP plume ejection. A possible candidate mechanism for C is ablation of large particulates (also known as defoliational sputtering), which is due to atomic bond anisotropy of graphite and was reported previously [23]. For Sn and Pb it will be phase explosion.

PHASE EXPLOSION

The first results of our study of phase explosion (PE) in prospective ALP propellants were reported in Ref. 22, while a more detailed account is presented in Ref. 21. In the context of this paper, the discussion of phase explosions will be limited to its impact on efficiency.

As reported in Ref. 22, time-delayed phase explosions were observed in all tested materials. We use the term "time-delayed" to stress that onset of PE occurred ~ 30-50 μs *after* the plasma plume formation and cessation, which spanned over first 1 - 2 μs. Observations of PE were made with an intensified CCD camera described elsewhere [18], and the images clearly indicate that the observed time-delayed PE is different in nature from LIP [22].

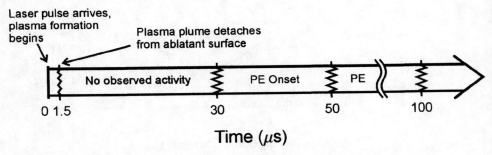

FIGURE 3. Timeline of events from laser-induced plasma to phase explosion.

The chronology of ablative events can be summarized in a timeline (Figure 3), where $t = 0$ corresponds to the arrival of the laser pulse at the target surface. Then, within 1 -2 μs laser-induced plasma completes its lifecycle, followed by ~ 30 μs long "dark" period, which separates LIP from PE. Then the PE event onsets and lasts for ~ 50 μs. In ~ 5% of observations the explosive removal of individual particulates could be traced to $t = 0$, however even in these cases the massive ejection always occurred at later times, as shown on Figure 3.

Images of PE showed discrete particulate traces extended over the exposure of the ICCD camera. The velocities of such particulates were estimated from these traces, and found to be ~ 100 - 200 m/s, *i.e.* ~ 2-3 orders of magnitude less than ion velocities measured by the TOF [22]. This indicates that the mass removed by PE would have a negligent contribution to the momentum transfer as compared to the energetic ionic component, unless it exceeds the mass removed with LIP by 2-3 orders. We found that removed mass partitioned about equally between LIP and PE for Sn and Pb, which means that the mass removed via PE indeed adds no momentum to the target.

The amount of mass removed during PE has been estimated by revisiting the analysis of mass removal rates obtained from the TOF study [8]. In order to accumulate measurable mass removal, determined by weighing the sample before and after ablation, we had to accumulate a large number of shots. In the course of this study, we introduced the "critical exposure η_{TOF}" as the number of shots providing steady-state TOF waveforms [8]. (Authors admit a poor choice of symbols, and hope that "η_{TOF}" will not be confused with the energy efficiency term). All measurements, including mass removal rates, were then performed by staying within the exposure limit established by η_{TOF}. When we turned to the study of PE, we found that PE has its own critical exposure. In other words, the number of consecutive laser shots leading to onset of PE on a given sample was finite, and we denoted this number by analogy with η_{TOF} as η_{PE}.

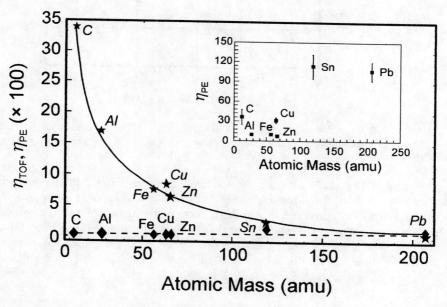

FIGURE 4. Critical exposures as functions of atomic mass. Stars and diamonds show η_{TOF} and η_{PE} data respectively. On the inset shows a close-up of data for η_{PE}.

Figure 4 presents both η_{TOF} and η_{PE} graphed as a function of atomic mass (the hyperbolic trend for η_{TOF} is explained in detail in [8]). This comparison shows that η_{PE} is significantly less than η_{TOF} in all tested materials except tin and lead, where $\eta_{PE} \approx \eta_{TOF}$. Thus, the measured mass removal rates for tin and lead are the ones most significantly affected by PE. The inset on Figure 4 shows how much η_{PE} for tin and lead exceeded η_{PE} of the rest of tested elements. Apparently, accumulated mass removals on other elements would have negligible fractions of mass removed by PE. This finding is supported further by the mass removal data shown in Figure 5.

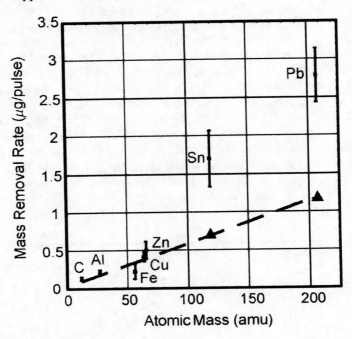

FIGURE 5. Mass removal rate as a function of atomic mass [8]. The dashed line represents a numerical fit to mass removal rates from elements where $\eta_{PE}/\eta_{TOF} \ll 1$, triangles denote extrapolated rates for tin and lead.

As seen in Figure 5, data for tin and lead diverges from the otherwise linear dependence for the mass removal rates presented as a function of atomic mass. Assuming that this deviation is due to PE, a linear extrapolation reveals the partition of removed mass between LIP and PE for tin and lead. As one can see from Figure 5, about ~60% of the total mass consumed per laser pulse for these two elements is removed inefficiently via PE. Hence, maximizing the efficiency of ALP will include choosing a propellant with $\eta_{PE} \ll \eta_{TOF}$, as indeed observed in Al, Fe, Cu, and Zn.

CONCLUSIONS

This paper summarizes first three years of research conducted by the UAH LP Group. The study is far from completion, but the first results are enough for us to believe that ALP will have its own respectable place in the Laser Propulsion field. The work reported in this review is being used to set system requirements for our upcoming flight demonstration tests.

ACKNOWLEDGMENTS

This work was greatly enhanced by discussions and assistance from all current and former members of the UAH LP Group: Wesley Swift, Jr., Dr. James Brasher, Ken Herren, Eric Broyles, Christian DelaCruz, Mary Breeden, Jun Lin, Casey Kemp, Jeremy Raper, Andrew Dollarhide, Carol Airhart, and John Outerbridge. Most of this work was performed in conjunction with Information Systems Laboratories, Inc., under NASA STTR Phase I (#NAS8-00185) and Phase II (#NAS8-02039) awards.

REFERENCES

1. Pakhomov, A.V. and Gregory, D.A., *AIAA J.* **38**, 725-727 (2000).
2. Kantrowitz, A., *Aeronautics and Astronautics* **9**, 40-42 (1972).
3. Mead, F.B., Jr., Myrabo, L.N., and Messitt, D.G., "Flight Experiments and Evolutionary Development of a Laser Propelled, Trans-Atmospheric Vehicle," in *High Power Laser Ablation I*, edited by Claude R. Phipps., SPIE Conference Proceedings 3343, Bellingham, WA: 1998, pp. 560-563.
4. Liukonen, R.A., "Laser Jet Propulsion," in *High Power Laser Ablation - 1998*, edited by Claude R. Phipps., SPIE Conference Proceedings 3343, Bellingham, WA: 1998, pp. 560-563.*Proc. SPIE* Vol. 3574, 1998, pp. 470-474.
5. Bohn, W.L., "Laser Lightcraft Performance," in *High Power Laser Ablation II*, edited by Claude R. Phipps. and Masayuki Niino, SPIE Conference Proceedings 3885, Bellingham, WA: 1999, pp. 48-53.
6. Sasoh, A., *Rev. Sci. Instrum.* **72**, 1893-1898 (2000).
7. Pakhomov, A.V., Roybal, A.J., and Duran, M.S., *Appl. Spect.* **52**, .979-986 (1999).
8. Pakhomov, A.V., Thompson, M.S. and Gregory, D.A, *AIAA J* **40**, 947-952 (2002).
9. De Young, R.J., and Situ, W., *Appl. Spect.* **48**, 1297-1306 (1994).
10. Weyl, G.M., "Physics of Laser-Induced Breakdown: An Update," in *Laser-Induced Plasmas and Applications*, edited by L.J. Radziemski and D.A. Cremers, New York: Marcel Dekker, 1989, pp.1-67.
11. Yabe, T., Nakagawa, R., Yamaguchi, M., Ohkubo, T., Aoki, K., Baasandash, C., Oozono, H., Oku, T., Taniguchi, K., Nakagawa, M., Sakata, M., Ogata, Y., and Inoue, G., "Simulation and Experiments on Laser Propulsion by Water Cannon Target", this volume.
12. Phipps, C.R., Seibert, D.B., II, Royse, R., King, G., Campbell, J.W., "Very High Coupling Coefficients at Low Laser Fluence with a Structured Target," in *High Power Laser Ablation III*, edited by Claude R. Phipps., SPIE Conference Proceedings 4065, Bellingham, WA: 2000, pp. 931-938.
13. Phipps, C.R., and Michaelis, M., *Laser and Particle Beams* **12**, 23-54 (1994).
14. Hettche, L.R., Tucker, T.R., Schriempf, J.T., Stegman, R.L., and Metz, S.A., *J. Appl. Phys.* **47**, 1414-1421 (1976).
15. Simons, G.A. and Pirri, A.N., *AIAA J.* **15**, 835-842 (1977).
16. Thompson, M.S., Herren, K.A., Lin, J., and Pakhomov, A.V., "Effects of Time Separation on Double-Pulsed Laser Ablation of Graphite", this volume.
17. Pirri, A.N., *Phys. Fluids* **16**, 1435-1440 (1973).
18. Pakhomov, A.V., Thompson, M.S., Swift, W., Jr., and Gregory, D.A., *AIAA Journal* **40**, 2305-2311 (2002).
19. Pirri, A.N., Monsler, M.J., and Nebolsine, P.E., *AIAA J.* **12**, 1254-1261 (1974).
20. Fajardo, M.E., "Velocity Selection of Fast Laser Ablated Metal Atoms by a Novel Nonmechanical Technique: Temporally and Spatially Specific Photoionization (TASSPI)", AFRL/PRS Final Report PL-TR-97-3051, Edwards AFB, CA, February 1998.
21. Pakhomov, A.V., Thompson, M.S. and Gregory, D.A, "Phase Explosion in Elementary Targets Exposed to Picosecond Laser Irradiation", submitted to *J. App. Phys*.
22. Pakhomov, A.V., Thompson, M.S. and Gregory, D.A, "Ablative Laser Propulsion Efficiency," presented at *33rd AIAA Plasmadynamics and Lasers Conference, 20 - 23 May 2002, Maui, Hawaii*, AIAA Paper #2002-2157
23. Iida, Y. and Yeung, E.S., *Appl. Spectrosc.* **48**, 945-950 (1994).

Effects of Time Separation on Double-Pulsed Laser Ablation of Graphite

M. Shane Thompson

Information Systems Laboratories, Inc.
Brownsboro, AL 35741-9455

Kenneth A. Herren

NSSTC
NASA Marshall Space Flight Center
Huntsville, AL 35812

Jun Lin and Andrew V. Pakhomov

Department of Physics
University of Alabama in Huntsville
Huntsville, AL 35899

Abstract. This work continues on previous investigations of elementary propellants for Ablative Laser Propulsion (ALP). The details on experimental methods used for alignment of a non-collinear pulse splitting apparatus are presented. Spatial and temporal coincidence of 100-ps wide pulses is demonstrated and the first data is reported on the pulse separation effects studied by means of time-of-flight (TOF) energy analyzer on graphite. The data includes ion velocity and number density, measured as functions of pulse separation. Possible models of observed phenomena are discussed.

INTRODUCTION

From the first formulation of Laser Propulsion (LP) by Arthur Kantrowitz in 1972 [1] to its latest achievements over the last decade, marked by successful test launches by Mead, Myrabo, et al. [2], LP has become the forerunner of future propulsion concepts. In a previous paper [3], Pakhomov and Gregory showed that perhaps the most efficient regime for LP is that in which direct ablation of a propellant dominates the momentum transfer; henceforth known as Ablative Laser Propulsion (ALP). The main advantages of ALP are discussed in detail in references [4-6], and of particular significance to this work, the effect of pulse width on energy transfer efficiency will be discussed in detail below.

CP664, *Beamed Energy Propulsion: First International Symposium on Beamed Energy Propulsion,*
edited by A. V. Pakhomov

The effect of plasma opacity on the energy transfer from laser pulse to target has been treated previously [4]. A plasma becomes purely reflective when its frequency exceeds the frequency of the incident light. When the critical plasma frequency, ω_{pc}, is reached, the electron density (n_{ec}) can be calculated from

$$n_{ec} = \frac{m_e \varepsilon_0 \omega_{pc}^2}{e^2},$$

(1)

where m_e is the electron mass, ϵ_0 is the permittivity of free space, and e is the charge of the electron. For the 532 nm irradiation used in this study, the corresponding critical electron density is then 10^{19} cm^{-3}. Assuming impact ionization as a leading source of electron density growth and neglecting the losses over first several hundred picoseconds, the following equation for electron density results [4]:

$$\frac{dn_e}{dt} = \nu n_e,$$

(2)

where ν is the ionization rate, taken as 6×10^{11} s^{-1} [7]. Solving Equation 2 for the critical time t_C, corresponding to the onset of the reflective regime, yields

$$t = \frac{\ln(n_{ec})}{\nu}.$$

(3)

Thus, for 532 nm laser wavelength, $t_C \approx 100$ ps. This critical time can be longer if two-photon ionization [Ref. 7, $\nu = 5 \times 10^{10}$ s^{-1}], and electron losses are taken into account.

As follows from Equation 3, a sequence of two laser pulses, temporally separated by intervals comparable to t_C, can be used for determination of the onset of the reflective regime in a laser-induced plasma. Preliminary results of such experiments will be presented in this paper. Ion velocity and number density are determined with an energy analyzer [4] and analyzed as functions of pulse separation (delay time); these serve as indicators for determination of t_C and, consequently, ionization rate ν.

EXPERIMENTAL TECHNIQUE

The experimental setup used in this paper is shown in Figure 1. The laser system (1) is comprised of a mode-locked Nd:YAG (Coherent Antares) used to seed a regenerative amplifier (Continuum RGA60). Output from the amplifier yields pulses approximately 100 picoseconds in duration at 532 nm with pulse energies at ~30 mJ. These pulses can be delivered in either repetitively pulsed (variable from 6-10 Hz) or single-pulse mode. The pulses are first expanded by the telescope (2), and then split into two parts by the custom-made cube mirror (3), made by Melles-Griot. The pulse splitting apparatus consists of a fixed and a movable arm, folded via retoreflectors (4a, b), it is used for achieving pulse separations over the range from 0.1 - 6.5 ns. For stability over large displacements of the movable arm (> 3 cm), this apparatus is mounted on an optical rail. For fine adjustment to the delay, the variable arm is attached to a stepping motor and controller (5) (Klinger UE30 and MC-4 controller) that provides 0.1 μm resolved movements.

After the pulses complete the delay stage, they are redirected by the cube onto the initial path towards the target. The pulses are focused by a 30 cm focal length achromatic doublet (6), and enter the ablation chamber through an optical window (7), which is attached at Brewster's angle for minimization of reflection losses. The pulses then ablate the portion of a target (8) placed at the focal plane of the lens. In this preliminary study we used graphite targets, the sources of highest ion velocities observed to date in our experiments [4]. Thus, the rest of the experiment is similar to the one described in Ref. 4. Carbon ions enter the time-of-flight tube and are sampled by the two copper plates in the energy analyzer (9). These plates are separated by 20 cm. A vacuum system (10), comprised of a roughing and diffusion pumps, maintains a ~3 mTorr pressure in the chamber. Voltages induced over the energy analyzer plates are read and stored by a digitizing storage oscilloscope (11). The oscilloscope is triggered by a photodiode (12) placed in the reflection path from the entry window (7). The waveforms from the oscilloscope are then transferred to a personal computer (13), where ion velocities and number densities are derived from the waveforms (see Ref. 4 for details).

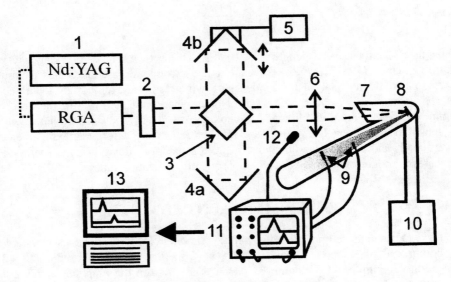

FIGURE 1. Experimental apparatus.

In order to achieve reliable and repeatable pulse separations, a temporal and a spatial calibration is performed. For calibration in the time domain, a two-photon photoconductivity (TPPC) autocorrelation technique is used [8, 9], implemented by inserting two optical wedges before the lens and crossing the diverted beams onto the surface of a SiC photodiode (Boston Electronics JIC1 EI17). The performance of such TPPC techniques have been recently reported to be comparable to second harmonic generation (SHG) methods, with the advantages of easier alignment and replacement of the expensive and usually prone to optical damage SHG crystal and photomultiplier tube by a single photodiode [8, 9].

For a given laser wavelength, the selected photodiode material should satisfy the following condition:

$$\lambda_g \leq \lambda \leq 2\lambda_g, \tag{4}$$

where λ_g is the wavelength corresponding to the band gap. As SiC has a band gap of 3.1 eV, it serves as an excellent material for autocorrelation of light in the visible part of the spectrum. The non-linear response of the photodiode is tested by variation of pulse intensity using neutral density filters. The response is quadratic, as expected for TPPC [8,9]. With this established, autocorrelation measurements of the beam are performed.

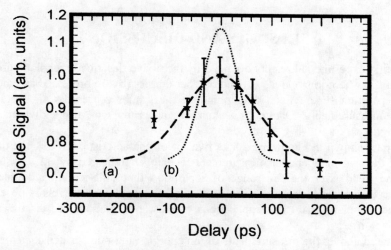

FIGURE 2. Autocorrelation trace obtained from SiC photodiode. Fit to the autocorrelation data is shown as (a), and (b) is the corresponding actual pulsewidth.

An autocorrelation trace obtained from the SiC photodiode is shown in Figure 2. The error in the pulse separation dimension is 1 ps, which is less than the width of the data points on the graph. It should be noted that the measured autocorrelation differs from the actual pulse width. For a Gaussian of full width 2σ, the autocorrelation $A(t_d)$ is:

$$A(t_d) = \int_{-\infty}^{\infty} \exp[-\frac{t^2}{2\sigma^2}]\exp[-\frac{(t-t_d)^2}{2\sigma^2}]dt = \sigma\sqrt{\pi}\exp[-\frac{t_d^2}{4\sigma^2}]. \tag{5}$$

Therefore, the profile obtained from the autocorrelation, denoted as (a) in Figure 2, is wider than the original beam by a factor of $\sqrt{2}$. The pulsewidth derived from the autocorrelation profile, indicated by (b) in Figure 2, is 103 ± 13 ps, in excellent agreement with the manufacturer specified value of 100 ps. The autocorrelation measurements yield not only the pulsewidth, but also the zero-delay point for the system. This point is used as the reference from which all delays are measured.

In addition to the temporal (or longitudinal) calibration, which sets the zero in the optical path difference (OPD), the non-collinear laser pulses must also be focused at the same point on the target. This "transverse" calibration, which insures spatial coincidence over the entire range of pulse separations, is achieved by making fine adjustments to the focusing lens.. The lens in the experimental setup is attached to a micrometer stage, which allowed for such fine movements along the propagation axis of the beam. Five laser shots were fired for each measurement from the fixed arm (variable arm blocked), the variable arm (fixed arm blocked), and the combination of the two arms. Using a magnifier, the spatial coincidence of the two beams is determined at the point where craters from each of the arms are equal to the crater from the combination.

RESULTS AND DISCUSSION

The first experimental results obtained by the above described technique are shown on Figure 3, where ion velocity (a) and number density (b) are presented as functions of pulse separation time. Each data point on the graphs represents an average of 5 separate measurements. Data points taken for single pulses are also shown on Figure 3 for reference.

The data shown in Figure 3 exhibits two major characteristics. First, the data from each of the two independent measurement techniques indicated by (a) and (b) shows the same trend over the full range of separation times. Second, a double-peaked structure that exceeds the signal from the single arm reference levels is observed for separations under 1 ns. The implications of these observations are discussed in this respective order.

As stated above, the measurements of ion energies and number densities are derived as independent observables. Ion yields are obtained by integration of the waveform recorded from the first plate of the TOF energy analyzer. Velocity, on the other hand, is found using the time separation of the peaks from both waveforms. Assuming that ion kinetic energy is directly proportional to electron thermal energy, the identity in behavior of electron density and kinetic energy can be directly deduced from the definition of the Debye length λ_D [10] as:

$$n_e = \left(4\pi e^2 \lambda_D^2\right)^{-1}\left(kT_e\right), \tag{6}$$

where T_e is the electron temperature and k is Boltzmann constant. Assuming the first term in parentheses in equation (6) is constant, the direct proportionality between the second term and n_e explains the similarity of Figures 3a and b.

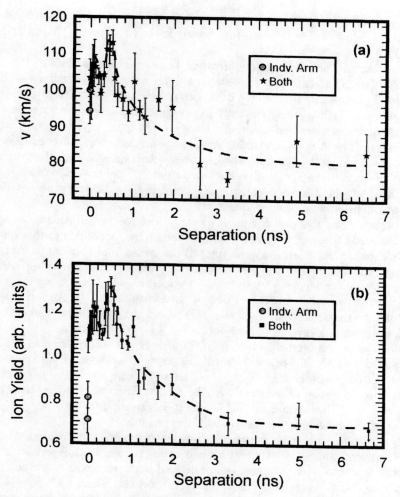

FIGURE 3. Ion velocity (a) and number density (b) as functions of time separation of the pulses. As indicated in the legend, the circles represent data from each individual arm, stars are from both arms combined. The dashed line shows the trend in the data.

The double-peaked structure of data profiles on Figure 3 is currently a subject to the study in progress. We will limit our discussion to a simple estimation. For the first minimum at the Figure 3, the Debye length can be found from Equation (6) under the assumption that $n_e \sim 0.1 n_{ec}$ (*i.e.* $\sim 10^{-19}$ cm^{-3}) and the electron energy is ~ 1.6 keV [4]. This would yield $\lambda_D \sim 100$ nm. Therefore, $\lambda_D k \approx 1.0$, where k is the wave number of incident EM wave. This meets the criterion for appreciable effects from Landau damping, given as $\lambda_D k \geq 0.4$ [10]. Thus, one may anticipate Landau damping as an essential mechanism of energy exchange between incident EM-wave and our predominantly collisionless plasma [4], which leads to observed data oscillations (Figure 3).

This assumption finds further confirmation from general solutions of the Vlasov equation [10]. The numerical simulations of the normalized distribution function performed by Dr. James Miller [11] demonstrated striking similarity to data profiles from Figure 3. This work is currently in progress, and ionization rates and other critical characteristics derived from this simulation will be reported elsewhere. In comparison to the Vlasov equation, which couples the continuity equation with Maxwell's equations, the approach used to obtain the expression for the critical time for onset of the reflective plasma regime in equation (2) can be viewed as a very first approximation. Still, it is worth noting that the first peak in the data profile is observed exactly at 100 ps, as predicted by equation (2).

The observed "double peaked" trend over the first ~600 ps of separation deserves some additional comment. It is important to mention that according to the solution to Vlasov's equation, the oscillatory character of the curves is not limited to short times. The long time tail in the simulation, while decreasing, still shows an oscillatory behavior. A similar tendency is seen in our experimental points; however, with the exception of the first two peaks, we do not have enough data to claim that we have clearly observed such an oscillation. For this reason the "data trend" line in Figure 3 goes down monotonically at $t_d > 1$ ns.

Finally, a comment on the data in Figure 3a for separations greater than 2 ns, where the measured velocity falls below the reference points from the single arm signals. It is believed that this result is due to general reduction in pulse energy, which could occur in continuous operation of the laser over the 5 hour time span required to collect the data. At least, this would be the simplest explanation to this feature. More experiments are underway in order to clarify this issue.

Ongoing studies are being performed to fully characterize all features of the data profiles, presented at Figure 3 and discussed here. For the next experimental step in these studies, electron temperatures will be derived from emission spectra, and they will be analyzed as a function of time separation between laser pulses. Numerical simulations based on solutions to Vlasov's equation will also support the experiments.

Although the study presented in this paper has rather general character, we would like to make a final remark on its application to ablative laser propulsion (ALP). ALP was introduced as a technique employing "short" (100 ps and less) laser pulses, with the justification for such a pulsewidth constraint built around equations (1-3) [3]. Figure 3 shows that the utilization of doubled pulses with separations under 1.0 ns can provide essential gains in both ion velocity and number density, as compared to the those quantities obtained from a single pulse from which the pair is produced. Conversion of these plasma characteristics into terms of propulsion means gain in specific impulse and coupling coefficient, *i.e.* better efficiency.

CONCLUSIONS

A method of two-pulse sequencing for studying plasma characteristics and first data obtained from the system is presented. This initial data verifies the calculation of the critical time for establishment of a reflective plasma at 100 ps under the given irradiation conditions. Two major observations are reported: (1) the independent observables derived from the TOF, ion velocity and number density, exhibit the same general trend over pulse separations ranging from 0.1-6.5 ns, and (2) an oscillatory behavior in measured parameters is seen over the first ~0.6 ps of pulse separation. The initial analysis suggests Landau damping as a major factor in the energy exchange between the laser pulses and the plasma at these conditions.

The graphite data presented here indicates the optimum pulse separation for maximum ion velocity and number density, proportional to specific impulse and coupling coefficient, is achieved at ~600 ps. It is projected that such a pulse separation will provide the most efficient regime for ALP.

ACKNOWLEDGMENTS

The authors are grateful to Wesley Swift Jr., who suggested the scheme of beam splitting / delaying apparatus, implemented in this work. We also would like to acknowledge the assistance and advice of the UAH Laser Propulsion Group members: Don A. Gregory, James D. Brasher, John Outerbridge, Carol Airhart, Jeremy Raper, and Andrew Dollarhide. This work was performed under NASA STTR Grant #NAS8-00185.

REFERENCES

1. Kantrowitz, A., *Aeronautics and Astronautics* **9**, 40-42 (1972).
2. Mead, F.B., Jr., Myrabo, L.N., and Messitt, D.G., *SPIE* **3343**, 560-563 (1998).
3. Pakhomov, A.V. and Gregory, D.A., *AIAA Journal* **38**, 725-727 (2000).
4. Pakhomov, A.V., Thompson, M.S. and Gregory, D.A., *AIAA Journal* **40**, 947-952 (2002).
5. Pakhomov, A.V., Thompson, M.S., Swift, W., Jr., and Gregory, D.A., *AIAA Journal* **40**, 2305-2311 (2002).
6. Pakhomov, A.V., Thompson, M.S., and Gregory, D.A., "Ablative Laser Propulsion Efficiency", AIAA Paper #2002-2157, May, 2002.
7. Weyl, G.M., "Physics of Laser-Induced Breakdown: An Update", in *Laser-Induced Plasmas and Applications,* edited by L.J. Radziemski and D.A. Cremers, New York: Marcel Dekker, 1989, pp. 1-67.
8. Feurer, T., Glass, A., and Sauerbrey, R., *Appl. Phys. B* **65**, 295-297 (1997).
9. Laughton, F.R., Marsh, J.H., Barrow, D.A., and Portnoi, E.L., *IEEE J. Quan. Elec.* **30**, 838-844 (1994).
10. Kruer, W., *The Physics of Laser-Plasma Interactions*, Redwood City, CA: Addison-Wesley, 1988, pp 5-15.
11. Miller, J.A. and Vinas, A.F., submitted to *Am. J. Phys.*

Characterization of liquid propellant for improved LOTV mission

Shigeaki Uchida, Masafumi Bato

Institute for Laser Technology and
Osaka University, Osaka Japan

Abstract   Laser thruster can improve the efficiency of laser orbital transfer vehicle (LOTV) in terms of energy and propellant mass. It is shown that by controlling specific energy according to the thruster operating condition, improvement of more than 60% is possible. Experimental investigation shows that those conditions necessary for the improvement is achievable by using liquid propellant.

INTRODUCTION

When used as an on-orbit thruster, laser propulsion system has many advantages over conventional means.

The first advantage is that it can separate energy source and propellant mass source. This means that energy efficiency and propellant mass usage can be better optimized compared with the conventional techniques. Since energy to laser propulsion system can be supplied from an external system, the optimization can be done by controlling the power input and propellant mass supply **separately**.

The second advantage is that it can cover a **wide** range of impulse parameter. One of the parameters that determines thruster performance is I_{sp}, specific impulse. It is an indicator of the thruster efficiency of propellant mass usage. The larger the I_{sp}, the higher the fuel mass efficiency, the system uses less propellant to accelerate a certain mass of payload. I_{sp} is directly proportional to the average exhaust velocity of propellant and thus the specific laser energy, Q^* absorbed by unit mass of propellant.

The importance of controlling I_{sp} can be seen by showing the optimum I_{sp} for energy consumption for specific velocity increment. Figure 1 shows an I_{sp} dependence of required energy to bring unit mass payload to the specific velocity increment. To achieve the optimization, I_{sp} have to be controlled over many orders of magnitude. A laser thruster can work with I_{sp} of a few seconds to a few thousands of seconds. This wide range can provide precise optimization of the thruster system to each operation requirement. For example, an orbit maintenance can be done with a velocity change of 100 m/s while orbit transfer operation requires a velocity change of more than 6 km/s. For the orbit maintenance operation, energy optimization can be done with I_{sp} of as low as a few tens seconds while the order of a few hundred of seconds is required for the orbit transfer operation. It is desirable that a single thruster

CP664, Beamed Energy Propulsion: First International Symposium on Beamed Energy Propulsion,
edited by A. V. Pakhomov
© 2003 American Institute of Physics 0-7354-0126-8/03/$20.00

system can provide these multiple parameters of propulsion to reduce the complexity and weight of the system and thus the cost of transportation in space.

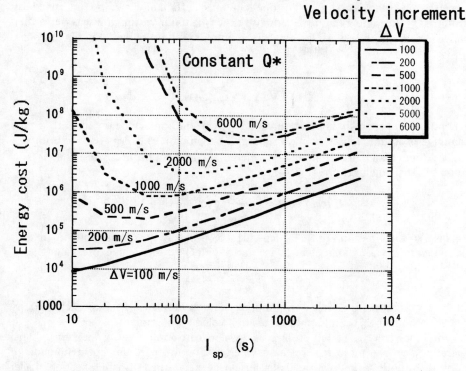

Figure 1 Specific impulse dependence of energy cost with velocity increment as a parameter. Parameters are from 100 m/s (bottom) to 6000 m/s (top).

In order to control the thruster parameter in such wide range of I_{sp}, Q* must be varied significantly. Conventional surface ablation cannot be used for the purpose since the amount of ablating material is predetermined by the skin depth of light into material and the intensity threshold for plasma generation. These physical constraints keep the amount of material absorbing laser light and energy flux impinging to the material very high. The typical corresponding I_{sp} is usually of the order of hundreds of seconds.

Simple method can be applied to control the Q* of laser irradiated material by impinging laser energy into layered target or propellant. Solid layered targets have been examined to expand the achievable parameter of laser ablating material [1, 2]. Liquid layered targets [3] gives better control over Q* and usability. Here viscous liquid propellant is proposed that is expected to give more compatibility in space environment

Once wide range of laser propulsion parameter has been obtained, the system can be used in various operation modes from orbital keeping to orbital transfer, where variable Q* might lead optimization of mission in terms of energy and propellant mass consumption as well as time duration.

215

In this paper, first energy and propellant mass consumption during orbital transfer operation from low earth orbit, LEO to Geo-Stationary Orbit, GEO was analyzed. It has been shown that energy and propellant mass efficiency can be improved by controlling Q* during operation. Secondly, experimental investigation was performed to show the applicability of laser thruster with liquid propellant to this purpose.

LOTV ANALYSIS

To optimize a propulsion operation, an analytical solution can be obtained if constant Q* is assumed. For example, energy cost, amount of energy used to bring a unit mass of payload to a certain velocity increment, ΔV can be calculated as a function of Q* using an expression[4],

$$Energy\,cos\,t = \frac{g^2 I_{sp}^2}{2\eta_{abs}}\left[\exp\left(\frac{\Delta V}{g I_{sp}}\right) - 1\right] \tag{1}$$

η_{abs} is laser absorption. In Fig. 1, I_{sp} dependence of the energy cost is plotted with a velocity increment as a parameter. I_{sp} is related with Q* through an expression

$$I_{sp} = \frac{\sqrt{2Q*}}{g}. \tag{2}$$

It can be seen that there is a cost minimum condition in terms of I_{sp} for each velocity increment. Smaller I_{sp} is preferable for a small velocity increment.

As mentioned above, laser propulsion system can control its propellant exhaust temperature. In the more general case where Q* is not constant but a function of velocity or altitude, numerical analysis has to be performed. In this section, Laser Orbital Transfer Vehicle [5] is analyzed numerically in order to calculate the energy efficiency and propellant mass usage. The model of the calculation is depicted in Fig. 2. For each laser irradiation or momentum bit, velocity change and mass of propellant exhausted are calculated. It is assumed that laser pulses carry constant amount of energy throughout the operation. Q* is given by a predetermined function and used to calculate the $\Delta m = \eta_{abs} E_{laser}/\Delta m$, amount of exhausted propellant mass at each laser irradiation. Laser absorption, η_{abs} is assumed to be constant. The velocity and altitude changes due to the impulse bit are then calculated using equation of motion and conservation of mechanical energy. The kinetic energy obtained from the impulse bit is redistributed into kinetic energy and potential energy at the new altitude. This completes the single cycle of calculation for an impulse bit.

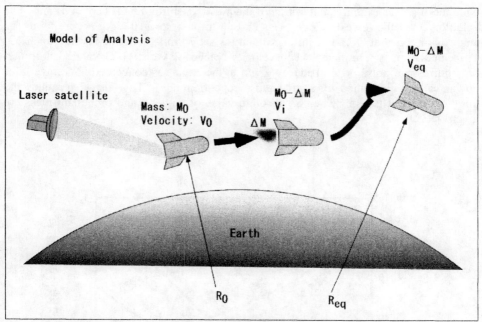

Figure 2 Model of calculation used for the LOTV performance analysis

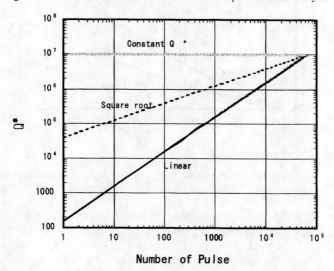

Figure 3 Three function types for Q* used in the numerical analysis

The choice of function type for Q* was made according to the previous I_{sp} optimization. For example, when velocity is small, a smaller Q* seem preferable. One can chose monotonically increasing Q* as velocity increases. Several function types were chosen for Q* variation that are shown in Fig. 3. Monotonically increasing and decreasing functions that vary like linearly and square root of the pulse number.

Number of pulses is equivalent to time if the pulse repetition rate is kept constant. The value of Q* at the lowest velocity was chosen to be below 10^4 J/kg according to the analysis of constant Q*. The initial velocity is set 7 km/s, the typical value for LEO. The calculation is terminated when velocity reaches 3 km/s, GEO value. Note that only number of pulse or total energy given to the engine is concerned. Average laser power can be calculated by setting pulse repetition rate. This also determines the duration of orbit transfer when an absolute value of payload and propellant mass are given.

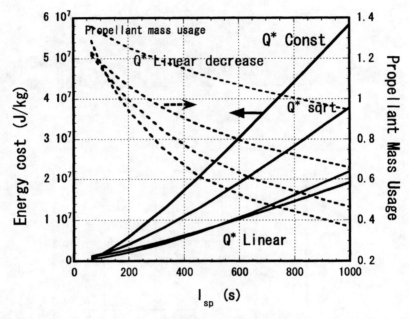

Figure 4 Numerically evaluated energy cost and propellant mass usage as a function of specific impulse. The solid lines correspond to energy cost. Broken lines correspond to propellant mass usage.

Two figure of merits, energy cost and propellant mass usage, have been calculated and plotted in Fig. 4. Since energy and mass are most precious resources in space, these two indexes have been chosen as most important factors that determine the performance of the propulsion system. They are plotted as a function of the maximum I_{st} value during the flight. For example, in the case of Q* being a linear function of time, the last I_{sp} value is the representative value. The solid lines and the broken lines correspond to the energy cost and propellant mass usage respectively.

Among the curves for the energy efficiency, the case of constant Q* exhibits the largest values. When Q* is varied, the cost of energy improves. The degree of reduction is largest when high I_{sp} values are taken. At the highest I_{sp} used in the analysis, the reduction is at least 30% in the case of square root function and as high as

65% when Q* changes linearly. Q* of decreasing function of time share the lowest energy cost with linearly increasing Q*.

As is expected, the propellant mass usage exhibits opposite trend as the energy cost. Case of linearly decreasing Q* shows the largest mass usage followed by linearly increasing Q* and square root cases. Constant Q* case is the lowest. It is worth noting that the linearly increasing Q* shows better mass usage than linearly decreasing Q* case even though they shares the similar lowest energy cost. It is rather surprising result since energy cost and the mass usage are inversely proportional to each other.

This numerical calculation also shows the total amount of laser energy is of the order of tens of MJ per kg of payload to GEO.

EXPERIMENTS

Experimental Setup

Preliminary experiments have been conducted to show the feasibility of the use of liquid propellant to give the wide range of Q*. Machining oil was chosen as the liquid propellant that seems to have proper viscosity for repetitive feeding to the engine. In the present experiment however, the propellant was supplied in a static condition, that is no feeding mechanism was used. The experimental setup is shown in Fig. 5. The propellant is placed on a small dish made of metal washer and glass plate. This is horizontally installed into a vacuum chamber with a mechanical rotary pump connected. The thickness of the oil is varied from 0.1 mm to 1.4 mm.

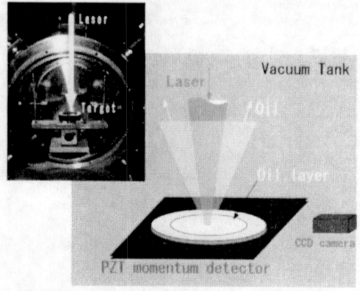

Figure 5 Experimental setup. Insert is the photo of the target chamber.

A Q-switched Nd:YAG laser with a maximum pulse energy of 500 mJ in 10 ns duration irradiates the target through the chamber top window. A 300-mm focal-length lens gives a focusing f number of 30. The target is placed slightly before the focus so that spot size on the target is of the order of a few mm.

Two diagnostics are used, momentum measurement and visual observation of propellant expansion. A piezoelectric plate was placed under the propellant dish. It measures instant force impinging on the target resulted from the propellant expansion. The output waveform is in voltage however, one can integrate it to obtain total impact given to the target during the reaction. Known mechanical impact is used to calibrate the momentum measurement. The characteristic response time of the device is about 300 µs.

A high speed CCD camera is used to obtain visual image of expanding propellant. The clock speed of reading out is accelerated so that the frame length is 2 ms. The obtained image is first stored into a memory bank and then transferred to a computer through SCSI interface. This diagnostic gives information of special and temporal profile of the propellant expansion.

In addition to the momentum and imaging diagnostics, the mass of blown off propellant is measured by measuring the mass difference between before and after the laser shot.

Experimental results

From the momentum measurements, C_m was calculated. Laser energy is monitored for every shot although absorption is not measured this time so that the calculated C_m's are underestimatd. Any quantities derived from laser energy such as specific energy are therefore, a relative quantity.

C_m is plotted in Fig. 6 as a function of incident laser intensity. Laser intensity is varied by changing laser pulse energy and the spot size is kept constant. Cases of oil thickness of 1.4 mm and 0.9 mm are shown. As a comparison, momenta from no oil layer targets are also plotted. When the laser intensity is below $3*10^8$ W/cm^2, there is a distinguishing difference in C_m depending on the thickness of the oil.

The mass measurement of the propellant gives that the order of propellant mass blown off after the laser irradiation is ten mg to a hundred mg. From this measurement, specific energy is calculated and plotted in Fig. 7. Q* also shows strong dependence on oil thickness at lower side of laser intensity. The difference is nearly an order of magnitude.

Figures 7 show the images of propellant expansion from two different oil thickness. When oil is 0.1 mm thick, expansion lasts about 4 ms (two frames) while 1.4 thick case is much slower and lasts more than 20 ms. The insertion is the image of propellant 26 ms after the laser irradiation.

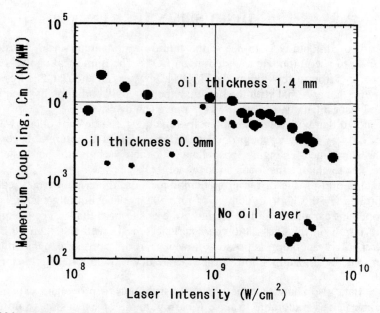

Figure 6 Momentum coupling coefficient as a function of incident laser intensity with the propellant thickness as a parameter.

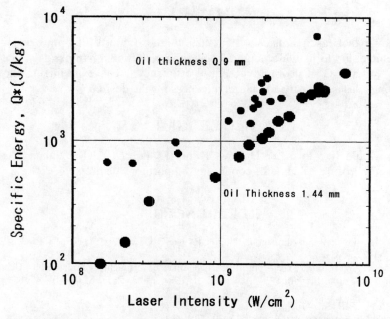

Figure 7 Specific energy as a function of incident laser intensity. Thickness of propellant is a parameter.

DISCUSSION

The purpose of this study is to see if the liquid propellant for laser thruster has sufficient ability to accommodate a wide range of Q*. The numerical analysis shows that a range of Q* from a few hundreds J/kg to a few tens of thousands J/kg will be required. Figure 6 shows that with the present experimental conditions of thickness of propellant and laser intensity, the required Q* range can be covered.

It is also shown that with low laser intensity region Q* can be controlled by a order of magnitude only by changing the propellant thickness. This is a preferable condition for laser thruster since changing propellant thickness seems to be easier than controlling optics to change the focus on to the target.

From the temporal profile of the propellant expansion, there is a certain repetitive frequency limited by the hydrodynamics of propellant. The limiting factors are the speed of propellant expansion and re-feeding time of the propellant. For example, in the case of 0.1 mm oil thickness, the characteristic time is well less than 10 ms and thus 10 Hz repetitive operation is possible however, in 0.9 mm the expansion time is much longer and the laser repetition rate has to be determined with this factor taken into account.

The image data also show that the angular distribution of propellant expansion is about 30 degree. It is preferable to have a narrow angle of expansion to utilize the propellant momentum. It is suggested to have certain kind of focusing structure on the laser irradiated surface.

SUMMARY

It is shown that laser thruster system has unique property of optimizing orbital transfer vehicle due to its wide range of operating conditions. Numerical calculation shows the variable Q* can improve energy efficiency. The experiments show the combination of laser intensity and target thickness for the desired Q*.

ACKNOWLEDGMENTS

Authors would like to thank Drs. M. Niino, K. Fujita and M. Nakano for their discussion and Prof. K. Tanaka for continuing support for this project.

REFERENCES

1. Laser-ablation-powered mini-thruster, C. Phipps, J. Luke,G. G. McDuff, T. Lippert, Proc. SPIE Vol. 4760, 833-842, (2002).
2. Characteristics of volume expansion of laser plasma for efficient propulsion, S. Uchida, K. Hashimoto, K. Fujita, M. Niino, T. Ashizuka, N. Kawashima, Proc. SPIE Vol. 4760, 810-820, (2002).
3. T. Yabe, Proc. SPIE Vol. 4760, (2002).
4. W. E. Mockel, J. Spacecraft and Rockets **12**, 700-701, (1975).
5. Laser Orbital Transfer Vehicle Feasibility Study Report, S. Nakai ed. Japan Aerospace Science and Technology, 2001.

Laser Plasma Microthruster Performance Evaluation

James R. Luke[*] and Claude R. Phipps[†]

[*]Institute for Engineering Research and Applications, New Mexico Institute of Mining and Technology, 901 University Blvd. SE, Albuquerque, NM 87106 USA

[†]Photonic Associates, 200A Ojo de la Vaca, Santa Fe, NM 87505 USA

Abstract. The micro laser plasma thruster (μLPT) is a sub-kilogram thruster that is capable of meeting the Air Force requirements for the Attitude Control System on a 100-kg class small satellite. The μLPT uses one or more 4W diode lasers to ablate a solid fuel, producing a jet of hot gas or plasma which creates thrust with a high thrust/power ratio. A pre-prototype continuous thrust experiment has been constructed and tested. The continuous thrust experiment uses a 505 mm long continuous loop fuel tape, which consists of a black laser-absorbing fuel material on a transparent plastic substrate. When the laser is operated continuously, the exhaust plume and thrust vector are steered in the direction of the tape motion. Thrust steering can be avoided by pulsing the laser. A torsion pendulum thrust stand has been constructed and calibrated. Many fuel materials and substrates have been tested. Best performance from a non-energetic fuel material was obtained with black polyvinyl chloride (PVC), which produced an average of 70 μN thrust and coupling coefficient (C_m) of 190 μN/W. A proprietary energetic material was also tested, in which the laser initiates a non-propagating detonation. This material produced 500 μN of thrust.

INTRODUCTION

The micro laser plasma thruster (μLPT) is a sub-kilogram thruster that is capable of meeting the Air Force requirements for the Attitude Control System on a 100-kg class small satellite. These requirements are 75 μN per axis with three thrust axes, and 320 N·s lifetime impulse. The μLPT uses one or more 4W diode lasers to ablate a solid fuel, producing a jet of hot gas or plasma which creates thrust [1]. The advantages of a laser-generated plasma are high thrust/power ratio, and a plasma heating mechanism which is physically independent of the ablation process itself. There are two modes of operation, referred to as "reflection" or R-mode, and "transmission" or T-mode. In the R-mode, the laser impinges on an opaque surface and creates a jet of ablated material. The disadvantage of this mode is that the ablated material can cloud the optics. A special grazing incidence optical system has been

CP664, *Beamed Energy Propulsion: First International Symposium on Beamed Energy Propulsion*, edited by A. V. Pakhomov designed and constructed, which will be discussed later in this paper. In the second operation mode, called T-mode, shown schematically in Figure 1, the fuel consists of two layers, with a transparent substrate and an absorbing coating. The laser light passes through the transparent material and is absorbed in the coating, which ablates and creates a jet. The performance is lower in T-mode, but since the optics are protected by the transparent substrate, all continuous thrust experiments to date have been done in T-mode. A pre-prototype continuous thrust experiment has been constructed and tested. The continuous thrust experiment uses a 505 mm long continuous loop fuel tape, which consists of a black laser-absorbing fuel material on a transparent plastic substrate.

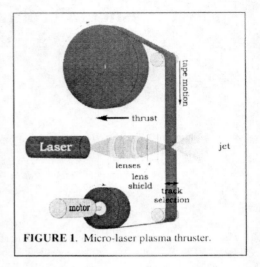

FIGURE 1. Micro-laser plasma thruster.

TESTING AND CALIBRATION OF THRUST STAND

The thrust stand consists of a torsion pendulum suspended on a steel wire. The thruster and electronics are mounted on the pendulum crossbar as shown in Figure 2. Power is supplied to the thruster through contacts in liquid mercury. The response of the pendulum was calibrated by applying a magnetic torque to the pendulum. This was accomplished by attaching a small coil of wire to the pendulum, and a larger coil to the support frame. The number of turns of wire in each coil is known. By applying a current to the coils, a known torque can be applied to the pendulum and its angular deflection measured. The pendulum response is 5 mN/rad at end of 15.5 cm lever arm. The pendulum is sensitive enough to resolve a 20 μN force. The oscillation period of the torsion pendulum is approximately 40 s. The mercury electrical contacts to the thruster are not entirely frictionless. Over time, a film of oxidation forms on the mercury, which can impede the motion of the pendulum. This situation can be remedied by causing the pendulum to oscillate, which keeps the film broken up. In addition to the torsion pendulum, thrust

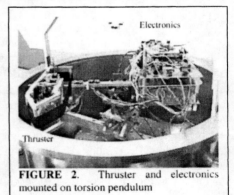

FIGURE 2. Thruster and electronics mounted on torsion pendulum

can be measured using a small flag pendulum placed so that it intercepts the exhaust plume. Although the flag pendulum misses some of the exhaust jet, and the fraction of the plume that sticks to the flag is unknown, it gives results which agree reasonably well with the torsion pendulum measurements. It has the advantage of much faster response, and therefore much higher data rate, than the torsion pendulum.

CONTINUOUS THRUST EXPERIMENTS

We have performed many continuous thrust experiments with the preprototype thruster. Several long (800 – 900 s) runs have been performed, and numerous shorter ones. Operating the tape drive and the track selection motors simultaneously, so that the laser path is helical, has worked well. The continuous thrust experiments have revealed some unexpected results, such as problems with the tape joint, outgasing of the fuel and steering of the plasma jet plume. During continuous operation, the exhaust plume is not perpendicular to the tape surface. It becomes deflected in the direction of the tape motion. As the fuel material is ablated in a continuous track, the end of the track forms a shape that steers the plasma jet. Depending on the conditions, the plume is steered by as much as $60°$ in the direction of tape motion. In order to avoid this plume steering, it is necessary to operate the laser in pulsed mode, rather than CW. When operated with short (2 ms) pulses, the plasma jet is well confined and perpendicular to the tape surface. By varying the pulse duration from 2 ms to 10 ms, it is possible to steer the exhaust plume by 50 to 60 degrees. This is a means of achieving thrust vector control instantly.

In order to restore the average laser power to the desired 2 to 4 W at a duty factor of 25%, it was necessary to increase the peak power. The lasers can be driven at twice their rated CW power of 2.5 W when operated in pulse mode. In order to obtain sufficient peak power, the output of four lasers was combined. Four JDSU 6380-A fiber-coupled lasers were obtained, and the four fibers were brought together in a bundle. The end of the bundle was successfully imaged on the fuel tape. The combination of the four lasers, electrically connected in parallel, required development of a new pulsed power supply. The new supply is capable of delivering 1 ms or longer pulses of up to 60 A, and the pulse waveforms are much cleaner than the previous power supply. Remotely controlled motorized focusing capability was recently added to the preprototype thruster. The JDSU lasers have proven to be very rugged. They have been inadvertently operated in air at 6 W CW without damage.

SUBSTRATE MATERIALS

Hundreds of substrate/coating material combinations have been tested over the course of this project. Substrate materials included Kapton polyimide, polyethylene terepthalate (PET), and cellulose acetate. Although acetate is somewhat hygroscopic and less strong than the others, it has a higher optical damage threshold. When the optical damage threshold is exceeded, or the tape burns through due to a tape drive malfunction or other reason, some of the plastic material is splattered back onto the optics. The foreign material on the optics absorbs the high intensity laser light, and destroys the lens. A thin glass shield was added to protect the lenses. The prototype thruster will employ a tape burnthrough sensor to avoid this problem.

CONTINUOUS LOOP FUEL TAPE

The preprototype thruster design uses a 505 mm long continuous loop fuel tape. Forming a reliable tape joint was somewhat difficult. Mylar adhesive tape makes a strong, flexible, long-lasting joint, but the mylar and/or the adhesive absorb the laser light and scatter material back onto the lens. A glued lap joint does not have that problem, but the thickness and stiffness of the joint area can cause trouble with the tape drive mechanism. Also, flexing of the tape material next to the stiff joint caused the tapes to fail in a short time. The most reliable tape joints are made using thin strips of mylar adhesive tape only at the edges of the fuel tape. The prototype thruster will use a reel-to-reel tape and so will not require joints to be made.

EVALUATION OF FUEL MATERIALS

One of the best passive fuel materials is black polyvinyl chloride (PVC). It has performed well in both static and dynamic tests. However, during the continuous thrust experiments, it was found to outgas significantly. As the PVC-coated tape flexes around the rollers of the thruster, jets of material come out which impart thrust that is comparable to the magnitude of the thrust from the plasma jet. The outgasing material is probably the plasticisers that keep the PVC flexible. Other black paints do not outgas, but they become brittle and crack. After aging for about one month, the PVC fuel is more stable, but still flexible enough to use. The coupling coefficient (C_m) measured for this material is about 190 μN/W has been measured, and the thrust is approximately 70 μN. In static tests, I_{sp} was 1800 s, and in dynamic tests it is 220 s. The lifetime impulse of the thruster using this fuel will be about 5 N·s/g of fuel, or 0.8 N·s for the 505 mm continuous loop fuel tapes used by the preprototype thruster. We have experimented with a non-plasticised PVC formulation with some success.

Aluminized kapton was tested as a fuel material. This material is stable and does not outgas in vacuum. Although the laser intensity is not high enough to ablate aluminum directly, reflection of the laser energy from the aluminum surface back into the kapton (in the T-mode) increases the intensity sufficiently to ablate the kapton substrate and blow the aluminum off the surface. Using this material, 26 μN of thrust was measured, with $C_m = 49$ μN/W. The amount of aluminum removed is not sufficient to produce this thrust, a significant amount of kapton is also ablated. Another low outgasing material is kapton with a proprietary thin black coating manufactured by Sheldahl Chemical Corp. It gave 6.7 μN thrust and $C_m = 26$ μN/W, with $I_{sp} = 34$ s. A special production run with the same black material in a thicker layer was also tested, but the performance was not improved.

Energetic Materials

In addition to the passive fuel materials, energetic materials have also been tested. Nitrocellulose lacquer with black iron oxide laser absorber has excellent adhesion to the acetate substrate and is flexible. It outgases less than PVC, and has good performance: $C_m = 124$ $\mu N/W$, 118 μN thrust, and $I_{sp} = 81$ s.

Our best fuel material is a proprietary energetic polymer developed by Thomas Lippert at the Paul Scherrer Institut in Switzerland [2]. In operation, the laser initiates a detonation, which does not propagate through the material. This may have other applications as well. In static tests this material has demonstrated $C_m = 500$ $\mu N/W$ and $I_{sp} = 500$ s. In continuous operation it produced 560 μN of thrust and $C_m = 320$ $\mu N/W$ at 120 s Isp. This Cm·Isp product is 80% of theoretical maximum for passive material (0.204). The continuous loop fuel tape provided for testing of this material had a very thick coating of the energetic absorber, 0.3 to 0.5 mm. This coating is an order of magnitude thicker than that of the other fuel tapes we have tested. With a thinner coating, this material may produce better results. We are in the process of acquiring the capability of making these energetic materials in-house.

R-MODE OPTICS

Because of the convenient geometry and the inherent protection of the optics by the transparent substrate, all continuous thrust tests to date have been done with T-mode. In static tests, R-mode illumination has resulted in greater thrust and greater I_{sp}. An R-mode test was performed using the pre-prototype thruster tape drive mechanism, and a laser and pair of focusing lenses mounted in front of the tape. This test was not successful due to rapid contamination of the optics by the exhaust plume. To avoid this problem, a unique optical system with grazing incidence R-mode illumination was designed to take advantage of the measured decrease of contamination flux with incidence angle. A schematic of this system is shown in Figure 3, and a photograph is shown in Figure 4. A small aspheric lens collimates the laser light, then the beam is expanded into a collimated annular beam by a pair of axicons. An axicon is a cone-shaped lens. The annular beam is focused to a point by another aspheric lens. This point is one focus of a toric ellipsoidal mirror. The other focus of the mirror is the fuel, which will be in the form of a 250 μm diameter fiber. Development of fuel fibers is in progress.

This optical system has now been constructed and evaluated. The optical system was manufactured at Applied Physics Specialties in Toronto, Canada, and was tested for quality of focus at their facility before taking delivery. It was determined that 80% of the laser energy is contained in a 200 μm circle, which is acceptable. In order to operate the R-mode optical system and measure thrust, it will be necessary to design and construct a fuel delivery system. Design is mostly complete, and construction of this fuel feed mechanism is underway.

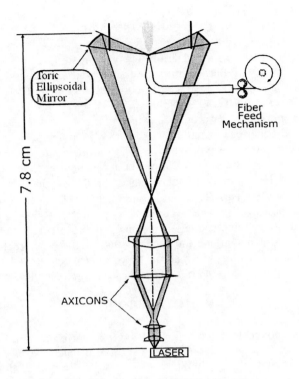

FIGURE 3. Schematic of R-mode Optical System

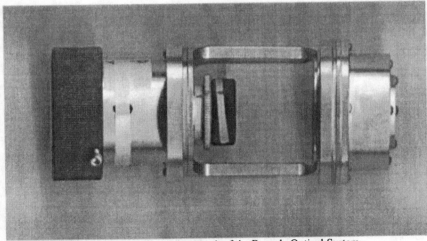

FIGURE 4. A Photograph of the R-mode Optical System.

SUMMARY

The micro-laser plasma thruster is recognized by the USAF as an alternative to micro-PPT's for attitude control on 100 kg class small spacecraft. Plans are underway for joint tests at AFRL/PRSS at Edwards AFB. The output of four lasers has been combined to obtain peak power of 14 W and average power of 2 W. By varying the pulse width from 2 ms to 10 ms, it is possible to steer the thrust vector by 50 to 60 degrees. Of the passive fuel materials, PVC has the best performance, achieving 70 μN of thrust, C_m of 190 μN/W, and I_{sp} of 220 s. With proprietary energetic materials, 500 μN of thrust has been measured. The R-mode optical system has been designed, constructed, and evaluated. Development of fuel fibers and a fiber feed mechanism is underway.

REFERENCES

1. Phipps, C and Luke, J., *AIAA Journal* **40**, 310-318 (2002).
2. Lippert, T., David, C., Hauer, M., Phipps, C., and Wokaun, A., *Rev. Laser Engineering*, **29**, 734-738 (2001).

Advantages of a ns-pulse micro-Laser Plasma Thruster

Claude R. Phipps[a], James R. Luke[b]

[a] *Photonic Associates, 200A Ojo de la Vaca Road, Santa Fe, New Mexico USA 87508*
Phone/Fax: 1-505-466-3877, Email: crphipps@aol.com
[b] *NMT/Institute for Engineering Research and Applications, 901 University Blvd. SE,*
Albuquerque, NM 87106-4339

Abstract. We describe a new project to develop an improved micro-laser-plasma thruster (LPT) using nanosecond pulses. As with the previous version driven by a ms-duration diode laser, the purpose of this microthruster is to position and attitude control of micro- and nano-satellites.

INTRODUCTION

In 1972, Kantrowitz suggested using laser power to heat a propellant sufficiently to produce a vapor or plasma jet for thrust [1], which is the principle of laser ablation propulsion. The principal advantages offered by this scheme are jet temperatures (and therefore specific impulse) which can be much higher than possible with chemical propellants, and, for certain missions, drastic cost reductions relative to conventional methods. However, practical applications have been realized only in the last decade [2, 3] as laser capabilities advanced to match Kantrowitz's vision.

THEORY

Pulsed lasers offer a much richer parameter space in which to work compared to CW lasers. The latter have been well treated elsewhere [4]. For pulsed lasers, the momentum coupling coefficient C_m is defined as the ratio of target momentum $m\Delta v$ produced to incident laser pulse energy W during the ejection of laser-ablated material (the photoablation process). For continuous lasers, it is the ratio of thrust F to incident power P:

$$C_m = \frac{m\,\Delta v}{W} = \frac{F}{P} \qquad (1)$$

In the ablation process, Q* joules of laser light (the asterisk is customary notation: Q* is not a complex number) are consumed to ablate each gram of target material:

$$Q* = \frac{W}{\Delta m} \qquad (2)$$

CP664, *Beamed Energy Propulsion: First International Symposium on Beamed Energy Propulsion*,
edited by A. V. Pakhomov

For the sake of discussion, we will consider a monoenergetic exhaust stream with velocity v_E. Momentum conservation requires

$$m\Delta v = \Delta m v_E,\tag{3}$$

so the product of C_m and Q^* is the effective exhaust velocity v_E of the ablation stream, independent of the efficiency with which laser energy is absorbed. This can be seen by writing:

$$C_m Q^* = \frac{(N-s)(J)}{(J)(kg)} = \frac{(kg)(m)}{(kg)(s)} = m/s\tag{4}$$

If for example, a significant amount of the incident energy is absorbed as heat in the target substrate rather than producing material ejection, Q^* will be higher and C_m will be proportionately lower, giving the same velocity in the end.

It is understood that real exhaust streams have velocity distributions, but we have shown [5] that the monoenergetic stream approximation will not introduce large errors $[<v^2>/<v>^2 \sim 1.15]$ for most laser-produced plasmas, and the principal points we want to make here will be made more easily using that assumption.

The specific impulse I_{sp} is simply related to the velocity v_E by the acceleration of gravity:

$$C_m Q^* = v_E = g I_{sp}\tag{5}$$

Energy conservation prevents C_m and Q^* from being arbitrary. Increasing one decreases the other. Using Eqs. (2) and (5), energy conservation requires that several constant product relationships exist:

$$2\eta_{AB} = \Delta m v_E^2/W = C_m^2 Q^* = g C_m I_{sp} = C_m v_E.\tag{6}$$

In Eq. (6), we introduce the ablation efficiency parameter, η_{AB} †1, the efficiency with which laser energy W is converted into exhaust kinetic energy. Choosing combinations of C_m and v_E that exceed 2 violates physics, since η_{AB} must be less than 1.

Since the maximum specific impulse of ordinary chemical rockets is about 500s, limited by the temperatures available in chemical reactions, larger I_{sp} values (exit velocity $v_E > 5km/s$) are accessible only by laser ablation, where temperatures can be many times 10,000K, or some other non-chemical process such as ion drives. Specific impulse I_{sp} up to 7600s has been measured using lasers with ns-duration pulses and non-heroic target intensities [6].

Ablation efficiency can approach 100%, as direct measurements with other types of lasers on cellulose nitrate in vacuum verify [6], but a value of 50%, or even less, is likely. The impact of $\eta_{AB} < 1$ is that the C_m value deduced from a given v_E may be less than the maximum permitted by conservation of energy. Exit velocity v_E is the fundamental quantity. The rate of mass usage is:

$$\dot{m} = \frac{P}{Q^*}\qquad\qquad kg/s\tag{7}$$

where P is laser optical power. When considering C_m and Q^* as design variables it must be kept in mind that the ablator lifetime increases with Q^* and decreases very rapidly with increasing C_m:

$$\tau_{AB} = |M/\dot{m}| = \frac{MQ^*}{\dot{P}} = \frac{2\,\eta_{AB}M}{P\,C_m^2} \tag{8}$$

For this reason, in laser propulsion applications, increasing C_m to get more thrust via the relationship

$$F = PC_m \tag{9}$$

from a given laser entails a serious penalty for ablator lifetime, because $\tau_{AB} \propto 1/C_m^2$ from Eq. (8).

It remains to see how laser intensity I (W/cm^2), pulsewidth τ (s) and wavelength λ(cm) combine to determine surface pressure. Vacuum plasma theory adapted from laser fusion [7] well describes the situation above plasma threshold. The principal results of that work which we will use here is, for the laser-initiated plasma-mediated pressure on a plane surface:

$$P_{AB} = 5.83\,\frac{\Psi^{9/16}}{A^{1/8}}\,\frac{I^{3/4}}{(\lambda\sqrt{\tau})^{1/4}} \tag{10}$$

and

$$T_e = 2.98\text{x}10^4\,\frac{A^{1/8}\,Z^{3/4}}{(Z+1)^{5/8}}\,(I\lambda\sqrt{\tau})^{1/2} \tag{11}$$

for the plasma electron temperature (K), where Ψ = the coefficient $(A/2)[Z^2(Z+1)]^{1/3}$, A is the plasma average atomic mass number and Z is the plasma average ionization state number.

The final constraint on the problem of laser-plasma generated thrust is defining the threshold for plasma onset, a critical element of the problem. It has been shown [8] that

$$\Phi_{th} = 240\,\tau^{0.45} \qquad\qquad \text{MJ/m}^2 \tag{12}$$

is an approximate expression for this threshold for ms to sub-ns pulse durations, independent of laser wavelength. This expression provides our main motivation for moving from ms to ns-duration pulses.

The vacuum coupling coefficient C_m is in the range 10 — 100 N/W for standard, front-illuminated, surface-absorbing materials [7]. Note that, from Eq. (6), $C_m*I_{sp} <= 2/g = 0.204$. In measurements with energetic target materials, products $C_m*I_{sp} = 0.18$ have been obtained, which is 90% of 0.204 [6].

232

In the laboratory, C_m and Q^* are relatively easy quantities to measure, and their product conveniently gives v_E which is a more difficult quantity to measure without, e.g., a laser-induced fluorescence setup or time-resolved shadowgraphy.

LASER PLASMA THRUSTER

Present Status

The micro-Laser Plasma Thruster (μLPT) [3] is a new micropropulsion option which is competitive with the micro-Pulsed Plasma Thruster (μPPT) on micro- and nano-satellite platforms. It takes advantage of the recent commercial availability of 4-watt diode lasers with sufficient brightness to produce a repetitively-pulsed or continuous plasma jet on a surface in vacuum. The diode is a low-voltage device with electrical efficiency in excess of 50%. A lens focuses the diode output on the ablation target, producing a miniature jet that provides the thrust. Single impulse dynamic range is nearly 5 orders of magnitude, with a measured minimum impulse bit of 1 nano N-s with 100μs pulses. Continuous thrust exceeds 75 N, and lifetime impulse of one fuel tape in the completed prototype (with 100g of ablatant) is expected to be 100 N-s. We have measured specific impulse as high as 1000s (greater than possible with chemistry), and thrust-to-optical power ratio C_m greater than 60 N/W, using these ms-duration pulses.

We have studied two modes of operation [Figure 1], which we term "Reflection" (R) and "Transmission" (T) modes. Each has advantages in performance. R-mode gives

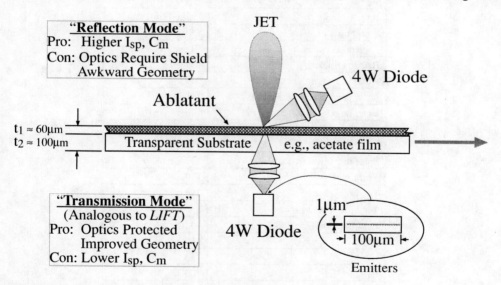

Figure 1. Illustrating two modes of operation of the μLPT

about 2 times better I_{sp} and 50% better C_m, but also offers significant design challenges to limit contamination of the illumination optics which occurs with the relatively cold jets produced by ms-duration pulses.

In T-mode operation, the focused diode beam passes through the transparent substrate and heats a specially prepared absorbing coating on the opposite side of the tape to a temperature sufficient to produce the ablation jet. The substrate is not destroyed in this process, but does contribute some mass to the ablation process due to heating of the substrate material which is in contact with the ablatant.

The LPT has the advantage over other micropropulsion techniques of being free of mysterious small-scale physics. Specifically, nothing erodes during operation except the ablation fuel. The LPT can operate pulsed or CW, and power density on target is optically variable, so I_{sp} can be adjusted in an instant to match mission requirements.

Construction of the long-pulse LPT

Figure 2 shows the existing "preprototype" LPT, which has been developed to test the basic components of the prototype. A small gearmotor drives the tape longitudinally at speeds up to 20 mm/s, and a stepper motor moves the laser transversely to illuminate one of 254 tracks.

Four JDS Uniphase 6380-A multi-transverse-mode diode lasers operating at 920 nm drive the long-pulse LPT. Standard 0.68NA aspheric lenses focus the 14W laser

TABLE 1. Preprototype specifications

Item	Value
Weight with fuel	850 g
Tape dimensions	50.5cm x 2.54cm
Backing thickness	100 m
Ablative coating thickness	90 m
Laser power (average, peak)	2 W, 14W
Laser target illumination area	500 m^2
Tape speed	20 mm/s
Pulse duration	2 ms
Pulse repetition frequency	100 Hz
Track width	100μm
Tracks	254
Tape lifetime	1.8 hours
Coupling coefficient C_m	80 μN/W
Force output/	150 μN
Q*	11 kJ/g
I_{sp}	400 s
Minimum impulse bit	0.6 μN-s

/ using C_m and P_{avg}, : at 1ms pulsewidth

output to a 500μm² spot, giving 2.8MW/cm² on target. These intensities are sufficient to form a plasma jet within 2ms. Figure 3 shows the plasma jet created by the μLPT. To avoid plume steering due to target tape motion, we must operate in repetitive-pulse mode. The preprototype thruster and its associated electronics unit each weigh about 425 g, for an 850g total mass.

Performance of the long-pulse LPT

Materials we have explored for the transparent substrate are cellulose acetate, PET and Kapton". Ablatants have been carbon-loaded PVC and a proprietary energetic coating. The ms-pulse LPT has been a highly successful program, exceeding requirements for 75 N thrust. It seems likely the finished prototype will meet requirements for 100N-s lifetime thrust per axis.

ADVANTAGES OF NS PULSES

Nevertheless, tremendous advantages will accrue from going to 6 orders of magnitude shorter pulsewidth, as summarized in Table 2. These benefits derive directly from Eq. (12). The immediate benefit is that 500 times less fluence on target is required to create a plasma. The impact of this on design is 15 to 20 times greater standoff distance for the target focusing optics from the target and its contaminating plume in R-mode. Since this contamination rate varies as $1/R^2$, the result is 400 times greater optics lifetime. These benefits are outlined in Table 2.

The new design will offer:

1. Freedom from thermal diffusion loss in targets. In our ms work, we have not been able to create plasmas on materials other than low conductivity organics with

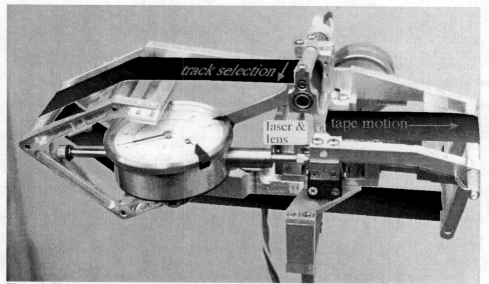

Figure 2. The preprototype ms-pulse LPT. Dial gage is used for setup, is not part of the unit.

the available intensity, even with very fast aspheric lenses and nearly diffraction-limited lasers. We cannot produce plasmas on metals.

2. Ability to access any target material, including metals. It will be a tremendous advantage in design to be able to use, e.g., stainless steel tapes as fuel for the thruster, avoiding the complexity of T-mode target design, as well as the difficult steps in their manufacture. This manufacture involves - in the case of our proprietary energetic coatings - a complex series of chemical synthesis procedures.

3. Freedom from outgassing organic target materials

Outgassing is currently a significant problem, although we believe we have reduced it to an acceptable level with new, proprietary target materials. However, with metal targets, all of the system but the electronics could be baked, if needed, to meet the most exacting spacecraft outgassing requirements.

CHALLENGES

Challenges that must be faced include:

1. Designing and building a 4-W, ns-pulse laser head weighing less than 200g

This will almost certainly require the design to be a diode-pumped fiber laser with ns-pulse injection and distributed saturable absorber.

2. Avoiding nonlinear optical (NLO) effects in the glass fiber

These include stimulated Brillouin and stimulated Raman scattering, and self-focusing due to nonlinear refractive index (n_2). Preliminary calculations indicate that we can avoid these problems, but more detailed calculations need to be done. We note that R. Waarts has succeeded[9] in getting 6.4 GW/cm^2 to propagate in a fiber laser,

TABLE 2. Comparison of near-term (1), developmental (2) and ms-pulse LPT's			
Laser version	ns-pulse (1)	ns-pulse (2)	ms-pulse
Pulse duration τ	500 ps	1 ns	2ms
Average optical power on target <P>	100 mW	4 W	2 W
Peak optical power on target P_{pk}	50kW	230 kW	12 W
NA of target optics	0.2	0.07	0.68
Lens diameter d_L	2 cm	2 cm	6 mm
Standoff distance s_2 (to target)	2 cm	15 cm	3 mm
Optics contamination time, R-mode	5 hours[a]	230 hours[b]	20s[a]
Repetition rate f	4 kHz	17 kHz	80 Hz
Pulse energy W	25 J	230 J	25 mJ
Spot size on target	40 m	100 m	25 m
Fluence Φ on target	2.0 J/cm^2	2.9 J/cm^2	4.9 kJ/cm^2
Intensity on target	4 GW/cm^2	2.9 GW/cm^2	2.8 MW/cm^2
Φ/Φ_{opt}	1.1	1.5	3.1

a) at near-normal incidence, b) at 45¡ incidence angle

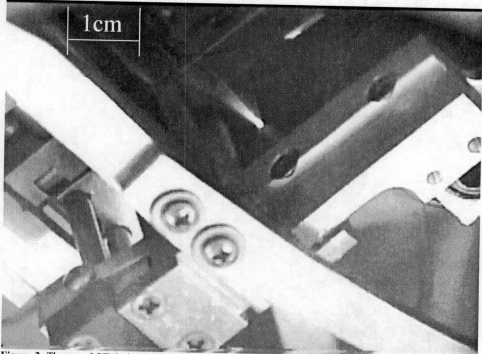

Figure 3. The ms-μLPT during operation produces a well-collimated, intense microjet.

whereas our application will only require 2.9GW/cm^2 in a 100 m diameter fiber. If there is a NLO problem, we can use a larger fiber. It is not necessary that the fiber laser be single mode.

3. Packaging the laser in a configuration with less than 0.2kg mass

A diode-umped fiber laser solution will satisfy the requirement. If we are forced to use standard, discrete-component solid state laser technology, meeting this requirement will be a challenge. However, we have one industrial supplier who believes they can reach this goal using their proprietary technology.

CONCLUSIONS

Present status and performance of the existing, ms-pulse LPT has been discussed and compared with performance advantages which we expect to achieve with ns-pulse duration. Challenges to be faced in that development will mainly be avoiding nonlinear optics limits in fiber propagation.

Acknowledgment. The microthruster work was completed with support from AFOSR contract F49620-00-C-0005

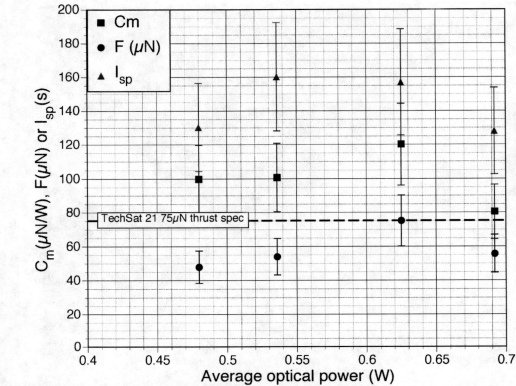

Figure 4. Typical measurements of thrust, thrust-to-power ratio and specific impulse for the preprototype in low-power tests.

REFERENCES

1. A. Kantrowitz, *Aeronaut. Astronaut.* **10**, 74 (1972)

2. F. Mead, Jr., L. Myrabo and D. Messitt, "Flight Experiments and Evolutionary Development of a Laser Propelled, Transatmospheric Vehicle", *Proc. High Power Ablation I,* SPIE **3343** pp. 560-563 (1998)

3 C. Phipps and J. Luke, "Diode Laser-driven Microthrusters: A new departure for micropropulsion," AIAA Journal, **40** pp. 310-318 (2002)

4 P. Loosen, "Advances in CO_2 laser technology for industrial applications", SPIE **1810** pp. 26-33 (1993)

5 C. Phipps, J. Reilly and J. Campbell, Optimum Parameters for Laser-launching Objects into Low Earth Orbit , J. Laser and Particle Beams, 18 no. 4 pp. 661-695 (2000)

6 C. Phipps and M. Michaelis, Laser Impulse Space Propulsion , *Journal of. Laser and Particle Beams* **12** no. 1, pp. 23-54 (1994)

7 C. Phipps, T. Turner, R. Harrison, G. York, W. Osborne, G. Anderson, X. Corlis, L. Haynes, H. Steele, K. Spicochi, and T. King, "Impulse Coupling to Targets in Vacuum by KrF, HF and CO_2 Lasers" , *Journal of. Applied. Physics.,* vol. 64, no. 3 pp. 1083-96 (1988)

8. C. R. Phipps and J. R. Luke, "Diode Laser-driven Microthrusters: A New Departure for Micropropulsion", *AIAA Journal* ,**40**, no. 2, pp. 310-318 (2002)

9. R. Waarts, "Fiber laser technology for commercial and DoD applications", *Proc. 13th Solid State and Diode Laser Technology Review, Air Force Research Laboratory, Albuquerque, NM, 5-8 June 2000*

Survey of Beamed Energy Propulsion Concepts by the MSFC Space Environmental Effects Team

P.A. Gray[*]. M.K. Nehls[¶], D.L. Edwards[¶], M.R. Carruth, Jr[¶].

[*]ICRC Huntsville, AL 35816

[¶]ED31 Environmental Effects Group
Marshall Space Flight Center, AL 35812

Abstract. This is a survey paper of work that was performed by the Space Environmental Effects Team at NASA's Marshall Space Flight Center in the area of laser energy propulsion concepts. Two techniques for laser energy propulsion were investigated. The first was ablative propulsion, which used a pulsed ruby laser impacting on single layer coatings and films. The purpose of this investigation was to determine the laser power density that produced an optimum coupling coefficient for each type of material tested. A commercial off-the-shelf multi-layer film was also investigated for possible applications in ablative micro-thrusters, and its optimum coupling coefficient was determined. The second technique measured the purely photonic force provided by a 300W CW YAG laser. In initial studies, the photon force resulting from the momentum of incident photons was measured directly using a vacuum compatible microbalance and these results were compared to theory. Follow-on work used the same CW laser to excite a stable optical cavity for the purpose of amplifying the available force from incident photons.

INTRODUCTION

Two techniques for using laser photons to produce propulsive forces are discussed in this paper. One technique is laser ablation, which uses short, intense pulses of laser light energy to ablate a small amount of material. The laser used in this investigation has a very short pulse, in the nano to picosecond range. This provides the instantaneous power necessary to ablate the surface of a propellant source without causing damage to the spacecraft [1]. This technique can also be applied to spacecraft that have an on board laser source. Typically, ground based laser systems have been considered for de-orbiting debris particles from low earth orbit. On board lasers have also been considered for ablative micro-thruster applications [2].

The other technique takes advantage of the momentum imparted to a surface when photons strike the surface of a material. This technique is applicable to laser sailing using a laser as a manmade source of photons. The laser has the advantage that large amounts of energy can be concentrated in a small area, thus reducing the size and mass of the sail [3].

ABLATIVE LASER INTERACTION WITH MATERIALS

Background

The Space Environmental Effects (SEE) team, at the Marshall Space Flight Center, began working in the area of laser interaction with materials by experimentally investigating ablative forces generated when lasers interact with materials. The SEE team designed a torsion balance and used a pulsed ruby laser to characterize the relationship between laser power density and ablative force for various materials representative of orbital debris. A natural transition of the SEE team was to utilize their experience with laser-material interactions and the existing test capability to investigate beamed energy propulsion. The most recent work for the SEE team is in the area of multi-layered thin films for micro-thruster applications. The concentration for the SEE team will be focused on optimizing the thin film composition for micro-thruster applications.

Apparatus

The laser used to produce the ablative force is a Lumonics QSR3, 3J pulsed ruby laser (694nm) with a Q-switch to create a narrow pulse. This laser can produce energies as high as 5 J with a 25 nanosecond (ns) pulse width. The torsion balance is located in a vacuum chamber operating in the 1×10^{-7} torr range. The chamber consists of a 13-inch conflat cross with a window at the top to view the torsion balance and a dial that is located beneath the torsion balance. A window on the side of the cross illuminated the torsion balance and the dial. The dial has a 3 degree per tick resolution. The motion of the torsion balance is viewed from above by a video camera and was recorded using a video recorder. The positive to negative displacement, $2\theta_{max}$, and the period of the pendulum are measured post test by replaying the videotape. The laser enters the chamber through an anti-reflection (AR) coated window that is optimized for 694 nm. The laser ablates material, producing an impulse to the torsion balance. The laser pulse is captured with a Molectron Model P5-01 laser pulse detector connected to a Tektronics model TDS460A, 500 MHz, digital storage oscilloscope. The pulse width is taken at full width at half maximum (FWHM). The total energy per pulse is measured using a Molectron Model EPM1000 pulsed laser energy meter with a model J25LP-RUBY energy detector. The energy detector samples 8.5% of the beam using a quartz beam splitter. The pulse detector was mounted so that it detected the diffuse reflection off of the ceramic diffuser on the energy detector. This eliminated the need for a secondary beam splitter for the pulse detector. The laser energy is attenuated by using a beam splitter stack and 0.4 and 0.8 neutral density (ND) filters. The $1/e^2$ beam diameter is measured with a Spiricon model LBA-300PC 8-bit beam profiler. The beam diameter is varied using different focal length lenses with the sample located at the focal length of each lens. Two lenses were used for these tests, a 350 mm and a 500 mm. The 500 mm lens produces a 0.41 mm dia. spot size, and the 350 mm lens produces a 0.3 mm dia. spot size.

Calculations

The coupling coefficient is defined as the momentum transferred to the pendulum divided by the laser power density. It is a measure of the efficiency of converting light energy to kinetic energy through the process of ablation. It is plotted as a function of the laser power density. Power density is defined as the laser energy divided by the pulse width and the beam area, which gives the familiar units W/cm^2. On a plot of coupling coefficient versus laser power density, the optimum coupling coefficient occurs at the highest point on the curve. The laser power density corresponding to the optimum coupling coefficient is the point of maximum efficiency in converting the laser energy to momentum. Increasing the power further will produce more momentum but the efficiency decreases. Laser power levels below that which corresponds to the optimum coupling coefficient produce force at less efficiency. In addition, the force produced decreases at a very high rate as laser power is decreased below that corresponding to the optimum coupling coefficient. Power levels should be equal to or above those corresponding to the optimum coupling coefficient in order to insure that sufficient ablation occurs to produce a significant propulsive force.

A torsion balance was used for measuring the momentum produced by the ablation process. The balance consists of a thin metal wire with a rotating pendulum suspended in the middle. The wire is fixed at both ends. Figure 1 defines the radius and angular displacement parameters for the torsion balance. Wire, of a known torsion spring constant, allows calculation of the maximum force or thrust produced by the ablation. The period of the pendulum is also measured. The momentum produced by the ablation is the product of the average force and the portion of the period from the zero position to the maximum displacement position. Since the force varies as a function of angular displacement, the force is zero at zero displacement and a maximum at the maximum angular displacement, θ_{max}. The average force over the time interval from zero to θ_{max} is simply the maximum force divided by 2. The relationship between the average force (F_{ave}) and the maximum angular displacement (θ_{max}) is shown in Equation (1).

$$F_{ave} = C\theta_{max} / 2r \qquad (1)$$

where C is the torsion spring constant, r is the distance from the center of the laser impact sight to the center of the tungsten wire, and theta is the angle of displacement in radians. The time interval from zero to max displacement is simply ¼ the period T. So the momentum, M, is given by;

$$M = F_{ave}\Delta T, \tag{2}$$

where $\Delta T = 1/4T$.

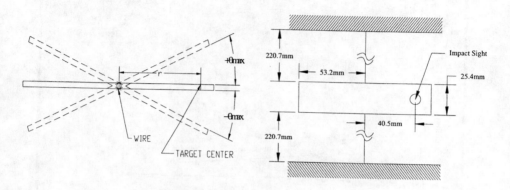

FIGURE 1. Schematics of torsion pendulum.

Results

The ablative coupling coefficients for several coatings and materials were measured. A graph of the coupling coefficient curves is located in Figure 2. On the white Z93 coating, the material darkened after the first hit. As a result, subsequent laser impacts had higher coupling coefficients, presumably due to increased absorption of the laser light. Also a muli-layer material, Aluminum – Teflon® (FEP) was measured and compared to bare aluminum without the Teflon® outer layer. The Aluminum-Teflon® material had an order of magnitude increase in coupling coefficient at the expense of a much larger damaged region. The outer Teflon® layer served to contain the gases evolved during the ablation process. The Teflon® layer formed a bubble containing the expanding gases and subsequently ruptured forming a crude nozzle which directed the high pressure gases in a direction approximately normal to the surface of the material. It is thought that the pressure resulting from containing the ablation products and the formation of the nozzle are responsible for the increase in the measured force. If this discovery is to be more than of academic interest as a propulsion technique, the amount of material used in each laser hit must be considered. A diameter, two or three times that of the laser beam, was damaged with each hit. However, multi-layer materials are still being considered for use in ablative micro-thrusters [1]. This technique used a clear material with the ablative propellant on the side away from the laser. The laser passed through the clear layer and ablated the propellant coated on the other side. Once the laser has hit an area, it is moved across the width of the tape. Once the entire width of the propellant is consumed, the tape is translated and a fresh area of propellant is exposed to the incident laser beam.

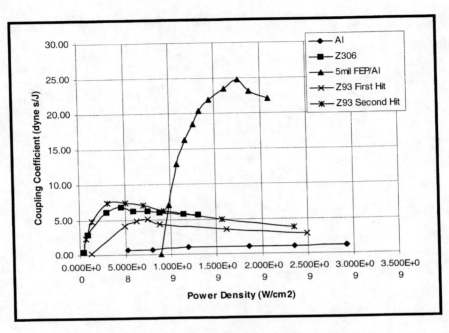

FIGURE 2. Coupling coefficient curves for various materials.

NON-ABLATIVE LASER INTERACTION WITH MATERIALS

Background

The SEE team has developed the capability to measure photon pressure. This capability was developed to experimentally verify the photon pressure theory and to provide solar sail designers the experimental validation required for mission scenarios. Originally, this photon pressure measurement capability was developed using a full spectrum photon source. The SEE team has expanded the application of this measurement capability to measure the photon pressure from a Continuous Wave (CW) laser source. This pressure measurement capability was further expanded to experimentally investigate the amplification of photon pressure [4]. This amplification process uses a semi-stable optical cavity to reflect laser photons back onto the primary photon-driven propulsion system, such as a laser sail.

Apparatus

The test chamber for this experiment consists of a 6 way vacuum cross mounted on an ion pump. The laser beam passes through a –250 mm focal length expanding lens, then through a standard glass view port. The divergence angle was less than 3°. Since the cosine of this angle is approximately one, the effect of angle was neglected. After the laser beam passed through the view port, a 45° turning mirror directed the expanded laser beam to the target.

The laser being used for this experiment is a Quantronix Model 118, 300W peak output, multimode, Nd YAG, CW laser operating at 1064nm. Figure 3 details the optical beam path for the photon force measurements and Figure 4 describes the photon amplification setup.

The force-measuring device was a CAHN model D-101 vacuum compatible microbalance. The sample was suspended from the balance with a long Nichrome wire. When the laser was directed onto the target and the photons impinged on it, they provided an upward force. This was measured by the microbalance as a decrease in target weight. The samples consisted of a thin stainless steel plate in the case of the photon force measurements and a 2" diameter mirror in the case of the amplified photon measurements. The thin stainless steel plate was chosen because of its lightweight and high damage threshold.

Laser power measurements were obtained, using a Molectron Model PM300F-50. The power meter was located after the expanding lens and before the chamber window.

The experimental setup for the photon amplifier consisted of an optical cavity of two 100 mm focal length (FL) concave mirrors. One of the mirrors (M1) has a small hole drilled in the center for the laser beam to pass through. The mirrors are mounted horizontally facing each other, one above the other, as shown in the schematic in Figure 4. The mirror with the hole in it (M1) is placed at the bottom of the cavity. The top mirror (M2) is suspended directly above the mirror with the hole (M1). The laser beam entered the cavity by reflecting off a 45° dielectric mirror.

As a comparison to the two-mirror system which provided an amplified photon force, the experiment was run with the lower mirror removed. In this case, incoming photons impinged on the mirror (M2) only one time.

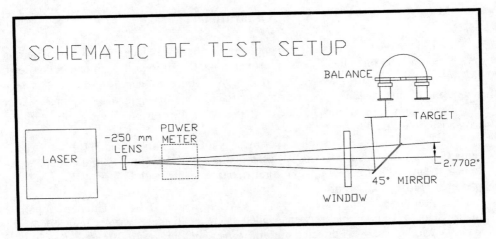

FIGURE 3. Photon force measurement optical setup.

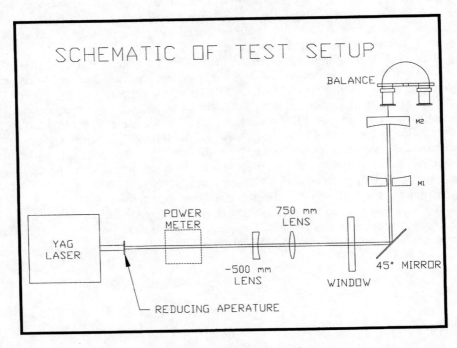

FIGURE 4. Photon amplification optical setup.

Calculations

Working formulas are presented in this paper. A detailed derivation of the formulas is located elsewhere [3]. The general equation for photon force is:

$$F = I/c, \tag{3}$$

where I is the total power from the laser in Watts and c is the speed of light in vacuum.

For a reflective surface, there is also a momentum contribution due to the reflected photons. A general equation for the photon force on a reflective surface is given by:

$$F_{total} = (I/c)(1 + R), \tag{4}$$

where R is the reflectance of the target material at laser wavelength of 1064nm. For the stainless steel plate R = 0.79.

A general equation that includes optical component losses is given by:

$$F_{calc} = (I/c)(1 + R)(\eta), \tag{5}$$

where η is an empirical term that takes into account measured losses through optical components. For the measurements of photon force with the stainless steel target η = 0.868.

Now consider multiple photon reflections onto the mirror. The result is that F_{calc} doubles with each reflection off the second mirror (sail). However, the losses increase with the square of the mirror reflectance, since each reflection cycle requires two reflections to complete. A general equation for total force that takes into account the loses dues to multiple mirror reflections is:

$$F_{calc} = (\eta_1)(I/c)(1+R)\Sigma(R^2)^{(i-1)}, \tag{6}$$

where n = number of reflection cycles, i = 1 to n, and where η_1 = 0.861 is an empirical term that takes into account measured losses through optical components for the amplification setup in Figure 4.

In order for equation (6) to be valid, the cavity must be stable. The relationship for a stable cavity is;

$$(1-d/2f)^2 <= 1, \tag{7}$$

where d is the distance between the two mirrors and f is the focal length of the mirrors. The assumption is made that both mirrors have the same focal length for this particular argument. This stability criterion limits the distance between the two mirrors. The separation distance must be less than or equal to 4 times the mirror focal length. The stability criteria is then:

$$d <= 4f .$$ (8)

Results

The laser was passed through an expansion lens to reduce the energy density. The lens focal length was −250 mm to make the divergence angle as small as possible to reduce off axis losses. The laser power was measured after the diverging lens and before the vacuum chamber window. As a measure of the validity of the photon force measurements, the measured force (F_m) was divided by that predicted from laser power measurements (F_{calc}). F_{calc} was calculated using the laser power measurements and equation (5). Since both numbers have error associated with them, a combined error for the ratio, of 4.1%, was calculated using the technique of the propagation of errors [5]. The value of the ratio was $F_m/F_{calc} = 0.986$ or 98.6%. Power measurements were taken periodically to compensate for power drift in the laser intensity.

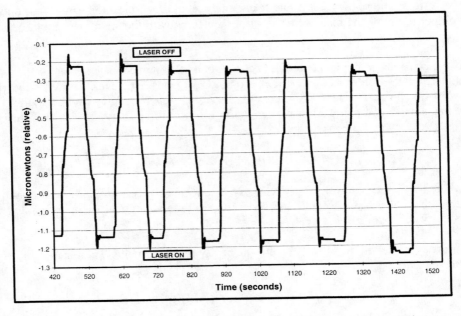

FIGURE 5. Laser force data obtained using a laser power of 200W on a stainless steel target.

An aperture was used to reduce the beam diameter of the laser to approximately 9.5mm diameter from 19 mm. This allowed the beam to pass through the bottom mirror (M1) that had a 9.5 mm diameter hole in it. A –500 mm and 750 mm lens were used in series to adjust the beam diameter slightly and also provide some beam steering so that the beam could be guided through the 9.5 mm diameter hole. This method did not harm the top mirror (M2) on the initial reflection, but the bottom mirror (M1) focused the beam back on to (M2) causing it to burn. This severely limited the amplification of photons within the cavity. Since alignment was difficult inside the vacuum chamber and space was limited, it is not known at this time if the low amplification factor was due to the mirror failure or the beam leaving the cavity through the 9.5 mm diameter hole in the lower mirror.

For the purposes of comparison, the geometry was retained, but with the bottom mirror removed. The laser power was increased to produce approximately the same force without the aid of the amplification. One data set at this power level was obtained before the top mirror burned. The measured force was 89% of the calculated force, which was based on power measurements and system losses. Prior studies have shown measured and calculated photon force to be within a few percent [3]. It is possible that the mirror degraded prior to failing, resulting in the large discrepancy between calculated and measured force.

The amplification factor (AF) for each experiment is defined by the ratio of measured to calculated force.

$$AF = F_m/F_{calc} \qquad (9)$$

The calculated force, F_{calc}, was calculated assuming no amplification in all experiments. An amplification ratio of greater than one indicates that force amplification did occur during the experiment. An average amplification factor for the two amplified force measurements yielded an amplification factor of 2.64.

Summary

The measurements indicate that the force from laser photons can be accurately measured with a vacuum compatible microbalance. This technique provided repeatable and reliable laser photon force data. Future work will involve the use of actual sail materials. With the current setup, the laser energy density is too intense for use on thin polymeric materials. Two factors that will improve this problem are increasing the beam diameter by the use of a larger vacuum chamber and sample, and changing the distribution of energy in the beam from a cone shape to a top-hat shape. Both of these techniques will reduce target hotspots.

This photon amplification experiment has shown that photon reflections within a stable cavity provided a means of optimizing laser photons and increasing the force they produce. An amplification factor of 2.64 was achieved. The theoretical number for this amplification was not achieved due to time and equipment constraints, but the concept has been demonstrated. In future experiments, a larger vacuum chamber and better dielectric mirrors should improve results considerably, with amplification factors approaching 50. Mirror reflectance is the driver for determining the theoretical maximum amplification factor. The higher the reflectance of the mirrors the more bounces can be obtained before reflective losses reduce the available momentum.

REFERENCES

1. Phipps, C. R., Luke, J. R., McDuff, G. G., and Lippert, T., "A Laser-Ablation-Based Micro-Rocket," *AIAA Plasmadynamics and Lasers Conference*, May 2002, AIAA 2002-2152.

2. Nehls, M.K., Edwards, D.L., and Gray, P.A., "Ablative Laser Propulsion Using Multi-Layered Material Systems," *AIAA Plasmadynamics and Lasers Conference,* May 2002, AIAA 2002-2154.

3. Gray, P.G., Edwards, D.L., and Carruth, M.R., Jr., "Laser Photon Force Measurements Using a CW Laser," *AIAA Plasmadynamics and Lasers Conference*, May 2002, AIAA 2002-2178.

4. Meyer, T. R., McKay, C. P., Pryor, W. R., and McKenna, P. M., "The Laser Elevator: Momentum Transfer Using an Optical Resonator," *38th IAF Conference*, Brighton, UK, October 11-17, 1987.

5. Bevington, P.R., *Data Reduction and Error Analysis for the Physical Sciences*, McGraw-Hill, 1969.

Optimization of Laser Ablative Propulsion Parameters: A Proposal

Bansi Lal*, Fang-Yu Yueh and Jagdish P. Singh

Diagnostic Instrumentation & Analysis Laboratory, Mississippi State University, Starkville, MS 39759
*on leave from Indian Institute of Technology Kanpur, India

Abstract. Laser ablative propulsion (LAP) is a useful technology in the launching of micro-satellites and in carrying out mid-orbit corrections. The thrust generated by a laser pulse is characterized in terms of momentum coupling coefficient. In this paper, we propose a simple experimental method to measure the momentum coupling coefficient (Cm) in ambient conditions. Cm for Al has been found to be equal to 12.7 dyn-s/J while for water it is 88 dyn-s/J. Results obtained for Al agree with the theoretical calculations.

INTRODUCTION

There being no traction in space, the thrust needed to move a space vehicle forward, is produced by ejecting the propellants out of it in the backward direction, necessitating the housing of both propellant as well as energy source in the space vehicle. This severely limits the vehicle performance, increases its cost and the risk of failure. All these problems can be eliminated with a remote power source capable of generating comparable thrust and specific impulse. Lasers which produce well collimated high intensity radiation, both pulsed as well as continuous wave (cw), from far IR to UV, can be employed as remote power source for space vehicles and the first proposal[1] in this direction came as early as 1972. There has been number of investigations covering various aspects of laser assisted propulsion since. The radiation from a remote laser (ground based or space based) is used in the following configurations for propulsion[2-4]:

(i)   Laser thermal propulsion: Analogues to chemical propulsion, the propellant housed aboard the vehicle, is heated by the laser radiation and the thrust is produced when heated propellant exits through a nozzle. Hydrogen is the most commonly used propellant which is heated either by solid heat exchanger or by particulate/molecular resonance /inverse Bremsstrahlung absorption. These systems are reported to deliver specific impulse in the range 875-1500 b_f-s/lb_m. The main disadvantage of this scheme is the need to store large amounts of hydrogen. Alternate schemes like (a) conversion of laser energy into electric energy to power an electric thruster, (b) the required thrust obtained from laser produced blast wave and (c) propulsion by inertial confinement fusion produced by laser energy, are being investigated actively.

CP664, *Beamed Energy Propulsion: First International Symposium on Beamed Energy Propulsion*,
edited by A. V. Pakhomov
© 2003 American Institute of Physics 0-7354-0126-8/03/$20.00

(ii)   Light sail: In this case photons act as propellant hence velocities >0.1c can be achieved by the vehicle. This makes it very attractive for interstellar missions. It is an interesting futuristic scheme and demands lot of work before a useful practical device can be built.

(iii)   Laser ablative propulsion (LAP): Radiation from a pulsed laser when made to incident on a solid target mounted on the back of the vehicle, produces plasma which generates the thrust as it expands rapidly in a direction opposite to the direction of laser pulse. Since the 'fuel' used in this case has very little weight, it can be used in the miniaturization of micro-satellites. Also, it is useful for mid-orbit manipulations. Recently there has been lot of interest in LAP and the present   paper reports our investigation in this field.

One of the important parameters of LAP is the momentum coupling coefficient (C_m) which is measure of momentum imparted on the vehicle by a laser pulse, has been subject of many experimental investigations[5-6]. However, most of these measurements have been carried out while irradiating the target in vacuum. This paper proposes an experimental arrangement to measure C_m in ambient conditions. The laser energy imparts linear motion to a target sliding over a surface and C_m is quantified in terms of the linear displacement of the target, mass of the target and the coefficient of friction of the sliding surface. The experiment can be carried out with a laser pulse of 100mJ or less.

Theory

When a laser pulse of energy W(J) falls on the target(Figure 1) mass m_A is ablated which ejects with velocity v_E imparting a velocity v_T to it. If the mass of the target is m_T (g) we have $m_A v_E = m_T v_T$. The momentum coupling coefficient C_m(dyn-s/J) is defined as[7] $C_m W = m_A v_E$. In terms of target mass and velocity $C_m = m_T v_T/W$ which

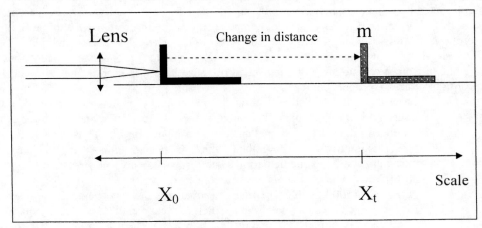

Figure 1.   Geometry of target motion under laser ablation.

252

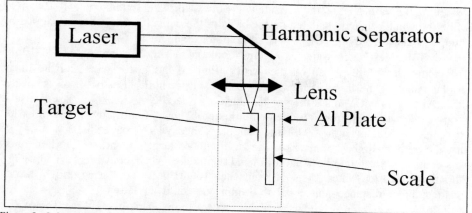

Figure 2. Schematics of the experimental setup

gives v_T = Cm (W/m_T). The target comes to rest after sliding over a distance x_T(cm) on the surface with coefficient of friction μ_s. Using elementary physics one can derive the following expression for Cm

$$Cm = [2\,\mu_s g(\,m_T)^2\,x_T/\,W^2]^{1/2} \qquad (1)$$

Equation (1) can be used to measure C_m for various materials under various conditions.

Experimental

Radiation from a frequency doubled Nd: YAG laser (Continuum SL1-10) operating in single shot mode is made to incident on a harmonic separator (Figure 2). 532nm pulse is focused on the vertical limb of an L-shaped target, using 50mm focal length fused silica convex lens. The target is made from 0.075mm thick Al foil. A sheet 25mm long and 8mm wide cut from the foil is L-shaped so that the vertical limb of the `L` is about 5mm. The targets with these dimensions weigh about 0.08g. The target is free to slide on a 0.1mm thick Al sheet (20cmX15cm) fitted with a steel meter scale of 0.5mm least count. This Al sheet is mounted on an X-Y-Z mount.

Results and Discussion

Targets in the mass range 0.01-0.1g have been investigated using 100mJ laser pulses. No reliable measurements with the targets having mass less or equal to 0.06g could be taken as most of the time instead of sliding on the surface, these have projectile motion. By trial and error method we found a target of mass 0.08g slides consistently on the surface and never takes projectile motion with 100mJ laser pulse. With the target of this mass the measurement of linear displacement was repeated 20 times on a single target while 3 targets were investigated. Average linear displacement of 5.8mm is measured with standard deviation of the data equal to 7%.

Assuming 0.3 values for μ_s the average value of C_m for Al is found to be 13.7 dyn-s/J while the value of Cm calculated on the basis of theoretical formulation developed by Phipps[7] is about 15 dyn-s/J. This is a fairly good agreement.

We also investigated water as ablation material. Using a micro-pipette (1-10µL) 1µL water drops are placed on the vertical limb of L-shaped target and the linear displacement of the target is measured on focusing a single laser pulse on the water drop. Al targets with mass 0.1g or less have a projectile motion when laser pulse is focused on the water drop. To overcome this problem a 35mm long and 8mm wide (0.3g mass) sheet cut from a 0.02mm thick brass sheet was L-shaped in such a way so that the vertical limb is 5mm, as is the case with Al targets mentioned earlier (brass was chosen as we found it difficult to bend the thicker Al sheet available with us). . This target has zero linear displacement when 100mJ pulse is focused on it while an average linear displacement equal to 13mm (Standard deviation 11%) is measured when a 1µL water drop is placed on its vertical limb. This gives the value of Cm equal to 88 for water, about 6 times more than that of Al. The increase in standard deviation in this case may be due to variation in the amount of water getting ablated as it is very difficult to place the water drop exactly at the focus of the laser pulse.

Conclusions

The present technique in which a laser pulse imparts linear momentum to an L-shaped target provides direct measurement of momentum coupling coefficient, hence can be used to optimize laser ablative propulsion parameters. Also, the momentum coupling coefficient of water has been found to be almost six times that of Al.

ACKNOWLEDGMENTS

This work is supported by the US department of Energy through Cooperative Agreement No. DE-FC26-98FT 40395.

REFERENCES

1. Kantrowitz, A., *Aeronautics and Astronautics* **9**, 40-429(1972)
2. Glumb, Ronald J., and Krier Herman, *J. Spacecraft* **21**, 70-79(1984)
3. Birkan, Mitat A., *Journal of Propulsion and Power* **8**, 354-360(1992)
4. Pakhomov, Andrew V., and Gregory, Don A., "Ablative Laser Propulsion: An Advanced Concept for Space Transportation" in *2000 Young Faculty Research Proceedings*, The University of Alabama in Huntsville, pp. 63-72
5. Nehls, Mary, Edwards, david and Gray, Perry, " Ablative Laser Propulsion using Multi-layered Material Systems" in *AIAA Conference* AIAA -2002-2154
6. Phipps, Claude R., Luke, James R., McDuff, Glen G., and Lippert, Thomas, " A Laser-Ablation-Based Micro-Rocket" in *AIAA Conference* AIAA-2002-2152
7. Phipps, Claude R, Reilly, James P., and Campbell, Jonathan W., *Laser and Particle Beams* **18**, 661-695(2000)

MICROWAVE PROPULSION

Conversion of Sub-Millimeter Waves to Gas Flow in Sonic Region

Donald G. Johansen

13450 Pescadero Creek Road, La Honda, California, U.S.A.

Abstract. Study shows potential for efficient energy conversion from sub-millimeter electromagnetic beam to hydrogen or helium gas at sonic and low supersonic speeds. Higher gas enthalpy is possible, compared with subsonic heating, leading to higher specific impulse for propulsion. Electrons driven by the beam E-field are excited to high temperature and transfer energy to the gas by collision. This process is limited to a narrow frequency range by gas breakdown at lower frequencies and low beam attenuation at higher frequencies. Propulsion system design factors, comparing IR vs. sub-millimeter beamed propulsion are discussed.

INTRODUCTION

Proposed microwave devices [13] for beamed energy propulsion (BEP) introduce beamed energy into a chamber via lens with gas heating and expulsion similar to chemical rockets. This is called a two-port engine [7] and contrasts with a device wherein the beam enters the gas exit [5]. This is called a single-port engine. Beamed energy propulsion (BEP) devices are also generally categorized as pulsed or continuous [10]. Pulsed devices [15] typically induce gas breakdown by creation of plasma with high electron density which absorbs the electromagnetic (EM) wave. This creates a laser-supported detonation (LSD) fluid wave which exits at high velocity.

We consider only continuous devices in this paper. Plasma creation greatly reduces engine efficiency. Therefore, only reversible processes will be considered. The heating mechanism of steady conversion is called laser-supported combustion (LSC) [7,8]. This type of heating assumes continuous wave (CW) beam generators or pulsed generators which operate at high repetition rate.

ENERGY TRANSFER

Beam attenuation requires ion-electron pairs in the gas. A two-step process transfers energy from the beam to the gas. First, the electron acquires kinetic energy from the beam E-field ($f=ma=Ee$). Between collisions, the electron oscillates about a constant velocity:

$$v = v_0 + (Ee/m\omega)\sin(\omega \cdot t) \qquad (1)$$

At each collision, excess electron energy is transferred to the gas molecule.

The process of the energy transfer is described by Ginzberg [3]. His model of energy coupling was done for gases in the ionosphere where pressure is less than one atmosphere. Free electrons are excited by electromagnetic waves extending from low-frequency radio to the microwave region. As will be shown, this work is also valid for pressures of several atmospheres and EM radiation at sub-millimeter wavelength.

Stallcop [6] uses quantum mechanics to model energy transfer from free electrons to neutral hydrogen atoms. The electrons are driven by IR radiation. This model was developed for astrophysical processes at low pressure.

The Ginzberg model is based on classical physics and is used to derive two equations which are important for beamed energy propulsion. These are used to calculate beam attenuation coefficient K (expressed in inverse meters) and free electron energy U (expressed in volts).

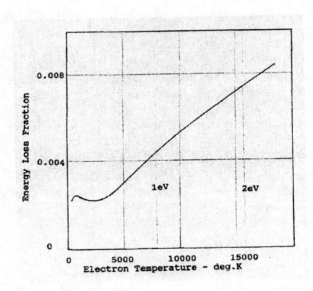

FIGURE 1. Fraction of electron energy lost per collision with molecular hydrogen.

Beam Attenuation

The equation developed for beam attenuation is:

$$K/nN = \sigma\sqrt{2/\delta} \cdot (e/m)^2 \cdot eEZ/\omega^3$$

(2)

In this equation, σ is the molecular cross-section. The parameter δ is the fraction of electron kinetic energy which is transferred to a molecule by a collision. These two parameters are properties of the gas molecule and vary with electron kinetic energy. Experimental values for the hydrogen molecule are shown in Figure 1. Values for hydrogen and atmospheric gases were calculated by Ginzburg [4]

Electron Energy

The equation developed for electron kinetic energy is:

$$U = (e/m) \cdot E^2 /(\omega^2 \delta)$$

(3)

Molecular energy levels are driven by electron collision and these are expressed in volts. Typical energy levels for molecules are in the 1 Volt region. Because the hydrogen molecule dissociates at 4.4 Volts [12] to create atomic hydrogen, electron KE must not exceed about 1 Volt.

Optimal Frequency

Equation 1) contains the number density of electrons and molecules. Given the gas temperature and pressure, N is determined from the gas law. The electron density, n, is

assumed as n/N=10e-4. The rms. field strength is related to the beam power density as $E^2 = SZ$, where S is the beam power density in Watts/square meter. Z=377 Ohms is the impedance of free space. Beam power is equated to stagnation enthalpy flow rate to determine E. Calculations are conducted at the sonic flow point based on equations of one-dimensional gas dynamics [1].

Attenuation and electron KE are shown in Figure 2a for sonic temperature of 2500 deg K and sonic pressure from 2 to 4 atmospheres. For this sonic temperature the gas expands to a theoretical exhaust velocity of 9245 m/sec. (Isp = 943 sec.). This is about twice upper stage performance using chemical propulsion. Attenuation length under 10 cm. is achieved near 200 microns beam wavelength for 2 atmospheres sonic pressure. The electron KE is below 1 Volt at this wavelength.

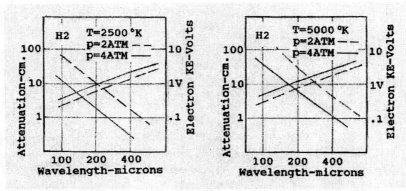

Figure 2. Hydrogen attenuation and electron kinetic energy in sub-millimeter wave region.

From Figure 2a it is seen that a 'window' for gas absorption exists near 200 microns. Below this wavelength, the attenuation length becomes large and a weight penalty occurs due to large nozzle size. Above 200 microns, electron KE becomes large and gas breakdown may occur.

The influence of sonic pressure is to shift the window to less than 200 microns as the pressure is increased to 4 atm. There is only a minor change in attenuation and electron KE within each window for 2 and 4 atm.

Figure 2b shows attenuation and electron KE for sonic temperature of 5000 deg K corresponding to a theoretical Isp of 1333 sec. which is about three times better than conventional propulsion. An upward shift in both attenuation length and electron KE occurs with increased temperature. Also, the window wavelength is increased.

Helium is also a candidate for BEP working fluid. Attenuation and electron KE for helium sonic temperature of 2500 deg K is shown in Figure 3a. The corresponding graph for 5000 deg K is shown in Figure 3b. Because helium is a monatomic gas and does not have vibrational degrees of freedom, less energy is lost per collision. This results in lower value of δ which causes increased attenuation length and electron kinetic energy. The value used for figures 3a and 3b for δ is 0.00027 which is based on kinetic theory. The theoretical value ($\delta=2m/M$) assumes elastic collisions [3]. Some energy is lost due to excitation of electron energy levels and actual values for δ will be somewhat higher due to this effect [11,12].

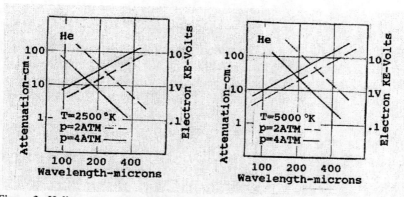

Figure 3. Helium attenuation and electron kinetic energy in sub-millimeter wave region.

Real Gas Effects

Figures 2 and 3 indicate favorable conditions exist for beam conversion in a short distance without gas breakdown. Calculated gas properties at the sonic point were used to determine the validity of the ideal gas assumption.

The gas is assumed to be weakly ionized at a constant ion fraction. A calculation can be made of thermal ionization effects based on the Saha equation [2]. At 5000 deg K the ion fraction is less than 2% of the assumed ionization.

The gas is assumed to be free of space charge effects. That is, the gas has no net charge at any location. A measure of space charge is the plasma frequency. A calculation of plasma frequency gives a value of 200 GHz. (1500 microns wavelength) which is an order of magnitude below the frequency of interest for favorable gas absorption. This means that the gas acts as a resistive medium as desired.

The beam frequency is assumed to be greater than the electron collision frequency. This means that an electron experiences one or more cycles in the average time between collisions. This assumption was made for Equations 1 and 2. The calculated electron collision rate is 0.146e12 collisions/second. This is almost two orders of magnitude below the beam radian frequency.

The gas is assumed to be weakly dissociated. The calculated dissociation fraction is 0.072 at 5000 deg K. At 2500 deg K., dissociation is negligible. The lower temperature is typical for chemical rocket design. The higher temperature is well above those encountered with chemical rockets. Chamber pressures for chemical rockets are of the order of 1000 psi. (68 atm.). This is an order of magnitude above the pressures considered here for conversion devices. The lower pressure will exaggerate dissociation effects as well as increase delay times for flow equilibrium.

Reaction Rates

Flow velocity is roughly 1 cm/microsecond Assuming 10 cm. attenuation length, equilibration times should be less than 1 microsecond to assure model accuracy. This is clearly the case for electron collisions. The molecular collision rate is 3e+09 collisions/second so roughly 3000 molecular collisions occur per flow centimeter.

Flow Stability

Two modes of beam entrance are possible. An intense radiation beam enters the heating chamber via an entrance window or the gas exit port and heats the working gas in the subsonic heating region. The gas exits through a supercritical nozzle and expands clear of the incident beam providing reaction thrust.

Proposed two-port BEP motors have been continuous devices. Heat generation is entirely inside the chamber and stability analysis is done by methods used for chemical combustion engines. In general, single-port devices have been pulsed. We consider here single-port devices which are continuous.

The flow geometry shown in Figure 4 is approximated by one-dimensional steady flow. The requirements of mass and energy conservation [1] give:

Mass: $\qquad \rho u A = W$ $\qquad\qquad\qquad\qquad\qquad\qquad$ (4)

Energy: $\qquad h + u^2/2 = h_r(1 + Q)$ $\qquad\qquad\qquad\qquad$ (5)

W is mass flow rate and h_r is reservoir enthalpy. $Q = P/Wh_r$ is the ratio of beam power to enthalpy flow rate from the reservoir. The gas is assumed to be thermally perfect and calorically perfect so that pressure is given by $p = \rho R T$ and enthalpy is given by $h = c_p T$.

The differential equations of momentum conservation and beam attenuation are:

Momentum: $\quad \rho u u' + p' = 0$ $\qquad\qquad\qquad\qquad\qquad$ (6)

Beam Power: $\quad P' = KP$ $\qquad\qquad\qquad\qquad\qquad\qquad$ (7)

K is a function of gas density and temperature as well as beam intensity.

Stability analysis using the above equations in unsteady form is used to demonstrate dynamic stability [9]. Here we derive the condition required for static flow stability. This is done by solving for the location of the sonic point.

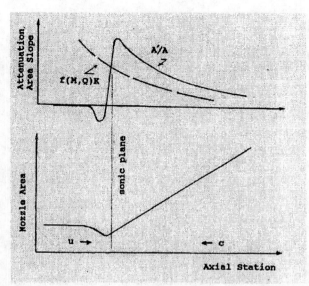

FIGURE 4. Sonic flow occurs at axial station downstream of throat with beam heating.

With heat addition, the sonic point is displaced downstream of the minimum diameter point of the throat. Consider the relation:

$$(M^2 - 1)u'/u = A'/A - \left(1 + \frac{\gamma - 1}{2}M^2\right)\frac{Q}{1+Q}K \qquad (8)$$

This equation is from the preceding equations for mass, energy, momentum, and beam power. At M=1 the right side of equation (8) must equal zero to maintain finite sonic velocity. Hence, the sonic point is specified by:

$$A'/A = \frac{\gamma + 1}{2}\frac{Q^*}{1+Q^*}K^* \qquad (9)$$

If Q>>1, the sonic point is determined by channel slope and attenuation coefficient. The sonic point is determined as the intersection of the area slope with curve of attenuation coefficient as show in Figure 4.

From the foregoing, it is seen that improperly chosen channel shape and attenuation profile can cause unstable flow. This type of instability, due to gas heating by the beam, is termed static flow divergence. Other types of instability may occur, due to acoustic delays in the subsonic region and upstream influence cause by beam heating in the supersonic region. A complete investigation of flow stability should include unsteady flow phenomena. Techniques are available for such analysis [9].

FLOW GEOMETRY

Conceptually, energy conversion for beamed energy propulsion might be described as gas flow in one direction absorbing energy from the beam directed from the opposite direction. This is not a useful model at low Mach Nr. as flow is governed by laws of gas dynamics while beam direction is simply described by reflection optics. At high Mach Nr. this model is approximately correct but low absorption is required in this region so that only limited benefit is derived.

Configurations where beam direction is opposite to flow are considered. Subsonic heating with beam entrance via the gas exit port is the principal source of gas enthalpy. Hybrid devices using lens or conduction to introduce beam energy into the chamber for preheating may also be considered.

Nozzle Design

Nozzle design is driven by the following considerations:
1) Nozzle throat diameter approximates absorption length.
2) No channel obstructions occur above M=1.
3) Absorption is rapidly reduced above M=1.
4) Expansion ratio of 25 to 100 is desired.

Also, the design must be insensitive to angular motion about axes normal to the beam. It is desired that the design have a positive static margin. This means that small angular motion about a transverse axis creates a restoring torque.

From the previous analysis, the design range for chamber pressures varies from one to ten atmospheres and sonic gas temperatures from 2000 to 5000 deg K. Because chemical rockets operate at several times this pressure and within this temperature range, it is expected that conventional cooling methods such as regenerative cooling are acceptable.

It is desired that the nozzle operate at high efficiency and low weight. Regenerative cooling not only maintains acceptable nozzle temperatures but also acts to pre-heat cryogenic propellants. This recovery increases efficiency.

It is desired that the nozzle design operate in a clustered configuration. This would shorten the thrust assembly and allow thin-wall construction to reduce weight.

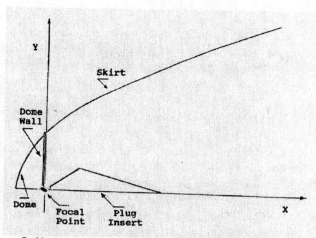

Figure 5. Nozzle cross section showing components significant to beam heating.

2-D Nozzle Design

A generic nozzle cross-section is shown in Figure 5. The wave front is parallel and extends approximately over the base. A parabola is the indicated shape to concentrate energy at a point within the chamber behind the throat. The parabola is described by the equation: $2x = y^2 - 1$. The parabolic arc is composed of a dome, described for x<0, and a skirt, described for x>0.

The dome and skirt contribute equal beam energy to the focal point if the base is extended to y=2. Beyond this value, energy from the skirt region is dominant. For this reason, the throat axis is oriented toward the skirt area. Orientation toward the dome area implies flow turning after throat passage and this is not feasible.

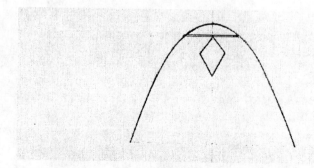

FIGURE 6. Bell nozzle created by rotation about longitudinal axis at dome center.

A dome wall is placed at x=0 between y=0 and y=1. This wall diverts flow toward the exit and away from the dome. A plug is inserted at the outer edge of the throat exit. The plug angle is chosen to align with the base edge so that beam entrance is not blocked. The plug is extended to a corner point in order to create an expansion fan which relieves the compression corner at x, y=0, 1.

3-D Nozzle Design

A 3-dimensional nozzle is created by rotation about the intended centerline. Rotation about the dome centerline (y=0) creates a bell nozzle. Rotation about a longitudinal axis which intersects the skirt creates a spike nozzle.

Figure 6 shows the bell nozzle. Because the focal point is on the rotation axis it remains the focal point. The bell nozzle skirt intercepts a larger fraction of the beam because it rotates at larger radius. For example, a base extended to x=2 intercepts 75% of the beam. If the base is extended to x=3, the skirt intercepts 89% of the beam energy.

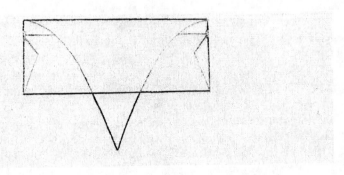

FIGURE 7. Spike nozzle created by rotation about longitudinal axis piercing skirt.

Figure 7 shows the spike nozzle. Rotation about a centerline which is outside the dome creates a focal circle. Because the dome sweeps through a larger radius, a larger fraction of beam energy is intercepted by the dome area. For example, for a centerline at x=2 only 25% of the beam is intercepted by the skirt. A centerline at x=3 intercepts 44% of beam energy.

The spike nozzle has an elongated throat in the x-z plane describing a circle. The radius of this circle is significantly larger than the corresponding radius of the bell nozzle. Because the two designs have throat areas which are approximately equal, a greatly reduced longitudinal size is indicated for the spike nozzle.

Misalignment with the beam centerline will cause the reflected beam to miss the throat. The dominant error will occur due to rotation about transverse axes which will cause y-axis motion of the focal circle. (This motion is not a problem for LPDE designs [15] which require beam concentration on a focal surface.) Parabolic surface tolerances will also cause defocusing and result in beam energy loss at the throat. This problem can be fixed by reshaping the parabolic surface to focus onto a discrete number of focal points evenly distributed along the focal circle.

The dome area introduces beam loss for the single-port engine design. The dome problem is more pronounced for the spike design. A lens might be used to port the dome energy into the chamber for gas ionization and heating.

DISCUSSION

Comparison with IR

Longer wavelength leads to improved efficiency for sub-millimeter waves. IR has been shown to work for laser supported detonation. However, much higher plasma content is needed to support LSD and this is a source of energy loss. It is also known that longer wavelength EM waves have reduced reflection loss so that material heating is reduced both at the generator and the terminal device.

A larger sending antenna is required for sub-millimeter waves. Diameters must be an order of magnitude larger compared with far infrared beaming devices. This applies also at the reception end. With longer wavelength, devices will be larger and heavier. This disadvantage is partly offset by dimensional tolerance relaxation and lower power density at longer wavelengths with large antenna size.

Comparison with Microwave

Higher efficiency is obtained from microwave devices as frequency is reduced. However, gas breakdown occurs because per cycle kinetic energy buildup for free electrons is increased as frequency is reduced. Proposed microwave BEP devices are based on electro-thermal heating inside the chamber.

Efficiencies above 80% are possible at 3.0 GHz [13]. However, efficiency does not scale favorably to higher frequencies.

Space vs. Ground-based BEP

Due to water vapor absorption, sub-millimeter wave transmission through the atmosphere is greatly attenuated. Reliable transmission is obtained at stratospheric altitudes and may be possible from one or more of the highest mountains if located in a dry region. High altitude aircraft (LTA or conventional) as support platforms are potential locations for BEP generation.

A BEP device will likely have very high peak-to-average power ratio. Intermediate energy storage and rapid dissipation of waste energy are factors which favor ground-based systems.

Reuse is an advantage of space-based BEP. An upper stage may be raised to LEO by the beaming device. After insertion and storage recharge of the beam generator, the stage is raised to higher orbit or escape trajectory. By this method, the velocity increment is doubled.

REFERENCES

1. Liepmann, H.W., Roshko, A. J., Elements of Gasdynamics, John Wiley, New York (1957).
2. Cobine, J.D., Gaseous Conductors, Dover Publications Inc., New York (1958) Ch. 4.
3. Ginzburg, V.L., Gurevich, A.V. Nonlinear phenomena in a plasma located in an alternating electromagnetic field, Soviet Physics-Uspekhi 3,147 (1960).
4. Ginzburg, V.L., Propagation of EM Waves in Plasma, Pergamon Press (1964) page 69.
5. Kantrowitz, A., Propulsion to orbit by ground-based lasers, Astronautics and Aeronautics, Vol. 10, (1972), pp. 74-76.
6. Stallcop, J.R., Absorption of infrared radiation by electrons in the field of a neutral hydrogen atom. Astrophysical Journal 187 (1974), pp.179-183.
7. Legner,H.H., Douglas-Hamilton, D.H., CW laser propulsion, J.Energy-Vol.2,No.2 (1978)
8. Kemp, N.H., Root, R.G. Analytical study of laser-supported combustion waves in hydrogen, J. Energy -Vol.3, No.1 (1979).
9. Johansen, D.G., Stability of radiation-heated flow, Acta Astronautica, Pergamon Press (1980)
10. Nebolsine, Pirri, Goela, Simons, Pulsed laser propulsion, AIAA Journal, Vol.19, No.1, (1984) pp.127-128
11. Csanak, G., Cartwright, D.C., Srivastava, S.K., Trajmar, S., Molecule Interactions and their Applications (ed. by L.G.Christoporou) , Academic Press (1984), pp.103-109.
12. Loeb, L.B., Basic Processes of Gaseous Electronics, Univ. Calif. Press, Berkeley, CA, p.393.
13. Frasch, L.L., Fritz, R., Asmussen, J., Electrothermal propulsion of spacecraft with mm. and sub-mm. electromagnetic energy. J. Propulsion-Vol.4, No.4, (1988), pp.334-340.
14. Keefer, D., Sedghinasab, A., Wright, N., Quan Zhang, Laser propulsion using free electron lasers, AIAA Journal, Vol.30, No.10 (1992), pp. 2478-2482.
15. Mead, F.B., Jr., Myrabo, L.N., Messitt, D.G., Flight experiments and evolutionary development of a laser propelled, trans-atmospheric vehicle, Proceedings of the SPIE, (1998) pp.560-563.

MW Energy Addition in Application to Propulsion

V. G. Brovkin, Yu. F. Kolesnichenko

Institute of High Temperatures of Russian Academy of Sciences,
Izhorskaya str. 13/19, 127412, Moscow, Russia

Abstract. Possibility of propulsion production by means of MW energy source disposed outside of an accelerating craft is discussing. MW energy supply conception is worked out in less degree in comparison with the laser one. Meanwhile, microwaves are attractive from the point of non-absorbing energy delivery through the atmosphere, relative simplicity of MW beam control and positioning, high efficiency of MW generators. The limitation for microwave energy application comes from the fact that the threshold of MW breakdown is sufficiently less than that for infrared or shorter wavelength lasers. The critical problem of MW energy transformation into heat is considered. Both experimental data and theoretical considerations for the efficiency of high-power MW beams interaction with gas, plasma and solid bodies are exposed. The cost of a thrust is shown to attain values up to *1N/kW* under atmospheric air pressure. Combination of laser and MW beams can also be regarded as prospective.

INTRODUCTION: THE SKETCH OF MBEP CONCEPT

Possibility of propulsion production with the aid of energy source disposed outside of an accelerated craft has received new investigation impulse during last years. The idea of energy addition in reactive thrusters through electromagnetic beams is attracting attention due to some obvious advantages. The main ones for the case of rocket engine are that the effective thermal power production of the working gas and its atomic weight are not limited by the chemical characteristics of fuels and, therefore higher values of the specific impulse can be achieved. In the case of air-jet engine the fuel is of no need at all. Despite of MW energy supply conception is worked out sufficiently less in comparison with the laser one, it was proposed several years before the well-known revolutionary article of Kantrowitz [1, 2]. Nevertheless, exactly the progress in laser technique renewed the recent splash of interest to the beamed energy propulsion. Meanwhile, the advantages of microwaves are still visible and it is reasonable to refresh our understanding of both advantages and those problems, which are critical for microwave beamed energy propulsion (MBEP).

Microwaves are attractive from the point of view of non-absorbing energy delivery through the atmosphere, relatively simple MW beam control for pointing and tracking, high efficiency of MW generators – these features stand better in comparison with laser ones. Another important advantage is that both MW generating and radiating systems can be realized in a block manner, i.e. they are scalable. Moreover, recent progress in fabrication of solid-state MW emission sources may lead to new appearance of these systems. At any case, producing and positioning of powerful beams of MW energy is not a limiting chain in MBEP progress. Rough estimations

CP664, Beamed Energy Propulsion: First International Symposium on Beamed Energy Propulsion,
edited by A. V. Pakhomov

show that the scale of power that might be used for purposes of through/in atmospheric flight is up-limited by the range of 1 – 10 GW in MW CW operation mode. These levels of MW power are extremely high, but not fantastic - MW installation of 20 Megawatt mean power was built in Russia about 30 years ago and was successively exploited for a long period of time [3]. And, finally, such MW beam-producing complex can be used in multi-task manner, say for delivering power to orbital stations and plants or, on the contrary, receiving it from an orbital solar power plants, different tasks of reconnaissance, etc.

Disadvantages of microwaves for BEP have the same rout as their advantages – the wavelength. Focusing of beams of MW energy at large distance demands rather extended radiating areas. Simultaneously, the lateral maximums in directivity diagram can cause undesirable consequences for environment. Large wavelength of microwave radiation leads also to the limitation of MW flux density by the value of MW breakdown threshold in the Earth's atmosphere, which is sufficiently less than that for infrared or optic-range lasers. But the critical problem for MBEP is organizing of the efficient process of on-board MW energy - heat conversion. This topic is closely linked with MW discharge properties and our paper is aimed mainly at lightening the physics of MW discharge phenomenon.

MW DISCHARGES: HISTORY

The idea of propulsion production by means of MW energy source disposed outside of an accelerating craft was announced about 40 years ago [1], just after appearing of the first experiments with MW discharge under the free-space conditions [4,5]. By that time the era of MW technique application went through a boom, the units with higher and higher power appeared, as well as the range of the mastered wavelengths was broadening. Experience in MW discharge investigation was also growing up. First microwave discharges were obtained and studied at 40-th, when the newly appeared radar technique actively developed. The discharges were burned in tubes, waveguides and resonators. Both electrode and electrodeless MW discharges were obtained. In spite of relatively small volumes of the examined MW plasma, many of important features of MW discharge were revealed and investigated. The most explicit statement of the results concerning MW breakdown was exposed in a well-known book of McDonald [6].

Probably be the first published investigations where the idea of MW energy concentration in space was realized and free localized MW discharge was obtained were the mentioned above works [4,5], appeared at the very beginning of 60-th. In [5] a striated MW discharge in argon was produced by high-power, continuous-wave 3.4cm wavelength radiation focused into the central region of a spherical 12-inch diameter discharge vessel. An X-band multi-cavity klystron was capable of delivering a CW output power of more than 2kW and fed a large 44-inch aperture parabolic mirror so as to produce a parallel beam of radiation. This beam of radiation was then focused to a small volume using a MW lens. This scheme of experiment was used from that time in many other MW installations.

Development of new powerful MW generators as well as MW technique as a whole, including the appearance of phased antenna arrays gave qualitatively new facilities for MW discharge experiment There appeared the possibility to create an intensive microwave beams in free space.

At the early of 70-th two Russian powerful microwave installations had been put in practice in Scientific Research Institute for Radio Device (in 1987 these installations were assigned to Moscow Radiotechnical Institute [3] together with the staff and collective of investigators). These installations were capable of creating a powerful microwave beams in virtually free space. Their possibilities and parameters noticeably exceed those of any other unit that has been built before or after. A phased antenna array formed a linearly or circularly polarized 7 GHz microwave beam and directed it into a vacuum chamber, which was hermetically closed from the ends by microwave transparent inputs. A special load absorbed microwave radiation that has passed through the chamber. Thus, a running wave regime was realized. In the region of the antenna geometric focus, a caustic was formed (shape beam), its half-depth and diameter being equal to 10 cm and 5 cm respectively (these limits are given according to the 3 dB level). In this focal region the regime of plane running EM wave occurred. This was the region in which the microwave gas discharge was investigated. The microwave flux density here attained 10^5 W/cm^2. Integral and speed photorecorders, electrodynamic and shadow facilities, and discharge emission spectra recorders were available during conduction of experiments. Several scientific organizations such as the Institute of General Physics of USSR Academy of Sciences, Moscow State University and many others took part in conducting investigations at these installations. It became clear that new horizons in gas discharge physics were opened.

GENERAL DESCRIPTION OF MW DISCHARGE PHENOMENON

It is well known that for self-ignition of a discharge by an electromagnetic (EM) impulse, an electric field intensity that is equal to or exceeds the breakdown threshold must be used. This is determined by the Paschen curve. Below this curve, there exists a curve for maintenance of a stationary CW discharge. The maintenance electric field of the stationary CW MW discharge is of order that at the Paschen curve minimum. The domain lying between these two curves belongs to propagating stimulated MW discharge (SMD). Thus, SMD exists in electric fields that are less than the breakdown field. Therefore, a stimulator is necessary for discharge ignition (this is the reason for discharges name - stimulated). One may use as a stimulator a pointed rod, a "whisk", made of fine wire, dielectric-metal plates, flame, aerosols, etc. During microwave beam illumination of the stimulator an initial discharge formation appears around it. After some time this formation "takes off" from the stimulator and propagates opposite to the electromagnetic radiation flux. In the discharge propagation regime, shape, structure, and other characteristics do not depend on the type of stimulator. The curve for the existence of stationary SMD lies much lower.

One general feature unifies self-ignited and stimulated MW discharges. Both are dynamic objects, i.e. they propagate towards MW radiation source. This phenomenon was revealed practically simultaneously with waveguide technique appearance as at transmitting of high power MW the breakdowns sometimes occurred and the originated plasma creation propagated in the waveguide. But the physical mechanisms for discharge propagation in super- and sub- breakdown fields may be quite different. The behavior of MW discharge in super-breakdown field is very sensitive to EM field and initial electrons distributions and its mechanism is the wave of breakdown. In principle, such a discharge may propagate at any defined velocity and it depends mainly on EM field distribution. The SMD is much less sensitive to this factor and is much more affected by its own self-organization processes. This MW discharge phenomenon will be presented here in a more detailed way as it becomes more and more evident that structural factor plays the most important role in MW field – gas interaction.

What are the possible mechanisms of MW field-plasma interaction? As it is well known, the most effective is the resonant interaction. At low pressures, when the angular MW frequency ω sufficiently exceeds the frequency of electron elastic collisions ν_m EM field interacts resonantly with plasma in a spatial region where Langmuir frequency is equal to ω (or, in the presence of magnetic field, to electron cyclotron frequency). This type of interaction has also been investigated at these installations (the so called "vacuum torch"). At high pressures, when $\omega \ll \nu_m$ the resonance is suppressed by dissipation and there can not be local plasma resonance's. But there may be resonance in interaction with plasma structure as a whole, if it forms, for instance, the appropriate slowing structure. Such structures are well known from theory of antennas.

MICROWAVE DISCHARGE IN SUB-BREAKDOWN FIELDS

Perhaps the most striking peculiarity of the SMD in free space is its structural variety. Here attention will be briefly paid only on the principles of structural organization of SMD caused in gas of middle and high pressure (p > 10 Torr) by action of an individual microwave radiation impulse.

Structural factor seems to play the most important role in MW field – gas interaction [7]. It was determined that there exist several qualitatively different SMD structures. The transition from one to another is rather sharp, i.e., takes place over a small variation of external conditions. Thus, the region for the existence of each SMD type is very definitely outlined. The most detailed investigation was made of the two fundamental SMD structures. Each separate region was named a *structural zone* (SZ).

The universality of the two investigated SZ is characterized not only by the commonality of their boundaries, but also by the facts that:

- The discharge propagation mechanism is the same within the SZ. The transition from one discharge structural type (SZ) to another is accompanied by a fracture in the logarithmic graph illustrating the discharge propagation velocity as a function of wave field intensity or gas pressure (see below). This is indicative of the change in the discharge propagation mechanism. The slope of the graph increases with increase of EM field strength in the subsonic and supersonic regions of SMD propagation velocities. This suggests that the transition from one discharge structure type (SZ) to another is caused by the appearance of a new propagation mechanism providing higher velocity of discharge propagation than the previous one (in other words, it confirms the principle of maximum propagation velocity). Thus, the discharge propagation mechanism and the discharge structure are closely linked.

- Structure formation in a SZ takes place at the base of a regular structural element, which was named the *base element* (BE). The form and dimension of the BE do not depend on the type of gas, and are determined by the microwave radiation polarization and wavelength (Fig.1).

Thus, the BEs, taken as structures, are *universal* and *stable*. Their shape and dimensions do not depend on the type of gas, the field strength and gas pressure, or the microwave beam width and convergence angle. Their shape does not even depend on the radiation wavelength, which determines only the dimension of the BE. The EM field polarization seems to be the most influential factor.

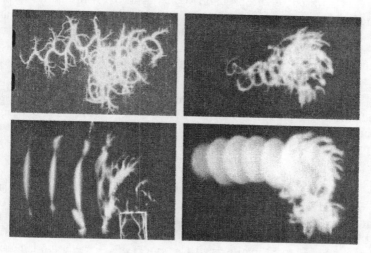

Figure 1. Base elements of MW discharge. Upper row – WSZ, lower row – DSZ. Left column – linear MW beam polarization, right column – circular polarization.

In the discharge structure, the BE reflects the wave character of the EM field in a most pronounced way. It may be regarded as a *self-organizing plasma running-wave antenna*, especially taking into account that its dimensions are **resonant** to the radiation wavelength, and that its polarization and axis of symmetry coincides with the polarization and wave vector of the EM radiation. All these conditions provide for very favorable reception of microwave energy. Depending on the SZ and radiation polarization, BEs in the form of dielectric, spiral, sinusoidal, director and other antennas are realized.

Different methods, using speed, shadow and integral photo-recording in the visible radiation range, were applied for measurement of the discharge propagation velocities [5, 6, 8]. The propagation velocity may be presented in the form: $v \sim E^{\mu} \cdot p^{-\nu}$, where μ and ν depend on the gas and are approximately constant for each structural type.

In SZ with sinusoidal base element a hierarchy exists not only in the dimensions, but also in the *velocities* of propagation of the discharge structure separate elements. The first velocity scale is of order 10^5 cm/s. This is the velocity of discharge propagation as a whole (more exactly, of the discharge front). The power indices vary: μ - from 1.7 in argon to 2.5 in air, ν - from 0.5 in air to 1 in carbon dioxide. The discharge propagation takes place with supersonic velocity. The second velocity scale is that of leader propagation (in a number of cases, the speed photo-recording made it possible to observe the bright leader propagating along the already paved way, the luminosity of the latter being very weak). This dependence corresponds to $\mu = 1$ and $\nu = 0$. Thus, we estimate the order of this velocity scale to be 10^6 cm/s. Let us point out that both the scale of the leader velocity and the dependence of velocity on E and p differ from those for the discharge as a whole. This fact confirms the pulsed, uneven, stepped character of discharge development in WSZ.

Measurements of discharge propagation velocities in SZ with dipoles have shown that values for μ and ν are two and four times greater than the values μ and ν in streamer discharge, and that they show a tendency to decrease with increase in pressure from 50 to 200 Torr. Thus the difference between the values for μ and ν is becoming small, and may be justified the approximation that the discharge velocity is a function of the parameter E/p: $\nu \sim (E_0 /p)^k$, where $k \cong 3...4,5$. This, along with the fact mentioned above of discharge structure reproduction (from pulse to pulse), leads us to conclude that in the process of discharge structure evolution, there appear regions of self-sustained breakdown in the sum field of the incident wave and the wave reflected from the dipole(s).

One of the distinctive characteristic features of the SMD is the existence of the extended UV hallo in front of the discharge. This phenomenon is closely linked with plasma characteristics in a channel, as namely channels are the sources of intensive UV radiation. The characteristic length of a hallo in air at atmospheric pressure is about 0.5cm and the mean electron concentration around 10^{12}cm^{-3}. At decreasing of pressure the length of a hallo increases and the mean electron concentration decreases. This phenomenon is observed in all gases. The model of UV radiation is absent up to now, and the question of the nature of UV emission that demonstrates absorption coefficient sufficiently less of that theoretically predicted has no answer since 1934, when Raether has discovered UV radiation from electron avalanche in air.

SELF-IGNITED MW DISCHARGE IN GAS FLOW

High flexibility of MW technique made it possible to create in supersonic flow two basic discharge configurations in respect to flow velocity - transversely and longitudinally oriented discharge structures. Investigation was fulfilled at specially created unique experimental complex, including wind tunnel and impulse 2µs pulse-duration 1kHz repetition-rate X-range MW generator [8]. Discharge properties in quiescent gas and in a flow turned out to be quite similar. Under the free flow conditions for linearly polarized MW flux separate plasmoid has a shape of elongated ellipsoid of revolution with maximal diameter of about *6mm* and total length about *20-*

22mm. Its shape and dimensions is independent of whether the discharge is longitudinal or transverse. Separate MW breakdown domain passes through several phases in its development and as a result of it the discharge plasmoid is not homogeneous, but have an internal structure. Inside relatively uniform bulk main domain of plasmoid, there appear very thin hot channel, or filament. The last can be characterized by high concentration of energy and high gas temperature, whereas the bulk domain demonstrates rather low temperature increase. The main domain can be characterized by specific energy input about $2J/cm^3atm$, electron concentration about $6 \cdot 10^{13} cm^{-3}$ and temperature rise *80-100K*. The rate of heating for the main domain of plasmoid is $\sim 100\ K/\mu s$. In the filament this parameter is sufficiently higher and attains ~ 2-$3\ kK/\mu s$. Gas temperature in the filament obtained for the moment of MW pulse termination turned out to be *2 800K* [9].

In spite of the filament can be observed by integral and high-speed photo devices, it is rather difficult to extract its parameters from the most commonly used experimental source – optic spectra. The reason is in sharply different dimensions of the bulk and filamentary domains – *0.5cm* and less than $10^{-2}cm$ in diameter, correspondingly. Therefore, the ratio of volumes of the filament to the main bulk domain of discharge is about *0.3%*. Under these condition processing of the optional second positive nitrogen system delivers only the temperature of the bulk component. Probably this is the reason for invalid interpretation of number of experiments in this area, when investigators attributed the obtained low temperature to the entire discharge volume and were forced to introduce the action of some phantom plasma effects.

Receiving of optic spectra information from the filament demands fixation of emission from species that appear in high-energy threshold processes. In our case the electronically excited nitrogen molecule positive ion is taken for this purpose. Comparison of the experimental results with those obtained in kinetic modeling let us determine the minimal ionization degree in a filament as $(3$-$5) \cdot 10^{-3}$. It means that electron concentration in the filament exceeds $10^{16}\ cm^{-3}$ at *70 Torr* air pressure.

The exact relation between energy deposited in the filament and the main domain of discharge is not determined yet, estimations show that energy dissipated in filament is less but noticeable of that in the main domain. Energy deposition in MW breakdown domain leads to shock wave formation and propagation from it. For carbon dioxide estimation of energy input by examination of shock wave created by separate discharge channel gives the value of energy deposition per unit length (in a form of instant heat release) of order *50-70mJ/cm·atm*. Such a value for air is not obtained directly, but have some indication to be found in this range.

For the both discharge configuration testing of AD body in air and carbon dioxide flows with Mach number about 1.7 revealed the 35% stagnation pressure decrease and of streamlining picture modification while interaction with discharge domains.

GAS DYNAMIC PHENOMENA IN QUIESCENT GAS NEAR-SURFACE MW DISCHARGE

Investigations of gas dynamic characteristics of MW stimulated discharge started simultaneously with the analogue experiments in laser area. While studying gas dynamic effects at MW irradiation of metal-dielectric targets in air of middle and high pressure, a new phenomenon – fast gas heating was discovered. This effect is also observed in SMD gas heating. Heating rates up to 50-100K/μs during 10...15μs were measured by the optical method, leading to gas temperature 1000...1500K by the end of this time interval. The time scale from several to tens microseconds completely excludes VT-relaxation, therefore the effect was named *anomalous heating*. The final temperature does depend on both gas pressure, and MW energy flux density, these parameters influence only the heating rate. The most popular nowadays explanation adds up to the assumption that a sufficient part of energy input in electronic states of molecules (up to 30%) converts to translational degrees of freedom during electronic states relaxation. Nevertheless, reliable kinetic schemes are unknown.

Thus, already first experiments have shown high rate of gas heating in near-surface discharge region and hence, the possibility of shock waves creation in the vicinity of a target [10,11]. The registered by optic emission spectral methods gas temperature reached 1800-2000K under the heating rate exceeding 30-40K/μs. The processes of shock waves formation and gas density evolution in the discharge region was registered by the shadow method. For discharge initiation the flat dielectric targets with metal insertions and the special antenna-type stimulators were applied. Both those and these were discharge stimulators with the lowered threshold of discharge ignition. As a result, formation of semispherical and cylindrical shock waves, propagating away from the target surface and discharge channel, correspondingly, was fixed.

These processes demonstrate their characteristic temporal phases and some peculiarities. The discharge near the surface starts its development firstly across the MW beam axis and only after reaching the beam boundaries begins its propagation towards MW source. Therefore, at the initial phase of energy supply the heating of a narrow flat gas layer with thickness about 2-3mm and transverse dimension about radiation wavelength takes place. This results in formation of a flat dense layer, which is gradually transforming in a semispherical shock wave. Simultaneously, the discharge front after some delay starts its movement in counter-MW-flux direction, catches and breaks the shock wave front. The shadowgram presents distinctly the small-scale fluctuations of gas density from a discharge channels. The velocity of shock wave under atmospheric pressure and MW energy density flux $(2-4) \cdot 10^4 \text{W/cm}^2$ did not exceed one and a half of sound velocity, i.e. Mach number M=1.5. The forefront of small-scale fluctuations (discharge front) is propagating with higher velocity, which reaches about 2.5-3 velocities of sound.

As a result of fast gas heating the pressure increases in the vicinity of a target surface. Thus, the fast-heated gas media exerts mechanical action on a target surface. The value of mechanical impulse J was measured by ballistic pendulum method – via registering of its deviation from equilibrium position. The dimension of the basic target – stimulator was about 2.5 of MW radiation wavelength λ. In a number of experiments the transverse dimension of a target was varied from λ/4 up to about 4λ. For experimental results analysis the dependencies of the specific impulse J/E over gas pressure, MW pulse duration and the total spent MW energy were plotted. The dependence of the specific mechanical impulse value over the gas pressure in a chamber is presented at Fig.2. This curve demonstrates non-monotonous character, reaching its maximal value of about 50 dn·s/J in the pressure range 70-100 Torr. The linear increase of the J/E quantity over radiation intensity is observed in the pressure range 50-760 Torr, and with increase of MW pulse duration J/E rises up to some stationary saturation value (under pulse duration exceeding 200-300μs). Experiments also showed that the rate of gas heating and, hence, the intensity of shock wave can be controlled by the regimes of MW energy supply of the discharge region. Variation of MW pulse duration, pulse repetition rate and the intensity of MW beam can localize the discharge at a desired domain.

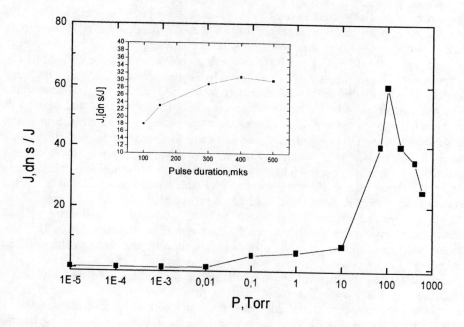

Figure 2. The dependence of the specific mechanical impulse value over air pressure in a chamber and MW pulse duration.

Application of impulse-periodic regime of gas media heating in a discharge region demonstrates the stable increase of J/E value practically in 2 times. Such a regime is mostly effective under 10-50μs pulse duration range. Subsequent increase of pulse duration leads to discharge run away from the near-surface domain and the effectiveness of energy transformation falls down dramatically. The increase of the mechanical impulse is also observed with increase of the target dimension from 2cm up to 10-12cm. The ultimate effect is attained in the case of target dimension about (2-3)λ.

Analysis of experimental result obtained for the low values of gas pressure in a chamber shows that in this case the main role in mechanical impulse production plays the vaporizing mechanism. However, the mass of evaporated substance of the target is not so large and the impulse value does not exceed several tenth units dn s/J.

CONCLUDING REMARKS

In the previous Sections we briefly presented the most important results concerning the properties of practically all types of MW gas discharge - self-sustained, stimulated, in vicinity of solid body surface, in quiescent gas and supersonic flow. How can these results be applied for MBEP? They definitely show what discharge types can be potentially used in MBEP, what are the ways for organizing the MW beam – flier interaction. Brief analysis shows that sub-breakdown fields are more effective for energy deposition process. Actually, the upper limit for the energy deposited per molecule (under the 100% absorbing coefficient) can be defined as $\varepsilon =$ S/pv, where S is the density of energy flux, p is gas pressure and v – discharge propagation velocity towards MW radiation source. As it was shown above, the velocity is a function of both p and S. For super-breakdown free localized beams this velocity is strongly dependent of S, such as the ε value tends to zero while S increase. For sub-breakdown fields ε-quantity also depends inversely of S, but sufficiently weaker: as $\sim S^{-1/2}$ for DSZ and S in about zero power for WSZ. Thus, inverse dependence of specific energy input is saturated in WSZ. Therefore, taking into account that absorption coefficient is rather high in this structural zone (of about 50%), it is natural to expect this type of discharge as prospective candidate for application. As its properties are investigated relatively well, it is possible to define the optimal parameters for irradiation of solid targets for mechanical impulse obtaining. This was fulfilled and the results of such optimization are partially presented in the previous Section.

At present level of knowledge we can suppose the radiation of the solid lower surface of the flier to be the basic way in propulsion production in the dense atmosphere. Atmospheric flight does not need in working gas. Disadvantage of relatively large dimension of MW beam spot turns as advantage, increasing the net force. Our investigation of MW discharge properties in wide beam, i.e. a beam with transverse dimension noticeably exceeding radiation wavelength, has shown that SMD in a wide beam looks like a structure that is "gathered" from the SMD structures in narrow beams. The characteristic dimension of the elementary cell is about radiation wavelength. Thus, the width of the beam does not change the basic structure of SMD. For this reason comparison of the discharge propagation velocity in wide and narrow beams at the same pressure and energy flux density demonstrates coincidence of these velocities in the bounds of measurement accuracy. It means that it is possible to scale the results of the previous Section to an arbitrary target – MW beam dimensions. As such, one can obtain that one Newton of thrust is provided by one kilowatt of MW power. It is seen that for practically reasonable demonstration the mean power of several hundreds of kilowatts is needed.

The flight in higher layers of atmosphere or low-orbit mission demands special treatment, as direct mechanism of solid body surface evaporation is not effective. The schemes of gas injection for creation of the near surface "atmosphere" can be proposed, as well as another ways of energy reception and conversion. Probably, for these technologies cooperative application of laser and MW beams will bring a breakthrough in beamed energy propulsion realization.

ACKNOWLEDGMENTS

At different stages of investigation authors collaborated with Dr.'s Igor Kossyi, Sergey Gritsinin, Dmitry Bykov and Dmitry Khmara. We express our deep gratitude to all these scientists.

Partially the results exposed here are obtained due to the support of the European Office of Aerospace Research and Development (EOARD).

REFERENCES

1. Shad, J.L., and Moriarty, J.J. Microwave rocket concept. – In: XVI Intern. Astronaut. Congress, Athens, 1965.
2. Kantrowitz, A. Propulsion to Orbit by Ground-Based Lasers. *Astronautics & Aeronautics*, May 1972.
3. Batzkikh, G.I., and Khvorostyanoy, Yu.I. Experimental unit for creation of powerful focused beam of microwave radiation in free space. *Radiotekhnika I Elektronika*, 1992, v.37, pp. 311-315. (in Rus)
4. Hamilton, CW *Nature*, **188**, p.1098, 1960.
5. Allison, S., Cullen, A. L., and Zavody, A. *Nature*, **193**, p.156, 1962.
6. MacDonald, A. D. *Microwave Breakdown in Gases*. New York, 1966.
7. Brovkin, V.G., and Kolesnichenko, Yu. F. Structure and dynamics of stimulated microwave gas discharge in wave beams. *J. Moscow Phys. Soc.* **5** (1995) 23 - 38.
8. Brovkin, V.G., Kolesnichenko, Yu. F., Leonov, S. B., et all. Study of microwave plasma-body interaction in supersonic airflow. AIAA Paper 99-3740, 30th AIAA Plasmadynamics and Lasers Conference, 1999, Norfolk, Virginia.
9. Kolesnichenko, Yu. F., Brovkin, V.G., Khmara, D. V., et al. Fine structure of MW discharge: evolution scenario. AIAA Paper 2003-0362, 41st AIAA Aerospace Sciences Meeting & Exhibit and 5th Weakly Ionized Gases Workshop, 6-9 January 2003 / Reno, Nevada.
10. Batanov, G.M., Gritsinin, S.I., Kossyi, I.A., et al. High-pressure microwave discharges. *Plasma Physics and Plasma Electronics* ed. L.M. Kovrizhnykh. New York: Nova, 1985, pp. 241-282.
11. Batanov, G.M., Gritsinin, S.I., Brovkin, V.G., et al. "Gas dynamic characteristics of stimulated MW discharge" in *Radiofizika*, edited by G.I.Batzkikh, Moscow: Moscow Radiotechnical Institute, 1991, pp.53-70.

MHD Augmentation of Rocket Engines Using Beamed Energy

John T. Lineberry*, James N. Chapman*, Ron J. Litchford[§]
and Jonathan Jones[§]

* LyTec LLC, Tullahoma, TN 37355
§ NASA Marshall Space Flight Center, MSFC, AL 35812

Abstract. MHD technology and fundamental relations that pertain to accelerating a working fluid for propulsion of space vehicles are reviewed. Previous concepts on MHD propulsion have considered use of an on-board power supply to provide the electric power for the MHD thruster which is accompanied by an obvious weight penalty. In this study, an orbiting power station that beams microwave or laser power to the spacecraft is considered which eliminates this penalty making the thruster significantly more effective from the thrust-to-weight viewpoint. The objective of the study was to investigate augmenting a rocket motor to increase the ISP into the 2,500 seconds range using MHD acceleration. Mission scenarios are presented to parametrically compare the MHD augmented motor. Accelerator performance is calculated for an array of cases which vary the mass throughput, magnetic field strength and MHD interaction level. Performance improved with size, magnetic field strength and interaction level, although lower interaction levels can also produce attractive configurations. Accelerator efficiencies are typically 80-90%. The results display a large regime for improved performance in which the extent of the regime is critically dependent upon the weight of the power receiving equipment (rectenna). It is concluded that this system has potential when used with an orbiting power station that transmits power to the space vehicle by microwave radiation or laser beams. The most critical technology improvement needed is a reduced weight rectenna system but more development is also needed on the MHD accelerator, which is currently underway with NASA sponsorship.

INTRODUCTION

The possibility of augmenting a chemical rocket engine to increase the exit velocity of the working fluid and thus the ISP of the engine has been discussed for years and was explored quantitatively in reference 1. In this reference and in the NASA Affordable In-Space Transportation Study (concluded in 1998) consideration of MHD propulsion that was given required carrying an on-board power generation system to generate the power needed for the MHD augmentation. The most attractive power generation option seemed to be an advanced nuclear power system but this is not only politically unpopular, but adds so much weight to the space vehicle that the advantage of using the MHD system is largely negated. In the past decade or so, a number of

CP664, *Beamed Energy Propulsion: First International Symposium on Beamed Energy Propulsion*,
edited by A. V. Pakhomov
© 2003 American Institute of Physics 0-7354-0126-8/03/$20.00

researchers have proposed that a large power station be placed in orbit. The use for this power station was in connection with a plan to beam electrical power to the ground for earth based electrical utility distribution. The technology is clearly available to deploy such a power station, either using large area solar panels or a fuel consuming power plant. The current area of interest is to use an orbiting power station to transmit propulsion power to spacecraft. This paper reports on an initial study to evaluate the potential of this concept for space MHD propulsion assuming that the power station is available.

The missions of primary interest were interplanetary missions and orbit raising. The system concept was to use the MHD accelerator coupled to a rocket engine exhaust, modified by add of some material that is partially ionized at rocket engine conditions. The rocket engine ISP is thereby increased by acceleration of the propulsive stream directly in the MHD accelerator by the action of Lorentz force.

The studies discussed herein were an assessment for NASA on the technical and feasibility of the potential use of a beamed energy driven MHD Chemical Rocket Motor (MCRM) to enhance the mission performance in orbit-boost-to-escape for deep space missions [2,3,4]. A type of mission scenario that is currently being considered by NASA for the pioneering transport and manned missions to the planets. An artist's rendition of the mission is shown in Figure 1.

The mission scenario encompasses the use of beamed energy (e.g., microwave) from an orbiting power platform to an orbiting deep space vehicle/probe. Considerable studies and research are underway within NASA on both the potential, needs and operational requirements for an orbiting power station as well as methods to beam energy to space probes and process this power for propulsion.

This mission includes beamed energy from the power station being captured by a receiving antenna (rectenna) on-board the space vehicle. This beamed energy is directly converted to electric power through microprocessors with rectifier arrays

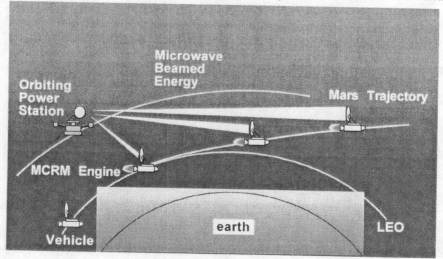

Figure 1. Space Mission Scenario - MCRM Driven By Beamed Power from an Orbiting Station

integral to the rectenna. The electric power produced is used to directly drive the MHD accelerator (magnet and accelerator channel). Beamed energy systems in the gigawatt's level have been projected which exceeds that required to accelerate a 80,000 kg craft at over 1.0 g based upon use of a very high power density receiving antenna. An increase in the LEO orbital velocity of 6,000 to 10,000 m/sec is considered that level needed for achieving a Mars trajectory with viable mission time for transport and/or manned spacecraft. Consequently, accelerations in the "g's" range are sought when this is accomplished across a single orbital path as shown in Figure 1

The MHD assisted rocket augments the impulse of the rocket motor utilizing energy (fuel/power supply) for the MHD accelerator that is not on-board. This concept is a vision by NASA as a prospective means for enhancing vehicle acceleration to acquire higher velocity and improved trajectory - to reduce deep space mission time.

THE MCRM CONCEPT

Our concept for the MCRM is sketched in Figure 2. The MCRM utilizes a standard rocket motor as a driver for the propulsion system burning conventional rocket fuels such as liquid hydrogen and oxygen. In addition, a trace amount of ionization seed (0.05 to 1.0% of K or Cs by mass) is added to the combustion stream to provide the ionizable material for achieving a thermal plasma

The combustion stream is accelerated through a nozzle to achieve a pre-defined optimum state at the MHD accelerator entrance. The MHD accelerator consists of a confinement channel with electrode and insulating walls surrounded by a high field strength magnet. Electric power is supplied to the MHD accelerator from the external power source.

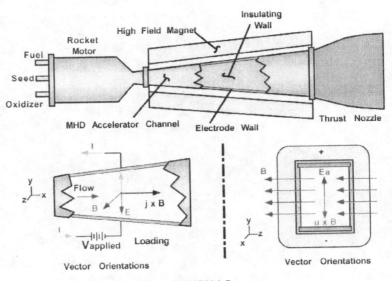

Figure 2. MCRM Concept

The major components of the beamed energy driven MCRM include the magnet, the MHD accelerator channel, the power conditioning equipment, and the rectenna/power conversion system. General evaluations of these systems in terms of technology readiness for optimization to the MCRM were considered in the NASA studies [2, 3, 4]

It is apparent that a superconducting magnet will be required for any airborne or space based propulsion system due to the weight and power requirements of conventional electromagnets. Our studies assumed the availability of room temperature superconductors. However, we applied a 10^5 Amp/cm^2 current density limit and imposed a safety factor of three in the coil thickness to provide for any structure that may be needed to support the coil internally.

Other technology issues relative to the magnet are contended to be engineering design matters. These include loads due to high accelerations during launch, magnet current control and shielding of instrumentation that is sensitive to magnetic fields and we profess that this element of the MCRM will not be the pacing element to its fielding.

The MHD accelerator utilizes the force created by an electrical current and a magnetic field applied mutually perpendicular to create a body force on ions in a plasma working fluid. The accelerating force is mainly transmitted to the neutral particles in the gas. By this process, it is possible to provide a direct and efficient increase in the working fluid velocity to propel the vehicle.

Power for the operation of the MCRM system under consideration comes from an orbiting power station that transmits the power to the spacecraft by microwaves. This concept has been studied for a variety of situations. [6-9] The transmission of electrical power from earth-to-space and space-to-earth by microwaves has been widely studied. The latter application was, and still is, being considered to evaluate the feasibility of a system of solar gathering satellites in orbit, beaming the electrical power to earth for distribution to electrical users on earth. In this application the frequency for the microwave power transmission is planned at about 2.45 Ghz, wavelength 0.122 m. This frequency is chosen rather than a higher one to minimize losses in transmission through the atmosphere. A space-to-space power transmission system will have no such consideration and a higher frequency should be considered. The performance of a high gain antenna applicable to the transmission and reception of microwaves is an inverse function of wavelength, λ. That is, ideal antennas having identical D/λ, have identical gain. Thus, higher frequencies favor smaller antennas for comparable power transmission. The limitation on how high the frequency can be depends upon the technology for generating, transmitting and receiving the microwave power.

Reference 8 suggests that a 300 Ghz system is possible in the near future, although some component development will be required. At these frequencies, integrated circuits are proposed for rectifying and processing the received power. Such a system should be diversified into multiple series/parallel circuits to produce the voltages and current needed to power the magnet and accelerator electrodes. The weight of these rectenna systems subject to contemporary technology is estimated [10] at between 0.5

to1.0 kg/kW. At this level, <u>the weight of the magnet and accelerator components are well below 1% of the total weight and are dwarfed by the rectenna weight</u>. Thus, the weight of this rectenna appears to be the key pacing issue for the beamed energy driven MCRM as well as for other space technologies applications.

An approximate equation that expresses the relationship between antenna sizes and range, as a function of wavelength, λ, is given below [10,14],

$$\eta = 1 - \exp\left(\frac{A_t A_r}{\lambda^2 L^2}\right) \tag{1}$$

where, λ is the wavelength of the microwave power, η is the fraction of the transmitter power captured by the receiving antenna, A_t, A_r are the transmitter and receiver array areas, respectively, and L is the distance between the transmitter and receiver. This equation is plotted in Figure 3 for an efficiency of 75% and assumes that the transmitting antenna dimension is ten times the receiving antenna dimension. For the 300 Ghz case, the antenna sizes seem to be manageable for powering over a significant range, certainly to 1,000 Km and beyond.

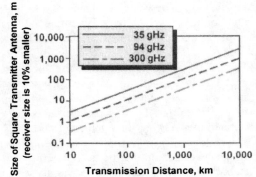

Figure 3. Required Antenna Sizes as a Function of Range to Space Vehicle

There is a considerable knowledge base derived from past work on MHD generators with regard to power conditioning and current control for MHD devices. This knowledge base can be brought to bear for the MCRM application. From the weight viewpoint, the integrated circuit rectifier design should be configured to produce the currents and voltages required by series and parallel combinations of individual rectifier stages. Koert and Cha [8] have outlined the requirements for such a system but additional development work will be needed. In this regard, a disadvantage to going to the high frequencies for power transmission is revealed in that the voltage received on each dipole is scaled down with wavelength and more dipoles are needed in series to drive each rectifier.

ANALYSIS

The first characteristic needing definition for the general boost from orbit-to-escape mission is total power requirement. If one considers boost from LEO with an initial orbiting velocity in the 4,000 to 7,000 m/s range (variation from perigee to apogee) and escape velocity to deep space from 10,000 to 13,000 m/s, then the fundamental kinetic energy imparted to the vehicle is:

284

$$KE = m_v \frac{V_{esc}^2 - V_{LEO}^2}{2} \tag{2}$$

where m_v is vehicle mass, Vesc is escape and V_{LEO} is initial velocity in LEO. As an example, for the velocity increase range required ($\Delta V \sim 6,000$ m/sec) and considering a mean vehicle weight of 100 mTons with a mean acceleration over a period of 60 minutes, one concludes that approximate power requirements for this mission are well in the gigawatt range; i.e.,

$$0.7 \text{ gW} < \text{Power} < 2.2 \text{ gW}$$

with accelerations between .08 and 0.25 g's and distances covered from 25,000 to 36,000 kilometers.

The extreme amount of power required to perform this mission is a <u>fact of nature</u> that has to be dealt with regardless of how the mission is approached or the propulsion system of choice. A general view of the power requirement is provided in Figure 4.

The mission concept under study of using beamed energy implies that the beamed energy system must be capable of producing, transmitting, receiving, conditioning, and managing both electrically and thermally, this level of power. This as the major engineering task that faces this specific mission and other beamed energy propulsion applications as posed by NASA. Furthermore, for the missions to prove viable, the beamed energy system components that are

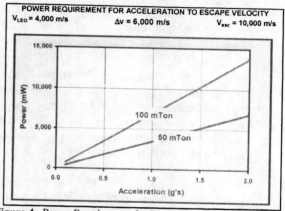

Figure 4. Power Requirement for Boost-to-Escape Mission

intrinsic to the vehicle must be of very high specific power to minimize their parasitic weight and enable high acceleration.

In our evaluation of the mission, the governing equations were parametrically solved on a spreadsheet to view mission characteristics and requirements. In this exercise, we considered combination variations in rocket motor controlling parameters including specific impulse, Isp, vehicle/payload mass, m_{veh}, throughput m_{dot}, and velocity increase to escape, ΔV.

For specification of the rectenna weight, a specific power parameter, ξ, was defined as

$$\xi = \left[\frac{\text{Power}}{\text{Weight}} \right]_{rectenna} \tag{3}$$

This parameter was varied over a broad range in our studies, i.e. $2 < \xi < 1,000$ kw/kg. based on information provided by NASA, literature, and personal communications with experts.[1] The total power required was derived considering the total beamed energy input with representative electrical efficiency, ηe. Electrical efficiency is,

$$\eta_e = \frac{\text{MHD Push Power}}{\text{Total Beamed Power}} \tag{4}$$

where ηe accounts for both power conditioning losses and MHD accelerator electrical efficiency. The MHD Push Power represents the total increase in kinetic energy of the rocket throughput across the MHD accelerator. With this formulation, the rectenna weight is computed as,

$$m_{\text{rectenna}} = \frac{\text{MHD Push Power}}{\xi\,\eta_e} \tag{5}$$

A series of plots were derived from calculations on the dynamics of the mission. The cases calculated and their range included: 80 metric ton payload: Isp from 500 to 3000 s; mass flow rates from 0.1 to 100 kg/s; and specific power parameter for the rectenna, ξ, from 1 to 1000 kW/kg. Unfortunately, the space limitations of this paper preclude inclusion of all these plots.

The dominant controlling factor for the system that is revealed is the rectenna power density. The dominance of rectenna power density on the performance arises from variation of rectenna weight. We note the following,

$$m_{\text{rectenna}} \propto \text{Power} \propto KE_{\text{exhaust}} \propto V_e^2 \propto \text{Isp}^2$$

i.e., the rectenna weight increases with the rocket exhaust velocity, Ve, squared; whereas, the specific impulse is directly proportional to the exit velocity. This variation lends non-linearity to the performance characteristics and produces the relationship that limits size of the range for which the MCRM advantage exists

In conclusion, the performance of the beamed energy driven propulsion system is paced by the power density of the rectenna.

- At the lower rectenna power densities, the rapid increase in the weight of the vehicle due to the rectenna size places limitation on the overall size of the MCRM system for which a significant increase in Isp over that of the fundamental rocket motor can be achieved.

- At very high rectenna power densities, a significant/revolutionary improvement in performance can be achieved by the MCRM with plausible Isp's in the 2,000 to 3,000 range.

Figure 5 summarizes these conclusions. This figure shows two maps of vehicle acceleration versus mass flow rate (size) for varying level of Isp and two distinct levels (bounds) of rectenna specific power. The lower plot represents a rectenna specific power which has been cited in literature as being optimistically achievable for the first generation receiver, i.e., 4 kW/kg (or 0.25 kg/kw). The upper plot represents an extremely optimistic (futuristic) rectenna power density of 1Gw/kg. This level is well beyond the state-of-the-art for this device but has been quoted as the goal "range" for beamed powered earth-to-orbit vehicles.

---

[1] Private Communications, Dennis Bushnell, NASA LaRC, Leik Myrabo, RPI, Richard Dickerson, JPL, Jonathan Jones, MSFC ASTP

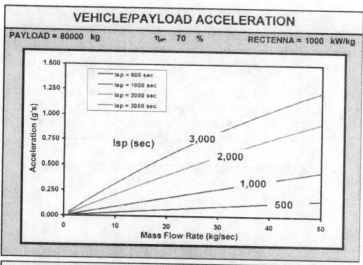

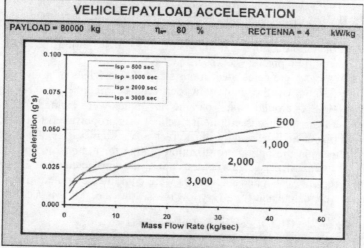

Figure 5. Vehicle Acceleration vs Size for Bounding Rectenna Specific Powers

At the lower level of rectenna specific power in Figure 5, the Isp lines cross over the 500 second line (representative of rocket motor only) in the lower left portion of the map. To the left of the cross over point represents the size of system for which the beamed energy propulsion system offers improved performance. As an example, an Isp of 3,000 seconds improves performance over the standard rocket motor for systems with throughput less than about 7.5 kg/sec. Consequently, rectenna specific power places restriction on system size. In this upper range, it is seen that significant performance enhancement is plausible with the MCRM lending vehicle accelerations in the "g" range for large vehicles.

MCRM Performance

The MHD accelerator calculation methodology used in this study is LyTec's one-dimensional MHD accelerator computer code which numerically solves the coupled the plasmadynamic equations. Detailed description of this code is contained in reference materials.[3-5]. The code has been used in past studies on MHD accelerators in application to both wind tunnels and small scale MHD thrusters. The code has provisions to model plasmadynamic and electrical wall losses with friction and heat transfer as well as flow field non-uniformity and boundary layer voltage drops.

In this study, only the Faraday configuration was used. Loading was imposed by specification of the load factor, k, defined according to:

$$k = \frac{E_y}{uB} \qquad (6)$$

Moreover, the load factor was specified as a function of distance along the accelerator (x coordinate) and a maximum current density was imposed. (That is, the load factor is used to calculate the required applied voltage until the resulting current density equals/exceeds the maximum specified; past this point, the applied voltage is calculated to give the specified current density.) Although the computer code has provision for accommodating real magnetic field distributions, a flat (constant intensity) magnetic field profile was specified.

A myriad of cases were calculated in our study. Table 1 lists of a few typical cases and quantifies some of the important performance parameters. The base case (suggested by NASA) around which parametric studies were formed was for a mass flow rate of 2.2 kg/s. As discussed, the Isp achieved is proportional to the square root of the applied power and the weight of the rectenna system varies as the first power. This fact implied that the optimum utilization regime for the beamed power, MCRM system is in lower size systems based upon the projected specific power for contemporary rectennas. Higher mass flows were computed as noted in the table. The last two cases were calculated for a very low mass flow rate, 0.022 kg/sec.

TABLE 1. Cases – MHD Accelerator Calculations

Case	m kg/s	Bmax T	Power MWe	Efficiency %	Jymax, Amp/cm^2	Velocity,m/ s	Isp	Length m
1	2.2	2	777.3	80.8	30	24,496	2,500	1.2
2	2.2	4	725	88.6	16	24,648	2,515	1.2
3	2.2	6	702.1	90.8	12	24,451	2,495	1.2
4	2.2	8	740.7	92.3	10.2	25,216	2,573	1.2
5	5	4	1,650	88.7	16	24,680	2,518	1.2
6	5	2	1,776	80.6	31	24,529	2,503	1.2
7	10	4	3,230	88.5	16.3	24,491	2,499	1.2
8	2.2	4	707.3	90.5	9	24,476	2,498	2.4
9	0.022	8	6.7*	89	5	23,970	2,446	1
10	0.022	4	6.8*	85.7	10	23,743	2,423	1

* Add 350 kW if water is used for fuel instead of H2-O2

It is noted that none of these calculated results are unique. For example, a longer length can be used with a lower current density to achieve the same acceleration at a slightly reduced efficiency. In all cases, choosing different inlet conditions to the accelerator will affect the detailed results. There is a high probability and need for further optimization of the accelerator, depending upon the optimization criteria one selects. In the first four cases of Table I, the magnetic field flux density was varied from 2.0 to 8.0 Tesla for a fixed mass flow rate of 2.2 kg/s. The accelerator was loaded to achieve an Isp of approximately 2,500 seconds over a 1.2 meter total length (including exhaust nozzle). As tabulated, the efficiency increases with magnetic field. This behavior is because the accelerating force is proportional to current density times magnetic field flux density. Therefore, at a higher magnetic field, proportionally less current is required and the resistive power dissipation in the plasma decreases as the square of the current density. The efficiency shown in Table I is the 'electrical efficiency' which represents the ratio of work done on the gas by the MHD force to the electrical power input. It should be pointed out that the electrical losses go to heating the gas which also has a second order propulsion benefit.

The estimated weight of the magnet increases from 845 kg for the 2 Tesla case to 988 kg for the 8 Tesla case. Weight of the entire MHD accelerator including magnet, channel and subsystems for these two extreme power density cases varied from 1,209 kg at 2 Tesla to 1,252 kg at 8 Tesla. The heaviest MHD system calculated was the large 10 kg/sec flow rate (Case 7) at 2,115 kg. The non-linear character in system weight arises from variations in MHD accelerator size and power density (i.e., the same power level for acceleration can be achieved in a smaller size accelerator as the magnetic field increases).

In cases 5, 6 and 7, higher mass flows were explored to view scale-up issues. These cases show how power requirement varies with mass flow rate, reaching more than a gigawatt for the 5 kg/sec case. The electrical efficiency does not reflect overall thermodynamic efficiency that improves with size by reduced surface to volume ratio.

Case 8 at 2.2 kg/s was chosen to show the design option of operation at half the power density by making the accelerator twice as long. Although the wall losses increase and the overall weight of the MHD propulsion system increases, it is clearly feasible to utilize this design trade-off to scale down the interaction level as needed for reliability and still maintain a very effective system.

The last two cases (9 and 10) are at a very low mass flow rate, 0.022 kg/s. These cases were calculated because system/mission analyses based on high rectenna weights scaled on a linear basis with power, indicate that a near term attractive regime for MHD augmentation is at small scale. These cases were expanded further in the rocket motor nozzle before entering the MHD accelerator in order to achieve an increase the physical size of the flow channel. However, more detailed calculations, especially in treatment of boundary layer growth and losses, are needed to be assured that this small size is feasible.

The footnote of Table 1 indicates the additional power required if direct heating of water (actually ice) is used as the working fluid for the propulsion system instead of hydrogen/oxygen rocket engine.

CONCLUSIONS

The MHD augmented rocket propulsion system (MCRM) is promising for boost from orbit of a deep space vehicle that can be powered by microwaves or lasers from an orbiting power station. Very high specific impulse and simultaneous with high thrusts are plausible with MHD accelerator propulsion.

Further work is needed to fully qualify this mission in terms of size, scope and detailed design needs. The beamed energy driven MCRM represents an advanced propulsion system that could significantly improve space transportation capability by lending high performance thrusters for deep space and inter-planetary missions.

Current and predicted technology advances in materials and subsystems needed for the MCRM are most promising. From our literature reviews, we contend that light weight, high temperature superconducting magnets could well be as close as a decade away from being a reality. Similarly, light weight, high strength materials based on carbon nanotube technology and composites are near term. In general, the principal facet that has plagued used of MHD propulsion in the past "Weight" will soon come to pass lending this system a leader for advanced space propulsion.

The pacing technology for the beamed energy driven MCRM (and other beamed energy advanced propulsion concepts under study by NASA) is the rectenna specific power. The weight of the rectenna based on contemporary estimates of these devices power density (1.0 to 4.0 kw/kg) far exceeds that of the total MHD system. The rectenna specific power is the controlling factor on what scale the MCRM is plausible – with small scale applicable to contemporary rectenna specific power projections. For the beamed energy driven MCRM propulsion system to be applicable for large scale inter-planetary missions, improvements in rectenna specific power of an order of magnitude or more are needed.

MHD acceleration scales to higher efficiency, lower weight per unit of thrust and lower cost per unit of thrust with larger size. The efficiency of the MHD accelerator depends on design parameters such as magnetic field strength, length, scale and operating regime. A conservative estimate of efficiency is 80%, with the electrical losses going into heating the working fluid which also benefits the accelerator.

Effective utilization of an MHD accelerator in space leads to very low static pressure in the downstream end of the accelerator, high Hall parameters and high axial potential. These conditions favor/require the utilization of a diagonal or Hall configured accelerator. These configurations can be designed to produce substantially the same performance as calculated in this study. Consideration need to be given to non-equilibrium MHD accelerator designs in recognition of the low pressure operating regime in which the space based system will operate. Taking advantage of non-equilibrium ionization in the extremely low density plasma that occurs with high expansion, provides a potential means for reducing power requirement and further enhancing the acceleration performance of the MCRM.

The MHD system is amenable to different working fluids, including H_2/O_2 engines, solid fuels with or without separate oxidizing agents and water. Different propellants and combinations need to be explored for MCRM application. The reliability and lifetime of the MHD accelerator is a function of power density, operating potentials, materials and cooling methods. All of these variables can be designed to be within limits needed for required reliability.

There is a deficit of knowledge and experience in MHD accelerator technology that plagues design issues related to the MCRM. No experimental work on the MHD accelerator has been done in the U.S since the 1960's. NASA ASTP initiated an effort in 2000 to begin some experimental MHD accelerator research directed at MHD space propulsion. This project is currently underway and a description of that effort and its goals are given in reference 11.

ACKNOWLEDGEMENT

This work was sponsored by NASA Marshall Space Flight Center, Advanced Space Transportation Programs under contract Order No. H-30549D and an SBIR Phase I under Contract No. NAS8-01147.

REFERENCES

1. Schulz, R. J., Chapman, J. N., and R. P. Rhodes, "MHD Augmented Chemical Rocket Propulsion for Space Applications," AIAA-92-3001, AIAA 23rd Plasmadynamics and Lasers Conference, July, 1992.
2. "A Beamed Energy Driven MHD Chemical Rocket Motor for Advanced Space Propulsion," Final Report, SBIR Phase I, NAS8-01147, LyTec-R-020-014, LyTec LLC, May 2002.
3. Lineberry, J. T. and Chapman, J. N., "MHD Augmentation of Rocket Engines for Space Propulsion," AIAA-00-3056, 35th IECEC, Las Vegas, NV, July 2000.
4. "Assessment of MCRM Boost Assist from Orbit for Deep Space Missions," Final Report, NASA Order No. H-30549D, LyTec LLC, May 2000.
5. Chapman, J. N., J. T. Lineberry, and H. J. Schmidt, "Application of MHD Accelerators to Hypersonic Ground Test Facilities", *Proceedings of the 28th SEAM*, June 1990.
6. Power, J. L., "Microwave Electrothermal Propulsion for Space", *IEEE Transactions on Propulsion in Space*, Vol. 40, No. 6, June 1992.
7. Sercel, J., C,. and Frisbee, R. H., "Beamed Energy for Spacecraft Propulsion", *Proceedings of the Princeton/AIAA/SSI Sixth Conference on Space Manufacturing*, pp252-265, 1987.
8. Koert, Peter and James T. Cha, "Millimeter Wave Technology for Space Power Beaming", *IEEE Transactions on Microwave Theory and Technology*, Vol 40, No 6, pp. 1251-1258, June 1992.
9. Brown, W. C. and Eves, E. E., "Beamed Microwave Power Transmission and its Application to Space", IBID, pp. 1239-1250.
10. Sercel, J. C., "Microwave Electric Propulsion for Orbital Transfer Applications," in JANNAF Joint Propulsion Conference Transactions, San Diego, California, April 9-12, 1985.
11. Litchford, R. J., Cole, J. C., Lineberry, J. T., Chapman, J. N., Schmidt, H. J., and Lineberry, C. W., "Magnetohydrodynamic Augmented Propulsion Experiment. I. Performance Analysis and Design," AIAA 2002-2184, 33rd AIAA Plasmadynamics & Lasers Conference / 14th International Conference on MHD Power Generation and High Temperature Technologies, Maui, HA, May 2002.

A 35 GHz Extremely High Power Rectenna For The Microwave Lightcraft

A.Alden

Communications Research Centre, Ottawa, Canada

Abstract. A rectenna has been designed to provide DC energy for the propulsion engines of the Microwave Lightcraft. It uses very high density sub-arrays to convert the microwave energy in a beam of power density 4 kW/cm^2 with an efficiency of 56 %. Each sub-array consists of an ensemble of very short dipoles and Schottky diode elements deposited on and within a semiconductor substrate. The substrate sits down on a low-loss wafer carrier which also performs the function of water cooling. The rectenna design permits the periodic reflection and focussing of the beam to maintain a plasma nose-cone in front of the vehicle. In addition, the rectenna is capable of transmitting stored energy to this 'air-spike' region to maintain that plasma in the event of a loss of beam power.

INTRODUCTION

A microwave-powered vehicle meeting a number of inter-continental and space transportation needs has been proposed [1]. A microwave beam is transmitted from a phased array of size 1 km by 1 km to the Lightcraft at a distance of about 1200 km. The DC power for the oscillators of this transmit array is generated by a large solar cell array, both solar cells and oscillators being co-located on a solar power satellite (SPS) in a low earth orbit (LEO) at an altitude of about 500 km. The 35 GHz beam is focussed on the rectenna of the 15 m diameter spacecraft, as shown in Figure 1. The primary function of the rectenna is to collect the microwave energy in the beam and convert it to useable DC electrical energy for the propulsion engines located around the periphery of the lenticular-shaped vehicle. The beam would power the Lightcraft from its liftoff to various altitudes up to 100 km, depending on the mission. The phasing of the beam and the trajectories of the vehicle and SPS are such that the beam remains at broadside to the major dimension of the vehicle.

Broadside travel (in the direction of the beam) is made possible by the generation of a plasma nose-cone or 'air-spike' in front of the vehicle, replacing the mass of a traditional physical conical forebody normally used for streamlining an aerospace vehicle. The central (1.5 m diameter) portion of this air-spike will absorb and reflect the microwave beam, leaving the outer section of the beam incident on the rectenna. The air plasma is induced and supported by a portion of the microwave beam energy incident on the Lightcraft. A secondary function of the rectenna is thus the periodic reflection and focussing of the beam to this air-spike region. In addition, the rectenna is capable of transmitting stored energy to the air-spike to maintain that plasma in the event of a loss of beam power.

CP664, *Beamed Energy Propulsion: First International Symposium on Beamed Energy Propulsion*,
edited by A. V. Pakhomov

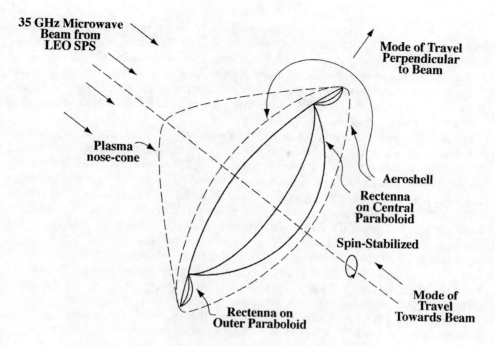

35 GHz Microwave Beam from LEO SPS

Mode of Travel Perpendicular to Beam

Plasma nose-cone

Aeroshell

Rectenna on Central Paraboloid

Spin-Stabilized

Mode of Travel Towards Beam

Rectenna on Outer Paraboloid

Figure 1   Microwave Lightcraft Powering System

RECTENNA DESIGN

Primary Function Design Considerations

For this extremely high power (EHP) application, normally-used rectenna designs, as shown in Figure 2, run into trouble. In these designs, each rectenna element collects the energy incident over an area a little more than a half-wavelength square surrounding that element. It accomplishes this using a resonant half-wavelength dipole antenna at the centre of the area allotted to that element, and a reflector plane behind the rectenna array.

For the Lightcraft application, the incident power collected by such a rectenna element would be around 3 kW. Because of the small active semiconductor areas necessary for 35 GHz operation of the rectenna diodes, these normal concepts place extremely high electrical power requirements on the rectification needed at each element.

In the present application, the dipole length and element separation are reduced from the order of a half-wavelength (4 mm) to around 50 μm. This high density of dipoles allows for a decrease in the power handling requirements of each diode by 4 orders of magnitude.

Rectennas normally use input and output circuit filters on each side of the diode for improved conversion efficiency, as shown in Figure 2. Owing to the difficulties involved in their implementation in the restricted area of the EHP design, the circuit filters have been relinquished. Output filter resonance of the diode junction capacitance is now provided by the appropriate spacing of the reflector plane behind the rectenna. The absence of input circuit filtering results in a reduction in rectification efficiency of less than 10 %.

Figure 2 shows a linearly-polarized (LP) rectenna. Such a design has serious disadvantages when motion of the transmitter or rectenna occurs. The rectenna would have to 'polarization-track' any rotation of a moving LP power beam, or alternatively, accept a 50 % power loss in a circularly-polarized (CP) beam. For the EHP Rectenna [2], a dual polarization format is employed, enabling all the power in a CP beam to be collected. The rectenna elements are laid in bands on paraboloidal surfaces within the Lightcraft, as shown for the central paraboloid in Figure 3. The x-polarization rectenna element at any location is oriented up the surface towards the rim, while the y-polarization element lies around the surface. Detailed modelling of this configuration shows that if the reflector and inter-element spacings are made specific functions of band position, matching of the beam to the rectenna can be achieved over the entire paraboloid with a single impedance level design. The rectenna does, however, require alignment between the beam direction and the axes of the paraboloids and manufacturing process variations in dipole and reflector spacings over the surfaces.

A 'sub-array' approach, Figures 3 and 4, is proposed to obviate part of the difficulty in manufacturing a large, curved-surface rectenna at 35 GHz. Each rectenna sub-array consists of an ensemble of antenna and Schottky diode elements deposited on and within a semiconductor substrate (wafer). Planar diodes are proposed to keep heat

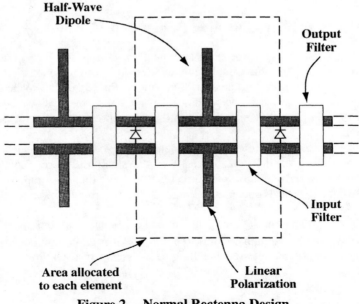

Figure 2   Normal Rectenna Design

dissipation close to the surface of the wafer. Each sub-array is surrounded by lossy material reducing stray reflections, diffractions and transmission of the high power beam in the regions between sub-arrays. This lossy material forms part of the rectenna water-cooling 'jacket' and is itself water-cooled. The space between sub-arrays is kept to a minimum to prevent a significant reduction in power collection.

Low-dielectric constant, low-electrical loss enclosures carry filtered de-ionized water along narrow channels at the surface of the wafer. The use of microchannel liquid cooling [3] has been proposed for the rectenna [4]. In the variant proposed here, these channels of water would flow directly over the anode metallization of each diode active area. The channels would traverse the entire sub-array and carry away the heated water for discharge from the vehicle. The serpentine channels pass at right angles to the dipoles and across the diodes they cool, to minimize the effect of the high-dielectric water on antenna performance. The width of each microchannel would be around 10 μm, the length of the diode active area. The height of each channel would also be small, between 10 and 20 μm. This would reduce the volume of coolant needed from normal microchannel requirements, and keep the effect of poor-thermally conducting channel walls to a minimum. Reliability problems related to the direct contact of water with the anodes would be limited by the use of a top gold metallization. If migration of the water into the passive regions of the wafer is a problem, an inert liquid coolant could be used. Viscosity considerations may also preclude the use of water in such narrow microchannels.

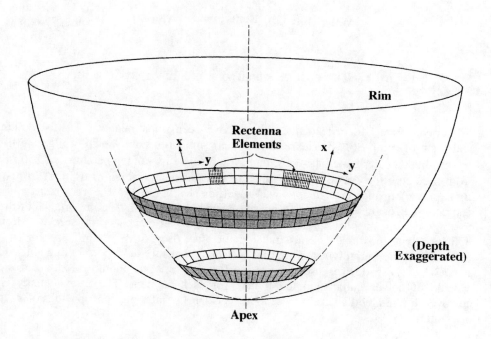

Figure 3   Bands of Rectenna Elements on Central Paraboloid

295

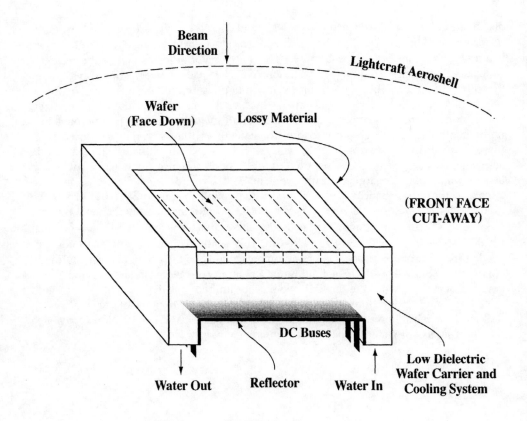

Figure 4   Three-Dimensional View of Sub-Array Structure

Electrical models for a rectenna element have been implemented using the circuit analysis program p-SPICE. The results of these simulations are shown in Figure 5 for two candidate wafer materials, GaAs and SiC, at active region temperatures of 250 °C. From the curve for GaAs, it is seen that an inter-element spacing of 61 μm (75 mW input power) results in a microwave-DC efficiency of 56 %. The rectenna circuit designs used for these calculations are not considered optimal. The importance of the cooling system is illustrated in Figure 6 where efficiencies are seen to drop significantly as the diode / surface metal temperature rises.

First-generation prototypes of the EHP Rectenna structure have been built and tested at 6 GHz. These devices use discrete (surface mount) Si Schottky diodes and are capable of power outputs up to 50 mW for each element. The array density is, however, limited by the size of these diode packages to inter-element spacings of 1 cm or greater.

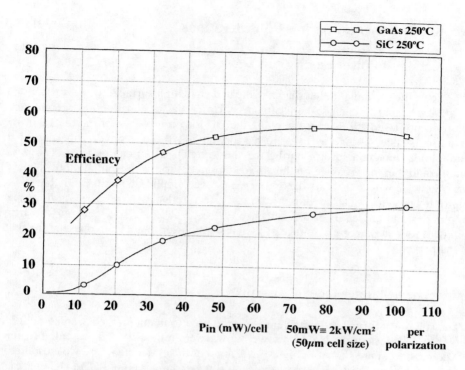

Figure 5   Rectenna Efficiency With Power Level

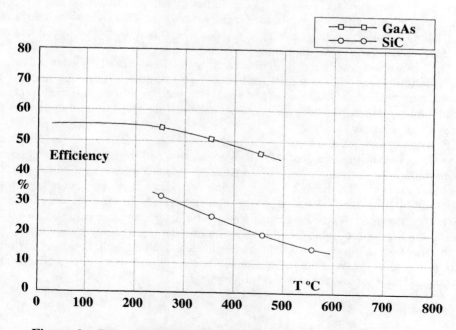

Figure 6   Rectenna Efficiency as a Function of Diode Temperature

Secondary Function Design Considerations

Redirection of Power Beam

The use of the rectenna, rather than a dedicated switching surface, for the reflection of the power beam to the air-spike is attractive for reasons of construction, weight etc. It has been proposed [5] that by switching the DC load impedance of the rectenna from its normal value (engine as load) to a short circuit, the reflective properties of the paraboloidal rectenna surface during the short-circuit period can be used. The power beam could then be focussed on the air-spike region. From exponential diode theory, it has been shown [2] that the reflected wave amplitude ratio and phase are approximately identical to those from a perfect reflector, over the varying beam amplitudes and angles of beam incidence of the paraboloid. The rectenna sub-arrays will then reflect a large percentage of the incident power to arrive in phase at the focus of the paraboloid.

Emergency Maintenance of Air-Spike

The sub-array format of the rectenna may be used to maintain the air-spike region in the event of a loss of the incident powering wave from the SPS. The essential features of such a system are shown in Figure 7. A number of solid-state microwave oscillators is located around the periphery of the central rectenna paraboloid. The DC energy to drive these sources is obtained from the stored energy in superconducting field coils around the outside of the vehicle [1]. The normal function of these coils is to provide the DC magnetic field required for the propulsion engines. This DC power is fed to the oscillators through power processing and protective switch gear running continuously under low-load conditions and monitored during flight.

The microwave oscillators are frequency- and phase-locked to each other and provide the source of injected RF power to each sub-array. They also run and are monitored continuously. A plurality of sources is used to provide sufficient power to feed the large number of sub-arrays, facilitate the routing of that primary power and to provide redundancy.

If the powering beam is lost, each sub-array switches immediately from its rectenna function to a transmit function. DC/RF switches, powered from the DC energy source, connect the sub-array to a DC voltage and the injected microwave signal. The DC voltage reverse-biases the rectenna diodes into avalanche breakdown, creating a set of RF transmission lines shunt-loaded at regular intervals by antenna elements with Schottky diodes across their terminals, reverse-biased into avalanche breakdown. Each shunt diode then acts as an IMPATT amplifier on the injected microwave signal. The sub-array then performs as an antenna of equivalent aperture with, however, additional RF power gain [6]. When each band of sub-arrays is correctly phased, a concentrated beam at the paraboloid focus will result, Figure 7.

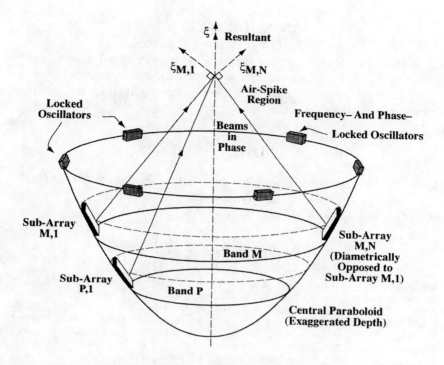

Figure 7   Paraboloidal Phased Array for Emergency Maintenance of Air-Spike

It is estimated that about 10^7 watts of CW power directed to the focus, are needed to maintain the air-spike. This represents about 80 mW per radiating amplifier-dipole element, for a 35 GHz transmit frequency. With IMPATT DC / RF conversion efficiency around 15 %, the DC requirement for this emergency function is around 7 x 10^7 watts. The energy stored in the Lightcraft field coils is estimated at around 10^8 joules and can thus only support the air-spike for 1.5 seconds. For the rectenna to perform this emergency function for the estimated 2 to 10 seconds necessary for speed reduction, the energy-storage capacity of the Microwave Lightcraft must be increased or some of the kinetic energy of the vehicle must be recoverable as DC electrical energy during the emergency.

CONCLUSIONS

A 35 GHz Extremely High Power Rectenna has been designed. Simulations suggest that an efficiency of 56 % or more is achievable for a monolithic GaAs rectenna at 250 °C in a 4 kW / cm^2 beam. Lower-power, lower-frequency hybrid prototypes have been built and tested successfully. Other applications for the technology are envisaged.

ACKNOWLEDGMENTS

The author wishes to acknowledge helpful communications with G. Ackerley. The assistance of P. Bouliane in the construction and testing of the lower-frequency prototypes is also gratefully acknowledged.

REFERENCES

1. Myrabo, L. N., 'Microwave-Boosted Spacecraft', *Presentation Notes, 5th Advanced Space Propulsion Research Workshop,* Pasadena: JPL, May 18-20 1994.
2. Alden, A., 'A 35 GHz Extremely High Power Rectenna For The Microwave Lightcraft', *CRC Contract Report No. CRC-VPRS-00-03,* Ottawa: Communications Research Centre, March 2001.
3. Tuckerman, D. B., and Pease, R. F., 'Ultrahigh Thermal Conductance Microstructures for Cooling Integrated Circuits', *32nd Electronic Components Conference,* San Diego, May 1982, pp. 145-149.
4. Communications from E. Somerscales and L. Myrabo, January 1999.
5. Communications from L. Myrabo, January 1999.
6. Bayraktaroglu, B., and Shih, H. D., 'Millimeter-Wave GaAs Distributed IMPATT Diodes', *IEEE Electron Device Letters,* Vol. EDL-4, No. 11, Nov. 1983, pp. 393-395.

MICROWAVE SAILS

Flight of Microwave-Driven Sails: Experiments and Applications

James Benford[1] and Gregory Benford[2]

[1]Microwave Sciences, Inc., 1041 Los Arabis Lane, Lafayette, CA 94549, [2]Department of
Physics, University of California Irvine, Irvine, CA 92697

Abstract. We have observed flight of ultralight sails of carbon-carbon microtruss material at several
gees acceleration. To propel the material, we sent a 10 kW, 7 GHz beam into a 10^{-6} Torr vacuum
chamber and onto sails of mass density 5-10 g/m^2. At microwave power densities of ~kW/cm^2 we saw
upward accelerations of several gees: sails so accelerated reached >2000 K from microwave absorption.
Data analysis and comparison with candidate acceleration mechanisms shows that photonic pressure
can account for 3 to 30% of the observed acceleration and that the remainder comes from desorption of
embedded molecules. This is a useful propulsion mechanism: A microwave beam source in orbit could
illuminate a sail, provoking desorption and enhancing thrust by many orders of magnitude over solar
sails, shortening the escape time to weeks (compared with years for solar sails).

MICROWAVE–DRIVEN SAIL FUNDAMENTALS

Robert Forward first proposed the microwave-driven sail as an extension of his laser-
driven sail concept[1,2]. No experiments and no sail flights by any method, including
laser and solar photon pressure have been done previously. The essential reason for
this is the lack of a material which could allow liftoff under one earth gravity. The
invention of strong and light carbon material has made sail flight possible, because
carbon sublimes instead of melting, so can operate at very high temperature. This is
because acceleration is strongly temperature limited.

The *acceleration from photon momentum* produced by a power P on a thin film of
mass m and area A is

$$a = [\eta+1] \, P/MA \, c \qquad (1)$$

where η is the reflectivity of the film of absorbtivity α, M is the mass per unit area
(m=MA) and c is the speed of light. (The carbon fiber sail material we use has ~1%
transmissivity.) Of the power incident on the film, a fraction αP will be absorbed. In
steady state, [which will be achieved in ~3 ms for 1 mm thickness sail] this must be
radiated away from both sides of the film (which may be of different temperatures)
which we describe with an average temperature T and emissivity ε by the Stefan-

CP664, *Beamed Energy Propulsion: First International Symposium on Beamed Energy Propulsion*,
edited by A. V. Pakhomov

Boltzmann law

$$\alpha P = 2A \, \varepsilon \, \sigma \, T^4 \qquad (2)$$

where σ is the Stefan-Boltzmann constant. Eliminating P and A, the sail acceleration is

$$a = [2 \, \sigma/c] \{ \varepsilon \, (\eta+1) \, /\alpha \} \, (T^4/M) = 2.27 \times 10^{-15} \, [\varepsilon \, (\eta+1)/\alpha] \, T^4/M \qquad (3)$$

where we have grouped constants and material radiative properties separately. *Clearly, the acceleration is strongly temperature limited.* This fact means that all materials considered previously in the literature (Al, Be, Nb, etc.) cannot be used for liftoff on earth, which requires acceleration greater than one gravity.

Another mechanism is *acceleration from sublimation pressure* due to mass ejected from the material downward to force the sail upward:

$$F = v_T \, dm/dt \qquad (4)$$

where v_T is the thermal speed of the evaporated carbon. The magnitude of this effect can vastly exceed the microwave photon pressure if the temperature is high enough. The essential factor is that this force must be *asymmetric*, i. e., if the sail is isothermal there is no net thrust. Thermal analysis shows that the skin effect produces a substantial temperature difference between front and back of our sails. Of course, for such thin material thermal conduction reduces such differences greatly.

SAIL MATERIAL—MICROTRUSS CARBON

Our carbon sails microtruss fabric 3-D architecture resembles a truss structure made of short discontinuous discrete carbon fibers with facings of long discontinuous discrete carbon fiber. Bonding the points where two fibers touch ("nodes") rigidizes the sail. The resulting sail material is stiff and lightweight, strong, conductive and creep-resistant. Most important, it is capable of high temperatures well above 2000 K, meaning sails of C-C could fly at high acceleration.

We measured the microwave properties of the C-C microtruss in a waveguide, so that the sail intercepted the wave completely. We measured the power transmitted through and reflected from the material and deduced the power absorbed. The result is very good: near-total 90% reflection, very low transmission (<1%), ~10% absorption (Figure 1).

From ESLI observations, the sheet resistance of sail material to DC is ~ 10 ohms. From the model of EM wave interaction with thin films of Cravey[3] et. al., the microwave properties vary with sheet resistance. It predicts 90% reflection, 10%

absorption for 10 ohms. So our experimental measurement of ~90% reflectivity at ~10 ohms fits this model very well.

EXPERIMENTAL APPARATUS

The configuration of the experiment is shown in Figure 2. An existing 25 kW X-band (7.1675 GHz) transmitter, using a Varian model VA-876 klystron, is connected by a waveguide to a water-cooled, stainless steel vacuum chamber with an internal diameter of 1.2 m and a length of 2.1 m. The chamber is pumped by a cryogenic-pump (10 inch) and a turbo-molecular pump (12 inch) backed up by Roots, Stokes and mechanical pumps in series. Heaters attached to the interior walls of the chamber bake out the chamber prior to the sail tests. The base pressure for the chamber is 2 x 10^{-7} Torr. The tube operates continuously, but for these experiments typically were pulsed for 0.2 sec. Safety requirements meant we had to place screens on all windows to avoid microwave leakage. Overhead silicon carbide absorbers reduce microwave reflections

The waveguide (WR-137 transitioning to WR-112) makes several turns and ends oriented vertically on the centerline of the chamber. We did not flare the waveguide into a horn because that would reduce the power density. The launcher geometry was a sail suspended over the waveguide. The sail had a grommet on axis which allowed it to slide on the support rod made of either carbon or alumina tube. This launcher geometry can been seen in some of the camera frames in Figure 3.

The pattern of the microwave beam was that of an open-ended waveguide. The angular distribution is given by $\cos^m \theta$, with m varying between 1 and 2.5 in the plane perpendicular to the vertical axis. We measured the profile by suspending Kapton sheet above the waveguide and heating it with microwaves. The overhead IR pyrometer/camera gave an image of the beam with ~1mm resolution, which showed that it was smooth and azimuthally symmetric. A line scan across the beam profile also showed the expected and predicted shape.

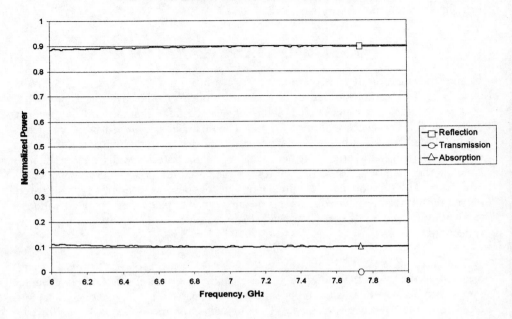

FIGURE 1. Microwave Properties of C-C Sail Material Measured in Waveguide.

FLIGHT OF SAILS

The most interesting aspect of the experiment is that the microwave beam lifted and flew sails in the chamber. Figure 3 shows four frames at 30 frames/sec; the data are given in Benford[4]. We saw the sail fly rapidly upward and strike the top of the chamber (55 cm). The video shows: Frame 1: quiescent sail, frame 2: sail lights up from microwave heating, frame 3: sail tilts, moves vertically 0.2 cm, velocity is 0.06 m/s (from framing interval), frame 4: sail has left the frame by moving at least 3 cm. Minimum velocity is 3 cm x 30 frames/sec=0.9 m/s, minimum acceleration a=0.9 x 30 frames/sec = 0.29 m/s^2. From the minimum height of 0.55 m, kinematics gives $v=[2gh]^{1/2}= 3.28$ m/s, and from the framing interval a = 98.5 m/s^2.

To summarize the flights, we see velocities from 0.3 m/s to 4.08 m/s, total accelerations from 10.9 m/s^2 to 132.3 m/s^2, or 1.1 gees to 13.5 gees. Many of these are lower bounds established by the framing interval of the camera. Temperature of the sails varies from 1700 K to 2300 K.

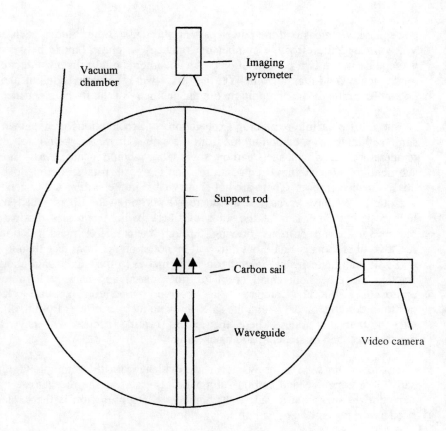

FIGURE 2 Schematic of the Apparatus.

INTERPRETATION OF PROPULSION MECHANISMS

We have analyzed two potential causes of the flights in detail and analyzed several others briefly. We conclude that both microwave photon pressure and asymmetric carbon sublimation pressure are viable mechanisms for explaining the data. But both fall short in explaining the data quantitatively. We have also concluded that neither asymmetric out-gassing nor electrostatics are credible mechanisms.

Photon pressure must of course be occurring. However, to fit the observations, it must be enhanced by a factor of 3 to 30. Additional lift could come from wave energy reflected by the sail and again by the microwave waveguide. To test for such resonance amplification of microwave power density in the sail region, we measured the effective cavity Q between the sail and the waveguide, and found it to be essentially unity. In the experiments, two disks of Al and strong microwave absorber were alternately positioned over the waveguide end. With 200 W emitted from the waveguide, we measured the power passing through a 3 mm hole at each disk center. We compared the effective number of "bounces" a photon made between disk and waveguide in the two configurations. An absorber should prevent reflection, while a metal plate would maximize it. Yet the fluxes were the same. This means there was no amplification of the photonic flux in the geometry of the flight experiments.

The most plausible remaining explanation for accelerations greater than gravity is some sublimation of carbon atoms from the sails. However, we do not measure sail temperatures at which such carbon sublimation should occur: Analysis shows the sublimation rates at these temperatures don't give thrusts sufficient to explain the observed acceleration. There are hot spots visible in the carbon sails, seen by eye and videotape. We have speculated that there is a portion of the fibers, which have locally high resistivity, perhaps at the joins of fibers, which increases dissipation of the microwave–driven currents flowing through them. The mass fraction of these resistive fiber joins is small, but they will be hotter on the front face than the back face and will sublimate faster. Comparing this model to our flight data shows that the fraction must be both small (<1% of fiber mass) and at very high temperature, approaching 3000 K. Compare to the observed temperatures about 500 K lower. A constraint on this model is that the sails show no melting effects from the flights. One expects that the nodes where the excess heating occurs will have a changed appearance, such as singeing and blackening.

However, no such signs whatever are evident from the microphotography of the sails. We conclude that carbon sublimation pressure is the mechanism only if such very hot regions exist in sufficient numbers. If sublimation is the explanation, we need a new generation of carbon sails without hotspots.

Possibly there is yet another unsuspected mechanism at work. The most plausible explanation is evaporation of absorbed molecules. This serious candidate is evaporation of physisorbed water vapor, chemisorbed hydrocarbons (C_n, H_m) and chemisorbed hydrogen. Although we used a bakeout procedure which heats the sail to ~1000 K for 100 seconds, this may not be enough. Our sails are driven to very high temperatures, exceeding 2000K, where evolution of absorbed molecules is greatly enhanced. Elevating the sail temperature for longer periods, to drive off water and hydrocarbons and to suppress subsequent adsorption rates, should reduce such contaminants. In principle, this can be accomplished through either an extended bake-out of the entire apparatus, or by heating the sail alone. The two essential conditions for this latter approach are that

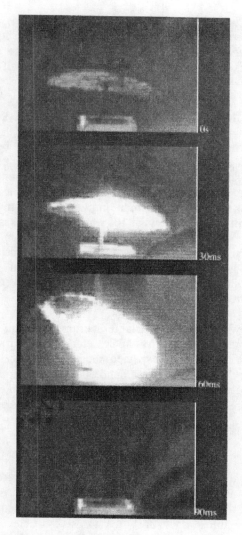

FIGURE 3 Liftoff of C-C Sail Under 10 kW Microwave Power.

• sufficient time at high temperature must be allowed to effect an initial decontamination. The more tightly bound chemisorbed hydrocarbons and hydrogen molecules require sail temperatures exceeding 400°C to be driven out.

• the microwave beam must be generated while the sail is too hot to allow re-condensation.

The most plausible explanation for the observed accelerations greater than gravity is evaporation of absorbed molecules (water, hydrocarbons and hydrogen) that are very difficult to remove entirely. The implication is that this effect will always occur in real sails and must be understood and anticipated in future sail experiments and missions. This includes not only microwave-driven sails, but also laser-driven and solar sails. Those planning sail demonstrations in orbit should be aware that the sun striking a sail could evolve molecules from the surface as well as impart photon momentum. This may be a serious source of error in all sail thrust experiments.

IMPLICATIONS OF HIGH-TEMPERATURE SAILS

A few years ago mission analysis of possible beam-driven mission was concentrating on metal coated sails. Typical performance characteristics are shown in Figure 4 which shows three kinds of high velocity missions of scientific interest that might be attempted in the next five to 10 years. Distance after ten years is shown since ten years is the current maximum mission duration that is likely to be chosen by NASA. Transmitter power starts at 10^7 watts and increases to 10^9 watts. Distance in AU and sail diameter in meters increase in roughly the same way. Sail size is chosen to produce maximum velocity for a fixed transmitter aperture. Figure 5 shows the same calculations for a carbon-carbon micro-truss sail. Notice that for a fixed aperture, the significant difference is that *with C-C material the required sail size is much smaller for a given mission.*

Figure 6 shows a hypothetical large aperture (1000km transmitter) that might be used on an unmanned mission to Alpha Centauri, similar to the one proposed by Dr. Robert Forward. in 1984. As expected, the sail diameter decreases with temperature, and the 10-year distance peaks at around 500 K sail temperature. This peak is about 1/3 the distance to Alpha Centauri. This example is power–limited. With higher power, higher operating temperature and greater distances are achievable.

CONCLUSIONS

An intense microwave beam has driven an ultralight carbon sail to liftoff and flight against gravity. Photon pressure, while certainly present, is insufficient to account for all of the observed effect. The acceleration calculated from sublimation at sail observed temperatures does not give thrusts sufficient to explain the observed acceleration unless there are very small hot spots in sufficient numbers. Other mechanisms have also been disproved. The most plausible explanation for the observed accelerations greater than gravity is evaporation of absorbed molecules. This is a useful propulsion mechanism (see J. Benford paper later in this session): A microwave beam source in orbit could illuminate a sail, provoking desorption and enhancing thrust by many orders of magnitude over solar sails, shortening the escape time to weeks (compared with years for solar sails). The sail returns to near the beam source on each loop of a steepening ellipse, enabling repeated accelerations. Plausible scenarios using ~ 100 MW microwave beam powers allow fast beam-plus-solar sailing missions to the outer solar system. For further discussion, see reference5.

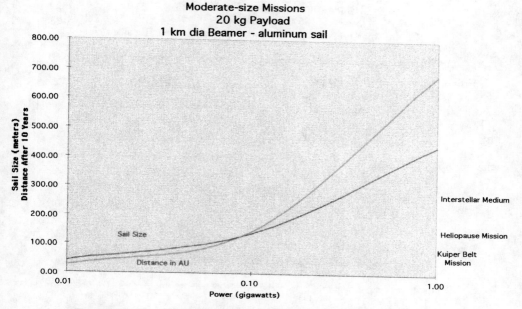

Figure 4 Missions for Fixed Aperture, Variable Sail Size, Aluminum Sails

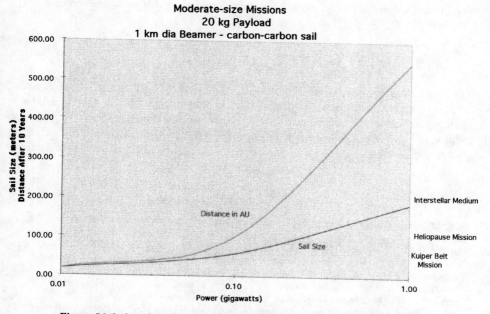

Figure 5 Missions for Fixed Aperture, Variable Sail Size, Carbon-Carbon Sails

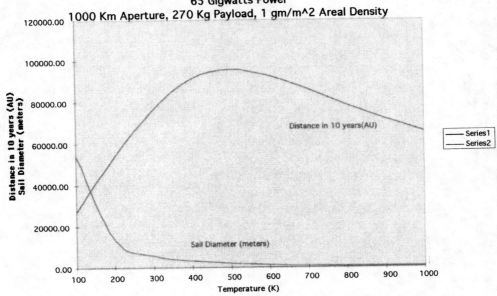

Figure 6 Distance traveled in 10 years and sail diameter as a function of sail operating temperature. Note that transmitter power is fixed.

ACKNOWLEDGMENTS

This work was supported by the Advanced Concepts Office, Office of Space Science, National Aeronautics and Space Administration through the Jet Propulsion Laboratory; Advanced Concepts—Technology Innovations project office. We thank Bob Forward, Richard Dickinson, Henry Harris and Neville Marzwell for helpful comments and support.

REFERENCES

1. Forward, R. L., "Roundtrip Interstellar Travel Using Laser-Pushed Light Sails", *J. Spacecraft*, **21**, 187, (1984).
2. Forward, R. L., "Starwisp: An Ultra-Light Interstellar Probe," J. *Spacecraft*, **22**, 345 (1985).
3. Cravey R. L. et. al., "Electromagnetic Losses in Metallized Thin Films for Inflatable Radiometer Applications", AIAA 95-3741 (1995).
4. Benford, J., Benford G., Goodfellow K., Perez R., Harris, H. and Knowles, T., "Final Report, Flight and Spin of Microwave-Driven Sails", Microwave Sciences (2000).
5. Benford, G. and Benford, J., "Desorption Assisted Sun Diver Missions", *Proc. Space Technology and Applications International Forum* (STAIF-2002), Space Exploration Technology Conf, AIP Conf. Proc. 608, ISBN 0-7354-0052-0, pg. 462, (2002).

Spin of Microwave Propelled Sails

Gregory Benford, Olga Gornostaeva

*Physics Dept
Univ. California, Irvine
Irvine, CA USA 92697*

James Benford

*Microwave Sciences Inc.
1041 Los Arabis Lane, Lafayette,
CA 94549 USA*

Abstract. It is not widely recognized that a circularly polarized electromagnetic wave impinging upon a sail from below can spin as well as propel. Our experiments show the effect is efficient and occurs at practical microwave powers. The wave angular momentum acts to produce a torque through an effective moment arm of a wavelength, so longer wavelengths are more efficient in producing spin, which rules out lasers. A variety of conducting sail shapes can be spun if they are not figures of revolution. Spin can stabilize the sail against the drift and yaw, which can cause loss of beam-riding. So, if the sail gets off-center of the beam, it can be stabilized against lateral movement by a concave shape on the beam side. This effect can be used to stabilize sails in flight and to unfurl such sails in space.

SPIN DRIVEN BY CIRCULARLY POLARIZED ELECTROMAGNETIC WAVES

Circularly polarized electromagnetic fields carry both energy and angular momentum. A polarized electromagnetic wave impinging upon a sail can spin a sail For a sail in space, this effect allows 'hands-off' deployment and control of the sail spin at a distance.

Some theory has treated wave angular momentum coupling to objects. (See references 1-6.) The wave angular momentum L acts to produce a torque through an effective moment arm of a wavelength, so $L = N \hbar k \lambda$, with N the photon number. The wave energy is $E = N \hbar \omega$, so the ratio of L/E imparted by a wave is $L/E = 1/\omega$. Therefore, longer wavelengths are more efficient in producing spin. *This effect also allows unfurling of sails by driving spin-up.*

With an eye toward eventual deployment in orbit, we study unfolding of sails by spinning them up. The wave angular momentum imparted to the sail scales as λ/D, $L/L_Z = \lambda/D$, where D is the transverse scale of the power beam, which will be close to the sail size for efficient propulsion. This arises because a wave focused to a finite lateral size D generates a component of electric field along the direction of

CP664, Beamed Energy Propulsion: First International Symposium on Beamed Energy Propulsion,
edited by A. V. Pakhomov

propagation of magnitude (λ/D) times the transverse electric field, as long as D exceeds the wavelength. So a sail beginning at small diameter can be spun open with an electromagnetic wave of wavelength a fraction of the beginning diameter. (Torque = wavelength × thrust, so wavelength on the order of the sail diameter should give maximum torque.) The spin S of a sail with moment of inertia I grows according to

$$\frac{dS}{dt} = \frac{2P(t)}{I}\frac{\alpha}{\omega}.$$

where P is the power intercepted by the sail. As the sail deploys, a shift to longer wavelengths could maintain the efficiency of transfer.

Clearly, there will be spin if the sail is an absorber of microwaves, $\alpha > 0$. This has been demonstrated for C-C microtruss sails by our team (Reference 5), and is explored further in the experiments described below.

Axisymmetric perfect conductors, for which $\alpha > 0$, cannot absorb or radiate angular momentum when illuminated. However, any asymmetry allows absorption. We have investigated the conditions under which a circularly polarized wave field transfers angular momentum to a macroscopic object, using exact electromagnetic wave theory. (See the Appendix for a discussion of the theory.) A rigorous solution of the boundary value problem for reflection from a perfectly conducting infinite wedge shows that waves convey angular momentum at the <u>edges</u> of asymmetries. Such absorption or radiation depends solely on the specific geometry of the conductor. We term this "*geometric absorption*". Conductors can also radiate angular momentum, so their geometric absorption coefficient for angular momentum can be negative! The geometric absorption coefficient can be as high as 5, much larger than typical simple material absorption coefficients, which are ~1 to 0.1 for absorbers and ~ 0 for conductors. In the appendix, we apply the theory results to recent experiments which spun roof-shaped aluminum sheets with polarized microwave beams (Reference 5). Below we describe tests of such techniques for spinning the sail on a thread suspended in the laboratory frame. The variables explored are various shapes (discs, 'roofs', strips) with 'cuts' which are chosen to optimally interrupt currents on the sail surface to produce maximized absorption.

To use this effect to spin objects in space, both a circularly polarized microwave beam and a local weak transient gas 'atmosphere ' would be required. This means producing a gas density with a mean free path of a tenth of the sail size or less. At ~100 km altitude, this would demand only 0.1 gm/s of gas, for 10 m sails. For spin-up taking minutes, only a few tens of grams are needed. A full discussion of our results is in Ref. 7, available on request.

SPIN EXPERIMENTS—APPARATUS AND MATERIALS

A schematic of the experiment is in Fig.1. The major systems are: High power microwave generation system; Microwave waveguide connection assembly and input into the experimental chamber; experimental chamber with sail-propeller support system; sail or propeller; vacuum and data acquisition system.

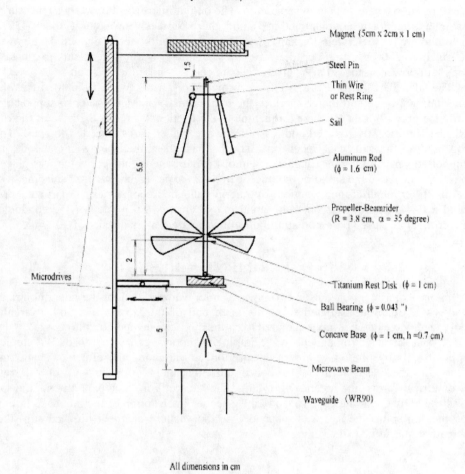

Propeller Assembly

Magnet (5cm x 2cm x 1 cm)

Steel Pin

Thin Wire
or Rest Ring

Sail

Aluminum Rod
(φ = 1.6 cm)

Propeller-Beamrider
(R = 3.8 cm, α = 35 degree)

Microdrives

Titanium Rest Disk (φ = 1 cm)

Ball Bearing (φ = 0.043 ")

Concave Base (φ = 1 cm, h = 0.7 cm)

Microwave Beam

Waveguide (WR90)

All dimensions in cm

FIGURE 1. This is the Style for Figure Captions. Center this text if it doesn't run for more than one line.

The microwave power source is a CW klystron amplifier (Varian VA-864) with maximum power of 10 kW at 10.6 GHz. The klystron has a 17 kV power supply and water cooling system (8 gal/min) with heat exchanger and water pump. The waveguide assembly is built using WR90 waveguide and waveguide components in X-band (23mm x 10 mm). All the flanges of the waveguide connections are sealed with copper tape. In our first experiments we used the open end of a waveguide with flange as a microwave input in the experimental chamber. A high power microwave pressure window manufactured by CPI (Communication and Power Industries) was used for atmosphere - vacuum separation. The output microwave power was measured using a directional coupler (-60 dB) and thermistor power meter (model 432A) manufactured by Hewlett Packard Co. The experiments were conducted at power levels up to 2300 W. In order to get 10 kW of output power it is necessary to increase the input power up to 180 mW, given the gain coefficient of the klystron.

The idea here is to use a lift force provided by a magnetic field to offset the gravitational force and thereby reduce the total normal force necessary to support the sail assembly. In this case, the torque due to the frictional force at the base of the rod will be significantly reduced and less microwave power will be needed to spin. The rod holding the sail was placed on a special base, described below. The rod was supported vertically by the magnetic field of the permanent magnet, which was placed on supports above the sail. For the microwave spin experiments we used sails made of carbon fiber material manufactured by Energy Science Laboratories. In particular, we used a C—C microtruss cone with areal mass density 19.6 g/m^2. For the conductor experiments we used common aluminum foil with areal mass density of 30 g/m^2.

SPIN EXPERIMENTS—RESULTS

In our experiments, we used two types of microwave radiation: the linear polarized wave of TE_{10} mode radiated from the open end of the waveguide, and a circular polarized wave of TE_{11} mode radiated from the circular waveguide or horn.

A polarized electromagnetic wave impinging upon a sail can spin it. We studied spin of sails by photon geometric absorption, investigating the effect of shape and 'cuts' in the basic form to influence current patterns on the sail of different geometrical forms like a cone, roof, strip, and square. For some experiments the sail was specially cut at the edge to study the effect of interruption of the induced currents on the sail spin. The sail was suspended in the vacuum chamber on a carbon fiber thread or on a kevlar fiber thread

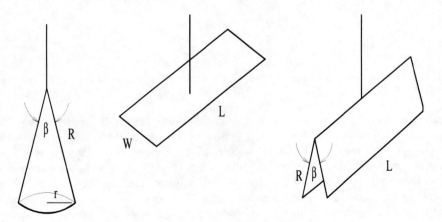

FIGURE 2. Sails shapes studies with circular polarization: cone, strip, and roof shapes.

The fiber thread was attached to the sail using high vacuum epoxy (Torr seal, Mechanical Laboratories Inc.). The carbon fiber was very difficult to handle due to its extreme fragility and stiffness. Each sail and its attached fiber were used only once and repositioning was difficult. Bringing the chamber up to atmospheric pressure completely destroyed the sail–carbon fiber connection. Kevlar fiber thread is much easier to work with, it is more flexible, strong and not as brittle as the carbon fiber. The sail-Kevlar fiber connection 'survived' several experiments. The sail was mounted so that it was centered with respect to the peak of the microwave power distribution and normal to the microwave beam. We tried to handle the sail material carefully to avoid possible surface contamination. Some experiments were performed with the support system described above. Each sail was irradiated by the microwave beam from below and spun in the direction of the wave polarization. The spin direction reversed if the direction of beam polarization was reversed. We studied sails made of different materials including carbon fiber material (carbon-carbon microtruss) (Energy Science Laboratory Inc., CA), aluminum foil and aluminum coated Kapton (courtesy of ORCON Corp., CA). Most of the experiments were performed with aluminum foil sails. All experiments were conducted in the evacuated chamber under vacuum conditions of 2 to 5×10^{-5} Torr to avoid the thermo-molecular effects described above.

We used the following model of rotational motion of the sail in order to estimate quantitatively the angular momentum transfer. We assume that only reflection takes place and no absorption of the microwaves by the sail material occurs, a very good assumption for aluminum.

$$I \frac{\partial^2 \theta}{\partial t^2} = T_{thread} + T_{rf} + T_{damping} .$$

$$T_{thread} = -k_\theta \theta ; \quad T_{damping} = -v \frac{\partial \theta}{\partial t}, \quad T_{rf} = \alpha \left(\frac{P_s}{c} \right) \left(\frac{\lambda}{\pi} \right).$$

where I is the moment of inertia of the sail, T_{thread} is the restoring torque of the fiber, T_{rf} is the torque due to microwaves, $T_{damping}$ is the damping torque, k_θ is the torsional spring constant, v is the damping constant, α is the coupling coefficient relating the microwave power on the sail, P_s to the applied torque, c is speed of light, and λ is the wavelength of microwave beam. At a certain total power of the beam P_t and respectively, power on the sail P_s, the sail will rotate to a maximum angle θ_m. Then the coupling coefficient is

$$\alpha = \left(\frac{2\pi}{\tau} \right)^2 \left(\frac{\theta_m}{P_s} \right) \left(\frac{cI\pi}{\lambda} \right).$$

We define the efficiency coefficient as the total use of all beam power, including that power which misses the sail entirely.

$$\varepsilon = \alpha \left(\frac{P_s}{P_t} \right) = \left(\frac{2\pi}{\tau} \right)^2 \left(\frac{\theta_m}{P_t} \right) \left(\frac{cI\pi}{\lambda} \right).$$

This coefficient gives the efficiency of using the total beam power to spin the sail.

Aluminum Strip Sail Tests

Strip sails made of aluminum foil (areal density $\sigma = 3$ mg/cm^2, sail length $L \gg$ sail width w) were investigated. The maximum angle of displacement, θ_m, from the initial position of the sail was determined experimentally for each power level of the microwave beam P_t. The rotation of the sail was recorded by a video camera and then displayed on a TV screen. The angle of displacement for each power level was determined through measurement of the projection of the strip on the screen. The microwave power intercepted by the sail P_s was calculated for each particular case, accounting for sail size and its position with respect to the horn. After power was turned off the period of oscillations of the sail τ was measured. We determined the value of the coupling coefficient α using the above equation. The coupling coefficient α and efficiency coefficient ε are shown to an accuracy within 20% or less.

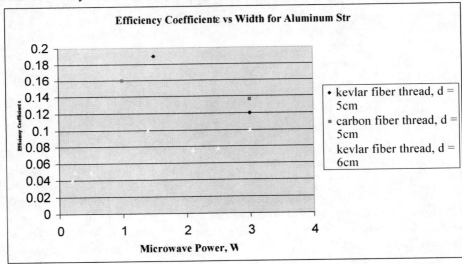

FIGURE 3. Coupling coefficient α vs. width of the strip for various horn distances.

A Scaling Rule for Coupling to Conducting Strips

We conducted our experiments on the microwave angular momentum transfer using aluminum shapes hanging from thin fibers of carbon or Teflon. The coupling coefficient between the beam and the sail was often quite high, ~1.

How can α exceed unity? A geometric argument may explain this. We know from theory that angular momentum is conveyed at the *edges* of conductors. As a wave diffracts around a conducting sail it is affected to a distance ~ λ laterally away from the sail. The diffraction pattern we observed (bright spots on the hanging thread, spaced about a wavelength apart) implies that wave energy within an area *larger than the sail area dL* (with L the length) can convey angular momentum. This is because

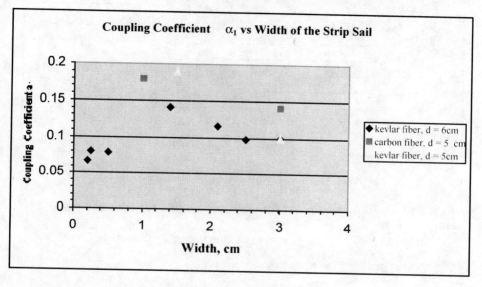

FIGURE 4. Efficiency coefficient ε vs. width of the strip.

the wave is disturbed over the distance λ, so the apparent area dL is in fact larger by a factor ~ $(1+2\lambda/w)$, taking into account the diffraction of the wave at the edges, where w is the width of the strip. We define the adjusted coupling coefficient α_1 as

$$\alpha_1 = \frac{\alpha}{1 + 2\lambda/w}.$$

where α is the experimental coupling coefficient. Fig. 5 shows enhanced coupling, so includes the wave energy conveying torque beyond the edge of the strip. This produces a peak between 1 and 2 cm, or about half a wavelength.

In order to investigate the effect of cuts in the sail surface, which can interrupt surface current patterns, on the beam–sail interaction, we conducted several experiments with strips that were slotted along the width or along the length of the strip.

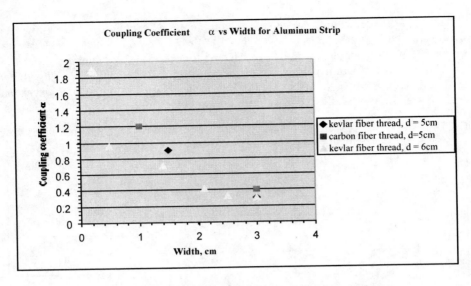

FIGURE 5. Coupling coefficient α_1 vs. width of the strip.

TABLE 1. Coupling coefficient for aluminum strip sails with and without cuts.

Strip (Al foil)	Moment of Inertia, $kg \times cm^2$	Coupling Coefficient
3 cm $\times$ 8 cm, without cuts	4.38×10^{-8}	0.3
3 cm $\times$ 8 cm, 2 cuts longitudinal (2.5 cm each)	4.38×10^{-8}	1.1
3 cm $\times$ 8 cm, 6 cuts transverse (1 cm each)	4.38×10^{-8}	0.68

The transfer of angular momentum from the microwave beam to the strip sail depends on the width of the strip and its relative size with respect to the wavelength of the beam. The smaller the width the higher is the coupling coefficient. However, the efficiency of using the beam power decreases with smaller width and has a maximum for a strip with width equal to about half the wavelength. Slotting the strip sail leads to an increase of the coupling coefficient. The increase depends on the size of the cut, the number of cuts and the spacing of the cuts. See Table 2 for some examples.

<u>Aluminum square sail tests.</u> Some tests were conducted with sails square in shape. The results of the experiments are in Table 2. In this case slotting the sail decreased the coupling coefficient and the smaller sail has a higher coefficient.

TABLE 2. Coupling coefficient for aluminum s square ails with and without cuts.

Strip (Al foil)	Moment of Inertia, kg×cm^2	Coupling Coefficient
3 cm × 3 cm, without cuts	4×10^{-9}	0.21
3 cm × 3 cm, 4 cuts (1.5 cm each)	4×10^{-9}	0.07
1.25 cm × 1.25 cm, without cuts	1.2×10^{-10}	0.54

<u>Aluminum roof-shaped sail tests.</u> We investigated aluminum foil sails in the form of a roof. Experiments were conducted for a 'long' (L= 6 cm) roof and for a short roof (L = 3.4 cm). We tried to determine the dependence of the coupling coefficient for the roof sail on its opening angle. This will determine the effect of multiple reflections within the roof.

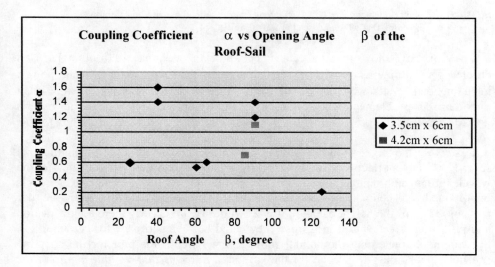

FIGURE 6. Coupling coefficient α vs. opening angle β of roof-shaped sails.

It is interesting to note that for the small roof (1.5cm × 3cm) with an opening angle of 2° the coupling coefficient is very high and equals 29, which is almost 30 times larger than the coupling coefficient for the roof with an opening angle of 90°. It seems that the behavior of the aluminum roof sail is similar to the behavior of a strip sail that has a size equal to the base of the roof sail. Apparently, for these geometries internal reflection (the influence of the opening angle) does not play a significant role in the sail dynamics.

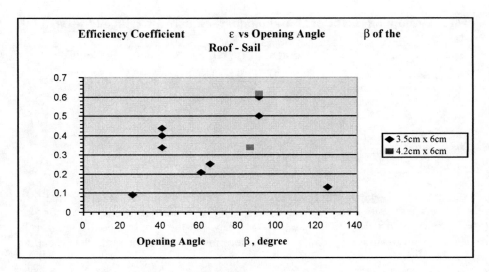

FIGURE 7. Total efficiency $\varepsilon = \alpha(P_s/P_t)$, which measures the total use of all beam power, including that which misses the sail entirely.

The plot above shows the total efficiency ε of beam power use, vs. roof angle. This efficiency we define as $\varepsilon = \alpha(P_s/P_t)$. This measures the total use of all beam power, including that power which misses the sail entirely. For a real system, this is an important design element. The two peaks occur where the roof width at its base is λ (120 degrees) or $\lambda/2$ (50 degrees). *This implies a cavity-like resonance effect for the roof.*

Results for the larger and long roofs ($r = 3.5$cm, $L = 6$ cm) have prominent maxima so the refractive effects are more expressed (the ratio of the roof height to the wavelength is important and the ratio of the width of the roof base to the wavelength is important also). The maxima correspond to the roof angles 40° and 90°, where the roof at its base is λ or the height of the roof is λ (within 20%). Clearly there are diffraction and reflection effects and it looks like they could be 'competitive ' for certain cases. For example, concerning the shorter roof (3.5cm x 3cm), the plot for the coupling coefficient is similar to the one for the strip sail. *It seems that the diffraction of the wave by the short roof (L = 3cm) overshadows the reflective effects due to the roof angle (height of the roof) and the roof sail 'behaves' as a strip sail.*

Carbon cone sail test

The carbon fiber material (carbon-carbon microtruss) cone should spin under the influence of a polarized microwave beam, due to both absorption of the microwaves by the carbon fiber material and to the geometric absorptivity, if it is not symmetric. The experiment was conducted with a carbon cone (a conical beamrider, JPL PO#1212552, mass is 0.056g) hanging on a kevlar fiber thread. The cone base had a radius $R = 2.4$ cm and a cone angle of 90°. The cone was placed at a distance of 5.5

cm from the horn of the polarizer. When the cone was irradiated with the polarized microwave beam, it rotated in the direction of the wave polarization.

FIGURE 8 Conical sail-beamrider made of carbon fiber material.

Taking parameters of the motion from the linear regime, we can determine the coupling coefficient for the carbon cone. For example, given that the moment of inertia of the cone is $I_c = 1.6 \times 10^{-8}$ kg cm^2 and the maximum displacement angle is $\Theta_m = 50°$ at beam power $P_t = 50$ W ($P_s = 15.3$ W) and the period of oscillation without microwave power is $\tau = 107$ s, the calculated coupling coefficient $\alpha = 0.1$. Since the measured absorptivity of the carbon material in the microwave is 10%, the rotation is due to simple absorption of the microwaves by the carbon fiber material. *This agreement checks our basic* picture.

We conducted many tests with cone sails made of *aluminum* foil. Theory shows that completely symmetrical objects (bodies of revolution) should not spin when irradiated by circularly polarized wave. *We did not observe any movement when we increased the beam* power *up to 1100 W, confirming that symmetric objects cannot be spun.* For a given cone, an areal cut does not significantly increase the coupling coefficient, even if it significantly destroys the shape and consequently the symmetry of the sail. But several cuts (just slotting the surface) will increase the coupling by a factor >10.

CONCLUSIONS

We conducted experiments demonstrating that a sail could be induced to spin when irradiated by a circularly polarized microwave beam. The sails spun due to transfer of the angular momentum of the incident microwave beam to the sail. The sails were suspended on either a carbon fiber thread or on a kevlar fiber thread. Spin properties were investigated as a function of the microwave power. We studied sails of different shapes and materials and calculated the coupling coefficient for each shape of sail. Our principal results

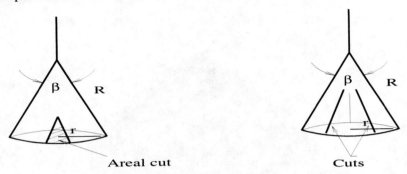

FIGURE 9. Types of cuts: Conical sails with an areal cut (left) and thin cuts (right). We find that the two cut types give very different couplings to circularized microwaves.

- As expected, a carbon fiber material cone spun under the influence of a polarized microwave beam with a measured coupling coefficient of 0.1, fitting independent measurements of carbon material absorption for microwaves in this frequency region.
- Coupling to conductors can be greatly enhanced, giving coupling coefficients of 1 or more, by the 'geometric absorption' effect (see Appendix), a previously unnoticed phenomena in classical electromagnetics.
- Transfer of angular momentum from the microwave beam to a strip sail depends on the width of the strip and its relative size with respect to the wavelength of the beam. The efficiency of using the beam power decreases with smaller width and has a maximum for a strip with width equal to about half the wavelength.
- *Slotting the strip sail leads to an increase of the coupling coefficient.* The behavior of the aluminum roof sail (of the size we studied) is similar to the behavior of a strip sail that has a size equal to the base of the roof sail.
- Slotting an aluminum cone sail can also significantly increase the coupling.
- Our experiments suggest strongly that cuts in good conductors (e.g., aluminum) constrain the circulating currents within it to the scale between the cuts. When this size is roughly a wavelength, maximum coupling occurs. Radial cuts give more coupling than azimuthal cuts.

REFERENCES

1. J. D. Jackson, *Classical Electrodynamics*, 1st edition, New York, J. Wiley and Sons, 1962, p. 201.
2. Christian Konz .and Gregory Benford,. to be published
3. Schindler, K., *Physica Scr.* **50**, 20 (1993)
4. Heitler, W., *The Quantum Theory of Radiation*, Clarendon, Oxford, 1954, p.401
5. Khrapko, R.I., *Am. J. Phys.*, **69**, 405 (2001)
6. Ohanian, H.C., *Am. J. Phys.*, **54**, 500--505 (1986)
7. *Deployment Of Ultralightweight Sails*, 2002 Final Report by James Benford, et. al.,available from jbenford@earthlink.net

Experimental Tests Of Beam-Riding Sail Dynamics

Gregory Benford, Olga Gornostaeva

Physics Dept.
Univ. Calif, Irvine
Irvine, CA USA 92697

James Benford

Microwave Sciences Inc.
1041 Los Arabis Lane, Lafayette, CA 94549 USA

Abstract. Stability is a neglected issue in proposals to propel light sails by beamed power. Whether the beam comes from a laser or a microwave antenna, power falls with angle from the beam center. This drives a sail sideways under any lateral perturbation—"tumbling down the hill.". While spin can help stabilize, the basic mechanics of pressures and sail averaging of them across its area remain unexplored in experiment, and only recently treated in theory. Here we report what is to our knowledge the first attempt to study beam-riding dynamics in the laboratory, using a slightly overweighted pendulum. A sail attached to the pendulum bottom can be made unstable by adding weight to the top end. Stability and oscillation are possible if this is corrected by electrodynamic beam pressure on the sail, directed from below, torquing the pendulum. We present both data and analysis. Our major points are: Microwave powers of a few hundred W can hold a sail steady. This is made possible because of the gradient in beam power with sidewise angle. Our experiments agree with the Univ. New Mexico numerical studies which show similar stability conditions. At higher powers, the sail can be oscillated in angle. Time-dependent feedback of beam power can manage a sail into stable motions across the beam. Theory shows this, but experiments are not yet done. Beam powers comparable to the strength of perturbing forces can plausibly achieve these effects in free sail flight.

INTRODUCTION

Generally, beam-riding sails are unstable, and will move away from the beam center in uncontrolled fashion. Here we describe experiments which explore the electrodynamic restoring forces acting on sails, which can in principle lead to stable sail flight.

To study stability in a g-field, one must compensate for gravity and yet give a source of instability. This implies the schematic design of Fig 1, where a sail hangs

CP664, *Beamed Energy Propulsion: First International Symposium on Beamed Energy Propulsion,*
edited by A. V. Pakhomov

from a rod that pivots about a bearing. The sail has mass M^* and the rod M'. Above the bearing is a counter-weight M which provides very nearly compensating mass, $M > M^* + M'$, making the whole assembly slightly unstable. The microwave source below provides a pressure which can make the sail return to vertical, hanging straight down, i.e., the angle θ goes to zero.

An important result emerges if the sail is displaced a distance d from the beam axis. It turns out that the *gradient* of the microwave power in angle is the stabilizing element, because the *difference* in torques about the sail center drives rotation of the assembly. Moving the sail slightly aside (~ 1 cm) helps minimize the needed power. Generally, the sail will fall to the side, increasing θ, and have some momentum as it passes through the ideal angle ~ d/H. It will go a bit further, then be arrested, where the mechanical torque equals the microwave torque averaged over the sail.

Still, there are considerable design requirements; one must fine-tune the lengths of the rod and counter-weight, and the counterweight mass. In our simple theoretical model, power P falls with angle as $\cos^2 \theta$, which helps off-center sail placement. A steeply falling beam would be much easier to stabilize. Experiment finds a gradient $\cos^m \theta$ with m ~ 5, promising ready stabilization. In a true, flying sail in zero-g, this may be crucial.

Flipping this arrangement, so the sail is upright and falls from vertical as it goes unstable, yields a *higher* power demand. This comes from geometric effects. Basically there are three geometric and physical effects affecting stability:

1. angle between wave and surface
2. area exposed
3. falloff of power with angle from the beam axis

The hanging sail mode (pendulum-like) wins overall for these factors.

Specifically, take $\theta = 12°$ and $d/R = 0.1$, with a 2 cm radius sail hanging 10 cm from the microwave source, on a rod 10 cm long. With the counter-weight to be 9 mm above the pivot (bearing), then the sail will be stable to sideways tumbling at power levels of 860 W or more. Roughly a kW should show beam-riding, then. This prediction we verified.

THEORY

The pendulum in Fig.1 follows the *difference in torques* from microwaves ("rf") and gravity ("gr") about the axis, which bisects the circular sail. Let us consider a pendulum with a counter mass M_c equal to the mass of the sail $M_c = M_s$. The pivot of the pendulum is at the distance d with respect to the *center line* of the waveguide.

M_s is the mass of the sail,
R is the radius of the sail (or 2R is a side of a square),
L_o is the length of the pendulum,
$2L_o$ is the length of the rod,
ΔL is the distance from the pivot to the center of mass,

326

$k_o = L_o / \Delta L.$,

$H_o + L_o$ is the distance from the pivot to plane of the edge of the waveguide,
L is the distance from the waveguide to the center of the sail,
θ is the angle of microwave beam propagation (see the figure), and
ϕ_1 is the pendulum decline angle from vertical position.

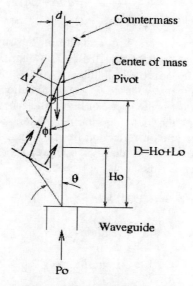

FIGURE 1. Pendulum with counter-mass off center

Static case: Pendulum with No Counter-Weight; Minimum Needed Pressure

To maintain the pendulum at an angle ϕ_1 the torque produced by the force due to the microwave field should be equal to the torque produced by the force of gravity. We consider the case of total reflection and no absorption of microwaves by the sail material. For a given ϕ_1, H_o, L_o we have

$$\theta = \arctan\left(\frac{\sin\phi_1 + d/L_\circ}{H_\circ/L_\circ + 1 - \cos\phi_1}\right).$$

$$1 = \frac{L_\circ (H_\circ/L_\circ + 1 - \cos\phi_1)}{\cos\theta}, \ \beta_1 = \phi_1 + \theta.$$

and the torque equilibrium equation is

$$T_{grav} = T_{rf}.$$

where T_{grav} is torque due to the gravitational force and T_{rf} is the torque due to microwave pressure, here:

$$T_{grav} = M_s g \Delta L \sin \phi_1,$$

$$T_{rf} = \frac{P_o A_k 2 \int rf(r) dS}{c} = P_o A - 2R(2K/c).$$

where S is the area of the sail, r is a coordinate on the surface of the sail, A_k is a normalization coefficient ($A_k = 0.38$ for Bessel $J_1(\theta)$distribution of the field in the plane XOZ and $A_k = 2\pi/3$ for $\cos^2 \theta$ distribution of the RF field in the plane XOZ), P_o is the microwave power in the waveguide, $f(r)$ is the function proportional to the force due to effect of microwaves, c is the speed of light. The function $f(r)$ is proportional to $J_1^2(\theta)$ or to $\cos^2 \theta$ due to the field distribution pattern and proportional to $\cos^2$(incident angle) due to the photon reflection from the sail. The K-integral assembles all the geometric factors governing electrodynamic pressures on the sail.

Then the *torque equation* is (mks units)

$$M_s g \Delta L \sin \phi_1 = P_o A_k 4R K/c \qquad (1)$$

and the microwave power in the waveguide is

$$P_o = \frac{M_s gcL_o \sin \phi_1}{4 A_k KRk_o} = \frac{M_s gc\Delta L \sin \phi_1}{4 A_k KR} \qquad (2)$$

where $g = 9.8$ m /sec^2, $c = 3 \times 10^8$ m /sec, $M_s = \rho \times 4R^2$, $\rho = 0.005$ kg/m^2 for our carbon sail and A_k is given above.

Dynamic case-- Pendulum with Counter-Weight

Consider a pendulum with a counter mass M_c equal to the mass of the sail $M_c = M_s$ and a pivot at a distance d with respect to the center line of the waveguide.

M_r is the mass of the rod,
$2L_o$ is the length of the rod,
The pivot point is at the center of the rod,
M_s is the mass of the sail, and
I is the moment of inertia of the system.
The equation of motion for the system is:

$$I \frac{d^2 \phi_1}{dt^2} = T_{grav} - T_{rf} \qquad (3)$$

where T_{grav} is the torque due to the gravitational force and T_{rf} is the torque due to the microwave pressure. In order to solve the motion equation we have to calculate the moment of inertia I for the system considered:

$$I = M_r (2L_\circ)^2/12 + M_s L_\circ^2 + M_s (L_\circ + \Delta L)^2 \approx M_r L_\circ^2/3 + 2M_s L_\circ^2 + 2M_s L_\circ \Delta L$$
$$= M_r L_\circ (L_\circ/3 + 2M_s/(M_r \times L_\circ) + 2M_s/(M_r \times \Delta L))$$

Then the equation of motion will be

$$I \frac{d^2\phi}{dt^2} = M_s g \Delta L \sin\phi_1 - P_\circ A_k 2R \frac{2K}{c} . \qquad (4)$$

Let us consider the value of integral K versus angle ϕ_1. The equation (1) for K gives us the direction of the torque due to the microwave pressure versus the angle of deflection ϕ_1 for given pendulum geometry. The sign of K is positive for counter clockwise rotation due to the RF torque and negative for clockwise rotation due to the RF torque.

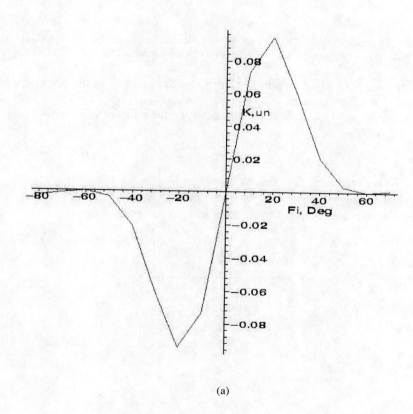

(a)

329

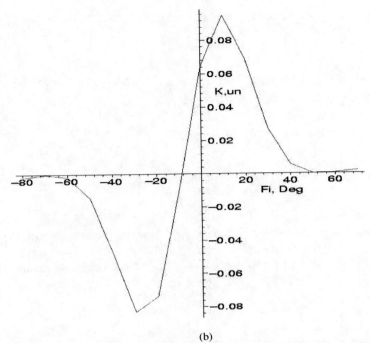

(b)

FIGURE 2. The value of integral K versus deflection angle ϕ_1 for centered pendulum (a) and off center pendulum (b), d = 1 cm. Note the asymmetry when d is not zero; this is the source of stabilizing torques.

The plot for $K\,(\phi_1)$ is shown in Fig.2 for a given geometry of the pendulum and its position with respect to the waveguide. As can be seen, we can use a linear approximation for the value of K for angles within the range of -23 deg to 10 deg:

$$K(t) = K_\circ + K\phi_1(t).$$

where $\phi_1(t)$ is in radians, $K_o = K\,(0) = 0.0643$ and k = 0.298 {1/rad}.

For small angles ϕ_1 ($\phi_1 < 17$ deg) we have sin $\phi_1 \sim \phi_1$

Then we have the following equation of motion for small angles ϕ_1

$$\frac{d^2\phi_1}{dt^2} = \frac{\left(M_s g\Delta L\phi_1 - P_\circ A_k 4R\left(\dfrac{K_\circ + k\phi_1}{c}\right)\right)}{I}$$

$$= \phi_1\left(\frac{M_s g\Delta L}{I} - P_\circ A_k 4R\frac{k\phi_1}{cI}\right) - P_\circ A_k 4R\frac{K_\circ}{Ic} = A + B\phi_1 \tag{5}$$

330

where

$$A = -P_\circ A_k 4R \frac{K_\circ}{Ic} \quad \text{and} \quad B = \frac{M_s g \Delta L}{I} - P_\circ A_k 4R \frac{k}{cI}.$$

In order to solve the equation analytically, perform the following transformation:

$$y(t) = A + B\phi_1(t).$$

then

$$\frac{d^2 y(t)}{dt^2} = By(t), \quad y(0) = A + B\phi_1(0).$$

This equation can be solved analytically, with solution

$$y(t) = (1/Ic)(- P_\circ A_k 4RK_\circ - P_\circ A_k 4Rk\phi_1(0) + M_s g \Delta Lc\phi_1(0)) \times$$
$$\cos\left(\left(\sqrt{(P_\circ A_k 4Rk - M_s g \Delta Lc)/cI}\,\right)t\right)$$

$$\phi_1(t) = \{(1/Ic)(- P_\circ A_k 4RK_\circ - P_\circ A_k 4Rk\phi_1(0) + M_s g \Delta Lc\phi_1(0))$$
$$\times \cos\left(\left(\sqrt{(P_\circ A_k 4Rk - M_s g \Delta Lc/cI)}\,\right)t\right)\} + \tag{6}$$
$$P_\circ A_k 4RK/cI \}/(M_s g \Delta L/I - P_\circ A_k 4Rk/cI)$$

We can write that

$$\phi_1(t) = \phi_1(0)\cos(\omega t) + A_0(1 - \cos(\omega t)) = A_0 + (\phi_1(0) - A_0)\cos(\omega t).$$

where

$$A_0 = (K_\circ/k)(P(\omega = 0)/P_\circ - 1) - 1.$$

$$\omega = 2\pi/T = \sqrt{(P_\circ A_k 4Rk/c - M_s g \Delta L)/I}.$$

critical stabilizing power:

$$P_\circ(\omega = 0) = \frac{M_s g \Delta Lc}{A_k 4Rk}.$$

Unbounded motion:
This solution will be a cosh function for power values satisfying the inequality

$$P_\circ A_k 4Rk - M_s g \Delta Lc < 0$$

Bounded motion:
Solutions are a cos function for power values satisfying the inequality

$$P_\circ A_k 4Rk - M_s g\Delta Lc > 0$$

with angular velocity

$$\omega = 2\pi/T = \sqrt{(P_\circ A_k 4Rk/c - M_s g\Delta L)/I}$$

(7)

$$\omega = 0, \text{ if } P_\circ(\omega = 0) = \frac{M_s g\Delta Lc}{A_k 4Rk}.$$

As we can see, period of oscillations does not depend on initial deflection angle and depends only on the microwave power and geometrical parameters of the system.

EXPERIMENT

The microwave source is a CW 10.6 GHz klystron with maximum power 10 kW. The waveguide assembly used WR90 and waveguide components in X-band. The microwaves are emitted from a horn with narrow power distribution pattern with a gradient $\sim cos^m \theta$ with $m \sim 5$.

Since the major part of the microwave power is reflected towards the bottom of the chamber, we covered the bottom of the chamber with silicon carbide tiles were used as an absorber. We used a 2.5-cm radius disk sail made of carbon fiber microtruss manufactured by Energy Science Laboratories Inc. A 3-cm long axle was placed through the center hole of an aluminum tube of 8-cm length and fixed with torr seal. A disk made of aluminum foil or a nut made of the nylon was used as a counterweight in the experiments. The sail-aluminum rod connection was made using thin silver wire. A digital video camera views the pendulum inside the experimental chamber through the glass window and hexcell screen in the viewing port.

We placed the same sail at the distance of 4 cm from the open end of the waveguide and off-center 1 cm from the center line of the waveguide ($d = 1$ cm in Figure 1). The initial angle of deflection from the position of unstable equilibrium was varied from 2 to 5 degrees, with accuracy +/- 0.5 degree. This is the range of angles the pendulum moves through, due to mechanical vibrations from coupling of the chamber to the pump. We irradiated the sail at each position with microwave power up to 30 sec.

The principal result is that the sail moved due to the microwave torque overcoming the gravitational torque, and passed through the (unstable) vertical position. The power level where this occurred appears in Figure 2 for each initial angle of deflection.

The movement of the pendulum we observed through the video camera taped and downloaded onto computer. We measured the power level corresponding to the movement of the pendulum, to within +/-50 W. The experiment divides into two separate tests:

Static Pressure Balancing (lower curve): Sail stability against gravitational torques occurs at the power levels predicted. This verifies the view inspired by the UNM numerical stability studies: *power gradients of form* $\cos^m \theta$ *can stabilize sail motion if* $m > 4.5$. *In our case,* $m \sim 5$, *so our experiment agrees with this feature of the UNM simulations.*

Power versus Initial Angle of Deflection

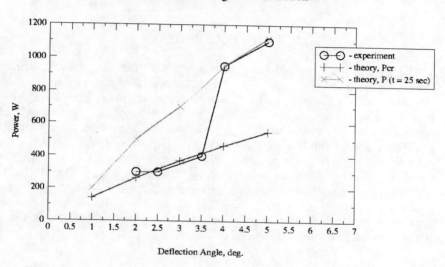

FIGURE 3. Power at Which the Sail Moved: Microwave torque overcame gravitational torque vs. initial angle of deflection. Points are data; curves are theory with no fitting parameters.

In Figure 2, the calculated value of P_{cr} ($\omega = 0$), the static case, predicts the power at which the pendulum remains at its initial angle of deflection after the pendulum is released by the actuator. The lower curve is theory for such static sail holding (with *no parameter fitting*) and fits well the power needed to just barely sustain the sail.

The calculations described in Section 2 suit the pendulum used in the experiment: pendulum off-center 1 cm from the center line of the waveguide, sail radius 2.5 cm, sail mass 18.5 mg, pendulum rod 8 cm, rod mass 397 mg, position of the center of mass of the system at the distance $\Delta L = L/50 = 0.8$ mm. The value of the center of mass (0.08-cm) is well within the adjustment capability of the countermass. Here L is the distance between sail and pivot (4 cm); distance between the sail and the end of the waveguide is 4 cm.

Dynamic Motions (upper curve): Vibrations in the system explain the transition from static force balance (lower curve of Figure 2) to the power needed to stabilize the sail against gravitational torques (upper curve of Figure 2). The lower angle cases are held in static equilibrium by the microwave power, but can be *driven by vibrations* to vertical (sail down).

At the higher angles, vibrations are insufficient to start the motion, so the full power calculated from the simple force equation must be used to move them to vertical and beyond. Therefore, the transition in Fig. 3 corresponds to the vibration-dominated limit of dynamics. The transition in Figure 3 at about 4 degrees shows that vibration accelerations can flip the sail to vertical for angles less than this, but the full power calculated for sail motion is needed at higher angles. That implies the vibrations give the pendulum accelerations of g (θ/56) =0.07 g. Reducing vibrations should then show a transition at lower angles.

CONCLUSIONS

The basic equations above are compatible with our static force measurements. Observing oscillatory measurements will demand a higher power level in future experiments. This implies that *achieving stability demands a sufficient gradient of microwave power across the sail diameter.* This is why placing the sail slightly off center from the microwave beam greatly aids stabilizing motion, since there the power gradient is larger. *This feature will be true of actual flying sails as well.*

Note that we have not studied the influence of *sail shape* on stability. The UNM simulation results to date say that a shallow "Chinese hat" sail is stable, but no other shape has yet. The class of unstable sails includes the simple flat sail we used in the pendulum experiments, which in the UNM simulations slides readily down the power gradient in angle. In the UNM work, spin seems unable to correct for this propensity. Further experimental work is needed to study sail shape.

NOMENCLATURE

R = radius of the sail ($2R$ is a side of a square sail) (m)
ΔL = distance from the pivot to the center of mass (m)
θ = angle of microwave beam propagation ($°$)
ϕ_1 = pendulum decline angle from vertical position ($°$)
K = integral that assembles all the geometric factors of the
  electrodynamic   pressures (m^{-1})
I = moment of inertia (kg-m^2)

ACKNOWLEDGMENTS

This work was supported by the Advanced Concepts Office, Office of Space Science, National Aeronautics and Space Administration through the Jet Propulsion Laboratory; Advanced Concepts—Technology Innovations project office. We thank Neville Marzwell for helpful comments and support.

REFERENCES

1. Benford, J., Benford, G., Goodfellow, K., Perez, R., Harris, H., and Knowles, T., "Microwave Beam-Driven Sail Flight Experiments," in proceedings of *Space Technology and Applications International Forum (STAIF-2001)*, Space Exploration Technology Conf, AIP Conf. Proceedings 552, ISBN 1-56396-980-7STAIF, 2001, p. 540.
2. Csonka, P. L., and Muray, J. J., "Radiation-Supported Space Mirror: Overview of Concept," *IEE Trans. Aerospace and Elec. Sys.*, AES-21, 320, 1985.
3. Genta,G., and Brusa, E., "Basic Considerations on the Free Vibrational Dynamics of Circular Solar Sails," in proceedings of *Missions to the Outer Solar System and Beyond*, July, 2000, pp. 81-87.
4. Schamiloglu, E., Abdallah, C.T., Miller, K.A., Georgiev, D., Benford, J., Benford, G., and Singh, G., "3-D Simulations of Rigid Microwave-Propelled Sails Including Spin," in proceedings of *2001 Space Exploration and Transportation: Journey into the Future*, Albuquerque, NM, February 2001.
5. Schamiloglu, E., Abdallah, C.T., Georgiev, D., Benford, J., and Benford, G., "Control of Microwave-Propelled Sails Using Delayed Measurements," in proceedings of *2002 Space Exploration and Transportation: Journey into the Future*, Albuquerque, NM, February 2002.

3-D Simulation of Rigid Microwave Propelled Sails Using Spin

D. Georgiev, E. Schamiloglu, C.T. Abdallah, and E. Chahine

Electrical & Computer Engineering Department,
The University of New Mexico,
Albuquerque, NM 87131

Abstract. This paper presents using Matlab, a software package developed for microwave-propelled sails. The software is user-friendly and may be used to design controllers, along with studying various sail designs and their stability.

INTRODUCTION

Microwave-propelled sails belong to a class of spacecrafts that promises to revolutionize future space travel. As an example, NASA's Gossamer Spacecraft Initiative focuses on developing spacecraft architectures for very large, ultra-lightweight apertures and structures. A goal of this initiative is to achieve breakthrough enhancements in mission capability and reductions in mission cost, primarily through revolutionary advances in structures, materials, optics, and adaptive and multifunctional systems. Solar and other types of sails will provide low-cost propulsion, station-keeping in unstable orbits, and precursor interstellar exploration missions. For a general introduction to solar sails and similar structures the reader is referred to [1]. For an introduction to the notion of beamed microwave power and its application to space propulsion the reader is referred to [2]. This paper is concerned specifically with the stability and control of carbon fiber sails propelled using microwave radiation.

The notion of *beam-riding*, *i.e.*, the stable flight of a sail propelled by Poynting flux, places considerable demands upon a sail. Even if the beam is steady, a sail can wander off the beam if its shape becomes deformed, or if it does not have enough spin to keep its angular momentum aligned with the beam direction, in the face of perturbations. Generally, sails without structural elements cannot be flown if they are convex toward the beam, as the beam pressure would cause them to collapse. On the other hand, the beam pressure keeps concave shapes in tension, so concave shapes arise naturally while beam riding. They will resist sidewise motions if the beam moves off center, since a net sideways force restores the sail to its position. Therefore, we concentrate on a conical shape for the sail and study its dynamics in 1-D and 3-D. We have previously shown using the Poincare-Bendixon theorem that such conical shapes will oscillate when a constant microwave power beam is used [3]. We present in this paper a simulation package that may be used to study the passive stability, along with various preliminary feedback controllers. Specifically, we have used the software to design controllers for 1-D motion using delayed measurements, and to design stabilizing controllers in three

CP664, *Beamed Energy Propulsion: First International Symposium on Beamed Energy Propulsion*,
edited by A. V. Pakhomov

dimensions.

MODELLING ASSUMPTIONS

The sail, which refers to the entire system, is composed of a reflector, a hollow mast, and a payload representing ball. The mast connects the payload to the reflector and it is attached to the reflector at the reflectors center of mass (C.M.). Although this study is a simplified look at the problem, the payload was included and displaced from the reflector (via the mast) for stability reasons. Without the payload system, which consists of the mast and the payload, the sail is inherently unstable to pitch and roll perturbations. Along with the basic configuration of the sail, certain assumptions, listed below, had to be made about the system.

1. No internal reflections are occurring within the reflector. This assumption may seem significant, however, after conducting some preliminary analysis, it was found that in fact only shallower reflector shapes are stable and only under minor angular perturbations.

2. The reflector's material has perfect reflectivity. The actual carbon mesh being considered has about 98% reflectivity, therefore, this assumption is valid.

3. The system is composed of either one or two rigid bodies, depending on the status of the hinge. This is perhaps the strongest assumption. The actual sail will be flexible, however, the purpose of this effort is to study the large scale control and behavior of the system. The structural as well as thermal analysis is left to future research.

4. The microwave source is a point source using a square wave guide. Other microwave models may be implemented in the future.

5. The payload and the mast do not interfere with the microwave power. This is a valid assumption since the payload configuration presented here is only a symbolic representation of the actual set-up.

6. The gravity vector is assumed to point in the negative Z direction of the inertial frame.

SIMULATION CODE

The unique features of this problem (i.e. photon propulsion) makes it difficult to study using commercial simulation software. Because of its wide availability and capabilities, MATLAB was chosen for the implementation of the equations of motions. One main advantage of using MATLAB is the availability of robust differential equation solvers. For example, the two-body sail system has twenty states, requiring an efficient integration method. In addition to the differential equation solvers, MATLAB also has convenient Graphical User Interface (GUI) structures, which makes the final package easier and intuitive.

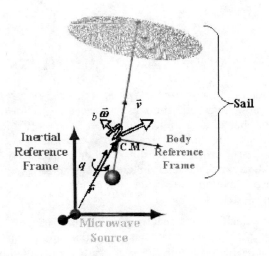

FIGURE 1. Single body coordinate systems and states.

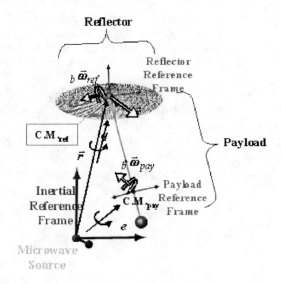

FIGURE 2. Two body coordinate systems and states.

Control Capabilities

An important of this code is the implementation of control algorithms. In the case of the one body system, the control inputs consist of the power coefficient P_t and the beam shape indices n_x and n_y.

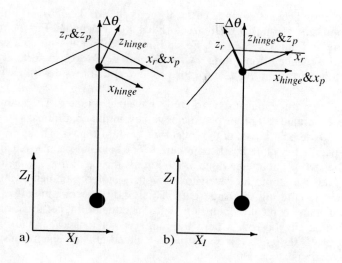

reflector/payload
hinge$_{eq}$ is perturbed $\Rightarrow$ perturbed
configuration

FIGURE 3. Controlling the reflector relative attitude by manipulating the neutral orientation of the hinge

The user has the option of adding delay to the system. Since the distance of the actual system from the control center will be significant enough to delay the control input, this feature brings this model and its control one step closer to reality. Delay is realized by keeping an ongoing global array of the state and the control inputs at every successful time step. A successful time step is located through the ode output print function. This is a function provided by MATLAB that prints the time and the state vector at all successful time steps. Therefore, a modified version of this function was created that not only prints to the screen but also stores the values to a global array.

Before the addition of delay, simulations consisted of the sail hovering above the source. At this point, however, the simulation of an actual sail flight can be done. Through the control capabilities the power of the source can be increased, causing the sail to accelerate in the inertial Z-direction. Furthermore, an inverse proportional relationship between gravity and the Z component of the position vector can be established, thereby simulating the sails departure of the Earth's gravitational field. Finally, delay can be gradually added to the system as the communication time between the control center and the sail increases.

Graphical User Interface

The simulation process is made easier through the MATLAB GUI utilities. Controlling the simulation through a GUI structure has many benefits. First, parameter values can be easily and quickly changed. Second, mistakes in entering data can be avoided, decreasing the chance of obtaining a wrong simulation. Most importantly however, the GUI helps to gain an intuitive feel for the problem and allows novice users the ability to experiment.

To aid the user, three main GUI structures have been created. The main GUI structure (see Figure 4) contains all of the parameter values, initial conditions, as well as an image of the sail and the beam front at the reflector's center of mass. The user is alerted when an invalid parameter value has been entered. The sail and beam front images are also updated whenever a parameter value or the initial attitude of the sail have been changed. Here, the user can store and revert to previously stored parameter values, so that no time is spent re-entering the same data. This GUI also contains all of the controller options, which are: controller inclusion, delay, and the initial perturbation of the hinge's neutral position (in the case where the hinge is not locked). Finally, other than the simulation itself, the eigenvalues can be calculated and an eigenvalue based stability region calculation can be conducted.

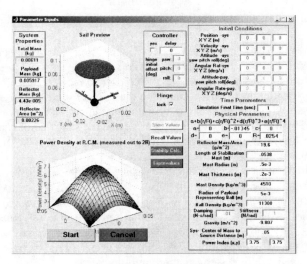

FIGURE 4. Main GUI: Model of Sail

The second GUI (see Figure 5) is called after the integration is complete. Here, the user is given access to some of the more common plots, such as displacement vs. time and phase portraits. All the parameter values can also be stored to a ASCII file for book

340

keeping. Finally, the user has the option of returning to the last simulation or recording an animation.

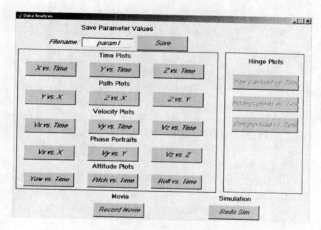

FIGURE 5. Main GUI: Data Entry

The last GUI (see Figure 6) is there to aid in the creation of the animation. Plots provide the means to compare two variales and analyze them quantitatively. An animation however combines all of the states and time, resulting in a more qualitative understanding of the data. MATLAB makes this animation possible with its capability of taking individual frames and converting them into an .avi file. The animation GUI makes the entire process very easy by substituting all of the syntax and formalities associated with the movie function, with more friendly parameters such as; the desired time segment, camera angle, movie resolution, and recording mode.

SAIL DYNAMICS AND PASSIVE STABILITY

Thus far, the control effort of this system has been emphasized as the main application of this code. However, due to the limits set on controllability or delay issues, the passive behavior of the sail system is of interest. In this section we look at the passive dynamics and stability through linearization and simulation.

Ideally, if the system is perturbed away from equilibrium, the sail's configuration will prevent it from going unstable. In this setting, Equilibrium conditions consist of the sail $z_b|z_r$ & z_p-axis being aligned with the Z_I-axes and the body origin coinciding with the Z_I-axes at a user-selected distance from the source. Perturbations from this equilibrium can occur in both translational and angular directions. The translational displacement

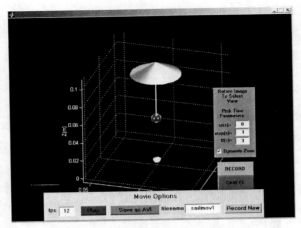

FIGURE 6. Main GUI: Movie Recording

can be represented with the cylindrical coordinates, Z_I & R_I and the angular with the Euler angles yaw, pitch, and roll.

The sail configuration described in this paper has been chosen to provide means of preventing instability for most of the possible displacement modes. The reflector shape is chosen such that the R_I component of the resultant force is inversely proportional to lateral displacement. In our case, where the source is a point source and the beam intensity falls off as the distance squared, a concave reflector is appropriate (see Figure 7).

Angular perturbation is more problematic. From figure 7 we can see that when the reflector shape is chosen to provide the 'restoring force' effect, the force is greater on the inner portion of the surface, causing it to rotate away from equilibrium, and therefore, making the system unstable to pitch and roll perturbations. A 'restoring torque', solving this problem, can be obtained with the addition of the payload (ball and mast). The payload provides a mass that is offset from the center of pressure. Therefore, when the system is accelerating in the Z_I direction, as it will be when exposed to the microwave rays, the payload will provide a torque counteracting the reflector torque. The magnitude of this torque is dependent on the acceleration, payload mass, and the payload C.M. to center of pressure offset.

Finally, the type of stability is dependent on the inclusion of the hinge. Without the hinge the system has no source of damping and therefore can at best be marginally stable. With the hinge, the sail can be asymptotically stable to some, but not all perturbations. For example, a vertical displacement will not produce relative rotation between the payload and the reflector and therefore will not result in asymptotic stability.

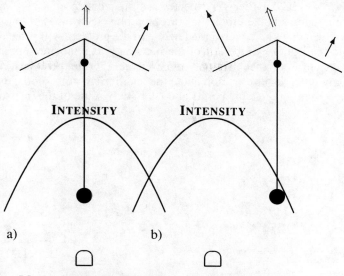

RESULTANT FORCE$_{mic}$ RESULTANT/RESTORING FORCE$_{mic}$

INTENSITY   INTENSITY

a)   b)

MICROWAVE SOURCEMICROWAVE SOURCE

FIGURE 7. A means of obtaining a 'restoring force' via reflector shape manipulation

Local Stability and Statistical Learning

We choose to illustrate the code using concepts of local stability and statistical learning. One method of analyzing the system stability is through linearization. The system is linearized by numerically evaluating the Jacobian,

$$\dot{\bar{x}} = \partial(\mathcal{M}^{-1}f)/\partial\bar{x}^1$$

Stability characteristics are then determined by the real parts of the Jacobian eigenvalues. A particular eigenvector is asymptotically stable when its corresponding eigenvalue has a real part less then zero, marginally stable when it equals zero and unstable when positive. Figure 8 illustrates the mode or eigenvector shapes present in the two systems. The mode shapes for the one-body system consist of various rotational and translational body modes. The two-body system also contains these modes, however, the majority of the modes describe relative motion between the two bodies. Finally, modes shapes 11-13 in figure 8a) and shapes 1,2&18-20 in figure 8b) require special attention. There are two types in this set. The first arises as a result of perturbation of the first quaternion element, which, when perturbed, returns to its original value of one after normalization. The second, corresponds to an attitude change about the Z_l axes. Although these modes

---

[1] In the one body case, $\mathcal{M}$ is simply the identity

do represent a valid perturbation, due to symmetry, the perturbed state is still part of the equilibrium specified earlier. As expected, a system displaced according to these modes will remain there, and therefore, the corresponding eigenvalues of these modes are usually zero. Even though, the numerical nature of the problem will place these eigenvalues slightly to the right or to the left of the imaginary axis they are not significant to any stability conclusions since they only represent equilibrium positions. Although this approach only provides information about the equilibrium point alone, the intent is to use it to locate a region of stable points and later study some of the interior points through simulation.

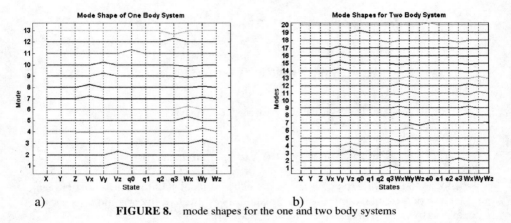

a)   b)

FIGURE 8.   mode shapes for the one and two body systems

The main problem with searching the parameter space however is the number of parameters. Altogether, the number of parameters describing the sail, the beam, and the equilibrium location is 17. Moreover, some of the parameters span 4 orders of magnitude. Therefore, if we wanted search the parameter space with a gridding approach approximately $400 \times 17 = 2 \times 10^{44}$ points would be required. A more efficient algorithm for searching through possible system configurations, plants, has been derived by Koltchinskii and Abdallah in [4]. According to [4] the number of <u>random</u> points required to confidentially approximate a region in the parameter space in which the plant will perform at a certain level and about which we can be confident is

$$m = \frac{\log(1/\delta)}{[1/(1-\alpha)]} \tag{1}$$

where

$$m = \text{number of points}$$
$$\delta = \text{confidence parameter} \in (0,1)$$

344

$$\alpha \;=\; \text{level parameter} \;\in (0,1)$$

When applied to our system, the confidence parameter represents the accuracy of the stability region boundary mapped by the points, and the level parameter designates the level of stability we can expect in that region. The only drawback is, the resolution of the stability boundary in the parameter space is dependent on the point density in that region. Therefore, the number of points required according to equation (1) is the minimum, but in order to get a better idea of the shape of the region the number of points in the vicinity will most likely need to be increased.

The first region in the parameter space studied represents reflectors of a similar size to the experimental reflectors already constructed for the lab experiments at the Jet Propulsion Laboratory, JPL. The JPL reflectors our approximately 5 centimeters in diameter and weigh 19.6 grams/m^2. Although, the dimensionality of the parameter space is not important with the above approach, for the sake of presentability a two dimensional parameter space was chosen. The two parameters picked are; shape parameter $b \in [-2cm, -1cm] =$ (-reflector height) and the ball radius$\in [1mm, 6mm]$. The space created by these two parameters is especially of interest because they represent the magnitude of the restoring force and the restoring torque respectively. The rest of the parameters were chosen to fit the JPL experiments [5]. Finally, the accuracy parameters, α and δ were chosen to be .005 and .01 respectively, which corresponds to a region of 99.5% stability in which we can be 99% confident. Figure 9 contains the results of this investigation.

Initially, the one body system was studied. However, as was stated earlier, no damping was included in this system, which means that we only observe marginal stability. Unfortunately, this data set is not adequate since marginal stability does not signify stability or lack their of. An answer to this question can be found by including damping. In the one body system damping can be caused by drag (during high altitude deployment) or reflector deflections. Modeling energy dissipation caused by reflector deflections would significantly increase the computational complexity. Therefore, for the purpose of this study a small drag like dissipative forces added to the microwave and gravity forces were adequate.

Clearly, in the one body system, the eigenvalues did drift towards the left hand plane. However, since the |maximum| of these values is very close to the absolute tolerance threshold (1×10^{-10}) one may dispute that the system is still marginally stable.

The two body system was found to be more suitable for the stability investigation. The addition of the hinge provides a significant and real damping force that effects all but one mode. As expected, the unaffected mode is the 'hopping' mode (displacement along the Z_I-axis), which, from test simulations, we know will result in periodic motion. However, in order to achieve complete asymptotic stability, the same small 'drag' force is added.

FUTURE WORK

The main future application of the code is to test a large variety of control algorithms. The effects of delay and the coupling of rotational/translational inputs are some of

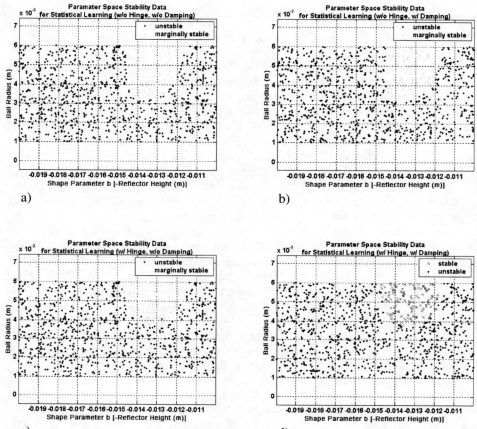

FIGURE 9. a)Stability region calculation for one body w/o drag forces, b) Stability region calculation for one body w/ drag forces, c) Stability region calculation for two body w/o drag forces, d) Stability region calculation for two body w/ drag forces

the issues which can be easily studied with the help of this code. Furthermore, minor modifications can easily be made to open up many more possibilities. The cylindrical structure of the mesh can be used to construct other common reflector shapes, such as the heliogyro. The simple one dimensional gravity term can be replaced with gravitational forces of different celestial bodies and by doing so an actual flight can be simulated. The effects of different wave guides can also be studied by replacing the intensity function. Finally, the parameter space analysis done thus far can be expanded to reveal other stable configurations.

REFERENCES

1. McInnes, C. R., *Solar Sailing: Technology, Dynamics, and Mission Applications*, Springer-Verlag, New York, 1999.
2. Benford, J., and Dickinson, R., "Space Propulsion and Power Beaming using Millimeter Systems," in *Intense Microwave Pulses III, Proceedings SPIE*, 1995, vol. 2557, p. 179.
3. Abdallah, C., Schamiloglu, E., Miller, K., Georgiev, D., Benford, J., and Benford, G., "Stability and Control of Microwave-Propelled Sails in 1-D," in *Proceedings 2001 Space Exploration and Transportation: Journey into the Future*, 2001, albuquerque, NM.
4. Kolthinskii, V., Abdallah, C. T., Ariola, M., Dorato, P., and Panchenko, D., *IEEE Transactions on Automatic Control*, **45** (2000).
5. Singh, G., Characterization of passive dynamic stability of a microwave sail, Tech. rep. (2000).
6. Kuipers, J. B., *Quaternions and Rotation Sequences*, Princeton University Press, Princenton, New Jersey, 1999.
7. Baruh, H., *Analytical Dynamics*, WCB McGraw-Hill, 1999.
8. Schamiloglu, E., Abdallah, C., Miller, K., Georgiev, D., Benford, J., Benford, G., and Singh, G., "3-D Simulations of Rigid Microwave-Propelled Sails Including Spin," in *Proceedings 2001 Space Exploration and Transportation: Journey into the Future*, 2001, albuquerque, NM.
9. Benford, J., Wireless power transmission for science applications, Tech. rep., Microwave Sciences, Interim Final Report Contract number NAS8-99135, Microwave Sciences, 1041 Los Arabis Lane, Lafayette, CA 94549 (2000).
10. Dickenson, R. M., Beam power density model (????), nASA JPL.
11. Shampine, L. F., and Reichelt, M. W., *SIAM Journal on Scientific Computing*, **18** (1997).

Dynamics and Control of Microwave-propelled Sails Using Delayed Measurements

C.T. Abdallah, E. Chahine, D. Georgiev, and E. Schamiloglu

Electrical & Computer Engineering Department,
The University of New Mexico,
Albuquerque, NM 87131

Abstract. This paper presents the microwave-propelled sail, its structure, and assumptions. We will first present its equations of motion, conduct stability analysis, then design a controller to make it asymptotically stable. Numerical simulations are included to illustrate our study.

INTRODUCTION

This paper will discuss a new generation of spacecraft, the microwave-propelled sail. The idea builds upon solar sails [4] which have been in the literature since the 1970's. The idea of microwave-propelled sails is very similar, but instead of the sun's photons hitting the solar sail at the right angle, the microwave-propelled sail alleviates that problem since we have "control" over the power source and its direction. The microwave sail architecture comprises very large ultra-weight apertures and structures. One of its distinguishing improvements is mission capability and reduction in mission cost, plus the ability of interstellar exploration missions. Microwave-propelled sails, along with solar and other types of sails will provide low-cost propulsion, and long-range mission. In [4], McInnes gives a general view on solar sails. Stability and control of carbon fiber sails propelled using microwave radiation in 1-D has been studied in [1, 2]. This paper will cover the analysis of the sail in 3-D, along with its equations of motion, and control design structure.

SAIL CONFIGURATION

The sail studied has an umbrella-like configuration with concave sides facing the radiation source and has a bounded motion behavior as described in [3]. The sail is composed of a reflector made out of a light-weight carbon fiber material, a hollow mast and payload represented by a ball. The mast is attached at the reflector center of mass (CM), and connects the payload to the reflector. The payload is not directly attached to the reflector for stability reasons. To obtain passive stability, earlier studies in [3]have shown that:

- The reflector must be located aft of the vehicle CM for rotational stability
- The reflector must be of a concave shape such that the concave shape faces the radiation source for translational stability.

CP664, *Beamed Energy Propulsion: First International Symposium on Beamed Energy Propulsion*,
edited by A. V. Pakhomov
© 2003 American Institute of Physics 0-7354-0126-8/03/$20.00

The notion of beam-riding, i.e. the stable flight of a sail propelled by Poynting flux caused by a constant power source, places considerable demands upon a sail. Even if the beam is steady, a sail can wander off the beam if its shape becomes deformed or if it does not have enough spin to keep its angular momentum aligned with the beam direction in the face of perturbations. The microwave beam pressure keeps concave shapes in tension, so concave shapes arise naturally while beam-riding. they will resist sidewise motions if the beam moves off-center, since a net sideways force restores the sail to its position (See Figure 1).

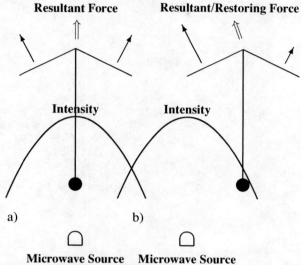

FIGURE 1. Beam riding ('restoring force') for Concave Sails

We list the assumptions needed to simplify our analysis of the microwave-propelled sail.

- The system is considered as a rigid body
- The reflector has full reflectivity. The actual carbon fiber used in some of our experiments has 98% reflectivity.
- There are no internal reflections.
- The payload and the mast do not block the microwave beam.
- There are no aerodynamic influences
- The microwave source is modelled as a point source with a square wave-guide.
- The gravity vector g, points towards the negative Z-axis of the inertial frame (See figure 4).

Since we have chosen our reflector to be of a conical shape, any cross-section orthogonal to the mast is a circle. The reflector surface is created by revolving a parameterized curve about the body z-axis. The following is a fourth order polynomial approximation of the parameterized curve:

$$f(r/R) = a_0 + a_1(r/R) + a_2(r/R)^2 + a_3(r/R)^3 + a_4(r/R)^4 \qquad (1)$$

where a_0, a_1, a_2, a_3, and a_4 are shape constants, r is the radial distance from the body z-axis, R is the radius of the circle. We obtain a conical shape when $a_0 \neq 0$, $a_1 < 0$, and $(a_2, a_3, a_4) = 0$, with concave facing-down shape. The circle is chosen because of its symmetry and its advantages to stability. For more details on the reflector shape design, the reader is referred to [3] (See Figure 2).

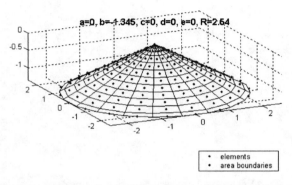

FIGURE 2. Representative sail shape illustrating elements and corresponding areas.

COORDINATE FRAMES & EQUATIONS OF MOTION

There are two coordinate frames defined for this system, as depicted in figure 3: the inertial frame and the body frame. The x_b, y_b, z_b axes of the body frame are attached to the vehicle CM with z_b aligned with the mast axis. The inertial frame $\{X_I, Y_I, Z_I\}$ has the gravity vector in the $-Z_I$ direction. The microwave source which is represented as a point source is located on the $\{Z_I\}$ axis at $\{0, 0, -D\}$ in the inertial frame (with $D > 0$). The microwave beam radiates in the $+Z_I$ direction with its maximum intensity aligned with the $+Z_I$. The offset between the vehicle CM and the reflector CM, defined as d $(d > 0)$. Since $D \gg d$ then we consider the distance from the source to the reflector CM to be D.

For a rigid body, the equations of motion are well established [3, 4].

$$\dot{\vec{r}} = \vec{v} \qquad (2)$$

$$\dot{\vec{v}} = \frac{\vec{F}}{m} + \vec{G} \qquad (3)$$

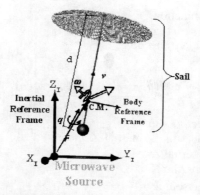

FIGURE 3. Microwave sail coordinate systems and states.

$$\dot{\vec{q}} = \frac{1}{2}\vec{q}\otimes\vec{\omega}$$ (4)

$$\dot{\vec{\omega}} = J^{-1}[-\vec{\omega}\times J\vec{\omega}+\vec{T}]$$ (5)

$\vec{r}$ is the coordinate vector in the inertial frame (m).
$\vec{v}$ as the velocity vector in the inertial frame (m/s).
$\vec{q}$ is the attitude quaternion that specifies body frame orientation in inertial coordinates and $\vec{q} = [q_1; q_2; q_3; q_4] = [q_1; \vec{\alpha}]$
$\vec{\omega}$ as the angular velocity vector in the body frame (rad/s).
m is the total mass of the system (Kg).
$\vec{G}$ is the gravity vector such that $\vec{G} = [0,0,-9.807]^T$ (m/s^2).
J is the vehicle moment of inertia (Kg/m^2).
$\vec{F}$ is the radiation-induced inertial force on the vehicle $(Kg.m/s^2)$.
$\vec{T}$ is the radiation-induced body torque on the vehicle $(Kg.m^2/s^2)$.

The force $\vec{F}$ and the torque $\vec{T}$ are given by [3]:

$$\vec{F} = \vec{q}^*\otimes\left[2\iint\limits_{ref} dA\rho_e cos^2\psi_e\frac{\vec{n}_{eb}}{\vec{n}_{eb}(3)}\right]\otimes\vec{q}$$ (6)

$$\vec{T} = \iint\limits_{ref}\left(\vec{r}_{eb}\times\left[2\iint\limits_{ref} dA\rho_e cos^2\psi_e\frac{\vec{n}_{eb}}{\vec{n}_{eb}(3)}\right]\right)$$ (7)

with $\vec{r}_{eb}$ is the vehicle CM to element location vector in the body-frame.
$\vec{n}_{eb}$ is the reflection unit normal in the body frame at $\vec{r}_{eb}$.
dA is the element area.
ψ_e is the angle between the element local normal and the direction of incident radiation.

351

ρ_e is the energy density function.

For a square wave-guide ρ_e becomes

$$\rho_e = P_t \frac{(cos^2\phi \, cos^{n_x}\theta + sin^2\phi \, cos^{n_y}\theta)}{4\pi s^2} \tag{8}$$

where P_t is the transmitted power.
n_x, n_y are the power indices in the inertial X and Y directions respectively.
θ is the angle with the inertial Z-axis.
ϕ is the angle with the inertial X-axis.
s is the distance from the source $= \sqrt{x^2 + y^2 + z^2}$

The physical control inputs to the system are therefore, P_t, n_x, and n_y but in the following, we will use the force $\vec{F}$ and the torque $\vec{T}$ as our control inputs.

STABILITY ANALYSIS

Let $\vec{x} = \{\vec{r}, \vec{v}, \vec{q}, \vec{w}\}$ be the state of the system. The equations of motion are then described by the nonlinear differential equation

$$\dot{\vec{x}} = f(\vec{x}) \tag{9}$$

The equilibria for the nonlinear system $f(\vec{x}) = 0$ are obtained as

$$\vec{x}_0 = \{(0, 0, z_{eq}), (1, 0, 0, 0), (0, 0, 0), (0, 0, 0)\}.$$

Since we do not have any source of natural damping, the system can be marginally or neutrally stable at best. Basically, equilibrium is achieved when the body-frame axes are aligned (parallel) with the inertial frames axes, and the origin of the body-frame is on the inertial Z-axis, at a desired distance from the source.

Perturbations occur in translational directions represented with cylindrical coordinates, R_l and Z_l, and in angular directions represented with the Euler angles, yaw, pitch, and roll. For most of the translational displacements, the reflector's concave shape will compensate and will bring the vehicle to equilibrium as discussed previously. The angular perturbations are more serious. When the reflector shape provides a "restoring force" effect, we notice that the force is greater on the reflector surface closest to the microwave beam leading to rotation away from equilibrium. This will cause the system to become unstable to pitch and roll perturbations. To compensate this effect, a stabilizing torque is induced by the addition of the payload. Next, we will attempt to get a more analytical understanding of stability through linearization.

Using the linearization technique as a way to analyse the stability of the nonlinear system, the linearized state equation becomes:

$$\dot{\vec{x}} = A\vec{x} \tag{10}$$

where A is the Jacobian evaluated at $\vec{x}_0$, $A = \left.\frac{\partial f}{\partial \vec{x}}\right|_{\vec{x}=\vec{x}_0}$. The stability characteristics of the linearized equations of motion are determined by the real parts of the eigenvalues of A. If these real parts are negative then the system is stable, unstable if they are positive, and marginally stable if the real part is zero [7]. We mentioned earlier that the system lacks natural damping, therefore the best performance that we hope to obtain is marginal stability. The vehicle has six degrees of freedom. One is a zero frequency mode which rotates the vehicle around the z_b axis. The other five are oscillatory modes. The first oscillatory mode is the bouncing or hopping mode that makes the vehicle translate up and down along the Z_I-axis. It is always neutrally stable. The other four are combinations of attitude and translation motion in the $Y_I Z_I$ and $X_I Z_I$ planes. They are a combination of pendulum and yo-yo modes. These four modes determine the neutral stability of the vehicle. Therefore, the system is usually unstable, and at best marginally stable [3]. In the 1-D case, we can stabilize the microwave-propelled sail using delayed measurements [2], and by feedback linearization [1]. In the following section, we will present a controller that will stabilize the sail in 3-D.

CONTROLLER DESIGN

Going back to the equations of motion and making the following changes in order to have the origin as the desired equilibrium. Let $\vec{e} = \vec{r} - \vec{r}_d$ and $\beta = q_1 - 1$. The new equations of motion become

$$\dot{\vec{e}} = \vec{v} \tag{11}$$

$$\dot{\vec{v}} = \frac{\vec{F}}{m} + \vec{G} \tag{12}$$

$$\dot{\beta} = -\frac{1}{2}\vec{\alpha}^T \vec{\omega} \tag{13}$$

$$\dot{\vec{\alpha}} = \frac{1}{2}(\vec{\alpha} \otimes \vec{\omega} + (\beta + 1)\vec{\omega}) \tag{14}$$

$$\dot{\vec{\omega}} = J^{-1}[-\vec{\omega} \times J\vec{\omega} + \vec{T}] \tag{15}$$

Using the nonlinear control law given in [6] and modified in [5].

$$\vec{F} = -m\left(\vec{G} + \vec{e} + \vec{v}\right) \tag{16}$$

$$\vec{T} = -\frac{1}{2}\left[\left(\tilde{\vec{\alpha}} + (\beta + 1)I\right)G_p - \gamma\beta I\right]\vec{\alpha} - G_r\vec{\omega} \tag{17}$$

where G_p and G_r are symmetric positive definite diagonal (3x3) matrices and γ is a positive scalar. Let us investigate the following Lyapunov function candidate [5].

$$V = \frac{1}{2}\vec{e}^T\vec{e} + \frac{1}{2}\vec{v}^T\vec{v} + \gamma\beta^2 + \vec{\alpha}^T G_p\vec{\alpha} + \vec{\omega}^T J\vec{\omega} \tag{18}$$

353

which is defined for all $\vec{x}$ such that $\vec{x} = [\vec{e}, \vec{v}, \beta, \vec{\alpha}, \vec{\omega}]$. The derivative of V is $\dot{V} = -2\vec{\omega}^T G_r \vec{\omega} - \vec{v}^T \vec{v}$ which is negative semi-definite. Let Ω be the set where $\dot{V} = 0$. The largest invariant set in Ω is the origin.

Replacing $\vec{\omega} = 0$ and $\vec{v} = 0$ in the equations of motion, we obtain the following. $\vec{e} = 0$, $\vec{\alpha} = 0$, $\beta I = -G_p (G_p - \gamma I)^{-1}$.

NUMERICAL EXPERIMENTS

The spacecraft model used in this simulation is a scaled version of the real microwave-propelled sail. The mass is 6.11345 g, the inertia matrix is given by 1.0e-006 *diag([0.3368   0.3368   0.0737]) Kg/m^2. The initial orientation of the sail is given by the $\vec{q} = [.85; .85; .85$ and $\beta = -0.004$. The gravitational vector is given by $G = [0; 0; -9.807]$. Using the above mentioned controller with the feedback gains chosen for $G_p = diag[100\ \ 100\ \ 200]$, $G_r = diag[100\ \ 100\ \ 100]$, and $\gamma = 100$. As

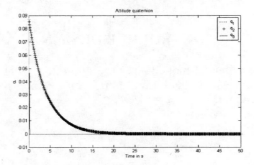

FIGURE 4.   Attitude vector $\vec{\alpha}$ of the sail.

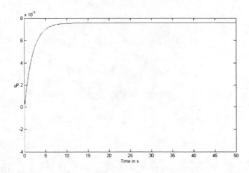

FIGURE 5. q_0 of the attitude vector $\vec{\alpha}$.

you see in Figure 5, q_0 converges almost to zero, while in Figure 4, the attitude vector converges to zero at different rates to zero, depending on the values of G_p.

We also tested the robustness of the controller when the sail is subjected to different physical variations: shape change, area variation and random disturbance. The de-

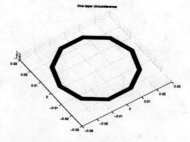

FIGURE 6.   One-Layer Flat Sail.

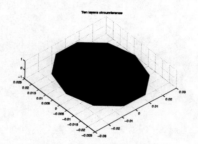

FIGURE 7.   Ten-Layer Flat Sail.

sired position we chose for the sail to converge to is $[x, y, z] = [1, 2, 3]$. Our initial conditions are: position (m) $[x_0, y_0, z_0] = [10, 20, 30]$, velocity (m/s) $[V_x, Vy, Vz] = [20, 30, 20]$, attitude $(°)$(yaw,pitch roll)$[yaw, pitch, roll] = [30, 40, 20]$, and angular velocity (rad/s) $[w_x, wy, wz] = [30, 20, 10]$ When we change the shape of our sail from a cone to a flat circumference and we vary the thickness of the circumference (see Figures 6 and 7), the controller still drives the sail to the desired position as shown in Figures 8 and 9 for one, and ten levels of thickness respectively, and with all other states reaching their equilibrium.

In the presence of random disturbances, we have investigated two cases. For a constant disturbance whose magnitude is random between 0 and 100, the maximum deviation from the desired position is 0.6115 as shown in Figure 10. For a random continuously changing disturbances, all states go to a different equilibrium position every time, as seen in Figure 11.

CONCLUSIONS

We have presented the general dynamics of a microwave-propelled sail, along with a controller that drives it to local stability, as was shown in our numerical example. More work is under way to have the power source and some of its parameters as the control inputs, along with accounting for delays in the input action.

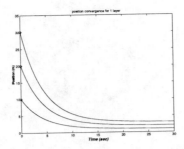

FIGURE 8.   Convergence of the sail to the desired positions for 1 layer.

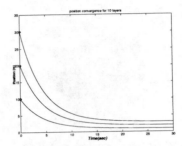

FIGURE 9.   Convergence of the sail to the desired positions for 10 layers.

REFERENCES

1. C.T. Abdallah, E. Schamiloglu, K.A. Miller, D. Georgiev, J. Benford,and G. Benford, *Stability and control of microwave-propelled sails in 1-D*, Proceedings 2001 Space Exploration and Transportation: Journey into the Future, Albuquerque, NM, pp.552-558,February 2001.
2. C.T. Abdallah, E. Schamiloglu, D. Georgiev, J. Benford, and G. Benford, *Control of microwave-propelled sails using delayed measurements*, Proceedings of the 19[th]Space Technology and applications international Forum, pp.463-468,February 2001.

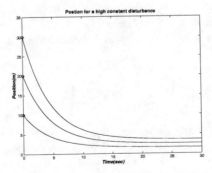

FIGURE 10.   Convergence of the sail to the desired positions after a constant disturbance of magnitude 99.

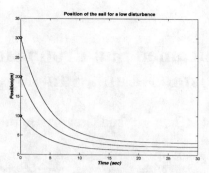

FIGURE 11. Convergence of the sail to the desired positions after a constant disturbance of magnitude 10.

3. G. Singh, *Characterization of passive dynamic Stability of a microwave sail*, Jet Propulsion Laboratory Engineering Memorandum EM-3455-00-001, 22 March 2000.
4. C.R. McInnes, *Solar sailing: Technology, Dynamics, and Mission Applications*, Springer-Verlag, New York, 1999.
5. S.M.Joshi, A.G. Kelkar, J.T.-Y. Wen,G. Singh, *Robust attitude stabilization of spacecraft using nonlinear quaternion feedback*, IEEE Transactions on Automatic control, Vol. 40, Issue 10, pp. 1800-1803, October 1995.
6. J.T. Wen and K.Kreutz-Delgado, *The attitude control problem*, IEEE Transactions on Automatic control, Vol. 36, Issue 10, October 1991.
7. T. Kailath, *Linear systems*, Prentice Hall Information and System Sciences Series, New Jersey ,1980.

Near-Term Beamed Sail Propulsion Missions: Cosmos-1 and Sun-Diver

James Benford[1] and Gregory Benford[2]

[1]Microwave Sciences, Inc., 1041 Los Arabis Lane, Lafayette, CA 94549, [2]Department of Physics, University of California Irvine, Irvine, CA 92697

Abstract. Next year the Planetary Society plans to launch Cosmos-1, the first solar sail. We are planning an experiment to irradiate the sail with the Deep Space Network beam from Goldstone. This can demonstrate, for the first time, beamed propulsion of a sail in space. The 450 kW microwave beam from the large 70-m dish can show direct microwave beam acceleration of the sail by photon pressure, and we can measure that acceleration by on-board accelerometer telemetry. In addition, we describe a mission scenario called 'Sun-Diver' using a powerful microwave beam on a solar-driven sail, to both heat and push the sail, accelerating by "boil-off" of coated materials. Sublimation and desorption work well with the new carbon sail materials which can take very high temperatures (>2000 K), can use promising new materials for mass loss, and promise new classes of missions. These missions make a close pass near the Sun, hence the name, to take advantage of high temperature characteristics of the sail by using the large solar flux at perihelion, yielding high velocities of ~50 km/s for >40 A.U. missions. Within ~5 years, the sailcraft flies beyond Pluto, giving high velocity mapping of the outer solar system, the heliopause and interstellar medium.

The ideas of sailing in space by various forms of photon pressure have captured the imagination of many for years. Solar sailing as been studied theoretically for 30 years, but has never been demonstrated. Beam-driven sail acceleration has been studied for 20 years and has recently been demonstrated in the laboratory. That research has suggested enhanced means of propelling sails. This paper describes our plans to detect solar and microwave acceleration in orbit, which would give a robust proof of photon-driven sailing, and outlines a new mission concept for sails.

BEAM-DRIVEN ACCELERATION OF COSMOS-1

The Planetary Society, in partnership with Russian laboratories, plans to launch the first solar sail in low Earth orbit in 2003 (http://www.planetary.org/solarsail). We have requested time on the Deep Space Network for two tasks: 1) To track the telemetry on the sail to determine its velocity and acceleration accurately under solar radiation (photon) pressure. 2) To irradiate the sail with a high power microwave beam from a large aperture to show direct microwave beam acceleration of the sail by photon pressure, and to measure that acceleration from receipt of on-board accelerometer telemetry. This would be the first demonstration in space of both solar sailing and microwave beam propelled sailing.

CP664, *Beamed Energy Propulsion: First International Symposium on Beamed Energy Propulsion*,
edited by A. V. Pakhomov
© 2003 American Institute of Physics 0-7354-0126-8/03/$20.00

The Cosmos 1 orbit will be polar, because the launch point is from a submarine in the Barents Sea. Its circular orbit is to be at 800 km. The orbital velocity is 7.4 km/s. The satellite moves across the sky at 0.06 deg/sec, so can be tracked readily. The orbital period is 101 minutes, so overhead time at most is 15 minutes. The orbital Doppler shift at the spacecraft telemetry carrier frequency (2.25 GHz) will be 56 kHz.

FIGURE 1. Cosmos-1 in orbit. The sail is ~100 feet across and has two independent sets of vanes which are to be steered to raise the orbit using solar pressure.

Solar-Driven Sailing

A clear measurement of solar thrust in orbit on a real Sailcraft will be a significant scientific contribution, the first step in the experimental study of solar sailing. Doppler tracking as a supplement to accelerometer measurements, adds confidence and robustness to the experiment, and gives a direct measure of velocity to use in modeling the sail orbit.

The sail diameter is 30m. The solar flux will give a maximum acceleration to the ~100-kg sail of 10^{-4} m/s^2. There are on-board accelerometers with sensitivity down to 10^{-7} m/s^2 and sample rates which can be varied from 10 to 100 Hz. Moreover, the delta-v from solar flux will produce a shift of the carrier of 10^{-3} Hz/sec. (This shift from solar acceleration should be detectable by the DSN: the DSN frequency resolution for carrier determination is 0.1 Hz, so tracking for ~ 100 sec is necessary for detection to be possible.)

We will model the rate of change of frequency for the few hundred seconds of track, incorporating orbital dynamics [including drag] and ionospheric propagation effects, and then see to what extent reality deviates from the model, thereby extracting the effect of solar pressure. The atmospheric and ionospheric calibration similar to that used for the Cassini gravity wave experiment may be needed. As both the shift of the carrier frequency and receipt of the accelerometer telemetry measure the acceleration of the sail, these can then be compared to each other and to modeling.

Microwave Beam-Driven Acceleration of Cosmos-1

There are two very different modes for accelerating the sail, the Tracking mode and the Impulse mode, with the former more attractive.

Impulse Mode

As the Cosmos sail passes overhead of Goldstone, the beam, directed to almost vertical, hits the sail at its minimum distance from the dish. The Goldstone 500 kW, 8 GHz microwave beam from its 70 m aperture will produce a beam size at the orbit about 0.5 km across at 800 km altitude. The impulse from the beam is applied in 0.07 seconds as the sail passes through the beam. The beam accelerates the sail at about 10^{-7} m/s2, giving a total impulse of 10^{-8} newton-seconds. For this to be detectable, sail electronics must be optimized to give the maximum sensitivity on the accelerometers.

The on-board accelerometers are derived from a gravitational wave experiment in Moscow and can measure accelerations to 10^{-12} m/s^2. We baseline for the Cosmos-1 accelerometers is to sample at 10 Hertz and integrate over 1 second. So the impulse will be sensed as a single pulse and then be integrated with lower accelerations, reducing its detectability. A faster cycle rate should be used to help the signal-to-noise ratio (S/N).

Tracking Mode

Consider an orbit taking the sail directly overhead Goldstone. As the sail rises above the horizon, its angular rate of motion is 0.13 degrees per second. The Goldstone dish has in a slew rate of 0.25 degrees per second, therefore it can track the sail until it reaches 23 degrees, where it moves faster than the dish can follow. The range at the horizon is 6270 km and the acceleration from the beam on the sail will be 10^{-8} m/s^2, for normal incidence on the sail. The sail will be inclined to the beam, with 30 degrees as a typical value. The acceleration on the sail will be reduced by the cosine of this angle. Although acceleration is lower than for the impulse mode, it occurs over a far longer time, about 200 seconds. Integrating over the trajectory, the total impulse is 2.5 x 10^{-6} newton-seconds, 250 times the impulse of other mode.

Again, the signal level will be much more detectable if the accelerometers are made more sensitive.

We can enhance the S/N ratio of the accelerometers by pulsing the beam (by modulating the input to the klystron amplifier), thus modulating the sail acceleration up and down, and use signal processing of the telemetry to bring the structure in the signals above noise. We can further improve S/N by choosing to modulate the beam amplitude at frequencies to excite resonant acoustic modes in the sail structure, then look for enhancement of those modes in the frequency spectrum of the accelerometer data. If the sail were rigid, the fundamental oscillation frequency would be about 10 Hz. Preliminary modeling of sail oscillations by our Russian colleagues say that the lowest-frequency oscillation mode is at 0.2 Hz, which corresponds to a gentle flapping of the sail. The fact that the sail, attached to its spars, is not a rigid body complicates attempts to make the system resonate acoustically. Therefore, we need to identify the resonant modes of the actual sail to understand what spectrum of oscillations we would be imposing our thrust upon.

So, calculations of resonant mode frequencies and their Q's are essential. We will attempt to measure the spectrum in flight by Fourier analyzing the data we get from the accelerometers. For this, the sampling time would have to be short to pick up the oscillations. We can oscillate the sail by varying be klystron power in amplitude between a fraction of a Hz to tens of Hz, depending upon which modes are typically excited in orbit and their damping rates.

Flight Preparation

From the above, since if the sources of noise are random (i.e., Poisson distribution), the signal scaling is $S \sim t/R^2$, noise $N \sim t^{0.5}$, so $S/N \sim t^{0.5}/R^2$, where t is the acceleration time and R is the distance from Goldstone to Cosmos. Calculations show that *the Tracking mode will have a S/N 15 times that of the Impulse mode.* Thus, we emphasize that mode.

We need observations of the amplitude and frequency spectrum of the sail every time it can be observed. We should gather data to understand modes of oscillation of the sail. The most useful measurement of the spectrum of oscillation will be as the sail is deployed. The typical case will be when the sail passes in and out of the Earth's shadow, and modes are excited and, eventually, damped. To see these acoustic oscillations of the sail, we need to have a sampling time closer to hundred Hz levels, compared to the present design level of 10 Hz.

Further, optimization of this experiment requires that the accelerometers be *as sensitive as possible.* A key factor for these accelerometers is the acceleration that gives an S/N = 1.

Another issue is the possible effect on electronics of the microwave electric field on the sail. The maximum field on the sail at closest range will be 30 V/m, which typically does not cause interference. Cosmos electronic equipment will be tested to this level at the X-band frequency of the Goldstone beam.

In summary, the power beaming experiment to the Cosmos sail will be optimized by

- Making the accelerometers as sensitive as possible,
- Operate the accelerometers at fast sample rate,
- Modulating beam amplitude to excite resonant modes in the sail.

The mission will last for about two months before drag-induced re-entry. During this period, when the sail passes over Goldstone we plan for a DSN 34-m antenna to measure the solar acceleration by Doppler tracking, and for the Goldstone Solar System Radar 70-m dish to test the predictions for microwave acceleration. (Since the Cosmos 1 orbit will be polar, tracking can be done in several, if not all, DSN locations.)

Implications

Demonstration of feasibility of a sail driven by the sun or a beam from a ground station will:

- Show the basic principle of solar and beam-driven propulsion in action in space, quantify the propulsion, and compare it with predictions.

- Demonstrate the potential of a higher-power DSN in future, when beam-driven sail experiments can be done at much higher accelerations.. The proposed demonstration is synergistic with ongoing investigations into DSN upgrades.

- Synergize with the Space Solar Power (SSP) Program. In current SSP concepts, microwave beams from SSP will be used to accelerate sails as space probes to very high velocities for outer Solar System missions. In the next section, we explore new ways to do this.

SUN-DIVER MISSIONS

In work reported in this session (see J. Benford et. al., this session) an intense microwave beam drove an ultralight carbon sail to liftoff and flight against gravity[1]. Although there was photon pressure, it wasn't strong enough to explain the observed accelerations. The most plausible explanation for the bulk of the observed accelerations greater than gravity is fast evaporation of heated absorbed molecules from the hot side of the sail on timescales short compared to that of thermal diffusion. This suggests development of sails that fly due to loss of "paint" from their illuminated side. Microwaves do not damage sail materials as short-range lasers do, and so can heat them less destructively. This approach promises to make microwave-riding sails greatly superior to both solar sails and laser-driven sails, because it uses the best features of both. After the coats desorb away, a sail can perform as a conventional solar sail, using an aluminum coat beneath. Solar sails are plagued in mission plans by low accelerations, which dictate long orbital times. Laser sails have problems with atmospheric distortion if the laser beam is fired from the ground, which microwave beams do not. *A natural collaboration emerges between subliming sails driven by beams in LEO, converting to greatly accelerated solar sails for the long mission.*

For a schematic of the approach, see Figure 2. For a full treatment, see Reference 5. This deployment takes advantage of high temperature characteristics of the sail to dive to within a few radii of the sun, where it achieves a high velocity by using the large solar flux at perihelion. The planned Solar Probe mission, flying to within 0.01 A.U., is an extreme example. For the near term use of beamed power, note that beam illumination at ~kW/cm^2 in LEO can simulate conditions any solar grazer mission will experience to within 0.01 A.U.

Conventional solar sail missions lower perihelion by adding and subtracting energy from the orbit over several revolutions around the sun[2]. Adding mass to a sail to be lost at the sun will generally lengthen this perihelion lowering time, because of lower accelerations. *Sublimation (or desorption) thrust from LEO into interplanetary orbit can omit the several-year orbits conventional solar sails need to reach ~0.1 AU. A second "burn" at perihelion, the highest available orbital velocity in the inner solar system, and thus optimum point for a delta-V, then yields high velocities for >40 A.U. missions.*

The mission phases are:

(1) Deployment in Low Earth Orbit by conventional rocket.

(2) Launch by a microwave beam from nearby in orbit. Beam heating makes a "paint" (*layer #1*) desorb from the sail. Under this enhanced thrust in repeated shots at perihelion in steepening elliptical orbits, the sail attains ~15 km/s velocity, canceling most of its solar orbital velocity, and so can fall edge-on toward the sun immediately. (This is far faster than using solar pressure to spiral down, which takes years.) It approaches the sun edge-on, to minimize radiation pressure on it in the inward fall.

(3) At perihelion, the spacecraft rotates to face the sun. Under intense sunlight ~20 times Earth insolation, the sail *desorbs away layer #2*, getting a ~50 km/s boost at its maximum (infall) velocity.

(4) It then sails away as a conventional, reflecting solar sail, with the final Aluminum layer revealed. Its final speed is ~ 10 AU/year. Within ~5 years, it sails beyond Pluto, giving high velocity mapping of the outer solar system, the heliopause and interstellar medium.

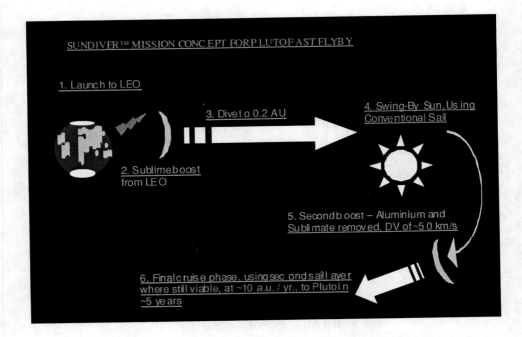

FIGURE 2. Phases of a Desorption-Assisted Sun-Diver Mission.

Obviously one needs a detailed orbital integration such as Sweetzer's[3] with plausible rates of mass loss gained from laboratory work, before judging the overall credibility of such a mission.

An Interstellar Sun-Diver

As a simple example, consider a sail falling sunward on a parabolic orbit. It will be accelerated by

- the ΔV imparted by desorption at perihelion

- ordinary solar sail acceleration on the outward-bound leg, once the desorped layer is gone, leaving a reflecting sail.

We can find an approximate expression for the final velocity V^F with respect to the sun, following energy analysis, as in Matloff[4]. The sail's parabolic velocity at distance R is

$$V = 1.4 \, (GM/R)^{1/2} = 93 \text{ km/s } (R/0.1 \text{ AU})^{-1/2} \tag{1}$$

At perihelion of 0.1 A.U. the sail reaches a temperature (for plausible values of absorption and emissivity)

$$T = 927 \text{ K } [(\alpha/0.3)(\varepsilon/0.5)^{-1}]^{1/4} (R/0.1 \text{ AU})^{-1/2} \quad (2)$$

For such temperatures, a considerable $\Delta V > $ km/s is plausible for a range of desorption materials. Losing its mass load at perihelion, the sail thereafter works as an ordinary solar sail, attaining a *final exit speed* from the solar system

$$V^F = 19.5 \text{ km/s } [(\Delta V/2 \text{ km/s}) + (3\sigma)^{-1}]^{1/2} \quad (3)$$

$$V^F = 3.9 \text{ AU/year } (\Delta V/2 \text{ km/s})^{1/2} [1 + 0.33 /(\Delta V/2 \text{ km/s})(\sigma')]^{1/2} \quad (4)$$

Here σ is the sail areal mass density in units of 100 gm/m^2. In the brackets, the first term comes from acceleration (a), the ΔV imparted by desorption at perihelion and the second from (b), ordinary solar photon acceleration on the outward-bound leg, once the desorped layer is gone, leaving a reflecting sail.

The sail's speed as it passes through the outer planets will exceed V^F. The linear sum of ΔV and the ordinary solar sailing momentum in the square root above means there will be a simple tradeoff in missions between the two effects, which are equal when the last term in brackets above is unity.

This is only a rough calculation, omitting many mission details, such as sail maneuvering near the sun. We assumed a perfectly reflecting sail on the outward leg, and that desorption would occur quickly at perihelion. For a full treatment, see Reference 5.

Conclusions

Using mass loss for thrust is not a new idea, but it is new to apply this idea, together with a powerful microwave beam, to both heat and push a sail. It is worth pursuing because sublimation and desorption

- work well with the new carbon sail materials, which can take very high temperatures (>2000 K),
- can use promising new materials for mass loss so far not studied for thrusting applications,
- hold the promise whole new classes of missions.

ACKNOWLEDGMENTS

We thank G. David Nordley for his calculations and Henry Harris for helpful comments.

REFERENCES

1. J. Benford et. al., "Microwave Beam-Driven Sail Flight Experiments", Proc. Space Technology and Applications International Forum, pg. 540, Space Exploration Technology Conf, AIP Conf. Proceedings 552, ISBN 1-56396-980-7STAIF, (2001).
2. Maccone, C., "Solar Focal Missions." IAA-L-0604, in *Proc. 2nd Intl. Conf. On Low Cost Planetary Missions,* Laurel, MD, (1996).
3. Sweetzer, T., "Advanced Propulsion Options for Pluto Flyby", JPL internal memorandum, (2001).
4. Matloff, G.L., *Deep Space Probes,* Chapter 4, Springer-Verlag, (2000).
5. Benford, G. and Benford, J., "Desorption Assisted Sun Diver Missions", *Proc. Space Technology and Applications International Forum* (STAIF-2002), Space Exploration Technology Conf, AIP Conf. Proc. 608, ISBN 0-7354-0052-0, pg. 462, (2002).

BEAMED ENERGY FOR
INTERSTELLAR MISSIONS

Space Based Energy Beaming Requirements for Interstellar Laser Sailing

Travis Taylor[1], R. Charles Anding[1], and D. Halford[1], Gregory L. Matloff[2]

[1]*Teledyne Brown Engineering, Inc., 300 Sparkman Drive, Cummings Research Park, P.O. Box 070007, Huntsville, AL 35805*
[2] *Bangs / Matloff Aerospace Consulting Co., 417 Greene Avenue, Brooklyn, NY 11216*

Abstract. This paper first presents a review of interstellar laser and maser sailing. Concepts for very large space based laser (VLSBL) architectures are discussed, which would be required for interstellar laser sailing. A discussion of the power, optical, and pointing requirements is given as well as a description of a diode based VLSBL architecture. Analysis of state of the art pointing jitter requirements for laser sailing is given. Statistical analysis of the beam angular distribution as a function of time is modeled and incident light pressure on a sail at large distances is calculated. Also given are concepts designs for very large aperture adaptive optics to improve laser sailing efficiency.

LASER / MASER PHOTON SAILING : A HISTORY

There is only one physically feasible mode of interstellar travel that is conceptually capable of two-way interstellar travel with travel times approximating a human lifetime—the beamed-energy photon sail. And the early development of this concept is largely the work of one man—the late American physicist Robert Forward. Although it is certainly true that others—notably Marx, Moeckel, and Norem [1-3] also considered some of these concepts in the same time frame, they were fully developed and integrated by Forward.

As discussed in the three major literature reviews of this topic [4-6], Forward began his examination of this concept in the early 1960's. [7]. The basic challenges for these early researchers was not dissimilar to that faced by more recent investigators:

(1) How to project a spacecraft large enough to carry humans to one of the near stars within a human lifetime?
(2) How to do this task with physics that is not impossible?
(3) How to reduce mission energy requirements and projected costs to believable levels?
(4) And finally, if the above is not daunting enough, can we return the crew (or their children) to Earth at the conclusion of their exploration?

Attempts to solve these problems certainly displayed creativity. But not all of them will prove to be technically feasible.

CP664, *Beamed Energy Propulsion: First International Symposium on Beamed Energy Propulsion,*
edited by A. V. Pakhomov
© 2003 American Institute of Physics 0-7354-0126-8/03/$20.00

One significant limitation to applicability of the laser or maser to interstellar propulsion was realized early on. That is the requirement to maintain beam aim and collimation to an accuracy defined by a 100-km or 1000-km sail size at the far end of a trillion-kilometer acceleration "runway."

Let the (disc) sail diameter normal to the energy beam by D_{sail} and the separation between beam-transmitter aperture and light sail be SEP $_{tran-sail}$. The angle (θ) subtended by the sail at a selected transmitter-sail separation is D_{sail} / SEP $_{tran-sail}$ radians. Assume, for example, that we wish to project a collimated electromagnetic beam against a sail 1000-km in diameter, at a distance of 10^{12} km. We must therefore maintain the aim of the beam to 10^{-9} radians. And that is not all of it—we must eliminate or compensate for beam drift, keep our transmitter perfectly aligned in the face of gravitational perturbations by solar-system bodies, etc. Such perfection must, in fact, be maintained for an acceleration period of decades duration and over such distances that the speed-of-light limitation renders feedback between ship and power station impossible.

Designers can improve things a bit by choosing a short beam wavelength (λ) and a large beam-transmitter aperture diameter D $_{beam-tran}$. Applying Rayleigh's criterion [6],

$$\theta = D_{sail} \ / \ SEP_{tran-sail} \ = 2.44 \ \lambda \ / \ D_{beam-tran} . \qquad (1)$$

One way of reducing the requirement for a long beam-collimation length is to design a spacecraft that can make several passes through the energy beam. One suggested approach is Lorentz-Force turning. If the sailcraft leaves the beam and then is charged to a sufficiently high electrical potential, its trajectory direction will be altered under the influence of the local galactic magnetic field [3,7]. Although magnetic alternatives to electrically charged spacecraft surfaces have been suggested [4, 6], Geof Landis has informed author Matloff that they are probably not feasible [6].

But even if thrustless turning is feasible, it may not prove to be a practical approach. Although it results in less demanding design parameters for the beam transmitter, we do not yet have good estimates of how long a large electrical charge can be maintained on a spacecraft surface in the interstellar plasma. Also, is the interstellar magnetic field constant enough over the light-months or light-years radius of a thrustless turn, so that the spacecraft can complete its maneuver and rejoin the beam.

Forward has suggested perforated light sails as a means of reducing light sail mass and thereby enhancing sailcraft acceleration in a power beam [8]. Unfortunately, the semi-empirical theory utilized by Forward to estimate sail optical parameters can only be used to calculate sail reflection, and only for the case of a superconducting sail. Even if we imagine very-high temperature superconductivity, we must conclude that a sail probably could not remain superconducting when pelted by megawatts or gigawatts of laser or maser radiation.

Matloff [6,9, 10] has applied the theoretical approach discussed by Driscoll and Vaughan to estimate spectral reflectivity, transmissivity, absorptivity and emissivity of a non-superconducting metallic perforated light sail [6,9,10]. Although an improvement, this theory is only applicable for very restrictive mesh-design parameters. Much work remains to be done on perforated light-sail optical theory and design before the true advantages of this approach can be ascertained. However, even our current theoretical understanding has been enough to demonstrate that, in the absence of very-high-temperature superconductors, metallic meshes may have less of an advantage over metallic thin-film sheet sails than had initially been assumed [11]. It has been suggested that dielectric thin-sheet sails may be superior to metallic sheet or mesh light sails for interstellar light-sailing application [12,13].

One method of maintaining beam collimation over interstellar distances is application of thin-film Fresnel lenses located in the energy beam between transmitter and sail [14]. Although such an approach is physically feasible, maintaining an optical link for three elements (transmitter, lens, and sail) for decades over trillions of kilometers presents a significant engineering challenge.

If all the engineering challenges can be solved, two-way interstellar travel might be possible using a laser-pushed light sail. Forward [14] has proposed an interstellar mission with sails and Fresnel lenses in the 1000-km range, and laser powers in the vicinity of 4×10^{16} wattts (about 1,000 x current terrestrial-civilization's power consumption). A spacecraft massing about 8×10^7 kg could be projected to epsilon Eridani at 10.8 light years.

Application of a multi-stage solar sail could allow two-way travel in the following fashion. One sail would be detached from the spacecraft upon its arrival at the epsilon Eridani system. Later, this sail would be maneuvered back into the power beam to reflect energy beamed from the solar system against the sail of the home-bound starship. Total-roundtrip travel time would be less than a human lifetime.

Many less-ambitious schemes have been suggested as alternatives. We might, for instance, accelerate micro-sails (1 meter or less in diameter) in the energy beam. After acceleration, these would be steered to impact against a much larger spacecraft, which would be accelerated by momentum transfer [15]. Such an approach might require development of intelligent micro-sails to home-in on and collide with the larger spacecraft [16].

Recent experimental work has included investigations of model spacecraft being pushed by power beams under vacuum conditions [17]. Other recent work has concentrated upon experimental tests of the stability of beam-riding spacecraft [18]. It seems that certain sail shapes may even be capable of automatically correcting for very small beam drift [19].

Most considerations of laser / maser – sailing assumes that the solar-powered beam-transmitting station is maintained in a constant inner-solar system position between the sun and spacecraft. Because this may be difficult to achieve, Matloff and Potter have considered the possibility of the power station following the much faster sailcraft on a barely hyperbolic trajectory relative to the Sun [20,21].

Many researchers in the field of beamed sailcraft propulsion assume that short-wavelength laser beams will always be superior to microwave maser beams. This may not always be the case.

First, discussions with G. and J. Benford and S. Potter revealed that operational microwave technology, currently applied to 1-cm wavelength microwaves, could be modified for application with millimeter-wavelength microwaves. The lower costs of microwave technology (as compared with similar-energy laser technology) could therefore be realized with shorter-wavelength microwaves.

But of greater significance, if the idea is feasible, is an application from general relativity. For many years, Claudio Maccone has been investigating the solar gravitational focus [22]. If an object occulted by the Sun is observed at a distance greater than 550 AU from the Sun, electromagnetic radiation from the object will be focused into a highly amplified and very narrow beam. The Sun-occulted object itself must be at least 550 AU on the far side of the sun from the receiver for gravity focusing of the object's EM emissions to occur.

But what if it is possible to tailor the wavefront of emissions from a solar-powered maser much closer to the Sun (say 30 AU) to have the same curvature at the solar limb as a point source 550 AU or more from the Sun? Then, beginning at 550 AU on the far side of the Sun from the power station, the maser radiation will be concentrated in a narrow beam, maintaining collimation for a very large distance [23].

Further discussions with Maccone reveal that many factors, such as variations in the solar coronal plasma, may render this idea unfeasible. But, it is certainly a most exciting concept and is well worthy of further study.

Very Large Space Based Laser Concepts

Current diode laser array technology can produce output irradiance of 10 MW/m^2 [24]. The diameter of these beams is small. Also, diode laser arrays are not very complicated and need only a radiator to remove heat from the diode material, collimating optics, and a power supply. The power supply for a space based application can be solar cells attached directly to the back of the laser diode substrate or radiator. The diode lasers are less efficient than, for example solar pumped lasers, but their simplicity in design make them an excellent candidate for laser sailing.

Assume a laser sailing mission requirement is to continuously deliver one solar constant of irradiance to a 1 km diameter sail and a 1000 kg spacecraft mass. The goal is to push the sail to 15% the speed of light within 10 years. In other words, the laser must supply about 2 x 10^{19} Joules of photonic energy to the sail. The total power delivered to the sail is approximately 64 GW. Each laser diode in the large array can generate 20 W. So, about 3.2 billion diode arrays would be required to generate this much power. This is a large number of arrays; however, physically it is not as extreme as it seems. Each diode array is roughly 1 mm by .5 mm by 3 mm in size (including heat sink). The surface area of the emitting plane of the diode array is 5 x 10^{-8} m^2. A circular array large enough to produce the power output requirements would be about 7.1 m in radius. Diode lasers operate at about 30 % efficiency, so the solar cell collector array would have to be able to generate 213 GW of power [25]. State-of-the-art solar cell technology is about 24% efficient for a cell approximately 1 cm in radius [26]. A cell this size at one AU from the sun can produce about 0.427 W. One hundred fifty six cells produce enough power to drive four diode arrays. Therefore, approximately 499 billion solar cells would be required to power the full diode array. The total size of the solar cell array would be about 15 km in radius. Placing the collector closer to the sun would decrease its size requirement.

Large Optics Concerns

Many papers have been published on the subject of laser sailing as previously discussed. Most of them, however, discuss the idea in general and give very little specifics of the large-scale optical components required. The type of optic usually discussed is the so-called O'Meara para-lens [14] shown schematically in Figure 1. It is basically a large Fresnel Zone lens made of concentric rings of a lightweight transparent material. The lens must be constructed in such a way that there are voids or free-space between the rings. Spars much like a telescope spider must be used to give the para-lens structural integrity.

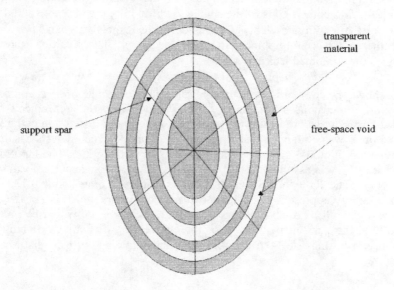

Figure 1. O'Meara para-lens

The para-lens design has been discussed in practically every paper and text written on laser sailing. However, no real diffraction analysis has been presented. Mallove and Matloff did a preliminary analysis and suggested that a reflective optic would be better [4]. It is the purpose of this section of the paper to show that a solid optic would have better properties (optically and structurally) than the para-lens.

The radial transmission function $t(\tilde{r})$ for a Fresnel Zone lens is written as [27]

$$t(\tilde{r}) = \sum_{n=0}^{R_{sail}} \frac{\sin\left(\frac{n\pi}{2}\right)}{n\pi} e^{\frac{in2\pi\tilde{r}}{X}} , \qquad (2)$$

where R_{sail} is the radius of the sail, n is the ring number, and X is the spatial frequency for the ring repetition. The aperture function for the sail circular hoop support and spider is

$$Aperture = circ\left(\frac{\tilde{r}}{R_{sail}}\right) - circ\left(\frac{\tilde{r}}{R_{hub}}\right) - rect\left(\frac{x-a}{L}, \frac{y-b}{W}\right)$$

$$- rect\left(\frac{x\cos\theta + y\sin\theta - a}{L}, \frac{y\cos\theta + x\sin\theta - b}{W}\right)$$

$$- rect\left(\frac{x\cos(-\theta) + y\sin(-\theta) - a}{L}, \frac{y\cos(-\theta) + x\sin(-\theta) - b}{W}\right),$$

(3)

where *circ* represents the circle function and *rect* the rectangle function per Goodman notation [28], R_{hub} is the radius of the spider's central hub, a and b are spatial offsets, L and W are the length and width of the spider arms respectively, and θ is the angle between the spider arms (assumed to be 120 degrees). The resulting transmission function for the combined lens and sail is

$$Optic = t(\tilde{r})Aperture. \qquad (4)$$

Figure 2 shows the combined Optic transmission function. The optic is considered to be the same dimensions as the sail (0.5 km in radius). Using Mathcad^{TM} and a numerical Fast Fourier Transform technique, the Fraunhofer diffraction pattern for the combined optic was computed and is shown in Figure 4. The analysis assumed 500 nm wavelength light at a distance of 2 light years from the optic. The input intensity of the laser beam was assumed to be 1360 W/m^2. The irradiance at the focal spot is ~ 2,500 W/m^2. The light in the central order is dispersed over about 0.125 km and significant energy is present in the higher diffraction orders which is not the case for reflective optics [28]. Analysis suggests that a reflective optic would be much more efficient than the para-lens [28]. Also, the engineering difficulties of building the large concentric rings for the para-lens and maintaining surface flatness seems daunting.

Figure 2. Transmission function for Fresnel Zone lens and aperture

Figure 3. Fraunhofer diffraction pattern of transmission function shown in Figure 2

Pointing and Tracking Concerns

A very precise control system will have to be utilized in order to keep the laser beam on the sail at large distances from the beam steering optic. Guidance and control of the beam steering optic could be via control vanes just as with the sail. The control vanes will allow for large-scale pointing adjustments. However, a method for making more precise pointing adjustments must be implemented. This control can be accomplished at the aperture of the laser by steering the diffractive optic mentioned

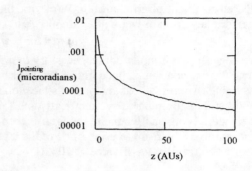

Figure 4. Pointing jitter versus distance in AUs

previously. All pointing and tracking systems have pointing errors due to system vibrations, imperfections in the optics, and many other factors. This so-called system "jitter" effectively tilts the beam at an angle to the optical path. The center of the beam is then moved away from the center of the target by [28]

$$\Delta \tilde{r} = z j_{pointing} \quad , \qquad (5)$$

where $j_{pointing}$ is the pointing jitter. The maximum jitter allowable for the center of the beam to remain on the sail is for $\Delta \tilde{r}$ to be equal to the sail radius. Figure 4 is a graph of the jitter versus distance. State-of-the-art in pointing and tracking jitter is little less than 0.1 microradians [29]. To simply reach 100 AUs would require a pointing jitter 4 orders of magnitude better than the state-of-the-art. Figure 5 shows the jitter versus distance in light years. Current pointing jitter technology must be improved 9 orders of magnitude in order to reach 5 light years.

Since the spatial distribution of the pointing jitter is a random Gaussian variable (RGV), the Central limit theorem requires that the

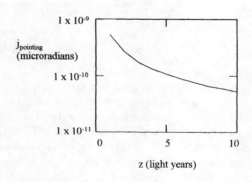

Figure 5. Pointing jitter versus distance in light years

irradiance distribution over a given period of time fill in a radial Gaussian distribution [30]

$$I(r) = I_o e^{-\frac{r^2}{2\sigma^2}} \qquad (6)$$

where I_o is the beam irradiance in W/m^2, r is the radial dimension in km, and σ is the standard deviation or beam jitter amplitude in km. The jitter amplitude is given as

$$\sigma = \Delta \tilde{r} = z j_{pointing} . \qquad (7)$$

We suggest that it might not be necessary to maintain the laser on the sail at all times since the distribution mapped out by the laser beam is radially symmetric. Over time, assuming that the sail can maintain its location at the mean or center of the laser beam, it will still be illuminated symmetrically and following the Gaussian profile. The loss of incident beam energy due to the jitter is then easily determined as a function of the distance from the main steering optic of the laser. Figure 6 shows the time integrated laser beam profile at a distance of 10 AU.

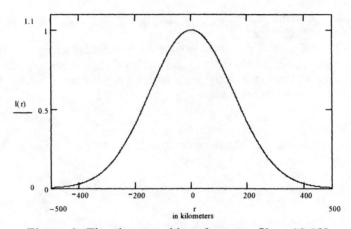

Figure 6. Time integrated laser beam profile at 10 AU

Assuming that the sail is 1 km in diameter, the average incident irradiance on it is just an integration of that region of the curve given in Figure 6. For simplicity, it is again assumed that the sail remains near the mean of the profile. Integrating equation (6) from 0 m to 500 m at a distance of 10 AU yields about 0.025% of the total beam incident on the sail. Assuming a 64 GW laser array, at 10 AU there would still be 20 W/m^2 incident on the sail or about 1.5% of the solar constant. Figure 7 shows a plot of percentage of power incident on the sail versus distance from the laser in AU. Realize of course, that improving the jitter will in turn improve the incident laser beam power.

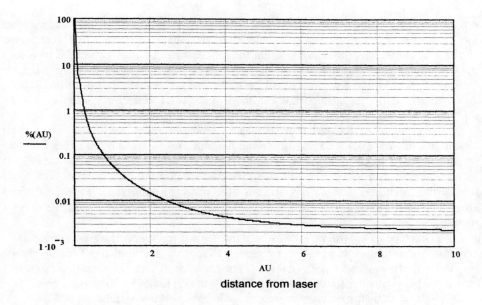

Figure 7. Percent of laser power incident on sail versus distance in AUs

Large Controllable Mirror/Antenna Concepts and Experiments

Thus far it has been assumed that the laser beam would be collimated by the optical system. However, the optical system should focus slightly beyond the lightsail so the entire beam will be incident on it. Forward suggested that this could be done with the para-lens [14]. Changing the focus of the para-lens would require changing the spacing and the width of the refractive rings. This is a difficult if not an insurmountable task. On the other hand, this task matches the characteristics of an inflatable membrane reflector or an electrically controlled membrane reflector almost perfectly. As the sail gets further away from the reflector, the inflation pressure or the electric field strength can be decreased. The decrease in inflation pressure or field strength makes the radius of the catenary curve larger and moves the focus further away from the reflector. The reflector is thus an active optical element. The electrically addressed optic as currently being developed by Taylor, Brantley, *et al*, [31] can also be pixilated which will allow for beam steering and wavefront correction.

The electrically addressed optic consists of two planes of conducting Al material a given distance apart. The surface reflector is contoured by placing a high voltage static field between the two conductive Al planes. Electrostatic forces pull the two surfaces together in a catenary like shape (which is parabolic). Control of the reflector surface flatness is accomplished by pixelating the control surface and varying the field strength spatially across it. This enables adaptive control of the reflector surface. Preliminary efforts suggest that the reflector can be maintained flat to within optical wavelengths for reflector diameters of larger than 5 meters with the adaptive control. Control of larger surface will be studied in greater detail in in future efforts. Also, control of larger surface flatness for a microwave reflector instead of visible light should prove much easier due to the longer wavelengths.

The figure of the reflector surface will be sensed by looking at light from visible or infra-red (IR) stars and observing the resultant Bessel sinc function (Besinc or Airy disk) on a ccd focal plane. This is a common telescope test know as the "Star Test". The Star Test will be used for adaptive control of the reflector. Deviations from the Besinc function represent deviations of the reflector membrane from the desired shape and will be used to apply the perturbation voltages to the tiles on the control membrane.

This type of control problem where the desired reference (the Besinc function) is a constant, is known as a regulator problem. Since this is a multi-input (many pixels on the focal plane) multi-output (many tiles on the control surface) we will employ a multivariable controller (multi-input, multi-output, coupled). Our solution to the surface figure regulator problem is an optimal linear-quadratic regulator (LQR) multivariable controller (system is linear, cost function which is optimized, is quadratic, and solution applies to the regulator problem). The LQR controller solves the regulator problem in an optimal fashion and yields a systematic method for the design of large multivariable controllers. This approach is common whereas adaptive optics are used to remove wavefront aberrations due to atmospheric distortion.

The LQR controller provides both the control for static deformation of the reflector surface to achieve the desired figure, and dynamic deformation to counter dynamic disturbances such as thermal cycling or structural vibrations.

A small-scale prototype of the optic/antenna was constructed and operated to demonstrate proof-of-concept of the approach. Lightweight, foam board was utilized as frame material to support the membrane, as shown in Figure 8. A 1-meter diameter piece of 1 mil thick aluminized KAPTON® polyimide film was bonded at its circumference to an elastic fabric. A circular hole was cut in the foam board frame and the fabric was secured to the frame creating a flat reflective surface. The elasticity of the fabric provided tensioning of the film keeping it flat in the de-energized state. Electrical connections were made to the aluminum coating at intervals around the perimeter of the film and this was connected to the positive terminal of a high voltage power supply. A second, solid sheet of foam board with an aluminum foil bonded to it was placed behind the reflective film. The negative side of the high voltage power supply was connected to the foil surface.

A standard, Dish Network™ satellite television LNBF (low noise block converter with integrated feedhorn) was mounted on an adjustable arm in front of the reflector and connected to a satellite receiver. The assembly was located such that it could view the sun and high voltage was applied, curving the reflector and focusing the sunlight. It was noted that due to imperfections in the film surface and curvature that the focused sunlight produced a spot size of about 8 to 12 inches in diameter. Adjustments were made to the support arm to place the LNBF at this focus. The

LNBF was connected to the satellite receiver and the antenna assembly was aimed at an Echostar satellite (Ku-band, 11.7 GHz to 12.2). Initial tests failed to acquire the satellite signal. The stock feedhorn for the commercial LNBF was approximately 2.5 inches in diameter and designed to operate with a more efficient reflector. To compensate for this, a larger conical feed horn was constructed of rolled sheet metal with an aperture of 4 inches. This was attached to the LNBF and the placement of the LNBF was re-adjusted.

With this larger aperture at the LNBF, the satellite signal was acquired. When high voltage was removed and the film relaxed, the signal was lost. High voltage was repeatedly applied and removed and consistently resulted in lock and loss of the satellite signal. This confirmed that the curvature produced in the reflector by the electrostatic forces formed a practical and useful antenna.

This prototype was constructed from readily available material and constructed with no controlled tolerances or precise mechanical adjustments yet it successfully demonstrated the concept. With optimized materials, controlled tolerances and accurate placement of components the efficiency of the antenna could be greatly improved.

It was also demonstrated that the curvature of film was proportional to the distance between the film and the ground plane and the differential voltage applied between the two. This was evident in the movement of the focal point and the diameter of the spot. This effort is in its preliminary stages and more data is forthcoming. However, we have witnessed the phenomenon.

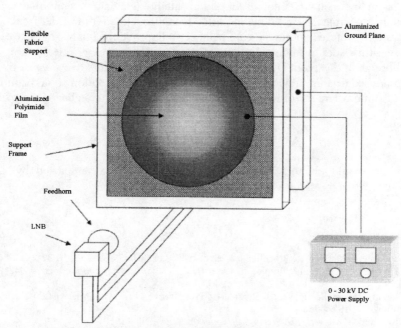

Figure 8. Schematic of Membrane Antenna Experiment

CONCLUSIONS

A history of laser sailing was given here that suggests the status of this field is still very immature. Which type of directed energy source, laser or maser, would be most effective remains to be seen and needs further investigation. Also, the power requirements of an interstellar mission is tremendous and would require very large space based solar collector arrays. The analysis given here, which did not take into account pointing jitter, suggests a minimum of about 15 km in radius for the collector. If pointing error is considered, it appears that the current state-of-the-art of jitter control is many orders of magnitude from enabling a laser sailing mission. Beam control is the largest obstacle for laser sailing.

Other obstacles such as very large space based lasers are merely engineering and deployment tasks. Although they are rather daunting tasks, there is no evidence that suggest it is beyond the current state of technology. Rather, the expense of the project would be the prohibitive factor.

Are there solutions to the pointing and tracking concerns? Possibly, new control algorithms and very large adaptive and steerable optics could offer hope. Preliminary analysis suggests that the standard O'Meara paralens would be inefficient and most likely difficult to control or steer. More analysis and experiment should be conducted, however, before the idea is totally discarded. On the other hand, the current status of large reflective optics may seems more likely to be technologically feasible in the near future.

Experiments conducted with aluminum coated polyimide materials show that the surface flatness of a microwave reflector can be controlled via static electric fields. Flatness was maintained well enough for digital satellite television reception using a commercially available LNBF with a modified feedhorn and an off the shelf satellite receiver system. Future experiments are planned to improve the quality of control and to enable real-time adaptive control and steering of a microwave or optical beam.

Although at this point laser sailing may seem very unlikely, we suggest that the book is not closed on this subject. Further study and experimentation should continue since laser sailing offers the only current likely candidate for interstellar space travel.

ACKNOWLEDGMENTS

The authors would lake to acknowledge that this research was supported by NASA MSFC Contract No. NASA # H-35191D.

REFERENCES

1. G. Marx, "Interstellar Vehicle Propelled by Terrestrial Laser Beam," *Nature, 211*, 22-23 (1966).
2. W. E. Moeckel, "Propulsion by Impinging Laser Beams," *J. Spacecraft and Rockets, 9*, 942-944 (1972).
3. P. C. Norem, "Interstellar Travel : A Round Trip Propulsion System with Relativistic Velocity Capabilities, " AAS 69-388.
4. E. Mallove and G. Matloff, *The Starflight Handbook*, Wiley, NY (1989).
5. J. H. Maudlin, *Prospects for Interstellar Travel*, Univelt, San Diego, CA (1992).

6. G. Matloff, *Deep-Space Probes*, Springer-Praxis, Chichester, UK (2000).
7. R. L. Forward, "Zero-Thrust Velocity Vector Control for Interstellar Probes : Lorentz-Force Navigation and Circling," *AIAA J., 2*, 885-889 (1964).
8. R. L. Forward, "Starwisp : An Ultra-Light Interstellar Probe," *J. Spacecraft 22*, 345-350 (1985).
9. G. L. Matloff, "An Approximate Heterochromatic Perforated Light-Sail Theory," IAA-95-IAA.4.1.01.
10. W. G. Driscoll and W. Vaughan eds., *Handbook of Optics,* McGraw-Hill, New York (1978), Chap. 8.
11. G. A. Landis, "Microwave Pushed Interstellar Sail : Starwisp Revisited," AIAA-2000-3337.
12. G. A. Landis, "Optics and Materials Considerations for a Laser-Propelled Lightsail," IAA-89-664.
13. G. A. Landis, "Advanced Solar- and Laser-Pushed Lightsail Concepts," NASA Institute for Advanced Concepts (NIAC) Phase I Final Report (May 31, 1999).
14. R. L. Forward, "Roundtrip Interstellar Travel Using Laser-Pushed Lightsails", *J. Spacecraft, 21*, 187-195 (1984).
15. J. T. Kare, "Interstellar Precursor Missions Using Microsail Beams," in *Space Technology and Applications International Forum-STAIF 2002*, ed. M. S. El-Genk, AIP (2002).
16. G. D. Nordley, "Momentum Transfer Particle Homing Algorithm," presented at NASA / JPL / MSFC / AIAA Annual Tenth Advanced Space propulsion Workshop, Huntsville, AL, April 5-8, 1999.
17. L. N. Myrabo, T. R. Knowles, J. Bagford, D. B. Siebert, and H. H. Harris, "Experimental Investigation of Laser-Pushed Light Sails in a Vacuum," AIAA 00-3336.
18. J. Benford, G. Benford, O. Gornostaeva, E. Garate, M. Anderson, A. Prichard, and H. Harris, "Experimental Tests of Beam-Riding Sail Dynamics," in *Space Technology and Applications International Forum-STAIF 2002*, ed. M. S. El-Genk, AIP (2002).
19. G. L. Matloff, "Self-Correcting Beam-Propelled Sail Shapes," in Research Reports—2000 NASA / ASEE Summer Faculty Fellowship Program, NASA / CR—2001-210797, L. M. Freeman , C. L. Karr, J. Pruitt, S. Nash-Stevenson and G. Karr eds. (Sept. 2001).
20. G. L. Matloff and S. Potter, "Near-Term Possibilities for the Laser-Light Sail," in *Missions to the Outer Solar System and Beyond, 1ˢᵗ IAA Symposium on Realistic Near-Term Advanced Scientific Space Missions*, ed. G. Genta, Levrotto & Bella, Turin, Italy (1996).
21. G. L. Matloff, "The Laser Light sail and Interstellar Colonization, in *Missions to the Outer Solar System and Beyond, 3rd IAA Symposium on Realistic Near-Term Advanced Scientific Space Missions*, ed. G. Genta, Levrotto & Bella, Turin, Italy (2000).
22. C. Maccone, "Interstellar propulsion by Sunlensing," in *Space Technology and Applications International Forum-STAIF 2001*, ed. M. S. El-Genk, AIP (2001).
23. G. L. Matloff, "Multiple Aspects of Light-Sail Propulsion by Laser / Maser Beams," IAA-01-IAA.4.1.03.
24. http://www.optopower.com, September 2000.
25. R.J. DeYoung, J.H. Lee, , et al., "One-Megawatt Solar Pumped and Electrically Driven Lasers for Space Power Transmission," *Proceedings of the Intersociety Energy Conversion Engineering Conference Proceedings of the 23ʳᵈ Intersociety Energy Conversion Conference,* Denver, IEEE Publishing, 1989, pp 709-714.
26. http://www.pv.unsw.edu.au/eff/eff_tab4.html, September 2000.
27. J.W. Goodman, *Introduction to Fourier Optics,* McGraw-Hill, New York, (1968).
28. T.S. Taylor, *Advanced Solar and Laser Sail Propulsion Concepts for Interstellar Travel,* Masters Thesis, University of Alabama in Huntsville, (2000).
29. W.H. Possel, "Lasers and Missile Defense: New Concepts for Space-Based and Ground-Based Laser Weapons," Occasional Paper No. 5, Center for Strategy and Technology, Air War College, internet copy, http://www.au.af.mil/au/awc/awcgate/cst/occppr05.htm, (July 1998).
30. S. Arnon, "Power versus stabilization for laser satellite communication," *Applied Optics*, Vol. 38, No. 15, 20 May 1999.
31. T.S. Taylor, L.W. Brantley, R.C. Anding, D. Halford, *Membrane Optic Research Program Internal Documentation,* Teledyne Brown Engineering and The National Space Science and Technology Center, Huntsville, Alabama, (2002).
32. T.S. Taylor, R.C. Anding, R. Boan, *Very Large Space Radio Telescope*, Teledyne Brown Engineering white paper, Huntsville, Alabama, (2002).

From the Sun to Infinity

Greg L. Matloff[1] and Travis S. Taylor[2]

[1]*Consulting Analyst. Also Adjunct Professor, New York City Technical College, Bklyn, NY and New School University, New York, NY.*
[2]*Principal Investigator, Advanced Projects, Teledyne Brown Engineering, Huntsville, Al*

Abstract. This paper first presents a review of interstellar solar sailing. Included is a treatment of close perihelion ("sundiver") trajectories, the effects of cable mass / strength and sail optical / thermal / structural properties. Also considered are suggestions to enhance performance by using cables that are partially supported by solar radiation pressure or to replace cables altogether as a means of affixing payload to sail. Various sail films that have been suggested are considered. Next, we present an approach to an Oort -cloud mission in the 2030-2050 time frame. Although "1000-year-arks" or probes to Alpha Centauri will not be feasible in this time interval, solar-system hyperbolic-excess velocities in excess of 300 km / sec may be achievable.

INTRODUCTION: HISTORY OF A NOVEL IN-SPACE PROPULSION CONCEPT

Although the solar sail is now in the flight-test and preliminary mission-planning phase, it has a long and venerable history [1-4]. Realization that spacecraft could accelerate by momentum transfer from reflected photons followed in the wake of early-20th century theoretical treatments of radiation pressure by Einstein and others. Perhaps the first to propose space travel by solar sails were Russian researchers, including Konstantin Tsiolovsky and Fridrickh Arturovich Tsander. Well before mid-century, these researchers and their Russian colleagues had performed preliminary studies of the application of solar-radiation-pressure to escape the solar system.

But the first rigorous considerations of the solar sail's applicability to extrasolar or interstellar space travel did not occur until the 1970's and 1980's. Although much of this research supported a British Interplanetary Society study of the feasibility of interstellar travel, most of the researchers who participated in this phase of interstellar solar-sail research were American [5,6].

Starting in 1974 and ending about 1990, the British Interplanetary Society (BIS) conducted work related to Project Daedalus, a study of a thermonuclear-pulse powered probe that could be accelerated to velocities as high as 0.15c (where "c" is the speed of light) and reach Alpha / Proxima Centauri (the nearest extra-solar stellar system at 4.3 light years) or Barnard's Star (which is 6 light years from the Sun) in one-way travel times less than a human life time [7].

One of the problems with the Daedalus concept was the fusion-fuel . To reduce neutronic irradiation, a mixture of helium-3 and deuterium was required. Since helium-3 is exceedingly rare on Earth, it was suggested that Daedalus could be fueled, at great expense, by helium-3 mined from the atmospheres of giant planets.

Editors of *JBIS* (*The Journal of the British Interplanetary Society*) acknowledged both the feasibility and difficulty of interstellar travel and scheduled up to four annual issues of *JBIS* to concentrate on "Interstellar Studies." Because of the difficulty and expense of obtaining helium-3 in quantity, socio-political issues relating to the acceptability of huge nuclear-propelled spacecraft, many *JBIS* authors considered alternatives to interstellar thermonuclear-pulse rocketry.

In the late 1970's, two American teams began to independently consider non-nuclear-propelled interstellar missions. On the west coast, a team of NASA Jet Propulsion Laboratory analysts directed by Louis Friedman [8], had analyzed the feasibility of exploring Halley's Comet in 1986 using a solar sail. One of the members of this study team, Chauncey Uphoff, then contributed to the TAU (Thousand Astronomical Unit) study at JPL [9]. This was a study of a probe to be launched in the first quarter of the 21st century that could reach 1000 AU (1.5×10^{11} km) from the Sun in 50 years, which implies a solar-system exit (or hyperbolic excess) velocity in nearly 100 km / sec.

TAU analysts concluded that only two propulsion systems would be capable of performing this mission during the projected time interval. The first was favored approach was the nuclear-electric or ion drive. As a back-up, Uphoff proposed a hyperthin (less than 1-micron) solar sail unfurled within the orbit of Venus. Unfortunately, Uphoff was only credited in the final TAU report with "unpublished calculations". His approach compares well with the predictions of the second group, whose results are available in the literature.

At about the same time as Uphoff was contributing to the TAU study, Gregory Matloff in New York collaborated with Michael Meot-ner on concept development for methods of propelling directed-panspermia paylads to interstellar velocities, on voyages with durations on the order of 10,000 year. [10]. The preferred propulsion approach was to fly a sailcraft with a lightness number (ratio of solar radiation pressure force to solar gravitational force) of 1. If the sail of such a craft is directed normal to the Sun, the spacecraft exits the solar system (following Newton's 1st Law) along a straight-line trajectory at its pre-sail-unfurlment solar orbital velocity. Mercury's orbital velocity is about 48 km / sec. Sail unfurlment at Mercury would result in a travel time to Alpha Centauri of about 27,000 years.

Although Meot-ner and Matloff realized that a sailcraft with lightness number greater than 1 would greatly reduce the duration of an interstellar expedition, analytical details of interstellar solar sails were published in the early 1980's by a team consisting of Matloff and Eugene Mallove [11,12]. Principal features of such a mission is an initial parabolic (or hyperbolic) solar orbit with a perihelion of a few mission kilometers (the so-called "sundiver" trajectory). At perihelion the partially-unfurled sail is exposed to sunlight. If the sail is highly reflective, very thin and heat-tolerant and the structure connecting sail and payload is strong enough, solar-system exit velocities of 1,000 km /sec or higher are possible, even for large payloads.

This approach renders robotic and peopled missions to the Alpha Centauri system of millennium-duration possible. In a land-mark paper, one of the final contributions of the Daedalus team, Alan Bond and Anthony Martin concluded that only one method of transferring human civilization to the stars would ultimately be feasible—the thousand-year ark or worldship—and only thermonuclear-pulse or the solar sail would be up to the task [13].

Since this early work, many papers have been published that examine methods of reducing interstellar-solar-sail voyage duration. These include hyperbolic pre-perihelion velocities, use of hyperthin or perforated sails, and cables so thin that they are affected by solar radiation pressure [14-18]. Computer simulations have revealed that various sail configurations are stable during the multi-g, hours-long near-Sun acceleration runs of solar-sail starships [19].

It has also been pointed out that for very close perihelions (within a few solar radii), the inverse-square law of solar irradiation needs correcting [4, 20]. Some solar-system-exit velocity advantage is obtained by optimizing sail aspect angle relative to the Sun during pre-and post-perihelion trajectories [21,22]. But all this work has not succeeded in decreasing interstellar-transfer time to much less than a millennium.

Recent Research : FOCAL, Aurora, the NASA ISP and Beyond

Starting in about 1990, many international researchers began to concentrate upon near-term interstellar solar-sail missions. Rather than being directed towards nearby stars on 1,000-year trajectories, these craft would study aspects of the near-Sun interstellar environment, out to a few thousand AU .

FOCAL (also called ASTROsail or SETIsail) originated through the efforts of Italian researcher Claudio Maccone and many colleagues [23]. This is a proposal to direct a sail-launched probe towards the Sun's gravitational focus at 550 AU. According to general relativity, the sun's gravitational field focuses radiation emitted by objects occulted by the Sun into a narrow, highly-amplified beam, at and beyond the solar gravitational focus. Consider a sailcraft that first makes a close flyby of Jupiter to direct it into a parabolic solar orbit with a perihelion near Mercury. The solar escape or parabolic velocity at the orbit of Mercury is about 67 km / sec. If the sailcraft has a lightness number of 1 and is oriented normal to the sun after perihelion, it will depart the solar system at 67 km / sec and reach the Sun's gravitational focus about 40 years after solar encounter. A long-lived spacecraft equipped with a modest suite of astronomical instruments could use the Sun's gravitational focus to make observations of interest to both the astrophysical and SETI (Search for Extraterrestrial Intelligence) communities.

Several of the participants in the international FOCAL study wondered whether a sailcraft could perform a scientifically useful function if directed towards targets less distant than the sun's gravitational focus at 550 AU. Further investigation led to Aurora, a sailcaft designed to explore the near interstellar medium, at a distance of about 200 AU from the Sun [24]. As well as trajectory analysis, some Aurora team members considered the design and stability of parachute-type sails with inflatable beam members [25]. One innovation in Project Aurora was the suggestion that the mass of a tri-layer Earth-launched sail (aluminum reflective layer, chromium emissive layer, and plastic substrate) could be reduced in space by utilizing UV-sensitive plastic that would evaporate in the space environment [26].

Beginning in 1998, NASA began to investigate the possibility of a near-term (2010-2020) solar-sail launched probe to the heliospause. Perhaps inspired by Aurora, the goal of the Interstellar Probe (ISP) was to project a suite of particle- and field-measuring instruments to the boundary of solar and interstellar space, which is projected to be about 200 AU from the Sun.

To perform its mission in 20 years, the ISP must depart the solar system at about 50 km /sec, about 3X the speed of the Pioneer 10 / 11 and Voyager 1 / 2 probes [27]. About 30 kg of scientific experiments are included in a total mission mass (excluding the sail) of about 150 kg. The sail has an areal mass density of about 1 gm / m^2 and a mass of 100 kg. To achieve its high hyperbolic excess velocity, the sailcraft departs Earth orbit and then proceeds to a 0.25 AU perihelion [28].

Much progress on sail films and structures to enable the ISP has occurred; much remains to be accomplished in this on-going research effort [29,30]. One aspect of recent research at NASA MSFC is testing of candidate solar-sail film materials in the simulated and real space environment [31].

In a consulting capacity, Vulpetti has applied his Aurora-derived software to ISP [32]. By using direct or retrograde perihelion -approach trajectories, there are two launch windows per year to reach any point in the near heliospause. Significant reductions in sail size and/or increases in areal mass density are possible if the spacecraft departs earth-space with a small hyperbolic excess velocity.

Some consideration has been given to an "Oort Cloud Trailblazer Mission" to be launched later in the 21st century. With a hyperthin sail (areal mass density about 0.1 gm /m^2), such a craft might attain speeds as high as 300 km /sec (0.001 c) Perhaps the ultimate earth-launched solar sail, this craft could travel more than 1,000 AU during its design lifetime.

A Proposed Oort Cloud Explorer

Created some 4.5 billion years ago from remnants of the solar system's development, the Oort Cloud consists of more than 12 billion comets that lie between 10^4 and 10^5 AU from the Sun [33]. It is of scientific interest to send a probe to the sphere of long period comets. As mentioned above, it is possible that a "sundiver" type solar sail could be used for a mission to the edge of the Oort Cloud within a reasonable mission time. We consider here one spacecraft architecture for such a mission.

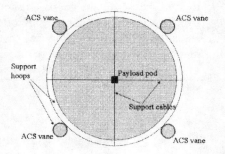

Figure 1. Oort Cloud Explorer (not to scale)

The solar sail spacecraft consists of a central disk sail 681 m in diameter, which is supported by an inflated torus or hoop. Four smaller hoop supported sails 5 m in diameter are attached to the main larger hoop at equidistant locations and are used for attitude control system (ACS) vanes. The mass of the hardware associated with the ACS vanes was assumed to be 50 kg. The payload of 150 kg is suspended from 4 Spectra 1000™ cables and is held in the center of the main hoop. Winch motors in the payload can reel in or out these cables to change the position of the payload thus changing the center of mass of the spacecraft. This allows for center of mass center of pressure steering as a redundant system to the ACS vanes. A depiction of the spacecraft is given in Figure 1. Assume that the support hoops are made of modern state of the art polyimide materials with an areal density of about 10 g/m^2 and that the sail material itself is on the order of 0.1 g/m^2. The support hoop for the large main sail is 0.5 m in diameter and the support hoops for the control vanes are 0.005 m in diameter and all of the hoops are 0.0025 m thick inflated polyimide. The polyimide can be thermally preformed so as constant inflation pressure is not required to maintain deployment [34], however, for this analysis it was assumed that all hoops remained inflated.

The total mass of the spacecraft is estimated as

$$m_{spacecraft} = \sigma_{sail} A_{sail} + m_{payload} + m_{(ACS)hardware} + m_{hoops} + \rho_{gas} * V_{hoops} \qquad (1)$$

where the masses are as specified by the subscripts, σ_{sail} is the areal density of the sail material, A_{sail} is the area of the sail, ρ_{gas} is the density of the inflation gas, and V_{hoops} is the inflated volume of the support hoops. The area of the sails is found as

$$A_{sail} = \pi \left(\frac{d}{2}\right)^2 N + \pi \left(\frac{D}{2}\right)^2 \qquad (2)$$

where d is the diameter of the ACS vane sails, N is the number of ACS vanes, and D is the diameter of the main sail. The mass of the hoops was found by

$$m_{hoops} = \sigma_{hoops} 2\pi^2 \left[\left(R_{outer}^2 - R_{inner}^2\right)\frac{D}{2} + \left(r_{outer}^2 - r_{inner}^2\right)\frac{d}{2} N \right] \qquad (3)$$

where σ_{hoops} is the areal density of the hoop material, R_{outer} and R_{inner} represent the outer and inner radii of the main hoop, and r_{outer} and r_{inner} represent the outer and inner radii of the ACS vane hoops. It should be noted here that a reflectance coefficient of ε=0.85 was assumed for the sail reflective coating. The above parameters and Equations (1) through (3) define the spacecraft architecture and drive the lightness number of the spacecraft to be

$$\beta = \frac{.00153\varepsilon A_{sail}}{m_{spacecraft}} = 2.003 \quad . \qquad (4)$$

Substituting the lightness number into the cruise velocity equation given by McInnes [4] enables the determination of the cruise velocity radially outward from the Sun as a function of the closest solar flyby distance (perihelion) of the spacecraft to the Sun. Figure 2 shows the cruise velocity in AUs per year as a function of the solar flyby distance in AUs. Note that the figure suggests that this spacecraft could achieve velocities approaching 125 AU/year with close solar approaches on the order of .01 AU. Figure 3 shows the trip time in years to reach the Oort Cloud (10^4 AU) as a function of the solar flyby distance in AU. This figure shows that mission times for the extreme close solar approach could be on the order of 100 years. Slightly better materials or a larger sail could possibly yield mission times of less than an average human lifespan!

It should also be noted that the same analysis was conducted for a square sail configuration with 4 triangular sail sections supported by 4 graphite booms extended from a central hub spacecraft. Similar results were achieved. However, slightly larger sail area was required in order to achieve comparable lightness numbers due to the mass requirements inherent to the square sailcraft's boom design.

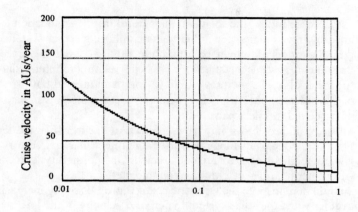

Figure 2. Cruise velocity in AUs/year as a function of solar flyby distance in AU

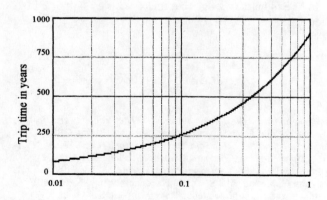

Figure 3. Trip time in years to reach the Oort Cloud as a function of solar flyby distance in AU

CONCLUSIONS

The solar sail has great potential for humanity's early forays into the galaxy, but it is important not to oversell this technology. Two decades of research has revealed that it will be very difficult for an interstellar solar sail to reach a solar-system hyperbolic excess velocity much in excess of 0.005c. Travel times to Alpha Centauri much less than 1,000 years may not be possible with the solar sail, which would severely limit our interstellar aspirations if faster propulsion options prove infeasible.

In the foreseeable future, the solar sail may therefore see application to the exploration of nearby interstellar space rather than neighboring stellar systems. As well as expanding the radius of human exploration to 1,000—10,000 AU, the interstellar solar sail can serve as a test bed for concepts relating to interstellar beamed sailing. Sailcraft materials and structures will have commonalities in both approaches and can be tested in solar sailors long before beam projectors and power stations are constructed and deployed in solar space.

Low-power beam projectors may have application to extra-solar exploration by solar sail long before the gigawatt-range power-beam projectors required for exploratory visits to neighboring stars are constructed. One possibility is to use a power beam for pre-perihelion acceleration of an interstellar solar sail. This could demonstrate the utility of beamed propulsion by increasing both the perihelion velocity and the solar-system hyperbolic velocity of the interstellar solar sail, thereby extending its exploratory range.

ACKNOWLEDGMENTS

We would like to acknowledge that this work was supported by NASA MSFC Contract Number NASA # H-35191D.

REFERENCES

1. E. Polyakhova, *Space Flight Using a Solar Satl – The Problems and Prospects*, Kosmichicheskiy Polet Solnechnym Parusom, Moscow, Russia (1986).
2. L. Friedman, *Star Sailing : Solar Sails and Interstellar Travel* , Wiley, NY (1988).
3. J. L. Wright, *Space Sailing*, Gordon & Breach, Philadelphia, PA (1992).
4. C. R. McInnes, Solar Sailing, Springer-Praxis, Chichester, UK (1999).
5. E. Mallove and G. Matloff, The Starflight Handbook, Wiley, NY (1989).
6. J. H. Mauldin, Prospects for Interstellar Travel, Univelt, San Diego, CA (1992).
7. A. Bond, A. R. Martin, R. A. Buckland, T. J. Grant, A. T. Lawton, H. R. Mattison, J. A. Parfatt, R. C. Parkinson, G. R. Richards, J. G. Strong, G. M. Webb, A. G. A. White, and P. P. Wright., "Project Daedalus : The Final Report on the BIS Starship Study," supplement to *JBIS, 31*, S1-S192 (1978).
8. L Friedman, "Solar Sailing : The Concept Made Realistic," AIAA-78-82.
9. L. D. Jaffe, C. Ivie, J. C. Lewis, R. Lipes, H. N. Norton, J. W. Sterns. L. D. Stimpson, and P. Weissman, "An Interstellar Precursor Mission," *JBIS, 33*, 3-26 (1980).

10. M. Meot-ner and G. L. Matloff, "Directed Panspermia : A Technical and Ethical Evaluation of Seeding Nearby Solar Systems, *JBIS, 32*, 419-423 (1979).

11. G. L. Matloff and E. F. Mallove, "Solar Sail Starships : Clipper Ships of the Galaxy," *JBIS, 34*, 371-380 (1981).

12. G. L. Matloff and E. F. Mallove, The Interstellar Solar Sail : Optimization and Further Analysis," *JBIS, 36*, 201-209 (1983).

13. A. Bond and A. R. Martin, " Worldships : Assessment of Engineering Feasibility," *JBIS, 37*, 254-256 (1984).

14. G. L. Matloff, "Beyond the Thousand Ark : Further Study of Non-Nuclear Interstellar Flight," *JBIS, 36*, 483-489 (1983).

15. G. L. Matloff, "Interstellar Solar Sailing : Consideration of Real and Projected Sail Material," *JBIS, 37*, 135-141 (1984).

16. G. L. Matloff, "The Impact of Nanotechnology Upon Interstellar Solar Sailing and SETI," JBIS, JBIS, 49, 307-312 (1996).

17. G. L. Matloff, "Interstellar Solar Sails : projected Performance of Partially Transparent Sail films," IAA-97-IAA.4.1.04.

18. G. L. Matloff, "An Approximate Hetereochromatic Perforated Light-Sail Theory," IAA-95-IAA.4.1.01.

19. B. N. Cassenti, G. L. Matloff, and J. Strobl, "The Structural Response and Stability of Interstellar Solar Sails," JBIS, 49, 345-350 (1996).

20. C. R. McInnes and J. C. Brown, "Solar Sail Dynamics with an Extended Source of Radiation Pressure," IAF-89-350.

21. G. Vulpetti, "3D High-Speed Escape Heliocentric Trajectories for All-Metal-Sail, Low-Mass Spacecraft, " *Acta Astronautica, 39*, 161-170 (1996).

22. B. N. Cassenti, "Optimization of Interstellar Solar Sail Velocities," *JBIS, 50*, 475-478 (1977).

23. J. Heidmann and C. Maccone, "ASTROsail and SETIsail," Two Extrasolar Missions to the Sun's Gravitational Focus," *Acta Astronautica, 37*, 409-410 (1994).

24. G. Vulpetti, "The Aurora Project : Flight Design of a Technology Demonstration Mission," in *Missions to the Outer Solar System and Beyond, 1ˢᵗ IAA Symposium on Realistic Near-Term Space Missions*, ed. G. Genta, Levrotto & Bella, Turin, Italy (1996), pp 1-16.

25. G. Genta and E. Brusa, "The Parachute Sail with Hydrostatic Beam : A New Concept in Solar Sailing," in *Missions to the Outer Solar System and Beyond, 1ˢᵗ IAA Symposium on Realistic Near-Term Space Missions*, ed. G. Genta, Levrotto & Bella, Turin, Italy (1996), pp 61-68.

26. S. Scaglione and G. Vulpetti, "The Aurora Project : Removal of Plastic Substrate to Obtain an All-Metal Sail," *Missions to the Outer Solar System and Beyond, 1ˢᵗ IAA Symposium on Realistic Near-Term Space Missions*, ed. G. Genta, Levrotto & Bella, Turin, Italy (1996), pp 75-78.

27. L. Johnson and S. Leifer, "Propulsion Options for Interstellar exploration," AIAA-2000-3334.

28. P. C. Liewer, R. A. Mewaldt, J. A. Ayon, and R. A. Wallace, "NASA's Interstellar Probe Mission," in *Space Technology and Applications International Forum— STAIF 2000*, ed. M. S. El-Genk, AIP (2000).

29. G. Garner, B. Diedrich, and M. Leipold, "A Summary of Solar Sail Technology Developments and Proposed Demonstrations," AIAA-99-2697.

30. G. Garner and M. Leipold, "Developments and Activities in Solar Sail Propulsion," AIAA 2000-0126.

31. R. Haggerty and T. Stanaland, "Application of Holographic Film in Solar Sails, in *Space Technology and Applications International Forum— STAIF 2002*, ed. M. S. El-Genk, AIP (2002).

32. G. L. Matloff, G. Vulpetti, C Bangs, and R. Haggerty, "The Interstellar Probe (ISP) : Pre-Perihelion Trajectories and Application of Holography," NASA / CR-2002-211730.

33. W.J. Kaufmann III, R.A. Freedman, *Universe Fifth Edition*, W.H. Freedman and Company, New York (1999), pp 406-407.

34. R. Bradford, United Applied Technologies Inc., Huntsville, Alabama, Personal Communication (2002).

The Application of Tension-Based Structural Design Concepts to Ultralightweight Space Systems

Glenn W. Zeiders

The Sirius Group, Huntsville Alabama

Abstract. Large ultralightweight structures for a variety of applications including transfer optics and sails for propulsion are an important supporting technology for space power beaming. Their specific requirements are, of course, quite different, the former generally being much smaller, typically no more than a few tens of meters in size and requiring relatively precise dimensional tolerances even with active optics, whereas the latter may be as large as a kilometer but with dimensional tolerances to only perhaps tens of meters, but the basic technologies required for launch/deployment and for maintaining shape can actually be remarkably similar. This paper presents a promising structural approach that has been developed and demonstrated during the past several years under a Phase II SBIR [1] from MSFC for large lightweight space telescopes but which, as will be shown here, may be quite applicable to other space missions as well.

BACKGROUND

Tension-based structures have long been recognized as being able to provide an outstanding combination of stiffness and low weight, and those characteristics have perhaps best been widely demonstrated in the masts and stays of sailing boats. NASA MSFC awarded a set of SBIR contracts to Sirius three years ago for the development of such structures for support of very large segmented space telescopes, and it became quickly apparent that tensegrities possessed unique qualities for meeting the needs.

That concept was conceived over fifty years ago by the artist Kenneth Snelson whose soaring masterpiece "Needle Tower" (Fig. 1) stands in the Hirshhorn Museum and Sculpture Garden in Washington D.C. Snelson called his design "floating compression", but it is better known today as "tensegrity", a term that was coined by his teacher, the architect/ mathematician R. Buckminster Fuller. The name refers to the structures' integrity under tension, and the concept has recently been widely publicized through articles in popular scientific journals and through a myriad of dedicated Web sites. While artists have been most attracted by the geometrical beauty of tensegrities, engineers have

FIGURE 1 Needle Tower

CP664, *Beamed Energy Propulsion: First International Symposium on Beamed Energy Propulsion*,
edited by A. V. Pakhomov
© 2003 American Institute of Physics 0-7354-0126-8/03/$20.00

come to recognize the significant value of the dimensional stability produced by the tensile "pre-load" under any orientation, the extremely low mass afforded by the absence of bending moments in the members, and by the potential ease of storage and deployment of the flexible tension elements and the slender compression ones. Tensegrities, however, remained only a curiosity for a long time and have, historically, been largely designed for visual appeal, and it is only now that useful design tools are beginning to emerge to enable their properties to be exploited for practical engineering applications.

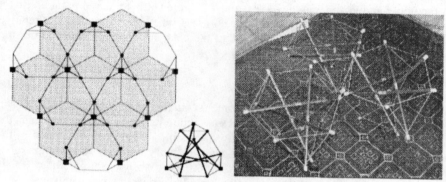

FIGURE 2 Six-Strut Tensegrity Modular Design

The first such attempt by Sirius was the six-strut "tetrahedron" (Fig. 2) whose four hexagonal faces and four triangular ones appeared well matched to the hexagonal shape that is typical of most segmented mirrors, and whose geometry permitted modular extension to very large arrays. Such a configuration is shown schematically to the left with mirror mount points designated by the black circles, and the photograph on the right shows a grouping using three individual tensegrities constructed from a commercial *Tensegritoy* kit. Deployment from a small stored volume could potentially be accomplished by using the tension elements that attach at each of the black squares, but further efforts to exploit the concept were seriously thwarted by the discovery that the hexagons were not planar and by inability at the time to properly analyze the details of the structure. It was realized, too, that the geometry of practical tensegrities with high-modulus tension elements would be defined by the initial lengths, not by the forces within them, and that the highly elastic nature of the *Tensegritoy* models could easily produce deceiving results.

There are a large number of possible tensegrity geometries, and an extensive search of the literature early in the Phase II program led to a little known journal where Furuya [2] described the three-strut octahedral tensegrity shown schematically in Fig. 3. It was readily apparent that the configuration was especially interesting for large space structures because of its capability to be merged with similar elements, its ability to be tightly stowed and later deployed, and its computational simplicity and physically understandable

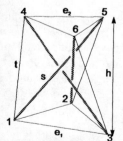

FIGURE 3 Octahedral Tensegrity

behavior. The importance of the latter aspects cannot be underestimated because they lend credence to the use of the concept and provide solid rules for realistic multi-element design, and the configuration was quickly adopted for use here.

CHARACTERISTICS OF OCTAHEDRAL TENSEGRITIES

Furuya concluded that the three-strut octahedral tensegrity of Fig. 3 is stable when the upper and lower triangles 4-5-6 and 1-2-3 are equilateral and parallel, that it is in equilibrium in the absence of external forces when the faces have a relative rotation of 30^0 about their centerline, and that the element lengths of the equilibrium structure were related by

$$s^2 = t^2 + \frac{2e_1e_2}{\sqrt{3}} \quad \text{and} \quad h^2 = t^2 - \frac{e_1^2 + e_2^2}{3} + \frac{e_1e_2}{\sqrt{3}} \; .$$

Sirius generated nodal force and moment equations to describe the behavior of the structure, but attempts to use NASTRAN finite-element analysis to model the response to external forces proved totally unsuccessful due to the combination of the prestress and the coupling between forces and geometry. The previous experience

FIGURE 4 Fundamental Two-Element Module

with the six-strut tensegrity had shown the absolute importance of detailed understanding, however, so a wood and wire model was constructed, and it quickly became apparent that the first-order reaction to external axial load or to a moment applied about the vertical axis was a height change accompanied by a proportional twist. The upper and lower faces remained essentially parallel even with nonuniform loading, and there was little visible effect of side load. It became equally clear then that the fundamental module should therefore be composed of two such octahedrons, rotated oppositely about a common triangular face as shown in Fig 4, and that adjacent modules could then be joined at the nodes of the upper and lower triangles to form a deployable structure that could be made arbitrarily large.

The empirical observations suggested that the tensegrity could best be modeled by perturbation from the basic equilibrium structure, and, after considerable effort with *Mathematica*, that indeed proved to be true, and Sirius has been able to develop linearized analytical models that enable the prediction of the behavior of the modules and of more complex assemblies to arbitrary loads and deformations. In particular, the analyses show in agreement with the earlier observation that each tensegrity element (two per module) behaves to lowest order like an adjustable coil spring, compressing $\Delta h = \frac{t}{36\sqrt{3}} \frac{e_1e_2}{h^2} \frac{F}{T}$ and internally rotating $\Delta\theta = \frac{t}{6\sqrt{3}h} \frac{F}{T}$ under an axial load F where e_1 and e_2 are the lengths of the lateral tendons, h is the element height, and t and T are the length and preload of a diagonal tendon. The lateral triangular faces of the octahedral tensegrity remain essentially parallel even with

392

nonuniform axial loading. The preload in the elements provides the resistance of the overall structure to axial load and is necessary as well to prevent the tensile elements from slackening. Equations from the linearized model predict the preload necessary to avoid such slackening as a function of known external loads. Resistance to lateral forces and moments is considerably greater and is produced by small strains in the elements, rather than by gross deformations as with axial loading, so it is the elasticity of the tendons -- not their preload -- that primarily determines the bending rigidity for more complex structures. Stable tensegrities are geometric in nature, and low-deformation structural elements with very large spring constants, sufficiently prestressed to avoid slackening, are essential for maintaining the dimensional integrity required by space optics.

The modeling of complex structures is facilitated if the problem can be cast into terms of elementary structures like thin plates and slender beams, and such forms have indeed been inferred from the linearized theory for octahedral tensegrities. Classical elastic theory predicts that the curvature of a beam at a plane with moment M is given by $1 / R = \nabla^2 w = M / D_b$, where the rigidity D_b is the product of the elastic modulus E and the moment of inertia, while the deflection w of a thin plate with distributed pressure load q is described by $\nabla^4 w = q / D_p$. The analysis of this work shows that the equivalent rigidity D_b for a typical tensegrity boom or tower with tendon elasticity E and cross-sectional area A is approximately $0.26e_1^2 EA$, while that for a plate is about $D_p = 3.6e_1 EA$. Note that the bending resistance of tensegrity structures does not have the characteristic thickness-cubed dependence of conventional ones, and that the rigidity is controlled primarily by the strength of the tendons. This is important because the overall weight of the structure tends to be dominated by that of the struts and the nodes, so a tensegrity structure can be made quite strong without a significant impact on the weight. The consequences of this will be explored later in the paper.

Although the elastic strength EA of the tendons is the defining quantity for the bending rigidity of tensegrity structures, it is the struts that most control the overall design, and they, because of the absence of bending moments in a properly designed tensegrity, are in turn defined by either column strength or by fabrication limitations. Buckling can occur if the axial load The compressive members (struts) of a properly-designed tensegrity are free of bending moments, but they could be subject to buckling if the axial load P were allowed to exceed the ability of the elastic bending curve to support it. The ultimate (not allowable) critical load for a column free to rotate at both ends is $P = \pi^2 EI/s^2$ where the moment of inertia $I = \pi HD^3/8$ for a thin-walled cylinder with diameter D and wall thickness H. Therefore, the maximum allowable slenderness ratio s/D is $\pi/\sqrt{(8\varepsilon)}$ where $\varepsilon = \sigma/E$ is the elastic strain in the column, and this is only 35.1 for typical values of $E = 30\,10^6$ psi and $\sigma = 30\,10^3$ psi. Note that the thin-walled cylinder is well known to be the most efficient shape for a column, and that a change of diameter, not tube thickness, is the most mass-efficient approach to preventing buckling.

DEPLOYMENT

Tensegrities are stable structures whose geometry is maintained by the lengths of the elements, and effectively shortening at least one strut or lengthening at least one tendon can collapse a single three-strut octahedron. However, the loss of symmetry would be undesirable for a multi-module configuration, and it would be best to adjust all three of a given class of element. The required length change would be much smaller, the packed system would be smaller, and no limitation would be imposed on the length of the elements, but the advantages would be obtained at the cost of two additional "mechanisms".

The first serious attempt at constructing a collapsing all-metal tensegrity is shown in Fig. 5 in various stages of its deployment. Toggled struts with three equal sections were joined by two pinned joints so that they formed one coaxial unit when extended. They took on the shape of a "Z" when partially collapsed, and lay closely packed and parallel to each other when fully collapsed. When the entire tensegrity is collapsed, all cables are still attached so there is only one way the unit can be redeployed. Unfortunately, the components were free to assume whatever position they chose and this, together with the tendency for tendons to occasionally loop around struts, required careful prepositioning to prevent tangling upon deployment.

FIGURE 5 Toggled Unit

The toggled approach was not suitable for autonomous deployment, so a pneumatic piston design was adopted with two-piece struts having inner and outer tubes. The gas pressure required to extend the struts and to provide the required pretension of the elements can be supplied by a small air compressor or with cylinders for terrestrial applications, or a solid-propellant gas generator can be used in space. No convenient means to automate strut collapse has been found, but it can easily be done manually where that is acceptable. The major issue is to provide a positive locking mechanism that will hold a deployed strut at the precise length required by the equilibrium structure, and the most attractive concept to date for doing so appears to be that shown in Fig. 6. The concept was used to produce the fully deployable tensegrity module of Fig. 4 with commercial aluminum tubes for the struts. The inner tubes have a piston and O-ring on the interior end, and each inner and outer tube pair essentially forms an air cylinder that can be extended pneumatically. About 42 lbs of extension force is available by using compressed nitrogen regulated to 300 psig. A needle valve controls deployment rate by producing a large pressure drop as the tubes fill, but produces full pressure in the tubes when needed for the final stage of deployment and tensioning of the elements. Tension forces of interest are so large in fact, that it has been found quite difficult to manually deploy the tensegrity, whereas the gas pressure does so smoothly and reliably. A groove in the cylinder piston and two cuts on opposite sides of the external

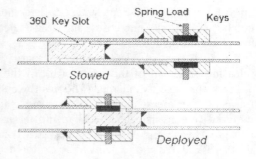

FIGURE 6 Pneumatic Strut Lock

aluminum tube have been machined to allow a stainless steel hairpin clip to latch into the piston independent of the relative rotation as the inner telescoping tube reaches full extension. The hairpin clip can be manually removed to collapse the telescoping strut back to the stowed position. The struts are shortened and lay parallel to one another when the unit is collapsed, the diagonal tendons remain straight and lay more or less parallel to the struts, but the lateral tendons tend to collapse to loops as their end points are brought together. The structure will fail to properly deploy if any of the loops encircles the end of a strut or catches on a hairpin clip, and that requires that the tendons be carefully arranged and secured before deployment. Experience has shown that the module consistently deploys from both vertical and horizontal starting positions when that is done. However, the locking mechanism represents a weak point of the present design, and care must be exercised, especially with manual deployment, to avoid bending -- and breaking -- the external tubes at the slots for the hairpin clips.

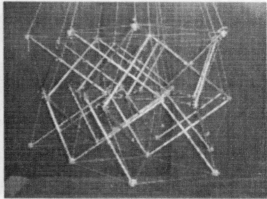

FIGURE 7   Seven-Module Deployable Tensegrity Assembly

Seven of the modules have been assembled side-by-side to produce a very impressive half-scale demonstrator of the base structure of an ultra lightweight space telescope for delivery to NASA MSFC under the contract. The assembly is shown in both stowed and deployed states in Fig. 7 prior to mirror attachment. As before, the structure consistently deploys from both vertical and horizontal starting positions when care is taken to arrange the tendons in the stowed state, but the plethora of wires does, however, cause deployment to become increasingly difficult to control as the number of modules increases, and quick-release Velcro straps are being used until a better solution is found.

Collapse of side-by-side tensegrity assemblies is best accomplished by releasing either all of the struts or all of the diagonal tendons as done above, but it appears that deployable masts/booms/towers may more efficiently be managed by releasing the lateral tendons in alternate planes, thereby minimizing the number of required "mechanisms" and collapsing the axial assemblies of modules into stacked planes. This should easily produce 100:1 compaction, but it was beyond the scope of the NASA SBIR, and remains to be explored.

The strut-and-tendon construction of tensegrities adapts them well to efficient storage and deployment, but, as with any other such scheme, that capability requires that concessions be made, and the costs of automated operation in particular should be carefully weighed against the use of human intervention. A suitable automatic system has not yet been identified for shortening (and pre-stressing) the tendons, and it is unlikely that one will be found for high-strength ground-based applications, partly because of the tendon bending stiffness associated with the required prestress. This may be less of a disadvantage than it would first seem, however, because stiff diagonal tendons for each tensegrity module could easily be swung into place, snapped into simple ball joints, tensioned to a preset level with toggle latches, and then assembled with adjoining modules using latches at the upper and lower nodes. This might even be done with a space-based structure because each tensegrity module could be assembled with its attachments (mirrors, antennas, etc.) inside the Space Station, and astronauts or robotic assemblers could then connect the modules outside.

APPLICATIONS

The original goal of the program was to develop ultralightweight primary mirror support structures for 20 meter and larger active space telescopes. Such systems use a secondary mirror with many small control zones to handle disturbances induced my maneuvers, debris encounters, etc., and the primary mirror profile should usually have an RMS error of no more than about 50 microns for such correction to be effective. The RMS deflection of a uniformly-loaded circular plate that is optimally supported at its three Abbe points is $6.35 \cdot 10^{-4} q A_p^2 / D_p$, so D_p must be 12.3 MJ for a 20 meter mirror with a mass loading of 10 kg/m^2 at 1g, and the requirement can be met by a tensegrity with $e_1 = 1.67$ m with 4 mm steel wires. These "wires" are quite thick (0.16"), but the loading is also rather extreme (Earth surface operation with NGST-level technology!), so it appears that a tensegrity structure can definitely meet the requirements of an advanced space telescope.

Fig. 8 shows the final telescope demonstrator built for the SBIR program using silicon wafers for dummy segmented mirrors and thermally formed plastic panels on magnetic mounts to support them. The seven-module assembly can be stowed in a cylindrical package at less than 5% of the volume of the deployed system, and it can then be deployed either manually or pneumatically to form a stable structure. The 2.5 m^2 demonstrator has a total mass of 8 kg for an overall areal mass density of about 6

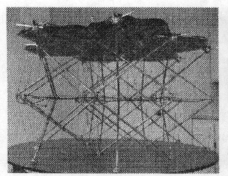

FIGURE 8 Seven-Module Space Telescope Demonstrator

kg/m², but it appears certain that the areal density could be reduced significantly, probably to about 1 kg/m², for a full-size operational system. Table I gives the mass breakdown on a per module basis, and it can be seen that the system is dominated by the tensegrity structure, followed by the dummy silicon mirrors. The tensegrity demonstrator suffers from several factors including the use of inexpensive commercial components instead of more exotic material, from the penalties associated

Plastic panel	67
Mirrors (7)	197
Magnets/Posts/Snaps	5
Wiffletree	47
Tensegrity	<u>750</u>
Module total (grams)	1066
Areal density (kg/m²)	6.1

Table I Module Mass Breakdown

with minimum available sizes (compounded by the half-scale nature of the demonstrator), and from the additional mass required by the pneumatic actuation scheme. The latter is a particularly important point because all automatic deployment schemes suffer significant mass penalties compared with manually-erected ones, and the overall cost of deploying a structure, whether autonomous or manual, must be carefully considered in an optimized system design. The commercial silicon disks used to represent the mirrors have a thickness of only 27 mils, but that is far thicker than needed (especially in space), and conventional lightweighting techniques would certainly be used should cost be less of an issue.

Another space application of considerable interest, a solar or laser sail, normally brings to mind the use of boom or mast supports, but the potentially low mass of tensegrities may make them interesting for use again as modules of a truss, especially if stress concentrations in the sails should prove to be an issue. Consider as an example a 100 km² circular sail with a total propulsive force of 450 newtons (solar), and, to provide consistency with the previous analysis (though not necessarily realism), let the sail be restrained again at the three Abbe points and let the RMS deflection be 100 meters. The combination of low loading and large allowable deflection -- even with the greatly increased area -- means that the required rigidity is then only 2.3% of that before. The modules are not therefore constrained by strength, but they are by mass because even a fabrication-limited one meter strut using an advanced material like BP Amoco's P120S would have a mass of almost 3 grams, and this would certainly exceed the NASA sail system long-term goal of 1 gm/m² if six struts and the accompanying wires and nodes populated an average of more than a small fraction of a square meter. However, the octahedral tensegrity modules are themselves each stable, so it should be possible to create a sparse truss at 10 gm/m^2 or better using an appropriate set of linked clusters of modules like those shown in Fig. 9.

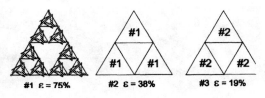

Figure 9 Sparse Tensegrity Clustering

Masts, booms, or towers can be constructed with equal ease with octahedral tensegrities by stacking them end-to-end in an axial configuration. This can be a very attractive way of making a weight-supporting tower, but the concentrated bending loads like those produced by a large solar sail do not produce a favorable comparison with techniques like the sparse truss described above. A square sail of the same area, for example, would require two 14 km booms, and if the sail were attached to the booms only at the ends and the payload were located at the center, the required rigidity for a peak deflection of 100 m would be $D_b = WL^3/48\delta = 1.29 \cdot 10^{11}$ J-m. This rather large value could be met with $e_1{}^2A = 0.6$ m^4 using a high-strength material such as BP Amoco P120S having an elastic modulus of 120 Mpsi, but that would correspond to e_1 = 87 m with 1 cm wires, and the resulting structure would be far too massive to be of interest. However, should the sail instead be 1 km^2 in size and if the allowable deflection were reduced to 10 m, the required rigidity would decrease by four orders of magnitude, and one could use a tensegrity with 3 mm wires and with $e_1 = 2.91$ m. There would be about 200 dual-element modules per boom at about 1 kg each, so the booms would have a total mass of about 400 kg or 0.4 gm/m^2 – certainly meeting the NASA mass goals and of a size easily permitting transport in the Shuttle bay.

Tensegrities clearly are an extremely attractive alternative for a variety of support structures for space applications, and they may also offer exciting commercial possibilities for ground-based structures such as easily-erectable domes and cell phone towers.

ACKNOWLEDGMENTS

This work was largely supported under NASA Marshall Space Flight Center Contract NAS8-00007 under the direction of Edward E. Montgomery IV.   Mr. Richard Cleve of Elk River Engineering was responsible for engineering development of the deployable tensegrity structure and provided custom fabrication and assembly, while Mr. Larry Bradford of CAT Flight Services was responsible for development of the plastic support panels of the telescope demonstrator.

REFERENCES

1. G.Zeiders, "Tension-Based Support Structures for Very Large Space Telescopes", The Sirius Group, Final Report for SBIR Phase II Contract No. NAS8-00007, April 2002
2. H. Furuya, "Concept of Deployable Structures in Space Application", *Int. J. of Space Structures* 7, 2 (1992)

Large Space Telescopes Using Fresnel Lens For Power Beaming, Astronomy and Sail Missions

James T. Early

Lawrence Livermore National Laboratory, Livermore California 94550

Abstract. The concept of using Fresnel optics as part of power beaming, astronomy or sail systems has been suggested by several authors. The primary issues for large Fresnel optics are the difficulties in fabricating these structures and deploying them in space and for astronomy missions the extremely narrow frequency range of these optics. In proposals where the telescope is used to transmit narrow frequency laser power, the narrow bandwidth has not been an issue. In applications where the optic is to be used as part of a telescope, only around 10^{-5} to 10^{-6} of the optical energy in a narrow frequency band can be focused into an image. The limited frequency response of a Fresnel optic is addressed by the use of a corrective optic that will broaden the frequency response of the telescope by three or four orders of magnitude. This broadening will dramatically increase the optical power capabilities of the system and will allow some spectroscopy studies over a limited range. Both the fabrication of Fresnel optics as large as five meters and the use of corrector optics for telescopes have been demonstrated at LLNL. For solar and laser sail missions the use of Fresnel amplitude zone plates made of very thin sail material is also discussed.

INTRODUCTION

The concept of using Fresnel optics as part of power beaming, astronomy or sail systems has been suggested by several authors[1,2]. The primary issues for large Fresnel optics are the difficulties in fabricating these structures and deploying them in space and for astronomy missions the extremely narrow frequency range of these optics. In proposals where the telescope is used to transmit narrow frequency laser power[1], the narrow bandwidth has not been an issue. In applications where the optic is to be used as part of a telescope, only around 10^{-5} to 10^{-6} of the optical energy in a narrow frequency band can be focused into an image.

This article addresses the fabrication of very large space optical systems with Fresnel lenses. The limited frequency response of a Fresnel optic is addressed by the use of a corrective optic that will broaden the frequency response of the telescope by three or four orders of magnitude. This broadening will dramatically increase the optical power capabilities of the system and will allow some spectroscopy studies over a limited range. Both the fabrication of Fresnel optics as large as five meters and the use of corrector optics for telescopes have been demonstrated at LLNL[3, 4]. For solar and laser sail missions the use of Fresnel amplitude zone plates made of very thin sail material is also discussed.

TRANSMISSIVE FRESNEL OPTICS

A key feature of this concept will be the use of a thin transmissive primary optic in the telescope design. The problem with membrane mirrors, of course, is that it is hard

CP664, *Beamed Energy Propulsion: First International Symposium on Beamed Energy Propulsion*,
edited by A. V. Pakhomov
© 2003 American Institute of Physics 0-7354-0126-8/03/$20.00

to control their shape to the $\lambda/20$ scale precision needed to attain $\lambda/10$ optical tolerances. The source of this difficultly is that mirrors reflect light, thereby doubling the effect of surface errors. In contrast, path-length errors caused by surface distortion of a thin, uniform thickness transmissive film are canceled (not reinforced) as light arrives at and then departs from the surface. In actual lenses, this path-length cancellation is not perfect because the incoming and outgoing rays are not traveling in quite the same directions. If the light is bent through angle θ, then the path-length error in a mirror is amplified by a factor $(1 + \cos\theta)$, whereas that in a thin lens is reduced by $(1-\cos\theta)$. Thus lenses have a $(1+\cos\theta)/(1-\cos\theta)$ advantage over mirrors. By making the lens weak, i.e., by keeping θ small, this tolerance gain becomes huge. In a very slow lens (f/100) visible-light tolerances increase up to nearly a centimeter, a tremendous practical advantage when trying to field a thin optical-quality membrane optic in space.

LLNL has also addressed the issue of fabrication of large meter scale optics utilizing techniques that can be scaled to much larger sizes. At LLNL we have facilities that were built for fabrication of meter scale diffractive optics that are needed for the new laser fusion laboratory currently under construction. We now have facilities, techniques and experience personnel needed to produce very large (up to one meter in a single piece) diffractive optics in large numbers for use as diffraction gratings, Fresnel phase plate lenses or optical control elements. We have also addressed issues of joining meter sized optics together to fabricate much large optical structures. This basic capability can now be used to address the issues of design and fabrication of Fresnel phase plate optics in glass or plastic and amplitude zone plates in sail materials.

DIFFRACTIVE OPTICS THEORY

The classic circular Fresnel pattern for an amplitude zone plate is illustrated in Fig. 1. For a phase plate the glass surface local slope results in a 2π phase shift in each zone with a 2π step at the zone edge. Just as with standard optics a point focus can also be formed with orthogonal cylindrical lenses (Figures 2).

Figure 1 – Circular Zone Plate   **Figure 2** – Linear zone plates:

                                   a) line focus   b) point focus with orthogonal lines

The total number of zones, N, are given by:

$$N = D^2/8\lambda F = D/8\lambda f_{number} \tag{1}$$

where λ, D and F are the wavelength, optic diameter and optic focal length respectively. The width of the outer zone (the smallest zone) is $2\lambda f_{number}$. When the f_{number} is very large, then these features become easy to fabricate using optical lithography technology.

Even with a normal lens the light distribution in the focal plane depends on the wavelength. For example the diameter of the central Airey spot, d, is proportional to the

$$d = (2.4F/D)\lambda \tag{2}$$

wavelength. In a broadband image the short wavelength light has a higher resolution. However all the wavelengths have the same focal plane for zero dispersion optics.

For a given Fresnel optic the focal length will also depend on the wavelength and

$$F = (D^2/8N)(1/\lambda) \tag{3}$$

will severely limit the useful bandwidth of a telescope. For good images:

$$\Delta\lambda/\lambda < 1/8N = f_{number}\ \lambda/D \tag{4}$$

If the f_{number} of the telescope is 100 and the diameter is 100m, then for 1μm light $\Delta\lambda/\lambda = 10^{-6}$. Unless the telescope features extraordinarily large f-numbers, the zone plate primary optics will be inherently micro-bandwidth optical elements. The bandwidth may be acceptable for laser communications or power beaming, but it is a major limitation for an imaging system. For an imaging system we must achieve higher optical bandwidth by the use of chromatic correction optics.

In the next section we discuss how to correct for this chromatic aberration associated with Fresnel optics. The intensity distribution in the focal plane will also be different for different types of Fresnel lens, i.e. phase plates, circular zone plates and orthogonal zone plates. We will not attempt to change these amplitude distributions in the focal plane nor the wavelength pattern variation found in standard optics. It is essential to note that the chromatic corrections will address only the variation in focal length for the Fresnel lens.

CORRECTOR PLATE FOR WAVELENGTH ABERRATIONS

To achieve high-precision chromatic correction for the zone plate, we try to cancel its chromatic aberrations with correcting optics in the optical train of the telescope. A technique was invented 100 years ago by Schupmann[5] who showed that any chromatic dispersion introduced at one optical element could be canceled by placing a second, inverse-power element with the same dispersion at an image site of the original element. The corrector optic's effectiveness will depend on its size, but increasing the bandwidth of the telescope by four orders of magnitude should be possible.

Physically, the reason Schupmann correction works for a diffractive telescope is clear (figure 3); light leaves each point of the Magnifying Glass's diffractive lens in an angular spray, each color being sent into a different direction. As the light from this site travels towards the Eyepiece it spreads apart, diverging both spectrally and physically; both effects must be corrected. The physical reassembly is achieved first, by making the light pass through a reimaging telescope as it enters the Eyepiece. This internal telescope focuses the surface of the Magnifying Glass onto that of the Fresnel Corrector, thereby physically recombining rays which left each site on the first diffractive lens to a matching site on the second one. Now each site on the Fresnel Corrector sees an incoming angular/color spray corresponding to that from the departure site on the Magnifying Glass; by employing an inverse (defocusing) diffractive profile, it can remove this angular/color spray. As a result, each ray bundle is now both physically and spectrally recombined; the set of bundles can then be brought to a common achromatic focus.

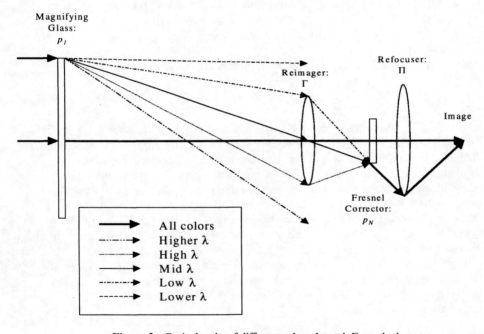

Figure 3 – Optical paths of different colors through Fresnel telescope

DEMONSTRATIONS AT LLNL

At LLNL we have fabricated Fresnel phase plates at diameters of 20 and 50cm. Using a corrector plate fabricated by JPL, we have demonstrated that the chromatic aberrations can be corrected in telescope configurations and that diffraction limited images can be formed with broadband ($\Delta\lambda/\lambda \sim 0.1$) light (figure 4). Both phase plates and amplitude plates have the same wavelength dependence in their chromatic aberrations, and broadband performance has been demonstrated with both types of optics.

To scale this technology to very large optics meter sized optical sections are made with current techniques joined to form a large optic. The joints must be designed to function in the space environment and to maintain the required optical position tolerances. Figure 5 shows a 5m Fresnel lens recently fabricated at LLNL. Most of the structure seen in this figure is bracing required to protect the lens from wind loads in this open air demonstration.

SPACE DEPLOYMENT

For a f_{number} 100 optic the outer (thinnest) ring is around 100μm for a visible optic. Fabricating the rings to a fraction of this dimension with standard optical lithography has been demonstrated with the LLNL equipment. Holding this in-plane precision in a deployed optic will depend on the material properties and the thermal control of the optic. If the optic is held at uniform temperature, then the only impact of a change in temperature is an easily corrected change in focus[4].

The optic can be held flat by spinning (the preferred method) or by applying radial tension at the circumference with external structures. For transmissive optics the out of plane optical tolerances are large and easily achieved. The stresses must however be distributed without causing local in-plane distortions in the optic. In the case of amplitude zone plates for sails, if the solar sail consists of metal deposits on plastic, then the difference in thermal expansion coefficients cannot be allowed to distort the lens. If the zone plate consists of holes in a uniform reflective or absorptive material, then the issue is the stress patterns created by the openings.

Thin glass optics are flexible enough to be rolled into configurations that can fit into the payload volumes of current launchers. Circular primary optics of 9m can fit in the Delta III fairing and of 20m can fit in the Delta IV. For a larger primary the optics must be folded in origami patterns (figure 5) or fan folded and rolled.

POWER BEAMING APPLICATIONS

Large low-mass Fresnel optics can enable power beaming over very long distances. Power beaming to and from geostationary orbits would be possible. Once long distance beaming is possible, then it no longer makes sense to locate the required power supplies and lasers in orbit. Placing these assets on the surface of the Earth will allow much simpler technology, practical maintenance and much lower costs. The laser power can be beamed up to a geostationary station using an Earth-based telescope with a large conventional primary optic and smaller adaptive optics in the optical train.

A satellite in geostationary orbit can capture the laser beam and redirect it to targets of interest using low-mass Fresnel phase plates. The Fresnel lenses will have transmissive efficiencies over 98%. Table 1 gives the typical ranges for several missions and the required lens size. The transmitting lens and the delivered spot size

Figure 4- Broad-band Diffractive Telescopes

- Two telescopes
  - 20 cm wide, 20 m long
  - 50 cm wide, 50 m long

- Each performed properly
  - Diffraction-limited focus
  - For broad-band (470 - 700 nm) light

Optics: 50 cm

Focal spots: 50 cm

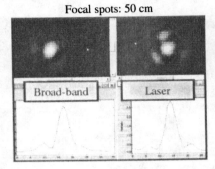

Lunar image: 20 cm

Fig. 5 -   5 m Fresnel lens made with 700μm glass

are assumed to be the same, but one can trade one against the other with their product being conserved.

Table 1 – Power beaming missions and required optic sizes

Mission	Range (10³km)	Visible light spot / optic diameter (m)
geo to leo	40	6
geo to geo	80	9
geo to moon	400	20
geo to L1 / L2	1500	40

NEARER TERM APPLICATIONS

Before large Fresnel optics can be used for power beaming, the technology probably needs to be developed and demonstrated on other nearer term applications such as astronomy or spacecraft optical communications. A 20m telescope based on Fresnel optics could be packaged in a Delta IV and deployed in geostationary or higher orbits. Such a telescope would provide eight times the resolution and sixty-four times the light gathering capacity of the Hubble telescope. This telescope could observe weather details on Mars with 10km resolution and on Saturn with 100km resolution. It could provide 1AU resolution of planetary accretion disks or planetary nebula out to 600 light years. If the light scattering properties are acceptable, then this telescope design could also be used for extra-solar planet finding missions.

For spacecraft communications the low mass and easy tolerances of thin membrane Fresnel optics makes optical communication with large transmitting apertures possible. For spacecraft communications:

$$\text{Data rate / spacecraft transmitter power} \sim d^2_{transmitter} D^2_{receiver} / \lambda$$

Table 2 compares potential laser and radio communication systems and show that 1μm lasers may have four or five orders of magnitude advantage in terms of higher data rates or lower power requirements.

Table 2 - Comparison of laser and radio communication systems

	λ(m)	$d_{trans.}$ (m)	$D_{rec.}$(m)	d^2D^2/λ
radio	0.01	1	10	10^4
laser	10^{-6}	10	3	10^9

SOLAR SAIL IMPACT: ENABLING FOR FLY-BY MISSIONS

The low payload capacity of solar sail spacecraft makes the inclusion of very large aperture imaging impossible unless the optics are extremely light. If the solar sail itself forms the primary optic, then very large aperture imaging systems may be possible. The solar sail must be reflective (or absorbing for carbon sails) and very thin. If the sail is used as a reflective optic, then the λ/10 optical precision requirement in a membrane structure will be extraordinarily difficult or impossible. The use of the sail as a transmissive optic greatly relieves the optical precision requirements, but one cannot make a transmissive Fresnel phase plate from this opaque material. The use of an amplitude zone plate optical design allows the use of reflective or absorbing sail material while still gaining the advantages of a transmissive membrane design.

For solar sails to be used in actual near and/or medium term missions, they must address requirements that are not easily satisfied with conventional rocket propulsion systems. This constrains potential sail mission opportunities to high Δv missions where conventional rocket systems are very expensive and have small payloads. For inner solar system missions a solar sail can be used for rendezvous or orbiting missions, but there are relatively few inner system mission opportunities. For outer solar system missions the solar sail becomes less effective at rendezvous or orbiting because of the low solar flux at the target.

There are numerous targets for high velocity fly-by missions in the outer solar system that potentially could be served with solar sail propulsion. The basic problem is that fly-by missions in the outer solar system have low scientific data returns due to the conditions at encounter. Any imaging systems will have very little time to take data, and the data is of low resolution. First the encounter relative velocity must be high. To reach outer system targets in reasonable times the outbound leg of the trajectory is highly elliptical. The velocity of the spacecraft is typically over 10-20km/sec (2-4 AU/y) relative to any target in a circular orbit[6]. Secondly with the exception of missions to Neptune or Uranus, the targets are typically small. Pluto and some satellites of the outer planets and potentially a few deep objects are of the order of 1000km in diameter. Most asteroids, comets, Kuiper belt objects and Oort cloud objects will be under 100km. To get any meaningful understanding of these objects we will want resolutions better than 1km. For deep space missions the problem is further exacerbated by the very low light levels that will push imaging systems to longer exposure times.

With the light payload capacities of deep space missions the typical imaging system will have an aperture of 10cm or less[6]. Table 1 gives the resolution of such a system for visible light as a function of range to target. Also listed is the time the target resolution exceeds a given value assuming the closing velocity is only 10km/sec. It can be seen that there is only a few hours of moderate resolution data and very little data with resolutions better than one kilometer. In this table it is assumed that the spacecraft has maneuvered close to the target and that the separation distance is primarily along the flight path. If the camera exposure times must be long due to low illumination levels, then the high resolution data could be quite limited.

If a large fraction of the solar sail is converted into a primary optic, then the imaging system capabilities can be dramatically different. If a 100m optic is part of a 200m diameter solar sail, then the imaging times and resolutions are given in Table 3. The long observation times and high resolution would enable significant imaging science. High resolution imagery of both hemispheres would be possible for even targets with low rotation rates. Dynamic events such as orbiting satellites could also be observed.

Table 3 - Imaging system performance assuming close fly-by at 10 km/sec

	10 cm optics			100 m optics			
Resolution	10.km	1.0km	100m	1.0km	100m	10m	1m
Distance to target (10^6km)	.7	.07	.007	70.	7.	.7	.07
Time to target	1 day	2 hr.	11 min.	3 mon.	1 wk.	1 day	2 hr.

For missions to the asteroid belt, Kuiper belt or Oort cloud the large solar sail optic would enable fly-by missions without the spacecraft being required to maneuver to the target. For these fields of small distributed targets the spacecraft would simply fly on a straight trajectory and targets would be imaged as they passed into range. The factor of 1000 increase in optic diameter would give a factor of 1000 increase in the range at which targets could be resolved. For example if a 100m optic was flown on a straight path through the Kuiper belt at 20 km/sec, in ten years based on current estimates of the belt population[6], it would be able to resolve hundreds of objects including approximately 100 objects with a 1000 line resolution across the image of the object.

One objective of the Gossamer Spacecraft Initiative for solar sails is to act as a precursor mission for laser driven lightsail for interstellar missions. While there are some speculative investigations on how to slow down a lightsail at the target star system, the basic concept of the lightsail is for a fly-by mission at speeds of 0.1c. The fundamental flaw in this mission concept is the high fly-by velocity's impact on data collection. While the target planets may be 10,000 km in diameter, Table 4 shows the limited observation times available at 0.1c. Laser lightsail sizes are quite large[7], so a one kilometer zone plate optic is assumed. The launch telescope for an interstellar lightsail must be around 100km. The use of this telescope for direct observation from the Solar System would provide resolutions near 1000km in the target system which are as good as the small optic fly-by. Very large optics are required to make the fly-by mission useful. If the optics investigated by this study are shown to be feasible, then the argument of using solar sails as precursors for interstellar or outer Solar System missions will be valid.

Table 4 - **Imaging system performance assuming fly-by at 0.1c**

	10 cm optics		1 km optics				
Resolution (km)	1000	100	1000	100	10	1	0.1
Distance to target(10^8km)	.7	.07	7000	700	70	7	0.7
Time to target	30 min.	3 min.	7 mon.	3 wk.	2day	5 hr	30 min

REFERENCES:

1. R.L. Forward, "Roundtrip interstellar travel using laser-pushed lightsails," J. Spacecraft & Rockets **21**, 187-195 (1984).
2. Y.M. Chesnokov and A.S. Vasileisky, "Space-based very high resolution telescope based on amplitude zoned plate," presented at the International Conference on Space Optics, Toulouse Labege, France, 2-4 December 1997.
3. R. A. Hyde, "Eyeglass: Very large aperture diffractive telescopes," Applied Optics **38**, No.19, 4198-4212, (1 July 1999).
4. R. A. Hyde, "Eyeglass: A Large Aperture Space Telescope," internal LLNL final report on LDRD Exploratory Research Project: 97-ERD-060
5. L Schupmann, *Die Medial Fernrohre: Eine neue Konstruktion for grosse astronomisch Instrumente* (B.G. Teubner, Leipzig, 1899)
6. P.K. Henry, R.J. Terrile, R.W. Maddock, and H.R. Sobel, "Exploring the Kuiper belt: an extended Pluto mission," 2nd IAA Symposium on Realistic Near Term Advanced Scientific Space Missions: Missions to the Outer Solar System and Beyond, Aosta, Italy, June 29 – July 1, 1998.
7. G.A. Landis, "Beamed Energy Propulsion for Practicle Interstellar Flight," J. British Interplanetary Society **52**, 420-423, (1999).
titute of Physics, 1998, pp. 651-654.

HYBRID SYSTEMS
AND NEW CONCEPTS

Fundamental Study of a Relativistic Laser-Accelerated Plasma Thruster

Hideyuki Horisawa[*], Hideaki Kuramoto[*], Keishi Oyaizu[*],
Naoki Uchida[*], and Itsuro Kimura[#]

[*]Department of Aeronautics and Astronautics, Tokai University, 1117 Kitakaname, Hiratsuka, Kanagawa, 259-1292 JAPAN
[#]Professor emeritus, University of Tokyo, 7-3-1 Hongo, Bunkyo-ku, Tokyo, 113-8856 JAPAN

Abstract. Potential concepts of laser-accelerators which can be applied to the spacecraft propulsion using a conventional technique were reviewed. Maximum ion energy of order tens of MeV through the acceleration was reported in recent studies. As for a proton beam accelerated up to 58 MeV, which was achieved in a recent study, its speed corresponds to 33 % of the speed of light and the specific impulse of order 10^7 sec. Also, a feasibility study of the utilization of these accelerators, based on the special theory of relativity, for space propulsion was conducted. It was shown that significantly high specific impulse can be obtained through this propulsion system for primary propulsion and attitude control applications. In addition, a preliminary experiment was conducted with an ultrashort-pulse laser system. Induced impulse in the forward acceleration mode was observed using a 12-μm thick Al foil as a target.

INTRODUCTION

The utilization of the extremely high gradients associated with plasma waves for the next generation of high energy linear particle accelerators was suggested more than a decade ago, and since then there have been fruitful discussion and theoretical investigations.[1-2] On the other hand, the peak power of artificial light sources has increased exponentially, from a kilowatt to a petawatt, over the past century, as shown in Fig.1.[1-3] This corresponds to an increase of more than a factor of ten each decade. Especially in recent years, there has been a tremendous progress in experimental accomplishment mainly due to the advances in the development of compact terawatt laser systems capable of exciting plasma waves with gradients as high as several tens of GV/m, which is approximately three orders of magnitude greater than that generated by the conventional radio frequency linear accelerators (RF linacs) that have an intrinsic limitation due to electrical breakdown on material surface.[1-2] Since the plasma is already ionized, there is no limitation due to electrical breakdown on the material surface. Therefore the utilization of laser-plasma interactions for particle accelerators can generate extremely high gradients. Moreover, interactions between the ultra-high field of the laser pulse and matter bring new

CP664, *Beamed Energy Propulsion: First International Symposium on Beamed Energy Propulsion*,
edited by A. V. Pakhomov

applications of plasma physics, and several techniques are under significant development.[1]

In 1990s, relativistic electrons accelerated close to the speed of light have become available in the distance of just the laser wavelength through laser-plasma interactions.[1-3] Later, fast ions have also become obtainable through laser-solid interactions in a short distance, [4-18] and the emission of relativistic ions will soon become possible.[14-18] In the case of propulsion based on the relativistic beams, extremely high values of the specific impulse can be expected through the relativistic effects.[19-22] In case of operations of this type of thrusters on earth and/or solar orbit, in which solar power is available, the merit of high specific impulse will bring on a significant advantage. The use of this type of propulsion system may also bring some solutions to the problem of the inherent penalty of extremely large mass of propellant required in interstellar flight missions.

In this study, potential concepts of the relativistic accelerator which can be applied to the spacecraft propulsion using a conventional technique are reviewed. In addition, fundamental analyses of thrust performance of relativistic plasma thruster are conducted. Also, a preliminary experiment was conducted with an ultrashort-pulse laser system.

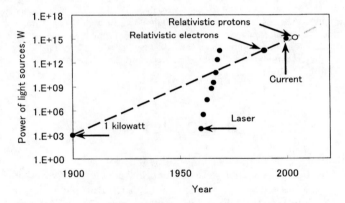

FIGURE 1. History of the peak power of artificial light sources.

Laser-Accelerated Plasma Thruster

Plasma-based accelerators[1-2] are of great interest because of their ability to sustain extremely large acceleration gradients, ~ 100 GV/m. In order to accelerate particles through the electric field of the plasma wave, following two conditions should be satisfied[1-2] ; (a) the wave proceeds longitudinally to the particle motion, and (b) the phase speed of the wave and the speed of the particle are the same, or their motions are in phase. One of the schemes to realize these conditions is to excite the plasma waves. The plasma based acceleration schemes which have received the most attention are the Plasma Beat Wave Accelerator (PBWA), the Plasma Wakefield

412

Accelerator (PWFA), and the Laser Wakefield Accelerator (LWFA), Self-Modulated LWFA (SM-LWFA), and $V_p \times B$ Accelerator (Surfatron Accelerator).[1-2]

The interaction of ultraintense laser pulses with solid targets leads to generation of fast particles, from x- and γ-ray photons to high energy ions, electrons, and positrons.[1-18] In particular, an interest has developed in ion acceleration by compact high-intensity femtosecond lasers with potential applications for the initiation of nuclear reactions on a tabletop.[4-18,23-25] Experiments now being carried out involve high energy ions generated in the interaction of laser pulses with solid targets and gas jets. In this case, strong electrostatic fields can be generated through charge separation. The efficiency of the laser energy conversion into a high energy electron component dominates this phenomena. Thermal expansion of the laser-driven plasma and ponderomotive electron expulsion constitute the most well-known examples of the electrostatic-fields production.[11]

As for the plasma, the ponderomotive force for unit volume F_p is,

$$F_p = -\frac{1}{2}\frac{\omega_p^2}{\omega_0^2}\nabla\frac{\langle E_0^2 \rangle}{4\pi}$$

$$= -\frac{1}{4}n_e m_0 c^2 \nabla a_0^2 \tag{1}$$

where $a_0^2 = (8.5 \times 10^{-13} I_0^{1/2} \lambda_0)^2$, n_e: electron number density, ω_p: plasma frequency, and ω_0: laser frequency, and electric field strength E_0 [V/m] of a focused laser beam of intensity I_0 [W/cm²],

$$E_0 = 2.74 \times 10^3 \, I_0^{1/2}. \tag{2}$$

For a laser pulse $I_0 = 10^{18}$ [W/cm²], and $\omega_p \sim \omega_0$, significantly high ponderomotive force is induced, $F_p \sim 20$ [TPa].

For propulsion applications, a collimated plasma beam would be preferable. The collimation requires a planer charge separation, which can be achieved by focusing an intense laser pulse onto the surface of a planar solid-density film. In this case laser light terminates at the target surface and drives high-energy electrons generated in front of it deep inside the target. Because of the planar charge separation these electrons produce a strong electrostatic field accelerating ions in a forward direction. In this case, high energy electrons expand faster and the ions form a well collimated beam confined in the transverse direction by the pinching in the self-generated magnetic field.[14] The ions in the beam expand in the longitudinal direction because the electric charge is not compensated inside. It has been shown that the mechanism of this anisotropic Coulomb explosion is at work in the case of the interaction of a petawatt laser with a thin slab of overdense plasma and accelerates ions up to relativistic energies.[14] The electromagnetic filamentation instability leads to magnetic pinching in the transverse direction and to collimated beam formation.

The reflective plasma acceleration (Fig.2(a)), is well-known,[26] in which the laser pulse is irradiated onto the target surface and in a short duration the plasma is induced on the target surface and accelerated against the incident laser pulse. However, there has been growing recognition of forward plasma acceleration (Figs.2(b) and 3).[4-17]

TABLE 1. Plasma accelerators by an intense laser pulse (λ_L : wavelength of the laser pulse, E_L: pulse energy, τ_L : pulse width, P_L: peak power, I_L: laser intensity, E_{ion}: accelerated ion energy, N_{ion}: number of ions, u_{ion}: velocity of ions, c: speed of the light, I_{sp}: specific impulse).

Laboratory	λ_L [μm]	E_L [J]	τ_L [psec]	P_L [W]	I_L [W/cm²]	Ions (Target)	E_{ion} [MeV]	N_{ion} /pulse	E_{ion}/E_L	$(u_{ion}/c)^c$	I_{sp}^c [Msec]
LosAlamos,[4] USA	10.60	–	1,500	–	~10^16	Proton (CH, Au)	~1	–	–	0.046	1.413
JAERI,[9] JAPAN	0.790	0.8	0.06	10^13	2~4 x10^18	He⁺, H⁺, F⁺ (PTFE)	1~	–	–	0.046	1.413
U.Mi.,[11] USA	0.5265	1	0.4	2.5 x10^12	3 x 10^18	Proton (Al foil)	1.5	>10^9	>2.4 x10^-4	0.056	1.731
RAL,[6] U.K.	1.053	30	1.8~4	–	2 x 10^18	Proton (Mylar disk)	4.2 (1.3)^b	–	0.1	0.094	2.899
RAL,[8] U.K.	1.053	30	~1	–	10^19	Proton (CH disk)	12 (1~)	–	–	0.158	4.910
RAL,[15] U.K.	1.053	50	0.9~1.2	–	5 x 10^19	Proton (Al foil)	18 (2~)	10^12	(0.06)^b	0.193	6.022
LLNL,[10] USA	1.053	<1k	0.45	~10^15	6 x 10^20	Proton (Metal foil)	50	–	–	0.317	10.12
LLNL,[16] USA	1.000	–	0.5, 5	~10^15	3 x 10^20	Proton (Plastic CH)	55 (10~)	3 x10^13	0.06	0.327	10.63
LLNL,[17] USA	1.000	–	0.5	~10^15	3 x 10^20	Proton (CH polymer)	58 (10~)	2 x10^13	0.12	0.336	10.92
U.Mi.,[18] USA	1.053	4	0.4	4 x10^12	6 x 10^18	He⁺ (He gas jet)	(0.5)^b,c	–	(0.05)^b	0.016	0.502
RAL,[13] U.K.	1.054	50	0.9	>5 x10^13	6 x 10^19	He⁺,He²⁺, Ne⁺ (gas jet)	3.6 (He)^c 6.0 (Ne)^c	–	0.0025 (He)	0.044	1.346
LLNL*,[5] USA	0.800	–	0.1	~10^15	10^20 ~ 10^21	C⁴⁺ (Carbon foil)	3 (1~)	–	0.0068~ 0.027	0.023	0.709
LLNL*,[7] USA	0.400	–	0.01~ 0.2	~10^15	10^18 ~ 10^22	Proton (Al, CH foil)	29	3 x10^13	0.2	0.243	7.667
JAERI*,[12] JAPAN	0.800	–	~1	–	10^16 ~ 10^17	Al⁶⁺ – Al⁷⁺ (Al foil)	~1	–	0.25	0.009	0.273
Osaka U.*,[14] JAPAN	1.000	–	0.018	~10^15	1.6 x 10^22	Proton (Plasma slab)	~1000	–	–	0.875	55.29

ᵃ results from PIC simulation, ᵇ average value, ᶜ estimated value, ᵈ radial direction

Recent experimental and theoretical results of laser plasma accelerators in various laboratories are listed in Table 1, in which estimated values of u_{ion}/c, i.e., fraction of the ion velocity to the speed of light, and specific impulse I_{sp}, based on the special theory of relativity, are also added.[21,22] Maximum ion energy of order tens of MeV through the acceleration has been reported in recent studies.[4-18] From the table, as for a proton beam accelerated up to 58 MeV, [17] its speed corresponds to 33 % of the speed of light and the specific impulse Isp of order 10^7 sec. Moreover, in theoretical studies, it was predicted that the relativistic ion beam, $u_{ion}/c \sim 87$ % and specific impulse $I_{sp} \sim 0.55 \times 10^8$ sec, is achievable with current laser facilities.[14]

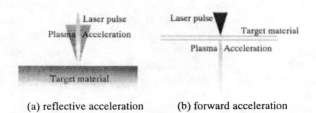

(a) reflective acceleration   (b) forward acceleration

FIGURE 2. Plasma acceleration through a laser-solid interaction.

Since no special nozzle nor channel but only a thin film target is needed for the collimated plasma beam formation as shown in Fig.3, the propulsion system employing this technique can be significantly simple and small. In this study, the fundamental investigation of the utilization of the forward plasma acceleration for space propulsion is conducted.

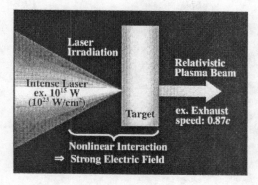

FIGURE 3. Schematic image of forward plasma acceleration through laser-target interaction.

PROPULSIVE PERFORMANCE OF A LASER-ACCELERATED RELATIVISTIC-PLASMA ROCKET

Potential concepts of the plasma acceleration techniques have been reviewed in the above section. In this section, to evaluate the applicability of these accelerators for the space propulsion, propulsive performances of the particles accelerated and exhausted with relativistic speed are investigated. In this case, significantly high thrust performance through the relativistic effects of the accelerated particles can be expected. Here, the particles to be accelerated are assumed as protons, and propulsive characteristics of the relativistic protons as the propellant are investigated based on the special theory of relativity, when the speed of the particle approaches the speed of light.

For the propellant, accelerated and exhausted at velocity, u_e', relative to a rocket, where the prime (') denotes the quantities measured by an observer on the rocket traveling at velocity v, momentum p_e, and specific impulse I_{sp} are given as,

$$p_e = \frac{m^{(0)} u_e'}{\sqrt{1 - u_e'^2 / c^2}},$$
(3)

and

$$I_{sp} = \frac{p_e}{m^{(0)} g_0},$$
(4)

where, $m^{(0)}$ is the rest mass, or the mass of a particle observed in the frame fixed to the particle, and c is the speed of light.

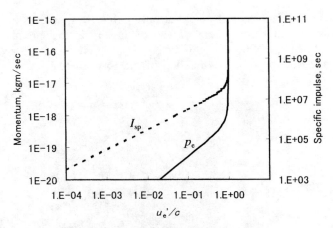

FIGURE 4. Momentum and specific impulse vs relative exhaust velocity of a relativistic particle.

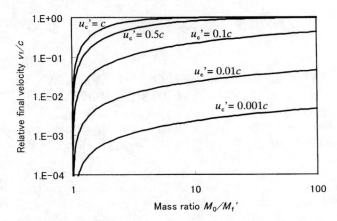

FIGURE 5. Relative final (terminal) velocity vs mass ratio for various exhaust velocities.

Relations of the momentum p_e, specific impulse I_{sp} with relative exhaust velocity u_e'/c are shown in Fig.4, for a proton as the propellant neutralized by a electron, where $m^{(0)} = m_e^{(0)} + m_p^{(0)}$ ($m_e^{(0)}$: mass of an electron, $m_p^{(0)}$: mass of a proton), and the speed neutralizing particles exhausted are assumed zero relative to the observer at rest. This figure suggest that the momentum and specific impulse of the particle increase abruptly and approach infinite values, as the exhaust velocity u_e' is nearly as great as the speed of light c. It is therefore expected that even a small amount of the propellant can generate the extremely large impulse, when the velocity is nearly as great as the speed of light. For example, for $u_e'/c = 0.1$, or the relativistic factor $\gamma = (1 - u_e'^2/c^2)^{-1/2}$ = 1.005, I_{sp} = 3.1 x 10^6 sec. This value is higher than the values of specific impulse estimated in typical design of the nuclear fusion thermal propulsions.

Denoting the initial rocket mass (at $v = 0$) by M_0 ($= M_0'$), the final (terminal) mass measured by the observer on the rocket by M_f and final (terminal) velocity by v_f, the relativistic rocket equation is given as,

$$\frac{v_f}{c} = \frac{1 - \left(M_0/M_f'\right)^{-2u_e'/c}}{1 + \left(M_0/M_f'\right)^{-2u_e'/c}}.$$ (5)

The relationship between the mass ratio M_0/M_f' and the relative final velocity v_f/c for various exhaust velocity u_e' is shown in Fig.5. From the figure, it is shown that larger values of v_f/c, or final velocity, can be obtained even with small M_0/M_f', as the exhaust velocity approaches the speed of light.

Comparison with a Photon Rocket

In this section, assuming the antimatter annihilation energy is utilized as their energy sources, comparison of the thrust performance of a laser-accelerated relativistic-plasma rocket to a photon rocket [27] is conducted. Based on the special theory of relativity, induced impulse of the laser-accelerated plasma (p_e, expressed as Eq.(3)), energy required to induce the impulse (T), and effective annihilation energy used for the laser-accelerated plasma (E_{eff}) are expressed as follows,

$$T = \frac{m^{(0)}c^2}{\sqrt{1 - u_e'^2/c^2}} - m^{(0)}c^2,$$ (6)

$$E_{eff} = \alpha \cdot m_{am}^{(0)}c^2,$$ (7)

where, $m_{am}^{(0)}$ is the mass of annihilated antimatter and matter pairs converted into energy with an energy conversion ratio α, which includes the efficiencies of laser system, laser-plasma energy conversion, etc., (for u_e' and $m^{(0)}$, see formar section).

Figure 6 shows relations of induced impulse ($p_e/ m_{am}^* c$) and propellant mass consumption ($m^{(0)}/m_{am}^*$), where m_{am}^* is a given value of antimatter and matter annihilation mass, or a given value of energy source. The ordinate is a ratio of induced impulse (p_e) to a given source energy ($m_{am}^* c^2$), normalized by multiplying c. As shown in this figure, the induced impulse for a given source energy increases with propellant mass consumption ($m^{(0)}/m_{am}^*$). So it is clear that higher impulse can be obtained for a given source energy in the laser-accelerated plasma rocket than the photon rocket, $m^{(0)}/m_{am}^* = 0$, in all cases.

Relationship between exhaust speed and propellant mass consumption of the laser-accelerated plasma rocket is shown in Fig.7 for given values of m_{am}^* and also for various α s. Maximum value of the exhaust speed can be achieved by the photon rocket, $u_e'/c = 1$. Higher values of exhaust speed can be obtained for smaller $m^{(0)}/m_{am}^*$ and also for $\alpha \rightarrow 1$.

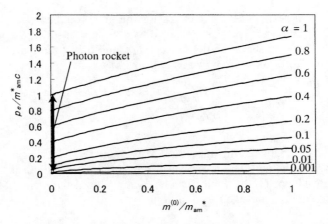

FIGURE 6. Ratios of induced impulse to a given source energy vs. propellant mass consumed.

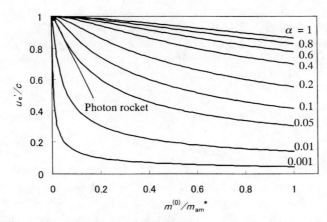

FIGURE 7. Relationship between exhaust speed and consumed propellant mass of the laser-accelerated relativistic plasma rocket.

Figure 8 shows a relationship between mass consumption of antimatter and matter pairs $(m_{am}^{(0)}/m^*)$ and induced impulse of the laser-accelerated plasma (p_e/m^*c) for a given value of m^* $(= m^{(0)} + m_{am}^{(0)})$ and also for various α s. It is shown that for α s below unity the impulse reaches maximum value $(\alpha/(2-\alpha))^{1/2}$ at $m_{am}^{(0)}/m^* = 1/(2-\alpha)$. Under these conditions, higher values of impulse can be achieved in the laser-accelerated plasma rocket cases compared to photon rocket cases in which $m_{am}^{(0)}/m^* = 1$.

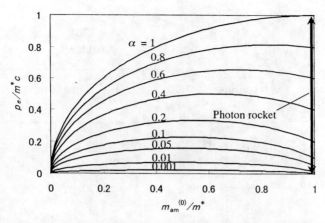

FIGURE 8. Relationship between consumed antimatter mass and induced impulse.

In cases of operations of these thrusters on earth and/or solar orbit, in which solar power is available, the energy can be supplied from the solar cells, or transmitted through the electromagnetic wave from the satellites. Regarding the combination of this almost infinite energy source with very small propellant mass required, significantly high specific impulse can be obtained through this propulsion system for primary propulsion and attitude control applications.

Moreover, for inter-stellar applications using a nuclear reactor or antimatter annihilation energy as the energy source, the higher specific impulse and final (terminal-) speed can be achieved compared to the nuclear fusion thermal propulsion which requires some more propellant mass, since this system requires a small amount of propellant mass. Therefore, comparing with conventional laser propulsion concepts, achievement of high thrust performance, such as higher specific impulse, can be expected.

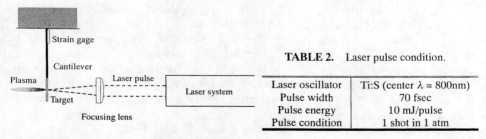

TABLE 2. Laser pulse condition.

Laser oscillator	Ti:S (center λ = 800nm)
Pulse width	70 fsec
Pulse energy	10 mJ/pulse
Pulse condition	1 shot in 1 atm

FIGURE 9. Schematic of experimental setup.

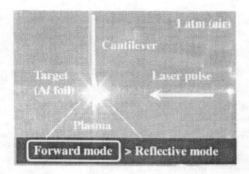

FIGURE 10. Photo of laser-induced plasma.

TABLE 3. Typical test result.

Target	A*l* foil (*t* =12μm)
*I*bit	5.7x10^{-7} Nsec (Forward)
*I*sp	13264 sec
*C*m	57 N/MW

PRELIMINARY EXPERIMENT WITH
AN ULTRASHORT-PULSE LASER

In order to evaluate thrust performance of the forward plasma acceleration through a laser-solid interaction, a thrust performance test, or impulse measurement, on a target irradiated by a femtosecond laser was conducted. A schematic diagram of an impulse measurement system is shown in Fig.9, and a typical condition of the laser pulse is given in Table 2. In this study, pulse energy of the laser was set 10 mJ/pulse with pulse width of 70 fsec and central wavelength of 800 nm, irradiated on a target in 1 atm air. An aluminum foil of 12 μm thick was used for the target material. As shown in this figure, the target is attached on a top of a calibrated copper cantilever. A strain gage is attached on the supporting edge of the cantilever, with which its displacement, or impulse, caused by induced plasma through the laser irradiation is measured.

A typical photo of the laser induced plasma is shown in Fig.10. It is shown that the plasma is induced from both sides, the front and rear surfaces, of the target to the incident laser pulse. Therefore, it can be seen that the plasma is accelerated to both sides, to the forward direction (forward mode acceleration) and to the reflective direction (reflective mode acceleration) in this case.

The impulse (I_{bit}), specific impulse (I_{sp}) estimated from removed mass, and momentum coupling coefficient (C_m), estimated in this case are given in Table 3. It can be seen that the net impulse is induced to the forward-mode side in this case. Although the plasma is accelerated to both the forward mode and the reflective mode, the impulse of the forward mode is larger, and the net impulse is induced to the forward direction as a consequence.

Since the laser pulse is irradiated in 1 atm air in this case, air (and the target surface) near a focal point of the target surface is probably broken down inducing the plasma on the reflective side (top surface). While in the bottom side, or the forward side, the plasma probably originated from the target material (and air) ablated and accelerated to the forward direction. If the reduction of the reflective-mode plasma is achieved by conducting the experiment in vacuum, it is expected that the thrust performance become significantly improved. This point is one of the next issues of our experiment.

CONCLUSION

Potential concepts of a relativistic laser-accelerator which can be applied to the spacecraft propulsion using a conventional technique were reviewed. Maximum ion energy of order tens of MeV through the acceleration was reported in recent studies, and as for a proton beam accelerated up to 58 MeV, its speed corresponds to 33 % of the speed of light and the specific impulse of order 10^7 sec. Also, a feasibility study of the utilization of these accelerators, based on the special theory of relativity, for space propulsion was conducted. It was shown that significantly high specific impulse can be obtained through this propulsion system for primary propulsion and attitude control applications. In addition, a preliminary experiment was conducted with an ultrashort-pulse laser system. Induced impulse in a forward acceleration mode was observed using a 12-μm thick Al foil as a target.

ACKNOWLEDGEMENT

Authors are grateful to Mr. Matsuoka, Dr. Deki and Dr. Arisawa of Advanced Photon Research Center, Kansai Research Establishment, Japan Atomic Energy Research Institute (JAERI), and Mr. Tamura of Tokai University for assisting the experiment.

REFERENCES

1. Esarey, E., et al., *IEEE Transactions on Plasma Science* **24**, pp.252–288 (1996).
2. Nishida, Y., *J. Plasma and Fusion Science Research* **73**, pp.411–414 (1997) (in Japanese).
3. Umstadter, D., *Nature* **404**, p.239 (2000).
4. Gitomer, S.J., et al., *Physics of Fluids* **29**, pp.2679-2688 (1986).
5. Denavit, J., *Physical Review Letters* **69**, pp.3052-3055 (1992).
6. Fews, A.P., et al., *Physical Review Letters* **73**, pp.801-1804 (1994).
7. Lawson, W.S., et al., *Physics of Plasmas* **4**, pp.788-795 (1997).
8. Beg, F.N., et al., *Physics of Plasmas* **4**, pp.447-457 (1997).
9. Zhidkov, A.G., et al., *Physical Review E* **60**, pp.3273-3278 (1999).
10. Roth, M., et al., *First International Conference on Inertial Fusion Science and Applications* (1999).
11. Maksimchuk, A., et al., *Physical Review Letters* **84**, pp.4108-4111 (2000).
12. Zhidkov, A., et al., *Physical Review E* **61**, pp.R2224-R2227 (2000).
13. Krushelnick, K., et al., *Physical Review Letters* **83**, pp.737-740 (1999).
14. Sentoku, Y., et al., *Physical Review E* **62**, pp.7271–7281 (2000).
15. Clark, E.L., et al., *Physical Review Letters* **84**, pp.670-673 (2000).
16. Hatchett, S.P., et al., *Physics of Plasmas* **7**, pp.2076-2082 (2000).
17. Snavely, R.A., et al., *Physical Review Letters* **85**, pp.2945-2948 (2000).

18. Sakisov, G.S., et al., *Physical Review E* **59**, pp.7042-7054 (1999).
19. Stuhlinger E., *Ion Propulsion for Space Flight*: McGraw-Hill (1964).
20. Kimura, I., *Rocket Engineering*: Yokendo (1993) (in Japanese).
21. Horisawa, H., and Kimura, I., *AIAA Paper* 2000-3487 (2000).
22. Horisawa, H., and Kimura, I., *AIAA Paper* 2001-3662 (2001).
23. Campbell, P.M., et al., *Physical Review Letters* **39**, pp.274-277 (1997).
24. Decoste, R., and Ripin, B.H., *Physical Review Letters* **40**, pp.34-37 (1978).
25. Cowan, T.E., et al., *Physical Review Letters* **84**, pp.903-906 (2000).
26. Phipps, Jr, C. R., et al., *Journal of Applied Physics* **64**, pp.1083-1096 (1988).
27. Mallove, E.F., and Matloff, G.L., *The Starflight Handbook*: John Wiley & Sons (1989).

Fundamental Study of a Laser-Assisted Plasma Thruster

Hideyuki Horisawa[*], Masatoshi Kawakami[*], Wun-Wei Lin[*],
Akira Igari[*], and Itsuro Kimura[#]

[*]Department of Aeronautics and Astronautics, Tokai University, 1117 Kitakaname, Hiratsuka, Kanagawa, 259-1292 JAPAN
[#]Professor emeritus, University of Tokyo, 7-3-1 Hongo, Bunkyo-ku, Tokyo, 113-8856 JAPAN

Abstract. In this study we propose a novel laser-assisted plasma thruster, in which plasma is induced through a laser beam irradiation onto a target, or a laser-assisted process, and accelerated by electrical means instead of a direct acceleration only by using a laser beam. Inducing the short-duration conductive plasma between electrodes with certain voltage, the short-duration switching or a discharge is achieved, in the laser-assisted thruster. Also, reductions of energy losses to electrodes, electrodes erosion, and an improvement of specific impulse through the intense current caused by the short duration discharge can be expected. Here, a fundamental study of newly developed two-dimensional laser-assisted pulsed-plasma thruster (PPT) and coaxial laser assisted PPT is conducted. A DC power supply (10 ~ 600 V) was used for the power source, and an Nd:YAG laser (wave length: 1.06μm, maximum pulse energy: 1.4J/pulse, pulse width: 10 nsec) was utilized. With this system, the peak current of about 500A with its duration of 3 µsec (FWHM) was observed in a typical case.

INTRODUCTION

There are growing interests on pulsed-plasma thrusters, PPTs, utilizing a solid propellant, usually PTFE (Teflon), for their system simplicity and advantages on miniaturization and mass reduction for the use of attitude or orbit control thrusters for small-sized spacecrafts, despite the low efficiency.[1-4] In their operation, short pulse discharges with several-microsecond duration are induced across the exposed propellant surface between electrodes, vaporizing and ionizing the surface, and also inducing the pressure force. Then the interaction of the discharge current (tens of kA) and its self-induced magnetic field results as the electromagnetic force, or Lorentz force, acting on the plasma and inducing a directed plasma beam exhaust, or thrust. In this electromagnetic acceleration process, it is necessary to complete phase changes of the propellant, such as vaporization and ionization, and the electromagnetic acceleration at the same time within a short duration of a pulse discharge. An improvement of the thrust performance can be expected with a shorter pulse duration case, since it is capable of higher current per unit time or higher power input, namely higher thrust, and of reducing loads on electrodes. However, it is difficult to complete

CP664, *Beamed Energy Propulsion: First International Symposium on Beamed Energy Propulsion*,
edited by A. V. Pakhomov

the process including the phase change and electromagnetic acceleration simultaneously during the discharge pulse, because there is a delay in the phase change of the solid-propellant after the pulse discharge initiation. The surface of the propellant will continue to evaporate long after completion of the discharge pulse, providing mass that cannot experience acceleration to high speeds by the electromagnetic and gasdynamic forces. The various masses including low-speed macroparticles can have quite different velocities. Since the residual vapor or plasma from the late-time evaporation of the propellant surface remains in the discharge chamber due to the delay after the pulse discharge completion, which cannot contribute significantly to the impulse bit, it has been difficult for the thruster of this type to improve the mass loss of the propellant and thrust efficiency.[1-4]

In order to reduce this late-time ablation and to improve thrust efficiency, effects of the utilization of the laser-pulse irradiation, or assistance, were studied, which can be expected to induce plasma from a propellant in a short duration, i.e., using a short-duration conductive region of the plasma between electrodes, the short-pulse switching or discharge can be achieved. Since the use of a shorter pulse of the laser enables a shorter duration of the pulsed-plasma in this case, the higher peak current and significant improvement of the thrust performance must be expected. In addition, depending on the laser power the laser-induced plasma occurring from a solid propellant usually has directed initial velocity, which should also improve the thrust performance compared to pure PPTs. In this study, a preliminary investigation was conducted on characteristics of pulsed discharges induced by laser pulse irradiations for various voltages applied to the electrodes including a low-voltage mode, which can be an electrically-assisted laser propulsion mode, and a high-voltage mode, a laser-assisted electromagnetic propulsion mode.

Laser-Assisted Plasma Thruster

Schematics of laser-assisted plasma thrusters developed in this study are given in Fig.1. The thruster utilizes the laser-beam assistance to induce plasma, ionized from a solid propellant between electrodes, and then an electric discharge is induced in this conductive region. Since the plasma is induced through the laser ablation of the solid propellant in this system, various substances, such as metals, plastics, ceramics, etc., in various phases can be used for the propellant. Therefore, this system must be effective not only for space propulsion devices but also for plasma sources in material processing. In addition, since the plasma has initial velocity through the laser ablation, it is expected that the thrust efficiency and propellant mass loss should be significantly improved. In this study, preliminary tests on two types of thrusters, (a) two-dimensional type thruster inserting a solid propellant between electrodes, and (b) coaxial type thruster using one of the electrodes (cathode) as the propellant, were conducted as shown in Fig.1.

424

EXPERIMENTAL

For the two-dimensional type thruster (Figs.1(a) and 2(a)), copper electrodes (5 mm in width, 20 mm in length) and an alumina propellant (3 mm in height) were used. While for the coaxial type thruster (Figs.1(b) and 2(b)), a copper tube (4 mm in inner-diameter) was used for an anode and a carbon rod (3 mm in diameter) for a cathode and also for a propellant. A schematic of experimental setup is shown in Fig.3. A Q-sw Nd:YAG laser (BMI, 5022DNS10, wavelength: λ=1064nm, maximum pulse energy: 1400 mJ/pulse, pulse width: 10~15 nsec) was used for a plasma source, or a laser assistance. The laser pulse was irradiated into a vacuum chamber (10^{-3} Pa) through a quartz window and focused on a target, or a propellant, with a focusing lens (f = 100 mm). Discharge current was monitored with a current monitor (Pearson Electronics, Model-6600, maximum current: 10 kA, minimum rise time: 5 nsec) and an oscilloscope (LeCroy, 9374TM, range: 1 nsec/div ~ 5 msec/div). In this study, preliminary experiments on switching, or discharge between the cathode and anode, and the current characteristics of the laser-induced plasma were conducted. At the first part of the experiment, lowered voltage conditions (~ 20 V) applied on the electrodes was selected to examine discharge current characteristics under the low current range (~ 10 A). In addition, discharge current characteristics for higher current cases, or higher voltages (i.e., up to 600 V), were also investigated.

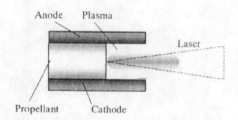

(a) Two-dimensional laser-assisted plasma thruster.

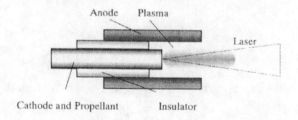

(b) Coaxial laser-assisted plasma thruster.

FIGURE 1.   Schematics of laser-assisted plasma thrusters.

Alumina-Insulator   Cu-Anode

Cu-Cathode

(a) Two-dimensional laser-assisted plasma thruster.

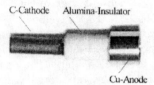

C-Cathode   Alumina-Insulator

Cu-Anode

(b) Coaxial laser-assisted plasma thruster.

FIGURE 2.   Photos of laser-assisted plasma thrusters.

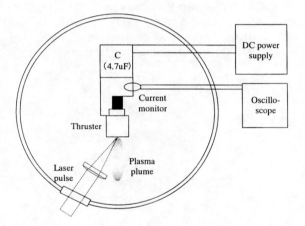

FIGURE 3.   Schematic of experimental setup.

RESULTS AND DISCUTTION

Discharge Characteristics through Laser-Induced Plasma

Low-voltage mode

At first part, lowered voltage conditions (~ 20 V), applied to electrodes, or charged to a capacitor, was selected to examine discharge current characteristics under the low current range (~ 10 A).

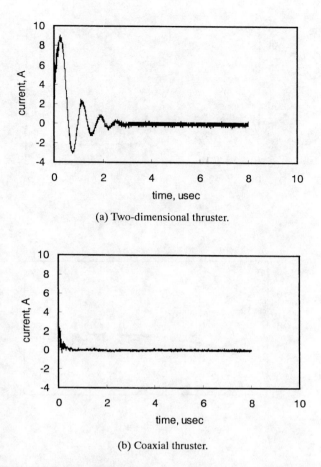

(a) Two-dimensional thruster.

(b) Coaxial thruster.

FIGURE 4.   Time variation of discharge current for low voltage conditions.

The temporal variation of discharge current for a two-dimensional type thruster is shown in Fig.4(a). It can be seen that the current abruptly rises up to its peak value (+9 A) within 200 nsec after the laser pulse irradiation, and drops down to minimum value (− 3 A) at 700 nsec. After oscillating with about 900 nsec period, its amplitude converges zero at about 3 μsec. Here, positive values on the ordinate mean the positive current from anode to cathode. It is confirmed that the electric discharge is achieved even under low voltage conditions (10 ~ 20 V).

The temporal variation of discharge current for a coaxial thruster is shown in Fig.4(b). Similar to a two-dimensional case, the current rapidly increases to its peak value (+3 A) within 10 nsec after the laser pulse irradiation, and drops down to minimum value (0 A) at 20 nsec. After oscillating with about 40 nsec period, its amplitude converges zero at about 400 nsec.

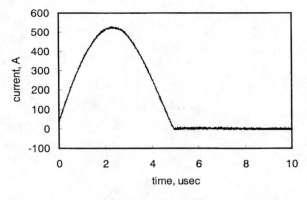

(a) Two-dimensional thruster.

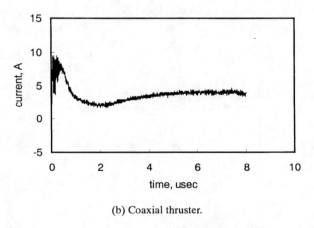

(b) Coaxial thruster.

FIGURE 5. Time variation of discharge current for high voltage conditions.

Although conventional pulsed plasma thrusters (PPTs) have been driven with several-kilovolts at several-microsecond pulse-discharges[1-4], it is shown that much shorter pulse discharges can be achieved under much lower voltage conditions with the laser-assisted discharges.

It has been reported that pulsed plasma with hundred-nanosecond duration is induced through a nanosecond laser pulse irradiation on a solid target.[5-7] So it is expected that within this duration a conductive region of the plasma enables short switching, or a short electric discharge, between electrodes. Therefore, it is presumed from these results that measured current patterns must depend on plasma behaviors, or pulses, which can be actively controlled with the incident laser pulses.

In laser ablation processes, it has been reported that the electron emission first occurs after laser pulse irradiation, followed by ions with higher degrees of ionization after several nanoseconds, after which those with lower degrees of ionization follow.[5-7] Therefore, the first positive peak of the discharge current in Fig.4(b) is probably due to the current by electrons. While the following negative peak of the current may be due to positive ions and/or to oscillation of the electrons, details of these points are not yet clear.

High-voltage mode

The time variation of discharge current of a two-dimensional thruster is given in Fig.5(a), for a high-voltage mode, where higher voltage (260 V) is applied to electrodes. From the figure, the current abruptly rises and reaches a maximum value, 520 A, at 2 μsec, after which it falls down to zero at 5 μsec. Comparing with the low current mode case shown in Fig.4(a), the maximum current increases significantly (50 times) and the pulse width, 5 times longer.

The plots of the current for a coaxial thruster in a high-voltage mode is shown in Fig.5(b). It is shown that the current increases rapidly and reaches 10 A at 0.5 μsec, and falls down to 2 A at 2 μsec, after which it rises again up to 3 A at 7 μsec. Although not shown in Fig.5(b), the current gradually decreases and converges zero at about 60 μsec. Comparing with the low-voltage mode case shown in Fig.4(b), higher maximum current and longer pulse width are obtained, which is followed by the second long pulse. A possible cause of the second long pulse in the coaxial thruster is probably due to residual plasma left in and/or around a discharge chamber with low current density, conducting small portion of the current, and also to plasma induced from a cathode surface at the first discharge, however, details of these mechanism are not clear at the moment.

From these results, it is found that the discharge duration at the low-voltage case is as long as the duration of laser-induced plasma. Therefore, the discharge in the low-voltage case must be controlled with the incident laser pulse, or laser-induced plasma. While in the high-voltage case, the discharge duration is much longer than that of laser-induced plasma. In this case, the laser-induced plasma should lead the main discharge from a capacitor, where some amount of the propellant surface must be vaporized through the main discharge.

The difference of current waveforms between high-voltage and low-voltage conditions is probably due to the difference of discharge processes in both cases. In the low-voltage case, discharge energy of the capacitor ($= CV^2/2$) is relatively small compared to the laser energy, however, in the high-voltage case, higher energy must be discharged. Considering ratios of the laser energy to these discharge energies, the discharge process in the high-voltage mode must be defined as the laser-assisted electric discharge, or the laser-assisted electric propulsion mode, while in the low-voltage mode with smaller electric energy, as the electrically-assisted laser-induced process, or the electric-assisted laser propulsion mode. As shown in these figures, there is a significant difference in the current waveforms between two types of thrusters.

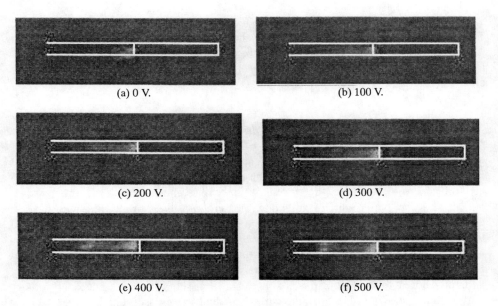

(a) 0 V.　　　　　　　　　　　　　　(b) 100 V.

(c) 200 V.　　　　　　　　　　　　　(d) 300 V.

(e) 400 V.　　　　　　　　　　　　　(f) 500 V.

FIGURE 6.　Photos of exhaust plasma plume from laser -assisted PPTs for various charged voltages.

Plasma Plume Observation

Photos of exhaust plasma plume of the two-dimensional type thruster are shown in Figs.6 (a)~(f) for various voltage conditions applied to electrodes, or a capacitor. In Fig.6(a) 0 V, the pure laser ablation plasma from the alumina (propellant) surface is shown, to which a laser pulse is irradiated. Cases for 100 V and 200 V are given in Figs.6 (b) and (c), respectively, where plasma expanding along a discharge channel from the alumina surface toward a channel exit can be seen clearly, compared to a case (a). It can be seen that the plasma emission in a near-surface region is stronger and decreasing with distance from the surface. On the other hand, the plasma emission distribution in a transverse direction is not changing significantly. In addition, it is shown that the extension and emission of the plasma plume increase significantly with voltages.

Photos of plasma plume of a coaxial type thruster for various voltage conditions are shown in Figs.7 (a)~(f). The plasma plume of a pure laser ablation process from a carbon rod (propellant and cathode) surface is shown in Fig.7(a), to which a laser pulse is irradiated. Larger extension of the plasma along the central axis, which is also confined to the center axis, is observed in a 100 V case (Fig.7(b)). In Fig.7 (c), a case of 200 V, larger distribution and stronger emission of the plasma plume, expanding significantly to an axial direction, is observed. With higher voltages, size and brightness of the plasma plume grow larger and its extension to an axial direction becomes significant.

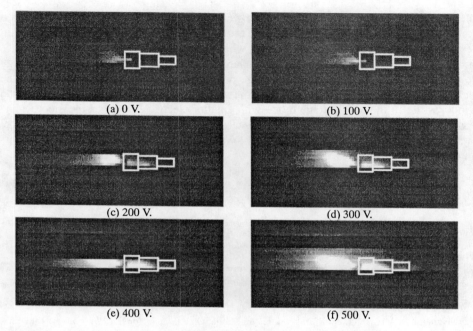

(a) 0 V.   (b) 100 V.

(c) 200 V.   (d) 300 V.

(e) 400 V.   (f) 500 V.

FIGURE 7. Photos of exhaust plasma plume from laser -assisted PPTs for various charged voltages.

From these results, it is found that a discharge mechanism of the low voltage mode is not similar to that of the high-voltage mode. As discussed above, the former case can be categorized to the electric-assisted laser propulsion mode, and another case, to the laser-assisted electric propulsion mode.

CONCLUSION

Novel laser-assisted plasma thrusters were developed and tested, in which plasma is induced through a laser beam irradiation onto a target, or a laser-assisted process, and accelerated by electrical means instead of a direct acceleration only by using a laser beam. A fundamental study of newly developed two-dimensional laser-assisted pulsed-plasma thruster (PPT) and coaxial laser assisted PPT was conducted. Inducing the short-duration conductive plasma between electrodes with certain voltages, the short-duration switching or a discharge was achieved.

At low-voltage conditions (~ 20 V), applied to electrodes, or charged to a capacitor, it was confirmed that the electric discharge, with the current up to 9 A with 500 nsec duration (FWHM), can be achieved even under low voltage conditions (10 ~ 20 V). From these results, it was found that the discharge duration at the low-voltage case was as long as the duration of laser-induced plasma. Therefore, the discharge in the low-voltage case must be controlled with the incident laser pulse, or laser-induced plasma. While in the high-voltage case (~ 600 V), the discharge duration was much longer than that of laser-induced plasma. In this case, the laser-induced plasma should lead the main discharge from a capacitor, where some amount of the propellant surface must be vaporized through the main discharge. Considering ratios of the laser energy to these discharge energies, the discharge process in the high-voltage mode must be defined as the laser-assisted electric discharge, or the laser-assisted electric propulsion mode, while in the low-voltage mode with smaller electric energy, as the electrically-assisted laser-induced process, or the electric-assisted laser propulsion mode. There were significant differences in the current waveforms between two types of thrusters in various voltage modes.

REFERENCES

1. Jahn, R.G., *Physics of Electric Propulsion*: McGraw-Hill, 1968, pp.198-316.
2. Martinez-Sanchez, M., and Pollard, J. E., *J. Propulsion and Power* **14**, pp.688-699 (1998).
3. Burton, R. L., and Turchi, P. J., *J. Propulsion and Power* **14**, pp.716-699 (1998).
4. Micci, M. M., and Ketsdever, A. D. (ed.), *Micropropulsion for Small Spacecraft (Prog. Astronautics and Aeronautics 187)*: American Institute of Aeronautics and Astronautics, 2000, pp.337-377.
5. The Institute of Electrical Engineers of Japan, *Laser Ablation and Applications*, Corona Publishing (1999) (in Japanese).
6. Ohyanagi, T., Miyashita, A., Murakami, K., and Yoda, O., *Japan J. Applied Phys.* **33**, pp.2586-2592 (1994).
7. Kokai, F., and Koga, Y., *Nuc. Instr. Meth. in Phys. Res.* **B121**, pp.387-391 (1997).
8. Phipps, Jr, C. R., Turner, T. P., Harrison, R. F., York, G. W., Osborne, W. Z., Anderson, G. K., Corlis, X. F., Haynes, L. C., Steele, H. S., and Spicochi, K. C., *J. Appied. Phys.*, **64**, pp.1083-1096 (1988).

Advanced Space Propulsion
with Ultra-Fast Lasers

Terry Kammash

Nuclear Engineering and Radiological Sciences
University of Michigan
Ann Arbor, MI 48109

Abstract. A novel propulsion system that is expected to evolve from promising worldwide research in which ultrafast lasers [with very short pulse length] are used to accelerate charged particles to relativistic speeds is presented. The LAPPS [Laser Accelerated Plasma Propulsion System] concept makes use of high power lasers with femtoseconds pulse lengths that are made to strike micron size focal spots in very thin targets giving rise to nearly collimated, charge neutral, proton beams with mean energies of several MeV. When utilized in a propulsion device these beams produce specific impulses of several million albeit at very modest thrusts, and require nuclear power systems to drive them. In this paper we examine the underlying physics issues and the technological problems that must be addressed to make LAPPS a viable propulsion device that could open up interplanetary and interstellar space to human and robotic missions in the not too distant future.

INTRODUCTION

One of the remarkable scientific developments of recent years is the demonstration, in table top experiments, that ultrafast (very short pulse length) lasers can accelerate charged particles to relativistic speeds. The University of Michigan and the Lawrence Livermore National Laboratory, among others, have produced, by this method, proton beams containing more than 10^{12} particles at mean energies of tens of MeV. In fact progress is being made so rapidly in laser technology that peak powers will soon be reached that can accelerate protons to rest mass energies. This means that these particles, when ejected form a propulsion device, will travel at 0.866 of the speed of light, and that translates to specific impulses of well over 10 million seconds. The implication of these facts for space propulsion are truly staggering especially when coupled to the fact that rep rates of kilohertz have also been achieved for high intensity lasers.

The underlying physics of the acceleration process is currently without an exact theory. Nevertheless a plausible, heuristic analysis consistent with sound physics principles can indeed be made to generate a mathematical expression that can predict experimental results with some measure of accuracy, consistency and reliability. When a high-intensity laser strikes a target, it produces at the surface a plasma with a size of about half a laser wavelength[1] due to the longitudinal electron oscillations resulting from the oscillating Lorentz force. In fact a free, stationary electron will

CP664, *Beamed Energy Propulsion: First International Symposium on Beamed Energy Propulsion*,
edited by A. V. Pakhomov

size of about half a laser wavelength[1] due to the longitudinal electron oscillations resulting from the oscillating Lorentz force. In fact a free, stationary electron will execute a "quivering" motion in the form of figure 8 when subjected to the electric and magnetic fields of the laser, and the spatial extent of the "quiver" is determined by the modified vector potential
"a_o" give by[2]

$$a_o = \frac{eA}{m_o c^2} \quad (1)$$

$$= 0.85 \times 10^{-9} \quad \sqrt{I(w/cm^2)} \quad \lambda \ (\mu m) \quad (2)$$

where e is the electron charge, A the standard vector potential, m_o the rest mass of the electron, c the speed of light, I the laser intensity and λ the laser wave length. It should be noted that a_o is related to the relativistic parameters "γ" through

$$\gamma = \left(1 - v^2/c^2\right)^{-1/2} = \left(1 + \frac{a_o^2}{2}\right)^{1/2} \quad (3)$$

hence the connection between the laser parameters and the velocity (or acceleration) of the electron with which it interacts. Twice in a laser period the electrons of the plasma re-enter the target while the ions remain virtually immobile due to their large mass. Returning electrons are accelerated by the "vacuum" electric field and subsequently deposit their energy inside the target. The electrons of the plasma become strongly heated by the laser light, penetrate deeper inside the solid target with relativistic speeds, and form a low-density, high-energy component of the entire electron population. These high-energy electrons create an electrostatic field, which accelerates ions in the forward direction while decelerating the electrons until both species drift out at the same rate. An electrostatic field near the target surface has a bipolar structure with the more pronounced component accelerating ions in the forward direction. If the laser pulse duration is longer than the ion acceleration time in the layer then the ions would acquire an energy equal to the electrostatic energy. Since this "ambipolar" potential causes both the electrons and ions to proceed at the same rate, they emerge from the back surface of the target in a perpendicular direction in a "neutral", nearly collimated beam form as shown in Fig 1. This emerging beam of charged particles is what provides the thrust in a propulsion device.

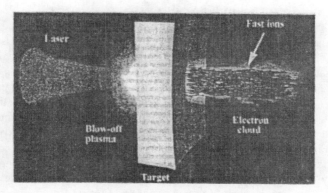

Figure 1. Ultrafast laser impinging upon target to produce fast ions.
(Courtesy of Scott C. Wilkes).

The electrons that are initially accelerated by the laser must overcome the Coulomb energy in the pre-formed plasma in order to penetrate the target to set up the electrostatic potential. Moreover, simple energy balance dictates that the energy imparted by the laser must appear in these electrons at some efficiency. Since these electrons create the potential, the electron energy must, therefore, equal that of the potential energy, and that in turn must equal that of the ions acted upon by this potential. When all these facts are put together, the energy of the ejected ions can be expressed by[3]

$$E_i = Z\sqrt{\eta IR\lambda} \tag{4}$$

where Z is the ion charge, η the efficiency of the energy conversion, R the size of the focal spot, and λ the wavelength of the laser. When the laser intensity is given in W/cm^2 and the spatial dimensions in microns, Eq (4) gives the ion energy in MeV. The above equation predicted reasonably accurately the one MeV ions accelerated by the 10 TW laser in the University of Michigan experiments [4], and the 5.3 MeV (mean) energy ions produced by the Livermore petawatt laser[5,6]. Clearly, several MeV ions produce specific impulses of several million seconds, more than required for almost all missions envisaged in the solar system and beyond. But, as we shall note shortly, it is the thrust which these present-day experiments can produce that is not adequate for most missions of interest. The thrust can be written in the form

$$F = \omega N_i M_i v_i \tag{5}$$

where ω is the rep rate, N_i the number of ions in the ejected beam, M_i the ion mass, and v_i the ion velocity. With the exception of the mass, the remaining parameters in Eq (5) lend themselves to change (increase or decrease as the situation may dictate) and present-day experiments are being analyzed and modified to address these changes[7].

THE LAPPS PROPULSION CONCEPT

The propulsion system addressed here is illustrated in Fig 2. It consists of a power supply which will most likely be nuclear, and the laser it drives which will be used to accelerate a beam of protons to relativistic speeds. These protons emerge, as noted earlier, in a nearly collimated form along with an equal number of electrons so that the beam is electrically neutral when it leaves the vehicle to provide the thrust. This is particularly significant since no space-charge effects will arise that might adversely influence the performance of the vehicle. Even present-day experimental data, if viewed from the standpoint of a propulsion system, will indicate a specific impulse of several million seconds which is significantly larger than any projected to be produced by "competing" advanced propulsion concepts such as those driven by fusion energy. Although a current Laser Accelerated Plasma Propulsion System [LAPPS] is capable of producing a very large specific impulse (I_{sp}), it does not produce large enough thrust (F) to make it suitable for, say, a manned interplanetary mission where a measure of balance between I_{sp} and F must be attained to result in an optimum travel time. This can be seen from the expression for the round trip time, τ_{RT} between two points separated the linear distance D, namely[8]

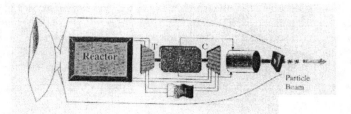

FIGURE 2. Laser-accelerated plasma propulsion system (LAPPS).

$$\tau_{RT} = \frac{4D}{gIsp} + 4\sqrt{\frac{Dm_f}{F}} \qquad (6)$$

where g is the earth's gravitational acceleration, and m_f the dry mass of the vehicle. The above equation is based on a constant thrust, acceleration/deceleration type of trajectory where it can be seen that the contribution of the two terms on the right-hand side are additive and must therefore be somewhat comparable in order to produce a reasonably optimum τ_{RT}. Eq (6) also reveals the need for large I_{sp} and F to obtain a small travel time but within the framework of the balance just alluded to. If a fly-by robotic mission[9] to an interstellar destination such as the Oort Cloud (10,000 AU) is contemplated then it can be shown that present-day LAPPS can accomplish such a mission in an acceptably short time. To put these predictions in proper perspective, we focus on some recent results produced at Livermore[5,6] and analyzed in detail elsewhere [3]. A summary of the experimental data, along with the characteristics of the propulsion system it may evolve into, is given in Table 1. In these experiments, a high energy conversion efficiency was observed in that half of the laser energy (500J) appeared in the ejected beam and at an assumed rep rate of one kilohertz this results in a 500 kW jet power. At this rep rate, a power source that delivers at least a megawatt of electric power would be needed, and based on present-day power conversion analyses, a mass to power ratio of 5 mT per megawatt electric is considered reasonable[10]. It is anticipated, however, on the basis of the progress made in the various

1. Proton Beam
    - i) Particle Population = 6×10^{14}
    - ii) Mean Energy = 5.3 MeV
    - iii) Beam Energy = 500 J
2. Laser Beam
    - i) Wavelength ≈ 1 μm
    - ii) Pulse Length = 500 fs
    - iii) Intensity = 3×10^{20} W/cm^2
    - iv) Energy = 1 kJ
    - v) Power = 1 Petawatt
3. Target
    - i) Material = Gold Foil
    - ii) Thickness = 125 μm
    - iii) Focal Spot Radius = 9 μm
4. LAPPS Propulsion System
    - i) Rep Rate = 1 kHz
    - ii) Specific Impulse = 3.2×10^6 s
    - iii) Thrust = 3.1×10^{-2} N
    - iv) Nuclear System = 1MW
    - v) Vehicle Dry Mass = 5×10^3 kg

TABLE 1. Present-day LAPPS Parameters

components of the power supply that a five fold improvement in the specific mass can be achieved in the near future. Since the LAPPS propulsion system described in Table 1 may be judged to be adequate for robotic interstellar missions[11], it is not particularly suited for manned interplanetary missions due to the smallness of the thrust as discussed earlier. For this reason, much of the research that should be done in the area of developing present-day laser accelerated plasmas into viable propulsion devices that could meet the challenges of missions within the solar system must focus on means of enhancing the thrust.

SOME MISSIONS WITH LAPPS

As a measure of the propulsive capability of present-day LAPPS we consider two missions, a fly-by interstellar mission to the Oort cloud, and a round trip journey to Mars. For the first case the equation of interest are[9]:

$$t_f = \frac{m_i - m_f}{F} v_e \qquad (7)$$

$$S_f = \frac{m_i v_e^2}{F} \left[1 - \frac{m_f}{m_i} + \frac{m_f}{m_i} \ln\left(\frac{m_f}{m_i}\right) \right] \qquad (8)$$

$$v_f = v_e \ln \left[\frac{1}{1 - \dfrac{F t_f}{m_i v_e}} \right] \qquad (9)$$

where t_f is the one-way travel time to destination, m_i the initial mass of the vehicle v_e the exhaust velocity, S_f the one-way distance to destination, and v_f the final vehicle velocity at destination assuming it started from rest. We apply these equations to a robotic mission to the Oort cloud for which S_f = 10,000 AU and the results are shown in Fig 3 where travel time is plotted against thrust for two values of m_f, namely 5000 kg and 1000 kg. The first mass is that which is given in table 1 which is based on present-day nuclear power designs and the scaling of 5 mT per MWe. The second takes into account research progress that is expected to take place in reactor design, radiator materials and design etc which will lead to a value of 1 mT per MWe. We note that present-day LAPPS at an m_f = 5mT will accomplish this mission in about 700 years, but if its thrust can be increased to 25 Newtons the same journey will take about 26 years, well within a scientist's lifetime. We note, further, that at the reduced mass of 1 mT, present-day LAPPS will accomplish the same mission in about 313 years, and at 25N thrust the travel time is reduced to about 12 years.

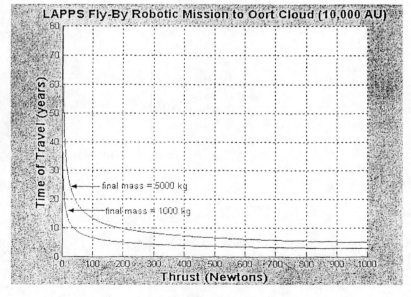

FIGURE 3. LAPPS Fly-By Robotic Mission to Oort Cloud (10,000 AU)

For the round trip to Mars we revert back to Eq (6) which can be cast in the following form

$$\tau_{RT} = 4 \left[\sqrt{\frac{Dg\,I_{sp}}{f\,\alpha}} - \frac{D}{g\,I_{sp}} \right] \tag{10}$$

where we have introduced the mass ratio

$$f = \frac{m_f}{m_i} \tag{11}$$

and

$$\alpha = \frac{P_j}{m_f} = \frac{Fg\,I_{sp}}{m_f} \tag{12}$$

which is the ratio of the jet power to the final (dry) mass of the vehicle. It can be shown[11] that the trip time can be optimized at $f = 1/4$ independent of the values of D and α. Although the travel time is minimum at this mass ratio, it is far from being economic since the propellant mass will be 3 times the dry mass, which in the case of LAPPS is 15 mT. For present-day LAPPS $\alpha = 1/10$, and at this value the optimum round trip to Mars will be about 350 days but at prohibitively large cost due to the large amount of propellant required. Going back to the un-optimized travel time, as given by Eq (6) and shown in Fig. 4 we find that the travel time by present-day LAPPS is 5193 days, which readily improves to 186 days upon increasing the thrust to 25 N. At the lower mass of 1 mT, the round trip time for current LAPPS goes down to 2323 days while at 25N that time is reduced to a mere 82 days-a very encouraging result indeed!

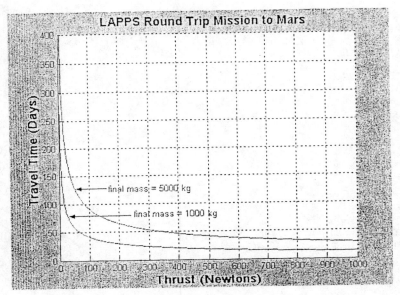

FIGURE 4. LAPPS Round Trip Mission to Mars

CONCLUSION

We have presented in this paper an advanced propulsion concept based on recent successful experiments in which ultrafast lasers were utilized in accelerating protons to relativistic speeds. We have shown that a LAPPS device based on present-day data is capable of producing several million seconds of specific impulse, albeit at modest thursts, and requires a nuclear power system to drive it. Our analysis also reveals that a modest increase in the thrust will allow such a vehicle to make a fly-by interstellar mission to the Oort cloud in a human's lifetime, and a round trip journey to Mars in about six months.

ACKNOWLEDGMENTS

This work was supported by the NASA Institute for Advanced Concepts (NIAC) of the University Space Research Association (USRA).

REFERENCES

1. Yu,W. et al, Phys. Rev. E **58**, 2456 (1999).
2. Sarachik E.S. and Schappert, G.T., Physical Review D, **1**, 2738 (1970).
3. Kammash, T. NIAC 01-01 Final Report, Dec (2001).
4. Maksimchuk, A. Gu, S. Flippo, K., and Umstadter D., Phys. Rev. Letters, **84**, 4108 (2000).
5. Snavely, R.A. et al, Phys. Rev. Letters, **85**, 2945 (2000).
6. Roth, M. et al, *High-Energy Electron, Position, Ton and Nuclear Spectrocopy in Ultra-Intense Laser-Solid Experiments on the Petawatt,* Proc. 1st Int. Conf. On Inertial Fusion Sciences and Applications, Bordeaux, France, Sept. 12-17 (1999), also UCRL-JC-135735, Sept. 16 (1999).
7. Kammash, T. Flippo K. and Umstadter D., 38th AIAA Joint Propulsion Conference, Indianapolis, IN July 7-10 (2002) paper AIAA 2002-4090.
8. Kammash, T. *Fusion Energy In Space Propulsion* AIAA progress in Astronautics and Aeronautics, vol. 167, Washington, D.C. (1995) p.69.
9. Kammash, T. J. Propulsion and Power **16**, 1100 (2000).
10. Mason, L. NASA Glenn. Res. Center, Private Communicate (2001).
11. Emrich Jr., W.J. 38th AIAA Joint Propulsion Conference, Indianapolis, IN, July 7-10 (2002) paper AIAA 2002-3931.

Near-Term Laser Launch Capability:
The Heat Exchanger Thruster

Jordin T. Kare

Kare Technical Consulting, 222 Canyon Lakes Pl., San Ramon CA 94583 jtkare@attglobal.net

ABSTRACT

The heat exchanger (HX) thruster concept uses a lightweight (up to 1 MW/kg) flat-plate heat exchanger to couple laser energy into flowing hydrogen. Hot gas is exhausted via a conventional nozzle to generate thrust. The HX thruster has several advantages over ablative thrusters, including high efficiency, design flexibility, and operation with any type of laser. Operating the heat exchanger at a modest exhaust temperature, nominally 1000 C, allows it to be fabricated cheaply, while providing sufficient specific impulse (~600 seconds) for a single-stage vehicle to reach orbit with a useful payload; a nominal vehicle design is described. The HX thruster is also comparatively easy to develop and test, and offers an extremely promising route to near-term demonstration of laser launch.

INTRODUCTION

Ground-to-orbit launch using a ground-based laser to power a rocket vehicle was originally proposed by Kantrowitz [1] and has been investigated by a number of researchers. The most widely-studied concept for ground-to-orbit laser propulsion uses pulsed lasers to convert an inert propellant to plasma, using either single laser pulses (e.g., [2], [3]) or double pulses (separate pulses to ablate a solid propellant and then heat the resulting gas) ([4], [5]). The main alternative approach has been laser-sustained plasma thrusters (e.g., [6], [7]) which would use a continuous-wave (CW) laser beam to create and maintain an absorbing plasma in a volume of flowing hydrogen gas, heating the gas to very high temperatures.

Both of these approaches have significant practical problems when incorporated into realistic vehicle and system designs, and are very difficult to develop beyond the scale of small demonstrations, such as those performed by Mead and Myrabo [8]. In particular, CW thrusters require large, precise focusing optics on the vehicle, and high-power, high-pressure windows that may not be feasible to build, while pulsed thrusters require very high peak laser powers, specific and precisely controlled pulse

CP664, *Beamed Energy Propulsion: First International Symposium on Beamed Energy Propulsion,*
edited by A. V. Pakhomov
© 2003 American Institute of Physics 0-7354-0126-8/03/$20.00

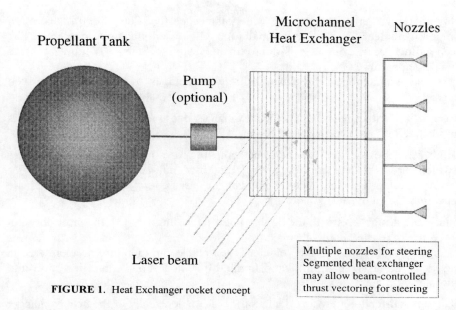

Propellant Tank

Microchannel
Heat Exchanger

Nozzles

Pump
(optional)

Laser beam

Multiple nozzles for steering
Segmented heat exchanger
may allow beam-controlled
thrust vectoring for steering

FIGURE 1. Heat Exchanger rocket concept

properties, and, for most concepts, focusing optics on the vehicle to produce sufficient flux at the propellant.

In the early 1990's, a third alternative for ground-to-orbit propulsion was proposed: the heat exchanger (HX) thruster. The original HX thruster, vehicle, and launch system concepts are described in [9], and a development program for the HX thruster is described in [10].

Recently, we have developed a more detailed analytical model and conceptual design for a heat exchanger vehicle. This paper briefly reviews the HX thruster concept and summarizes the current vehicle design, thruster and vehicle models, and modeling results.

THE HEAT EXCHANGER THRUSTER CONCEPT

The heat exchanger (HX) thruster concept is illustrated in Figure 1: a simple pressurized tank of liquid hydrogen is connected to a flat-plate heat exchanger, optionally through a pressure-boosting pump. The heat exchanger absorbs laser energy and transfers it to the hydrogen, producing hot hydrogen gas which is exhausted through one or more conventional nozzles. By using a laminar-flow microchannel structure, the heat exchanger can have low areal density (<10 kg/m^2), very high specific power (up to 1 MW/kg), and low pressure drop.

Key to the practicality of the HX thruster is that, with hydrogen, moderate temperatures will produce sufficient Isp for ground-to-orbit launch. The original HX thruster was assumed to have an output temperature of 1000 C (1273 K). This is low enough that the heat exchanger can be build from relatively common materials, notably nickel, rather than the exotic materials needed for 2000 - 3000 K operation assumed in nuclear- and solar-thermal (and CW laser plasma) thrusters. The corresponding specific impulse is approximately 600 seconds. (A more accurate value is given below.)

The heat exchanger rocket has multiple advantages over other laser-driven rockets:

- Compatible with any laser. The heat exchanger does not care what wavelength is used, and does not need a pulsed beam.

- High efficiency. The low operating temperature minimizes reradiation losses (which vary as T^4 and therefore as Isp^8), and moderate-temperature hydrogen is a nearly ideal propellant, so over 90% of the laser energy reaching the vehicle becomes useful exhaust kinetic energy.

- Flat-plate design. As with the double-pulse planar thruster [5], the laser does not need to strike the heat exchanger from a specific angle, as it does with designs using focusing optics on the vehicle. The vehicle can therefore accelerate at an angle to the laser beam, allowing flexible trajectories and direct insertion into circular orbit.

- The thrust direction (nozzle orientation) is also independent of the heat exchanger orientation. This allows the heat exchanger to be oriented for good aerodynamics and/or efficient beam collection, an advantage over even pulsed planar thrusters.

- **Simple development**. Except for the heat exchanger itself, all the components can be designed and tested using conventional aerospace techniques and facilities. The heat exchanger does not require any specific laser for testing, and can be tested with non-laser heat sources such as solar furnaces, arc lamps, or even electric heaters.

HX VEHICLE CONCEPT

The current HX vehicle concept is shown in Figure 2. This vehicle is sized for 100 MW of received power.

Relative to the original vehicle concept, this design has several new features:

1) "Side-fire" design: the heat exchanger forms one side of the vehicle, rather than forming the vehicle base ("tail-fire" design). This allows the heat exchanger area to be much larger (weight permitting) for a given overall vehicle size, while maintaining a relatively conventional low-drag shape. It also allows the vehicle to start considerably uprange of the laser and fly nearly horizontally over the laser (Figure 3), which extends the powered trajectory length and duration significantly.

2) Large heat exchanger area and reduced flux: The heat exchanger described in [9] was designed to handle 10 MW/m^2, in order to minimize the heat exchanger mass and maintain a reasonable base area for a tail-fire vehicle. However, current concepts for the laser system (using non-coherent laser diode arrays [11] make the flux at the

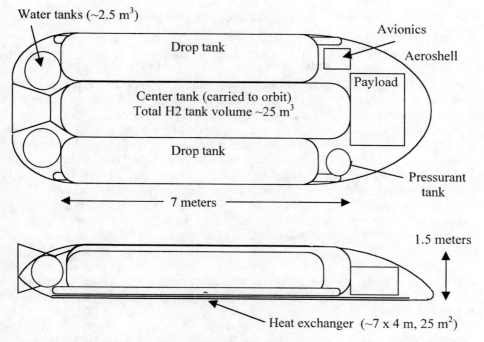

Water tanks (~2.5 m³)

Drop tank

Avionics

Aeroshell

Payload

Center tank (carried to orbit)
Total H2 tank volume ~25 m³

Drop tank

Pressurant
tank

7 meters

1.5 meters

Heat exchanger (~7 x 4 m, 25 m²)

FIGURE 2. HX vehicle concept, including droppable hydrogen tanks. Dimensions are approximate.

vehicle even more of a system design driver than for conventional coherent lasers. With the side-fire configuration, a larger heat exchanger area is acceptable, and the baseline heat exchanger is now approximately 25 m², operating at a maximum flux of 4 MW/m². Higher fluxes may still be desirable for higher-power launch systems; for fixed vehicle proportions and fixed trajectory, the heat exchanger area will scale as the mass (and therefore laser power) to the 2/3 power, so the heat exchanger flux would naively scale as (laser power)$^{1/3}$.

3) Improved tanks: The original vehicle assumed a spherical aluminum hydrogen tank and a very low tank pressure (500 kPa, ~70 psi) to minimize tank mass. The low tank pressure drove both the heat exchanger design (to minimum pressure drop) and the "chamber" (heat exchanger output) pressure; the low output pressure in turn drove the nozzle dimensions and made operation at low altitude highly inefficient.

Modern composite tanks offer substantially higher performance than aluminum, currently up to $P_b V/W$ (tank burst pressure x volume over weight) of 1.6 x 10⁶ inches (approximately 400,000 N-m/kg), vs. <3 x 10⁵ inches for aluminum, and optimize to a cylindrical, rather than spherical, geometry [12]. For large single-use tanks with no cycle-life requirement, even higher values should be possible [13], and the current model assumes $P_b V/W = 2$ x 10⁶ inches. The new nominal vehicle design therefore assumes a tank pressure of 1.73 MPa (250 psi) and a chamber pressure of 1.04 MPa

(150 psi). (This value has not been optimized, and the optimum for a real system may be somewhat higher or lower, but probably within a factor of 2).

4) Water injection. High exhaust velocities are inherently energy-inefficient at low vehicle velocity, i.e., at takeoff. Laser-launched vehicle designs also tend to be light and low density, with large frontal area, resulting in high drag forces and low velocities in the atmosphere. The HX vehicle also has a hot hydrogen exhaust -- problematic near a ground launch site -- and low chamber pressure.

In [9], it was assumed that some alternate propulsion mechanism would be needed to get the vehicle through most of the atmosphere before starting the HX thruster. Possibilities ranged from aircraft- or balloon-launched vehicles, to laser or chemical (hydrogen fueled) ramjets, to small solid-rocket first stages. These options add substantially to either the vehicle complexity or the system logistical complexity -- especially since any chemical jet or rocket would reintroduce range safety issues (the possibility of a powered, out of control vehicle) that pure laser propulsion eliminates.

In the current design, these problems are solved by a classic propulsion enhancement: water injection. By injecting water into the hydrogen stream downstream of the heat exchanger, the mean molecular weight of the exhaust is increased and the Isp lowered, without affecting the heat exchanger operation or properties. Nozzle efficiency is (to first order) unaffected, and the thrust increases in inverse proportion to the Isp.

The mass flow rate increases as the square of the thrust, but a vehicle can carry several times the mass of water as it carries hydrogen, and still maintain higher initial thrust-to-weight. Because water (1000 kg/m^3) is ~14 times denser than liquid hydrogen (70 kg/m^3), the water tank mass remains small compared to the hydrogen tank mass.

5) Drop tanks. The vehicle could be configured with a single large hydrogen tank, at some reduction in total tank mass. However, at the current system parameters there is a large benefit to stretching the definition of single-stage-to-orbit and dropping part of the propellant tank volume, along with the vehicle aerodynamic shroud (aeroshell) and possibly other hardware, once clear of the atmosphere. In the three-tank design shown, the outer two tanks are discarded with the aeroshell. The baseline vehicle is not designed to be reusable, so dropping tanks is not an additional hardware expense. As with multistage rocket systems, the dropped hardware does increase downrange safety issues, and may constrain the allowable launch sites and trajectories. However, the hardware in question is extremely light (typically 200 kg total) compared to current expendable stages or, e.g., the Shuttle external tank, so this may be less of an issue. Depending on the trajectory, the dropped components may also be designed to break up or burn up on reentry.

(Larger systems would be expected to have even "skinnier" tanks compared to the heat exchanger, since increasing the laser power by a factor f would increase the nominal heat exchanger dimensions by f$^{1/2}$, but increase the tank dimensions by only f$^{1/3}$. Conversely, at lower laser powers, and assuming constant heat exchanger flux, the propellant tanks tends to dominate the vehicle dimensions.)

TABLE 1. HX Thruster Nozzle Characteristics

Chamber pressure (Pc), MPa	1.04 (150 psi)
Throat diameter, cm	14.7
Max. area ratio	50
Exit angle, degrees	20
Length factor	0.7
Mean k (Ratio of specific heats) in nozzle	1.385
HX exit temperature, K	≥1273 (1000 C)
Chamber stagnation temp., K	1330
c star, m/s	3434
Vacuum Isp, s	591

SYSTEM MODEL

The thruster and vehicle models are components of an overall laser launch system model written in Mathematica. The model contains a trajectory integrator which permits simulating vehicle flight with user-settable control laws, to determine the actual trajectory and mass ratio required to reach orbit. The difference between the vehicle dry mass and the calculated final mass is the payload mass. The integrator allows a pseudo-staging event (based on time, velocity, or other factors) where the vehicle parameters and control laws can change discontinuously; this is used to simulate dropping tanks.

Details of the whole model are beyond the scope of this paper, but some of the factors included in the trajectory calculation are:

- Gravity
- Air drag vs. altitude and velocity, with velocity-dependent drag coefficient
- Spherical Earth with rotation
- Vehicle initial altitude, position (relative to the laser), and velocity
- Vehicle (and therefore heat exchanger) attitude relative to the laser beam
- Laser beam profile, including diffraction (Gaussian beam), source dimensions (for non-coherent sources) and atmospheric turbulence effects as a function of vehicle elevation
- Atmospheric attenuation

Unlike earlier approximations, this model calculates the actual power intercepted by the heat exchanger, rather than assuming a fixed power or simple function of range and elevation.

THRUSTER MODEL

The thruster model uses the formulas given by Sutton [14] to calculate thrust, I_{sp}, and nozzle size for specified input gas temperature, chamber and ambient pressure, and gas characteristics. Baseline nozzle characteristics are given in Table 1.

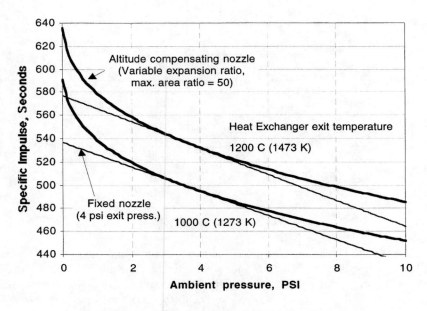

FIGURE 3: Calculated I_{sp} vs. ambient pressure for HX thruster

The chamber temperature is slightly higher than the heat exchanger exit temperature due to the kinetic energy of the flowing gas (Mach number M = 0.5) at the heat exchanger exit..

The thruster model allows either a fixed nozzle, with the expansion ratio set by ambient pressure at the (specified) minimum operating altitude, or an altitude-compensating nozzle with a fixed maximum expansion ratio, such as an aerospike or mechanically-variable nozzle. (Variable nozzles are difficult to build for chemical rockets due to the problems of cooling the moving elements; at the low operating temperatures and pressures of the HX rocket, cooling should not be an issue.) For the nozzle parameters given above, the calculated Isp vs. ambient pressure and Isp vs. altitude are shown in Figures 3 and 5. Note that a real vehicle may choose to use multiple nozzles for improved directional control or better aerodynamics.

Water injection is modeled simply as a multiplier in mass flow rate, and a proportional decrease in Isp.

HEAT EXCHANGER DESIGN AND FABRICATION

The heat exchanger design presented in [9] and [10] remains a baseline for the HX thruster. This design uses the laminar-flow microchannel concept, in the form proposed by Tuckerman [15] for liquid cooling of semiconductors: the heat exchanger consists of many narrow channels, nominally with high-aspect-ratio rectangular cross section, sized to maintain laminar flow and thus low pressure drop

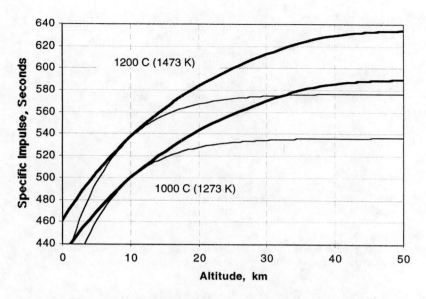

FIGURE 4. Calculated I_{sp} vs. Altitude for HX thruster. See Fig. 3 for curve details.

within the heat exchanger. The original channel dimensions were 200 um wide x 2 mm deep x 3 cm long, with 200 um wide fins separating the channels. The short channel length was required both to minimize pressure drop in the channel and to provide sufficient overall flow cross section to keep the flow in the channels subsonic, but required a sophisticated manifold structure on the back of the heat exchanger to interconnect channel assemblies.

The heat exchanger parameters have not been recalculated for the revised design, but qualitatively, higher hydrogen pressure and higher allowed pressure drop will allow longer channels. Lower flux will allow lower aspect ratio channels and thinner fins, both of which reduce the heat exchanger areal density, so the heat exchanger specific power (nominally 1 kg/MW) should remain roughly constant. The heat exchanger mass is modeled as 1 kg/MW + 1 kg/m^2 to allow for area-dependent support structure and manifolds.

The baseline fabrication process for the heat exchangers remains unchanged from the original concept [16]: the channels and fins are formed from from stacks of filler and nickel ribbons. The stacks are electroplated on the front and back surfaces to seal the channels and form inlet and outlet manifolds, and the filler material is dissolved or burned out.

HEAT EXCHANGER VEHICLE MASS

The vehicle mass model includes allocations for:

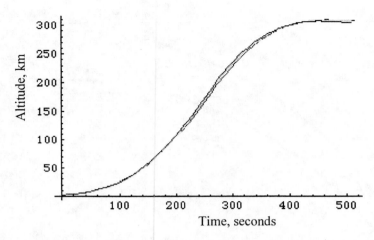

FIGURE 5. Sample launch model output: vehicle altitude vs. time. Note low velocity and acceleration at low altitude. (Plot shows 4 very similar trajectories overlaid.)

- Tank mass $M_{tank} = (M_{H2}/\rho_{H2}) * P_{tank} / (P_b V/W)$
- Pressurization hardware $M_{press} = 0.2\ M_{tank}$
- Residual pressurant gas   Tank volume of H_2 gas @ P_{tank} and 300 K
- Heat exchanger $1\ kg/MW + 1\ kg/m^2$
- "Plumbing" (pipes, valves, and other hardware)
- Aerodynamic shroud $2\ kg/m^2$ (1 mm thick composite structure)
- Avionics   6 kg for guidance, telemetry, and control
                                    5 kg for cables
                                    100 W-hr of Li primary batteries
- Structure   10% of other dry mass

Note that many of these are fixed by the initial vehicle parameters. The main exception is the hydrogen tank mass, which is calculated after a model trajectory has been flown and the amount of hydrogen expended has been determined. Separate hydrogen tank masses are calculated for the first and second "stage" trajectories, so that the effect of dropping empty tanks at the transition can be included correctly. Plumbing mass is calculated from propellant mass flow rate and heat exchanger size; pipes are assumed to have PV/W of 100,000 inches

MODEL RESULTS

Table 2 summarizes the input parameters for a nominal vehicle and trajectory. A sample plot from the corresponding trajectory is shown in Figure 5. Figure 6 shows the mass budget for this vehicle. Figure 7 shows the payloads achievable with some variations around the nominal trajectory. The strong effect of initial vertical velocity (which is present for any type of launcher) strongly suggests that some type of

TABLE 2. Baseline Launch Characteristics

Initial laser power at vehicle	100 MW
Heat exchanger exit temperature	1000 C
Nozzle type	Altitude compensating
Launch Altitude	2 km
Launch Velocity (V_0) Horizontal Vertical*	0 100 m/s
Water injection	Linearly decreasing, launch to t=150 s
Thrust vector control	Vertical, launch to t=70 s Fixed angle, t=70 s to tank drop Horizontal, tank drop to orbit
Flight time To tank drop To orbit	270 s 513 s
Tank drop altitude	188 km
Tank drop velocity	1.32 km/s upward, 1.23 km/s downrange
Maximum laser range**	495 km
Payload to orbit	122 kg

*For ground launch, V_0 is provided by a pneumatic catapult or similar mechanism
**Primary laser site 400 km downrange of launch site

catapult be used to give the vehicle an initial boost; a 100 m/s catapult requires only 50 meters of acceleration distance at 10 g's acceleration.

One factor in this model is arguably unrealistic, although not unphysical. To allow relatively long trajectories with moderate ranges, and without having the vehicle go over the laser's horizon, the laser site is located near the mid-point of the trajectory, 400 km downrange of the launch site. This requires a second laser (with much shorter-range optics) to power the vehicle from launch until it reaches a reasonable elevation as seen from the main laser. Such a double laser system may in fact be reasonable to build, as it divides the launch time between two lasers and increases the maximum launch rate nearly two-fold. A single laser site can be used, but tends to drive the system to shorter and higher "lofted" trajectories and impose much tighter requirements on beam divergence.

DISCUSSION AND CONCLUSIONS

The two largest uncertainties about the heat exchanger thruster concept have always been whether the heat exchanger itself could be built with the necessary performance, and whether the vehicle design would close -- that is, whether a realistic vehicle mass budget would leave a useful (or even positive) value for payload mass.

The first question can only be fully resolved by building and testing heat exchangers, but by assuming higher operating pressures and lower fluxes than earlier designs, the current design greatly relaxes the heat exchanger design constraints.

Dry mass

Heat exchanger	125.0
H2 tank	38.8
Press. hardware	42.0
Pressurant	5.1
Nozzle	5.4
Plumbing	24.1
Avionics	12.6
Structure	47.0
Payload	122.1
Tot. taken to orbit	422

Dry mass (dropped)

Water tank	108.0
H2 tank	63.1
Pressurant	12.4
Aeroshell	33.4
Tot. dropped	217

Liquid Hydrogen

In drop tanks	1277
In center tank	784
Total LH2 mass	2061

Water	2700
Gross LiftOff Weight	5400

FIGURE 6. Mass budget for baseline laser launch vehicle (all values in kg)

Based on the results of fairly detailed modeling, as presented here, the answer to the second question is a definite yes. With water injection to boost initial thrust, vehicles can carry payloads directly from near sea level to low Earth orbit.

Positive payload can be achieved without air launch, drop tanks, or other aids, although with current assumptions, the predicted payloads are small. With drop tanks, however, the predicted payloads can exceed 150 kg for a 100 MW vehicle, with a dry payload fraction of 25%. There is thus considerable margin for vehicle weight growth and other inevitable differences between model and reality.

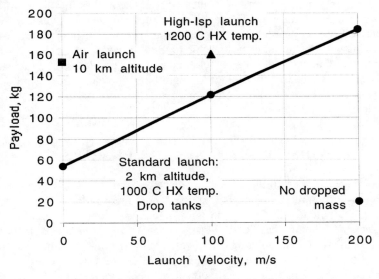

FIGURE 7. Payload for various HX vehicle launch options

There is also still room for extensive optimization of the vehicle design and trajectory, and improvements (such as higher operating temperature) in the system performance. Larger systems will achieve even higher payload mass fractions.

In the decade since the heat exchanger thruster was first proposed, its advantages over other types of laser-propulsion thrusters have remained unchanged. It should be seriously considered in any plan for development of laser launch.

REFERENCES

1. Kantrowitz, A., "Propulsion to Orbit by Ground-Based Lasers," *Astronautics and Aeronautics* **10**(5) 74 (1972).
2. Phipps, C. R., Reilly, J. P., and Campbell, J. W., "Optimum Parameters for Laser-Launching Objects Into Low Earth Orbit," *Lasers and Particle Beams*, **18**(1) 1-35 (2001)
3. Pakhomov, A. V. and Gregory, D. A., "Ablative Laser Propulsion: An Old Concept Revisited," *AIAA Journal* **38**(4), 725-727 (2000).
4. Douglas-Hamilton, D. H., Kantrowitz, A., and Reilly, D. A., "Laser Assisted Propulsion Research," in *Progress in Astronautics and Aeronautics*, Vol. 61, ed. by K. W. Billman, AIAA New York, 1978, pp. 271-278.
5. Kare, J. T., "Pulsed Laser Propulsion for Low Cost, High Volume Launch to Orbit", *Space Power J.* **9** (1), 1990. (Also UCRL-101139, LLNL 1989).
6. Krier, H. and Mazumder, J., "Fundamentals of CW Laser Propulsion," *Proceedings of the AFOSR Workshop on Laser Propulsion,* AFOSR-TR-88-1430 (February, 1988).
7. Komurasaki, K., Arakawa, Y., Hosoda, S., Katsurayama, H., and Mori, K., "Fundamental Researches on Laser-Powered Propulsion," presented at 33rd AIAA Plasmadynamics and Lasers Conference (Maui, HI, USA, May, 2002), AIAA-2002-2200, AIAA 2002.
8. Mead, F.B., Jr., Myrabo, L. N., and Messitt, D.G., Flight Experiments and Evolutionary Development of a Laser-Propelled, Trans-Atmospheric Vehicle," SPIE **3343**, pp. 560-563, 1998.
9. Kare, J.T., "Laser Powered Heat Exchanger Rocket for Ground-to-Orbit Launch," *J. Propulsion and Power* **11**(3), 535-543 (1995).
10. Kare, J. T., "Development of Laser-Driven Heat Exchanger Rocket for Ground-to-Orbit Launch", International Astronautics Federation IAF 92-0614 (1992), presented at the World Space Congress, Washington D.C., 1992. (Also UCRL-JC-111507, LLNL 1992)
11. Kare, J.T., "Laser Power Beaming with Non-Coherent Diode Arrays", UCRL-JC-116095 (LLNL 1994).
12. Mitlitsky, F. and Myers, B., "Development Of An Advanced Composite Lightweight High Pressure Storage Tank For On-Board Stoarge Of Compressed Hydrogen," UCRL-MI-123802, LLNL 1996.
13. Weisberg, A., LLNL/DOE tankage development program, private communication, 2000.
14. Sutton, G. P., *Rocket Propulsion Elements*, 6th edition, Wiley, New York, 1992, pp. 43-84.
15. Tuckerman, D. B., "Heat-Transfer Microstructures for Integrated Circuits," UCRL-53515, LLNL 1984.
16. Steffani, C., "Electroforming Thin Channel Heat Exchangers," in *Proceedings of the Electroforming Symposium, SURFIN 91,* Amer. Electroplaters and Surface Finishing Society (1991.

In-Tube Laser Propulsion Configurations

Sukyum Kim*, **, Naohide Urabe*, Hiroyuki Torikai*, Akihiro Sasoh*,
and In-Seuck Jeung**

*Shock Wave Research Center, Institute of Fluid Science, Tohoku University, Sendai, Japan
**Department of Aerospace Engineering, Seoul National University, Seoul, Korea

Abstract. Laser propulsion research activities at Shock Wave Research Center, Institute of Fluid Science, Tohoku University, focus themselves on 'in-tube' configurations. The thrust is enhanced in a confined acceleration region. Other advantages are obtained from the viewpoint of practical application. We are now investigating various extensions of the Laser-driven In-Tube Accelerator (LITA); (1) ablative in-tube propulsion, (2) thrust enhancement using applied magnetic field, (3) plasma pre-generation using a pilot laser irradiation, (4) demonstration of supersonic laser propulsion. The progresses in these subjects are presented.

INTRODUCTION

Laser Propulsion is a kind of propulsion system reducing the launch cost by the use of laser beam energy. Recently, Mirabo et al.[1,2] and Schall et al.[3] demonstrated vehicle launch of a projectile with a 10 kW repetitive-pulse CO_2 laser. In most of laser propulsion experiment, an object is accelerated in an open atmosphere.

Sasoh[4] developed the concept of Laser-driven In-Tube Accelerator (LITA). In LITA, an object is accelerated not in an open air but in a tube. The propulsion performance is enhanced by the confinement effect. Also, driver gas species is selectable, and the fill pressure can be tuned.

So far various experiments of LITA have been carried out at Institute of Fluid Science, Tohoku University. A 3.0 gram projectile has been launched vertically in a 25 mm-bore tube using a repetitive pulse CO_2 TEA laser. A momentum coupling coefficient higher than 300 N/MW has been obtained. The coupling coefficient is almost in proportion to the reciprocal of the speed of sound. This tendency is caused by the dependence of the duration time of the propulsive force on the speed of sound[5,6].

In order to improve LITA performance, further studies are currently performed or planned. LITA with laser ablation, thrust enhancement with applied magnetic fields, the plasma re-generation using a pilot laser irradiation, supersonic laser propulsion using a ballistic range, and numerical simulation of the above-mentioned operations.

In this paper, concept of advanced LITA configurations and progresses in experiments are reported.

CP664, *Beamed Energy Propulsion: First International Symposium on Beamed Energy Propulsion*,
edited by A. V. Pakhomov

LITA WITH LASER ABLATION

In laser propulsion, high coupling coefficient is needed for launching. Laser ablation is a possible method to enhance the performance. Our study utilizes ablation in the acceleration tube for propulsion (Figure 1). The projectile has ablator material placed in the shroud. Moreover, by means of setting vacuum in the tube condition, there is no fluid dynamical drag against launching.

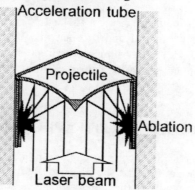

Figure 1. Schematic of ablative propulsion in tube

Experimental Apparatus

Currently, we are conducting preliminary experiment in the atmospheric air. Using a ballistic pendulum, we have measured the impulses obtained with and without ablating material placed in the projectile shroud (Figure 2). The impulse is approximately in proportion to the maximum deflection of the pendulum. The deflection was measured with a combination of a knife-edge, a He-Ne laser and a photodiode.

A CO_2 TEA laser is used as the energy source. The nominal output energy per pulse is 10 J at a maximum, and the pulse width is 2 μs. The highest repetition rate is 50 Hz. The output beam is introduced from the laser to the projectile using plane molybdenum mirrors.

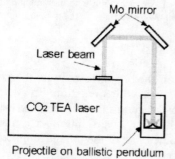

Figure 2. Experimental setup

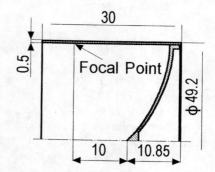

Figure 3. Cross-sectional view of projectile

The projectile is shown in Figure 3. It is made of aluminum alloy, A7075-T6, and comprises a projectile reflector and a shroud. The mass of the projectile is 12.24g. The projectile has a parabolic mirror, by which the laser beam is focused onto circular peripheral in the projectile shroud. POM (Polyoxymethylene) is used as the ablator. The ablator is placed on the inner wall.

Experimental Results

So far a coupling coefficient of 32 N·s /MJ ($\pm 6\%$) has been obtained with the ablator at the laser energy of 10 J and in open air. Without ablator, the coupling coefficient is 16 N·s /MJ ($\pm 11\%$). Although the impulse is doubled with the ablator, the coupling coefficient is not satisfactory. Future investigations are currently conducted.

THRUST ENHANCEMENT WITH APPLIED MAGNETIC FIELD

In magnetic field, the plasma get the force perpendicular to the magnetic field. By this force, plasma can be confined in the place around the focusing point. From the plasma concentration, it can be expected to create a stronger blast wave with an applied magnetic field and to make higher pressure behind the projectile.

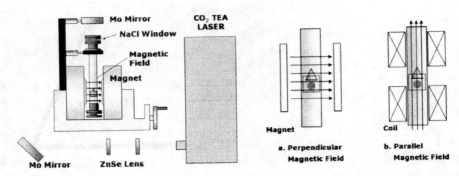

Figure 4. Experimental Set-up

Figure 5. LITA operation with a. perpendicular and b. parallel magnetic fields to the acceleration tube

Experimental Set-up

Figure 4 illustrates the experimental setup of the experiment with horizontal magnetic field. The CO_2 TEA laser can output 5 J/pulse with a repetition frequency of up to 100 Hz. After two ZnSe lenses and two molybdenum mirrors, the cross-section of the laser beam is reduced and the reduced laser beam introduced into the acceleration tube through the NaCl window. The other side of the tube is plugged with a brass flange. The acceleration tube is made of acryl.

In this experiment, a magnetic field perpendicular to acceleration tube axis (Figure 5a) is applied. This magnet can create 0.6 T at a maximum a magnetic field.

Experimental result

The experiment is carried out with the perpendicular magnetic field. (Figure 5a) The coupling coefficients with and without magnetic field are compared to each other. Xenon is used as the driver gas and the fill pressure is varied from 10 kPa to 300 kPa. But, the effect of magnetic field cannot be confirmed. Currently, experiment of the parallel magnetic field (Figure 5b) is prepared for.

PLASMA PRE-GENERATION USING A PILOT LASER IRRADIATION

In the plasma state, the energy of laser beam is easy to be absorbed. Absorption of laser beam energy by the plasma can reduce the unused laser beam energy. And absorbed energy can be used to make stronger blast wave. To make plasma state before the arrival of CO_2 laser, a pilot laser that is focused at the same focal point with the CO_2 laser can be used.

Operation

Figure 6 shows the operation of the pilot laser and the main laser schematically. First, a pilot laser beam from the rear side of tube is focused and the plasma is created around the focal point by the breakdown. A beam of main laser from the frontal side becomes focused in the area of plasma state and beam energy can be absorbed more efficiently.

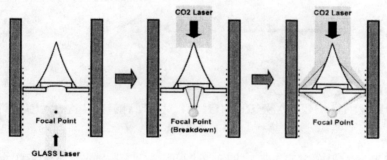

Figure 6. Operation of pilot laser and CO_2 laser

Experimental Set-up

This experiment will be carried out using ballistic pendulum. (Figure 7) And the impulse can be compared with that of normal LITA operation. Also visualization is planned. A glass laser will be used as a pilot laser. The output energy is 6 J/pulse at a maximum.

Figure 8 shows the projectile for this experiment. This projectile has two reflecting surfaces that have the same focal point.

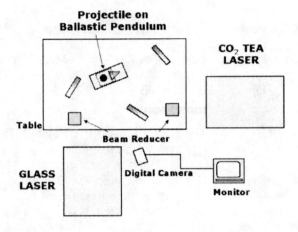

Figure 7. Experimental set-up using pendulum in order to measure the thrust

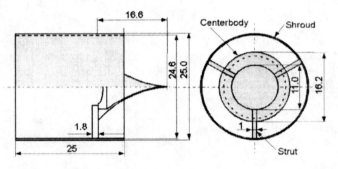

Figure 8. Projectile

SUPERSONIC LASER PROPULSION USING A BALLISTIC RANGE

At Shock Wave Research Center, a ballistic range which is utilized as a ram accelerator, is able to be operated as a single-stage light gas or powder gun. It has a bore of 25mm. With smokeless powder, it can launched a 12-gram projectile up to 1.3km/s, using the facility, demonstration of laser propulsion at a supersonic regime is planned to be conducted. A laser beam is supplied from a single-pulse CO_2 TEA laser. It is capable of producing a laser energy of 380 J. In order to introduce the laser beam into a vacuum chamber, a temperature controlled NaCl window is installed. This experiment is planned to be conducted early 2003.

NUMERICAL SIMULATION

Efforts for numerically simulating the experimental results are currently made. The interactions among the laser-induced blast wave, the projectile and the acceleration tube wall are main interests to study.

In this study, the viscous effects are assumed being small since the propulsion mechanism is governed by the shock wave dynamics. Therefore, Euler equation for an axi-symmetrical geometry is with ideal gas assumption and heat addition term to model the laser focus used as the governing equation[7].

$$\frac{\partial Q}{\partial t} + \frac{\partial F}{\partial x} + \frac{\partial G}{\partial y} + \frac{1}{y} H = W$$

$$Q = [\rho \quad \rho u \quad \rho v \quad e]^T , \quad W = [0 \quad 0 \quad 0 \quad q]^T$$

Since the LITA systems are governed by the unsteady propagation of a blast wave, the numerical method should be have a high accuracy with respect to space and time. A finite volume method is used for the discretization of the governing equations in space. The Roe's approximate Riemann solver is used for the evaluation of numerical flexes at computational cell interfaces with MUSCL extrapolation method and minmod limiter function to keep higher order spatial accuracy near discontinuities while maintaining the TVD characteristics[8].

Crank-Nicolson method is used for a second-order accurate time integration. The temporal stiffness of the present problem is given at the initial period of computation by the energy source function, modeled as a square pulse. Because the stiffness is posed by a given source function and is independent of fluid dynamics, it cannot be eliminated by only introduce the second-order time accuracy. Therefore it is inevitable to use small time step, especially for starting period of calculation. The Crank-Nicolson method permits a relatively-larger time step than second order explicit scheme, so the expense of additional computation load in comparison with a simplest first order explicit scheme is small.

SUMMARY

In the present study, experiment using laser ablation in LITA has been conducted. LITA with magnetic field perpendicular to the tube did not yield a significant performance improvement.

Experiment of LITA with parallel magnetic field and the plasma pre-generation using a pilot laser are currently in preparation. Supersonic laser propulsion is also planned. Numerical simulation for the experimental is now in progress.

ACKNOWLEDGMENT

This project was supported by Japan Society for the Promotion of Science as Grant-in-Aid for Scientific Research (S) # 13852014. The first author's stay in Tohoku University was supported by BK21 Program sponsored by Seoul National University.

REFERENCES

1. Mirabo, L.N., Messitt, D.D., and Mead Jr., F.B., "Ground and Flight Test of a Laser Propelled Vehicle" AIAA paper-98-1001, 1998.
2. Mead Jr., F.B., Mirabo, L.N., and Messitt, D.D., "Ground and Flight Test of a Laser-Boosted Vehicle" AIAA paper-98-3735, 1998.
3. Schall, W.O., Bohn, W.L., Eckel, H.-A., Mayerhofer, W., Riede, W. and Zeyfang, E., "Lightcraft experiment in Germany" *High power Laser Ablation III, proc. SPIE* 4065, pp472-481, 2000.
4. Sasoh, A., "Laser-driven in-tube accelerator (LITA)" *Rev.Sci. Instrum.*, Vol. 72, No. 3, 2000.
5. Sasoh, A., J.-Y. Choi, I.-S. Jeung, Urabe, N., Kleine, H. and Takayama, K., "Impulse Enhancement of Laser Propulsion in Tube" *Postcpy Astronautyki*, 27, pp.40-51, 2001.
6. Sasoh, A., Urabe, N., Kim, S., and Jeung, I.-S., "Impulse Scaling in Laser-driven in-tube" *Applied Physics "A"*, 2001, Submitted
7. Choi, J.-Y., Sasoh, A.,Jeung, I.-S.,Kim, S., Cho, H.-J.,"Numerical and Experimental Studies of Laser-Driven Ram Accelerator," AIAA paper-2001-3924(2001).
8. Choi, J.-Y., Jeung, I.-S. and Yoon, Y., "Computational Fluid Dynamics Algorithms for Unsteady Shock-Induced Combustion, Part 1: Validation," *AIAA Journal*, Vol. 38, No. 7, pp.1179-1187, July 2000.

An Experiment To Demonstrate Spacecraft Power Beaming and Solar Cell Annealing Using High-Energy Lasers

Capt Richard Luce, USAF and Dr. Sherif Michael

Naval Postgraduate School, Space Systems Academic Group
699 Dyer Rd, Monterey, CA 93943-5000

Abstract. Satellite lifetime is often limited by degradation of the electrical power subsystem, e.g. radiation-damaged solar arrays or failed batteries. Being able to beam power from terrestrial sites could alleviate this limitation, extending the lifetime of billions of dollars of satellite assets, as well as providing additional energy for electric propulsion that can be used for stationkeeping and orbital changes. In addition, laboratory research at the Naval Postgraduate School (NPS) has shown the potential to anneal damaged solar cells using lasers. This paper describes that research and a proposed Maui experiment to demonstrate the relevant concepts by lasing PANSAT, an NPS-built and operated spacecraft.

INTRODUCTION

Science fiction works as far back as *War of the Worlds* have vividly demonstrated the concept of directed energy. Perhaps the most visually stunning motion picture example is the Death Star from the *Star Wars* series of movies. In these cases, the directed energy was used as a weapon, transferring tremendous amounts of power for destructive purposes.

Humankind's ingenuity has imagined other applications for wireless power transfer, one of which is beaming power from the ground to spaceborne satellites. This idea, which seems at first glance to belong in the science fiction movies, actually belongs to the realm of science fact.

There are many reasons to implement such a seemingly far-fetched idea. This paper will explore the motivations for spacecraft power beaming, peruse the history of work done in the field, discuss the current and future status of power beaming programs, and describe a proposed experiment to demonstrate and validate the concept of laser power beaming.

There are two main missions which can benefit from beamed power from the ground: propulsion and electrical power supplementation. Electrical or thermal engines can be used to transfer satellites from low earth orbit (LEO) to geosynchronous orbit (GEO), to reposition GEO satellites, or for stationkeeping. In addition, beamed power can provide power to satellites during eclipse or after on-board batteries have failed, in essence replacing the sun's rays. Finally, lasers could be used to anneal radiation-damaged solar cells on orbit, extending the lifetimes of

CP664, *Beamed Energy Propulsion: First International Symposium on Beamed Energy Propulsion,*
edited by A. V. Pakhomov
2003 American Institute of Physics 0-7354-0126-8

these extremely valuable assets. Laser power transmission to space makes these applications feasible.

POWER BEAMING FOR ELECTRIC PROPULSION

Chemical rocket engines are powerful but heavy and have relatively low specific impulse (I_{SP} in the vicinity of 300 seconds). Electric thrusters such as plasma thrusters and ion engines produce low thrust but have much higher I_{SP} (around 1,000 to 5,000 seconds). Thermal thrusters, which function by having a large mirror that concentrates solar flux or laser energy to directly heat hydrogen fuel, fall somewhere in the middle, producing moderate thrust and moderate I_{SP} around 600 seconds [1]. Reference [2] also describes an ablation microthruster concept where a less than ten watt laser ablates material which is ejected to generate a small amount of thrust.

When it comes to orbit raising, that is transfer of satellites from LEO to GEO, this has become very big business. An average GEO satellite weighs 3,500 pounds whereas the chemical rocket to deliver it to orbit typically weighs four to five times that amount. Launching from Earth to LEO costs about five dollars per pound, but the cost to transfer from LEO to GEO is about one hundred million dollars! With an average of seventeen commercial and six military GEO launches per year, we easily spend over one billion dollars on orbit raising [3]. His paper asks the valid question, "Is this cost effective over the long run?" In proposing a laser power beaming infrastructure he also admits that the cost of a worldwide power beaming setup would be expensive, but proposes that international cooperation and sharing could make the cost feasible.

Reference [3] also discusses repositioning GEO satellites on orbit. Changing mission requirements often dictate a change in the subsatellite point, and both defensive and offensive space control missions also require agility. Repositioning existing assets is both faster and more economical than launching new satellites, and lasers powering an electric propulsion system can cut maneuver times in half.

The final application of laser power beaming for propulsion purposes is stationkeeping. Orbit perturbations require small amounts of thrust that are ideally suited for electric propulsion. Laser power can enhance solar flux and enable these thruster firings during eclipse or when the solar power is being tapped for other spacecraft uses.

POWER BEAMING FOR OTHER APPLICATIONS

Laser power can also enhance solar flux for other applications. The average commercial GEO communications satellite costs $250M and earns $25M to $50M per year. They are in eclipse for ninety days a year for up to seventy minutes at a time. Rechargeable batteries are used to power the spacecraft during eclipse, but they are the most common failure item. In addition, these batteries are heavy. Twenty percent of the mass of a typical GEO communications satellite is the power system and over half the mass of the power system has no other function than to provide power during eclipse. Eliminating the need for energy storage (which is needed less than one percent of the time) would reduce the satellite mass by ten percent, enabling a smaller launch vehicle or more mass for the payload. Reference [4] studied this approach, concluding that missile tracking sensors, space-based lasers, and a space-based radar would all see significant payload mass benefit from using a power beaming approach.

Although lasers can be used during eclipse to replace the power flux produced by the sun, they can also supplement the sun's energy when the satellite is not in eclipse. This can further reduce the time required for orbit transfers or enable high power demand applications. Of course satellites would have to be designed to take advantage of such higher power output, and they could also include photovoltaic arrays that are highly efficient at light wavelengths that can be easily produced on the ground.

In fact, one of the competing technologies to laser power beaming is microwave power beaming. The concept is exactly the same, except that satellites already have photovoltaic power conversion systems. Implementing a microwave power beaming apparatus would require a whole new satellite subsystem and R&D in as-yet unexplored areas. In addition, microwave power conversion systems are far less efficient than light to power converters [5]. Laser power beaming is clearly the way to go.

The bottom line is that laser power beaming from the ground will save satellite mass, complexity, and cost, resulting in lower launch costs, higher satellite reliability, and longer satellite lifetimes. Of course this must be traded off against the cost of developing, building, and maintaining the ground-based facilities. Even Hal Bennett, one of the pioneers of this research, estimates the cost of setting up a laser power beaming system at over one billion dollars, but he goes on to say that this money would be made back in less than a year as the cost per kilogram of useful payload is reduced by more than a third using laser power beaming instead of rockets for LEO to GEO transfer [6]. The history of power beaming theory is rife with studies of just this sort!

THE HISTORY AND FUTURE OF POWER BEAMING

As far back as 1965, NASA has been interested in laser power transmission, "quite simply because it may ultimately allow space missions which would be impossible by other means." The article goes on to say, "Of particular interest to NASA is the transmission of power over long distances for applications such as direct conversion to propulsive thrust or electrical power." [7]

More recently, the Air Force Research Lab (AFRL) determined that one could use 370 kW continuous wave lasers to deliver 110 kW of energy for orbit transfer applications. Four sites worldwide with 4 meter beam directors would enable the transit to be completed in about forty days. This is in comparison to a typical chemical thruster transfer which takes one to three days [8].

Some scientists have looked at novel ways to convert laser energy to spacecraft power without using photovoltaics. Reference [7] proposed some interesting optical rectification, laser-driven magnetohydrodynamic, and reverse free-electron laser concepts, and Reference [5] described the use of thermophotovoltaics which actually rely on the generation of heat rather than the absorption of photons to create electricity.

Sandia and NASA have done other studies into photovoltaic energy transfer of beamed laser energy [3]. Current satellites predominantly use silicon (Si) or gallium arsenide (GaAs) solar cells which exhibit peak efficiency to light at 950 nm and 850 nm respectively. They have found that solar cells' response to monochromatic light near these optimum wavelengths is roughly double that produced by sunlight. Reference [10] conservatively calculates that the equivalent of one sun of illumination would require 915 W/m^2 of 850 nm light. Reference [5] puts the figure at closer to 700 W/m^2. In any event, a 100 kW laser with realistic-sized adaptive optics can easily produce one or more suns of illumination. Some current HELs operate at 530 nm, the wavelength of a frequency-doubled Nd:YAG laser. The atmosphere transmits 67% of the energy at this wavelength and that GaAs operates at 25% efficiency at 530 nm [11]. One sun of illumination at that wavelength still only requires around a megawatt of laser output.

Looking even further into the future, Sandia and NASA have independently proposed 200 kW solutions. In each case, an array of lasers is coherently fused to produce the high-energy beam. These systems would produce 3 kW/m^2 at the satellite, essentially quadrupling the power output of the solar arrays.

There are inherent limitations to beaming laser energy from the ground into space. Atmospheric turbulence, thermal blooming, and plain old divergence all conspire to steal energy from the laser beam that is intended for the target. Although the use of large telescopes and adaptive optics has done wonders to ameliorate these concerns, the best way to minimize these detrimental effects is to simply have the laser beam path traverse less (or none) of the atmosphere. After all, if there is no beam distortion to correct then the maximum amount of power will reach the intended target.

Because of this, in the future airborne and space-based lasers will have a role to play in beaming power to satellites. Forcing the Airborne Laser (ABL) to perform this mission will be problematic, however, as the platform was designed primarily for missile defense and there will be both predictive avoidance and political concerns with using a defensive platform for what could be perceived as a potentially hostile act.

The Air Force is also actively pursuing a space-based laser (SBL) for missile defense applications. For beaming power to other satellites, it completely solves the atmospheric turbulence and thermal blooming problems.

TRW is the lead contractor for the SBL program, which is funded by the MDA and run by a program office at the Air Force's Space and Missile Systems Center (SMC), part of the Air Force Space Command (AFSPC). It is the culmination of thirty years' research into high-energy lasers, beam control systems, precision optics, and fire control systems.

SBL will be a 4 m aperture megawatt-class HF laser operating at 2.8 μm. The atmosphere is opaque to this wavelength, meaning that beam scatter will be absorbed and will not reach the ground. This is one specific reason why an HF laser was chosen for the SBL rather than a COIL.

Although SBL (like ABL) is designed as a missile defense program, current architecture studies are considering the possibility of additional missions. Offensive counter space is specifically mentioned (SBL FAQ), meaning the SBL could have the inherent capability to direct it's high-energy laser at other satellites. Power beaming would be a natural complementary mission.

The SBL program is currently building a $3B integrated flight experiment (IFX), scheduled for launch in 2012 with a lethality demonstration planned the next year. (IFX Fact Sheet, 2002) The operational SBL constellation is envisioned to have 20 satellites operating at a 40° inclination, providing full-time worldwide coverage [12].

Assuming the diffraction-limited case, SBL would be able to impart tremendous amounts of energy to other satellites. In fact, given notional numbers for orbit altitude and assuming a 1 MW output, it could beam 40 MW/m^2 to satellites in GEO orbit! Unfortunately, any current solar cell technology demonstrates 0% conversion efficiency at the long wavelength of an HF laser. To be feasible, new satellites would have to include completely dedicated power conversion equipment to use this laser light.

POTENTIAL GROUND-BASED POWER BEAMING SITES

The United States has the world's most advanced facilities for laser research and applications. Academic, government, and military organizations own and operate these sites for a myriad of applications and users. While building lasers, even high-powered ones, has become a commodity operation, propagating those laser beams through precise optics and into space and tracking moving satellites is still a tricky business. There are only a handful of places in the world that can do it reliably and regularly.

Firepond Optical Facility

The Firepond Optical Facility is located at 42.6°N, 71.5°W near Westford, MA and is operated by the Massachusetts Institute of Technology's Lincoln Laboratories (MIT/LL). MIT/LL is a Federally-Funded Research and Development Center (FFRDC) that is chartered to perform research and other functions that cannot better be performed by contractors or government laboratories.

Laser propagation to space at Firepond started with the Clemson University/MIT Light Detection and Ranging (lidar) project funded by the National Science Foundation (NSF). This lidar used a 25W, 31.7 Hz pulsed Nd:YAG laser to observe Rayleigh scattering in the atmosphere up to about 85 km [13].

The most modern instrument at Firepond is a lidar system designed, built, and operated by Thomas J. Duck and Dwight Sipler. This lidar was primarily used for day and night measurements of middle-atmospheric temperatures and gravity waves. "An important accomplishment was the first ever Rayleigh lidar observation of a mesospheric inversion layer in daytime." [14]

Duck and Sipler's work expanded on the experiments run by the Clemson/MIT team. Their higher power laser and cleaner optical path enabled more precise measurements and at higher altitudes than the earlier work. This laser is Nd:YAG, but it is a powerful injection-seeded frequency-doubled (532 nm) Spectra-Physics model with a 30 Hz pulserate that operates at 10 MW peak power. This laser was propagated using the Firepond 1.2 m aperture telescope, resulting in a narrow 0.1 mrad field of view [14].

Firepond is no longer considered an active lasing site. It is expected that it would take on the order of a few months to reactivate the site, making other locations more attractive for the purposes of the experiment described in this paper.

High-Energy Laser Systems Test Facility (HELSTF)

The U.S. Army Space and Missile Defense Command (SSDC) operates the White Sands Missile Range (WSMR) HELSTF "to support test and evaluation of high-energy laser systems, subsystems, and components, and to support the conduct of damage and vulnerability tests on materials, components, subsystems, and systems." [15]

HELSTF was established in the 1970's and first used as the test site for the Navy's SEALITE program. TRW, under contract to the SEALITE program, built the Mid-Infrared Advanced Chemical Laser (MIRACL), a continuous wave (CW) DF laser operating at 3.8 μm. It is the highest power continuous output laser in the United States, operating in the megawatt-class (Reference [16] puts the total output power at 2.2 MW).

"The laser is basically an exotic rocket engine composed of individual module assemblies each having many nozzle blades. The modules are fed from an upstream combustion chamber and are designed to produce an optically uniform downstream flow field as a lasing medium. A gaseous oxidizer is reacted with a fuel mixture and ignited in the combustor to produce fluorine. Deuterium is injected into the flow to chemically combine with the fluorine atoms and produce the required population of excited DF molecules upon which lasing is based." [17]

HELSTF is host to other laser programs, most notably the joint U.S.-Israeli Tactical High-Energy Laser (THEL) program which developed an operational system for theater missile defense. In addition, MIRACL was used in the fall of 1997 for a controversial ASAT research test. The laser was fired at MSTI-3, a U.S. satellite that was past its useful lifetime, to improve computer models used for planning protective measures for U.S. satellites [17].

HELSTF is nearly ideally located for high-energy laser testing. It is located in the south-central part of the WSMR, far from civilized areas or common air traffic routes. There is sunshine 350 days per year, and the dry desert climate provides for excellent seeing conditions most of the time [18].

From the perspective of this paper's experiment, HELSTF has two main weaknesses. First, the very high power of the site has embroiled it in political debate.

Second, it costs on the order of one million dollars per firing, making it well above the budget of any simple demonstration.

Starfire Optical Range (SOR)

The Air Force Research Laboratory (AFRL) SOR is located atop a small peak at Kirtland AFB in Albuquerque, NM. It is the nation's and most likely the world's premier location for ground-to-space laser beam propagation research and experimentation. Dr. Robert Fugate, the technical director of the site, is a world-renowned expert in adaptive optics and personally pioneered and oversaw the development of laser guidestar imaging.

"The primary mission of the SOR is to develop and demonstrate optical wavefront control technologies. The SOR houses a 3.5 m telescope (one of the largest telescopes in the world equipped with adaptive optics designed for satellite tracking), a 1.5 m telescope, and a 1.0 m beam director. In addition to its primary research charter, the SOR also supports field experiments by others within the research community." [20]

Dr. Fugate's work with laser guidestar adaptive optics has led to dramatic increases in the performance of ground-based astronomical telescopes. Normally in adaptive optics systems, bright natural stars serve as "beacons" for the adaptive optics system. By looking at these beacons, the system can measure and correct for atmospheric distortions. Unfortunately, there are a limited number of suitable stars which do not completely cover the sky. Dr. Fugate has used a laser to measure the scattering from atmospheric molecules or a layer of sodium atoms 60 miles above the earth's surface to create an artificial beacon or guidestar in the direction of interest [1].

SOR, while not at as high an elevation as AMOS, is still located at an altitude of over 8,000 feet. The thermal environment, however, is much messier due to the mountainous nearby terrain, and the light produced by the nearby city greatly increases optical noise.

The government and contractor personnel working at the site are extremely knowledgeable, competent, and dedicated. They are used to working very long hours, have the ability to rapidly reconfigure test equipment when needed, and have more experience with satellite tracking and laser beam propagation to space than anyone else.

In addition, there are a variety of lasers on site at SOR. Due to the dynamic nature of their mission and experiments, laser sources are constantly being added or removed. At the current time, there are no lasers of high enough power to overcome the fact that the satellite to be used in this paper's experiment only rises to fairly low on the horizon at SOR (~10°). In addition, SOR is constantly overbooked with high priority experiments.

Maui Facilities

The Air Force Maui Optical and Supercomputing Site (AMOS) is a set of locations on the island of Maui operated by the Air Force Research Laboratory (AFRL). The Maui High-Performance Computing Center (MHPCC) is located in Research Park in

467

the city of Kihei, while the optical station itself is located atop Mount Haleakala and is used for both research purposes and operational space situational awareness (SSA) for the Air Force Space Command (AFSPC).

There are many telescopes the site uses for optical tracking of space objects, most notably the 3.67 meter Advanced Electro-Optical System (AEOS) which became operational in 2000. There is also a 1.6 m telescope, two 1.2 m telescopes, two 1.0 m telescopes, a 0.8 m beam director/tracker, and a 0.6 m laser beam director. (Pike) Lasers can be fired at elevation angles as low as $-5°$ depending on other conditions, meaning there is no theoretical limit on what satellite orbits can be reached from the site [21].

There are several lasers on site at AMOS that are used for lidar research, active tracking, and other purposes. There is a 9 J CO_2 laser, a 0.35 J tunable alexandrite laser, a 6 W tunable argon ion laser, and a 4.7 W tunable krypton laser on site. In addition, the High Performance CO_2 LIDAR for Space Surveillance (HI-CLASS) laser operates at 30 J and 30 Hz [22].

In the past, a 500W Holobeam Nd:YAG laser was used for high-energy experiments. That laser has since been moved back to the mainland, but in summer 2002 a new Ytterbium in yttrium argon garnet (Yb:YAG) laser will be installed that operates in different modes from 500 W to 1000 W depending on desired beam quality. This laser operates at 1.03 μm and can be frequency doubled to operate at 535 nm [23].

Mount Haleakala offers the distinct advantage of being at 3,058 meters elevation, reducing the amount of atmosphere a beam must travel through to get to space. The climate is favorable and stable virtually year-round, and there is minimal scattered light from surface sources. In addition, the thermal soak of the Pacific Ocean for thousands of miles in every direction greatly reduces atmospheric turbulence, leading to extremely good seeing conditions (on the order of 1 arc second) for a terrestrial site [24].

AMOS is also well-suited to outside experimenters using the facility. AEOS in particular was designed with modularity in mind, such that multiple sets of research can take place simultaneously in rooms arranged under the telescope itself. Such visiting experiments pay a small fee for site support and overhead. Technical and administrative support is provided from offices at Premier Place, a technical park in Kihei, Maui. Mount Haleakala is also home to NASA's LURE (Lunar Ranging Experiment) facility, operated by the University of Hawaii.

AMOS and NASA are currently working on an aggressive series of experiments in the area of wireless power transmission for the Marshall Space Flight Center contribution to NASA's Space Solar Power program. They recently purchased a brand new Coherent fiber laser that will operate in the tens of watts at 850 nm. This source will not be suitable for lasing demonstrations to orbiting assets, but will enable ground demonstrations that can lead to that all-important next step.

Space Laser Energy (SELENE) Corporation

The most potentially visionary work done in this area appears to be that of the SELENE Corporation, a private company operating out of Lompoc, CA. Dr. Hal Bennett is the president of the company and is applying his Naval career expertise to making space power beaming a reality. SELENE has proposed a ground-based infrastructure beaming power to a reusable "space tug" that will repeatedly transfer satellites from LEO to GEO and, if needed, back again. The SELENE concept includes a $25M power beaming complex with a 12 m telescope and a Berkeley Lab-constructed, Russian-designed 200 kW free-electron laser in Ridgecrest, CA. The entire program, including the space tug, is estimated to cost $350M. Currently the company has spent about ten million dollars on studies but has not yet broken ground [25].

When considering ways to beam power to satellites or to perform laser annealing of radiation-damaged solar cells, ground-based sites offer many advantages. Weight and space considerations are not very important, maintenance of the laser and associated hardware is relatively simple, and the physical infrastructure is already in place in several locations.

Still, there are disadvantages associated with ground-based sites as well. Even the best ground sites have to propagate the laser beam through much of the atmosphere, forcing the use of adaptive optics and still resulting in tremendous losses. Predicative avoidance is complicated, and there are safety issues concerning people on the ground and aircraft in the vicinity. Raising the laser higher off the ground ameliorates some of these issues, but it also introduces additional challenges to overcome.

ANNEALING OF RADIATION-DAMAGED SOLAR CELLS

NPS has done extensive research into the damage caused by radiation in the space environment. Using an in-house linear accelerator that generates 27 MeV electrons, students are able to simulate aging various types of solar cell materials and measure the I-V curves showing their degraded output. In space, electrons, protons, neutrons, gamma rays, and alpha particles all conspire to degrade the crystal lattice of the semiconductor material. These effects are especially pronounced when traversing through the Van Allen radiation belts, which all low-earth satellites do and geosynchronous satellites so when in their transfer orbit. Mid-earth satellites such as the GPS constellation live in this harsh radiation environment, necessitating extra hardening of the solar cells.

Thermal annealing is a well-understood phenomenon, but is not especially useful for on-orbit solar arrays. The high temperatures needed to anneal the photovoltaic materials are impossible to attain, and would likely cause damage to other materials in the cell anyway. NPS students have demonstrated current injection annealing techniques that work at very realistic temperatures.

Reference [26] describes how forward bias injection involves the application of an increased voltage potential applied across the cell. Forward current flow increases exponentially as the potential increases, increasing the power dissipated in the cell proportionately. This technique has been successfully applied to both indium phosphide (InP) and gallium arsenide (GaAs) cells at temperatures well below 100° C. Silicon (Si) cells, which require much higher temperatures for thermal annealing, have not been tested as it is expected they would show a resistance to forward bias injection annealing. Still, a student recently noticed some annealing affects when illuminating simulated aged silicon cells.

This type of annealing is called photoinjection and is the second type of minority carrier injection annealing. Reference [27] describes that a laser is used to both force the production of a large current and to produce additional heat needed for realigning the crystalline structure of the solar cell. Since the response of solar cell materials to monochromatic light is not linear across the entire solar spectrum, it is possible to use a laser tailored to a cell's peak efficiency, thereby reducing the power needed to generate a large current. Students have completed thesis research using argon-ion lasers to repair most of the damage in GaAs cells.

Spacecraft are starting to be launched with multijunction solar cells which consist of several layers of different photovoltaic material. Each material reacts to light energy at a different energy level, maximizing the efficiency and power that can be generated from sunlight. Future multijunction cells could include a material that is highly responsive to monochromatic light at a wavelength where lasers can easily generate large amounts of power. Such foresight will enable power beaming for any of the aforementioned applications.

PROPOSED EXPERIMENT

We are obviously not yet ready to implement power beaming schemes. To be most effective, satellites will have to be designed from the start with power beaming in mind. This will mean including specific layers in multijunction solar cells that are highly responsive to monochromatic light, rethinking the idea of batteries and other power storage techniques, and even mission planning to illuminate various satellites during eclipses. If power system failure were eliminated as a satellite lifetime constraint, satellite systems engineers would have to consider the projected lifetimes of all other spacecraft components to ensure they can stand up to the longer anticipated mission durations.

As of today, we do not have high enough power lasers at the wavelengths needed to generate response in Si, InP, or GaAs solar cells. Still, an experiment using current technology that demonstrates all the ingredients needed to make power beaming a reality is possible. It would have to involve generating laser power, acquiring and tracking a satellite, beaming laser power to it during a pass, and verifying that laser energy was received at the satellite.

NPS is currently planning such an experiment, with the much-appreciated support of AFRL/DE leadership, SOR personnel, and AMOS site staff. On 29 Oct 98, the

Space Shuttle launched the Petite Amateur Navy Satellite (PANSAT) out of a get-away special (GAS) can. This small spacecraft (NORAD catalog number 2552) is in a 28.5° inclined, nearly circular orbit at 555 km. It consists of eighteen faces but can be thought of as roughly spherical with a diameter of about a half a meter. Sixteen of the faces are covered with silicon solar cells with one face composed of GaAs solar cells.

Telemetry from the satellite is collected at the school in Monterey, CA, and the spacecraft director, Dr. Rudy Panholzer, has already given permission for it to be lased. AMOS has verified that it is bright enough to be seen and tracked, so current discussions are revolving around what kind of laser will be required to cause a noticeable jump in the solar array current. It currently appears that either a high-power (2 kW or so) Nd:YAG or lower power (750 W) diode laser operating around 850 nm would be ideal for a power beaming demonstration using PANSAT.

Once a suitable laser is purchased, installed, checked out, and readied for use, the plan will be to illuminate PANSAT over a complete pass (approximately ten minutes horizon to horizon). Then, when the satellite gets into view of Monterey, it will begin downloading its telemetry data which includes output current from eight of the solar panels sampled at two-minute intervals. Historical data indicates that panels not facing the sun produce zero output, indicating that albedo does not generate enough energy to cause any photovoltaic generation.

It is possible to move the PANSAT ground station to Maui in order to get real-time verification of laser illumination. This has the added benefit that the satellite can be commanded to download status information every few seconds rather than relying on the two-minute snapshots. In the past the satellite could have been commanded, via software uplink, to operate in this mode regularly, however due to battery degradation the satellite now loses power during eclipse and dumps any software that has been uploaded [28].

For the purposes of a simple, short-term demonstration of the end-to-end concept of power beaming, the two-minute snapshot should prove sufficient. By scheduling multiple passes over different days, the chance of successfully illuminating the satellite and measuring the effect will be maximized. Once this demonstration is complete, it will provide conclusive proof that power beaming is both achievable and feasible.

CONCLUSION

Laser power beaming to satellites is an idea that is still just slightly ahead of its time. Researchers at SOR concluded that "…the technology for projecting tight laser beams to satellites is maturing rapidly…" and "…such strong progress in beam directing lays the foundation for beaming power to satellites in the near future." It will take the concentrated effort of scientists and engineers throughout the community to move this promising technology out of the province of the near future and into today.

REFERENCES

1. Fugate, Robert Q. and James F. Riker. "Beam Control for Ground to Space Power Beaming", *AFRL Briefing*, January 28, 2002.
2. Ziemer, John K. "Laser Ablation Microthruster Technology", *33rd AIAA Plasmadynamics and Lasers Conference*, May 20-23, 2002.
3. Lipinski, R. J. and D. A. McArthur. "FALCON Reactor-Pumped Laser and Applications for Power Beaming to Space", *Sandia National Labs Briefing*, April 28, 1994.
4. Bamberger, Judith A. and Edmund P. Coomes. "Power Beaming Providing a Space Power Infrastructure", *IEEE AES Systems Magazine*, November, 1992.
5. Michael, Sherif. Personal discussions, Naval Postgraduate School Department of Electrical and Computer Engineering, Space Systems Academic Group.
6. Bennett, Hal E. "DoD and Navy Applications for Laser Power Beaming," *SPIE Laser Power Beaming II*, February 8-9, 1995.
7. Nored, Donald L. "Application of High Power Lasers to Space Power and Propulsion", *2nd NASA Conference on Laser Energy Conversion*, January 27-28, 1975.
8. Lipinski, Ronald J. and Dorothy Meister, Phillip Leatherman, Robert Fugate, Carl Maes, W. Joseph Lange, William Cohan, Richard Cleis, James Spinhirne, Raymond Ruane, Janice Glover, Robert Michie, and Andrew Meulenberg, Jr. "Laser Beaming Demonstrations at the Starfire Optical Range."
9. DeYoung, Robert J. "A NASA High-Power Space-Based Laser Research and Applications Program", 1983.
10. Monroe, David K. "Laser Power Beaming to Extend Lives of GSO NiCd Satellites", *SPIE Laser Power Beaming Proceedings*, January 27-28, 1994.
11. Landis, Geoffrey A. "Space Power by Ground-Based Laser Illumination", *Intersociety Energy Conversion Engineering Conference*, September 15, 1991.
12. SBL FAQ, http://www.sbl.losangeles.af.mil/IFX_FAQ/FAQs.htm.
13. Meriwether, John and Dwight Sipler. "Millstone Hill Lidar", http://www.haystack.mit.edu/~dps/lidar0.htm.
14. Duck, Thomas J. "The Firepond Lidar at Millstone Hill / MIT Haystack Observatory", http://aolab.phys.dal.ca/~tomduck/milidar/.
15. HELSTF Home Page. http://helstf-www.wsmr.army.mil/index.htm. September 2001.
16. Talbot, John P. "Airborne Star Wars Laser", http://home.achilles.net/~jtalbot/history/starwars.html.
17. WSMR Public Affairs. "High Energy Laser Systems". http://www.wsmr.army.mil/paopage/Pages/laser.htm.
18. Pike, John. "Mid-Infrared Advanced Chemical Laser", http://www.fas.org/spp/military/program/asat/miracl.htm.
19. HELSTF Home Page. http://helstf-www.wsmr.army.mil/index.htm.
20. Starfire Optical Range Home Page, http://www.de.afrl.af.mil/SOR/.
21. Skinner, Mark, Boeing Maui Site Support. E-mail conversations, 2002.
22. "AMOS Maui Space Surveillance System", http://ulua.mhpcc.af.mil/AMOS/mission.html.
23. Snodgrass, Josh, Capt, USAF. E-mail conversations, 2002.
24. AMOS Users' Manual. http://ulua.mhpcc.af.mil/AMOS/manual.html.
25. Preuss, Paul. "Earth to Space: Powering Communication Satellites from the Ground", http://www.lbl.gov/Science-Articles/Archive/ground-satellite-power.html.
26. Cypranowski, Corrine, Sherif Michael, and Bruce Anspaugh. "Forward-Biased Current Annealing of Radiation Degraded Indium Phosphide and Gallium Arsenide Solar Cells", *IEEE Conference*, 1990.
27. Chase, Charles T. "Annealing of Defect Sites in Radiation Damaged Indium Phosphide Solar Cells Through Laser Illumination.", Master's Thesis, NPS, December, 1995.
28. Horning, James A. Personal discussions, Naval Postgraduate School Department of Aeronautics and Astronautics, Space Systems Academic Group

MISCELLANEOUS APPLICATIONS

Generation and Focusing of High Brightness Pulsed X-rays

-Toward the X-ray Driven Micro-Ship-

Makoto Shiho[1], Kazuhiko Horioka[1], Yuji Kiriyama[1], Sadao Aoki[2] and Takashi Yabe[3]

1 Department of Energy Sciences, Tokyo Institute of Technology, Nagatsuta 4259, Yokohama 226, Japan

2 Institute of Applied Physics, Tsukuba University, Tsukuba Ibaragi 305, Japan

3 Department of Mechanical Science and Engineering, Tokyo Institute of Technology, O-okayama, Tokyo 152, Japan

Abstract. A method for energetic micro-X-ray-beam formation is proposed using a pulse-powered radiation source and a grazing incidence mirror. The expected radiation sources for the micro-beam formation are X-pinch plasmas, bremsstrahlung point sources and capillary X-ray lasers. Radiations from the pulse power driven source are focused using a Wolter mirror. The Wolter mirror is made of pyrex-glass and has a capability to focus photons from a point radiation source with energy up-to a few keV. Based on the experimental results, potentiality of the pulse power point X-ray source as a driver of micro-ships is discussed.

1. INTRODUCTION

Recent developments in high power laser and radiation sources materialize the idea of laser driven propulsion of vehicles[1-4]. In fact, vertical launch of a 100-g rocket[3] and horizontal drive of mini airplane[5] has already been demonstrated, and momentum coupling efficiency; Cm of 5000N · sec/MJ has been achieved by experiment and simulation[5]. If we could extend this scheme to X-ray region, we might be able to manipulate an object in opaque materials such as human body in visible wavelength region by locally depositing the X-ray radiation energy.

CP664, *Beamed Energy Propulsion: First International Symposium on Beamed Energy Propulsion,*
edited by A. V. Pakhomov

To materialize the concept, we should investigate following issues;

a) Mechanism of plasma ablation and energy deposition process on to materials
b) X-ray light Source
c) Light focusing system in X-ray wavelength region

For a) several considerations are given in section 2. For b), several light sources, which our group (Tokyo Institute of Technology, TIT) has are briefly described in section3. For c), X-ray focusing systems are described and discussed in section 4. In section 5, results of a preliminary X-ray focusing experiments are described. In section 6, future experimental systems are described.

2. SEVERAL CONSIDERATIONS ON X-RAY MICRO SHIP'S DRIVE

In our previous demonstration experiments of micro-airplane flight[5], we used a 590mJ/5ns YAG laser for producing a driving force. The laser light was focused on to a spot of less than 0.1mm diameter. The target was thin aluminum plate with water overlay. The energy of the laser light made plasma on the surface of the aluminum   and plasma transfer momentum to the aluminum plate. When there is water over lay, the momentum transfer is enhanced as shown in Fig.1.

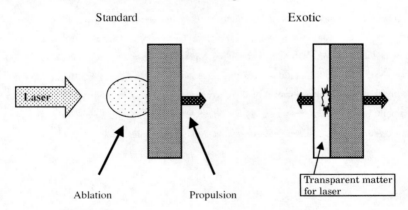

Fig.1 Structure of Standard Ablation Type (Standard) and Target with Water Overlay (Exotic)

We then consider the case of X-ray driven micro-ship. The micro-airplane used in the experiment was paper crafted plane, of which the weight was about 10g. When we consider X-ray driven micro-ship, size of the ship would be less than the order of 0.1mm. Even when the ship is made by steel, the weigh would be the order of the micro grams. Supposing   that the weight is 1μg, and the momentum coupling coefficients Cm obtained in experiment holds, the required X-ray power would be $590 \times 10^{-3} \times 10^{-7}$ / 5ns=5.9×10^{-8} /5ns= 59nJ/5ns.

It apparently seems that the value would be obtainable by suitable choosing of X-ray light sources and a proper focusing system. But we should notice the following facts; In case of usual laser propulsion, laser ablation plasma works as the first trigger of producing driving force, and this process is quite well understood by the knowledge from extensive research activities for laser driven nuclear fusion over a decade. On the other hand, absorption process of intensive X-rays might be quite different from the process based on the scheme of "visible-light and high- temperature-plasma interaction". Even more, we have no lasers in the X-ray region. To evaluate the driving ability with pulsed X-rays quantitatively, we have to explore scientifically the energy deposition process at high intensity level by focused X-ray beams to verify what the basic concept of manipulating an object is in opaque materials using an energetic X-ray radiation source.

3. LIGHT SOURCES IN TIT

Figure 2 shows an overview of various light sources from visible to X-ray region. In the VUV to X-ray region there almost no intense light sources except for synchrotron radiation source. Since the machine time of using synchrotron radiation source is

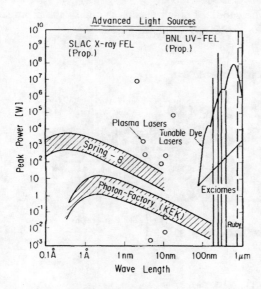

Fig.2 Various advanced light source and their power.

477

limited, and since considering the fact that high energy deposition on to a spot with about 0.1mm (present experimental target), we, TIT group, are now preparing another type of intense pulse X-ray sources for the study of X-ray drive micro ships.

High energy density plasmas and relativistic electron beams produced by pulse power generator are known as bright radiation sources at VUV ~ X-ray region. Typical light sources at TIT and their typical operating region are summarized in Table 1.

Table 1.   Specification of TIT Pulse Powered Radiation Sources

	Photon Energy	Output Energy	Pulse Width
VUV Laser	<100eV	≦mJ	≦10nsec
Pinch Plasma	≦keV	≦kJ	≦10nsec
Vacuum Spark	1eV~10keV	~J	~nsec
Bremsstrahlung	≦MeV	≦kJ	~100nsec

Brief descriptions for each of the light sources are given bellow.

3-1 Discharge Pumped VUV Lasers

Recently lasing by using fast discharge plasma is demonstrated[6,7]. By controlling the plasma parameters and their distribution we can increase the output energy of the discharge type laser[8]. The capillary pinch plasma is one of the promising candidates of the energetic "coherent" VUV~soft-X-ray drivers.

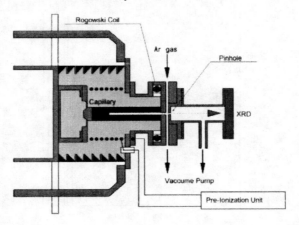

Fig.3 a) Schematic view of TIT VUV Laser.

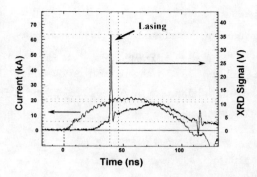 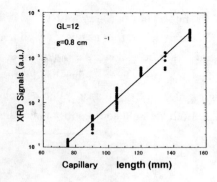

Fig.3 b) Time evolution of lasing signals.   **Fig3. c) Spatial growth of lasing signal**

3-2 Pinch Plasmas

Z-pinches including plasma-focus systems are the most powerful X-ray sources[9]. The relative simplicity of design and high power density make pinch sources attractive for applications.

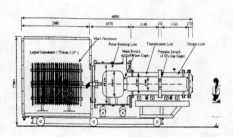 

Fig.4 Schematics and photograph of Z-pinch X-ray source. The stored energy is 10kJ, peak current is 200kA, and dI/dt>10^{12} A/sec.

3-3 Vacuum Sparks

In the first stage of vacuum breakdown, X-rays are generated.   Due to the simplicity of configuration, possibilities for producing high power pulsed X-rays based on the temporal mechanism of vacuum breakdown are attractive[10].

3-4 Bremsstrahlung X-ray Sources

Relativistic electron beams are well known energetic radiation sources at hard-X-ray regime. Figure5 shows the induction accelerator installed recently in Tokyo Institute of Technology. On the other hand, bremsstrahlung X-rays can be also produced by using a high voltage pulser. Photographs of the miniature X-ray tubes are shown in Fig.6. Both are light sources for rather hard X-ray region.

Fig.5 Induction accelerator LAX-1 of TIT. Energy is 1MeV, current is 3kA, and pulse width is 100ns. This accelerator would be used for flash X-ray.

 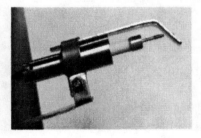

Fig.6 Photographs of miniature X-ray tubes. Current of 1A and the voltage of 100kV prototypes are under development.

4. FOCUSING SYSTEM

There are various type of light focusing methods in VUV to X ray region as: Kirkpatrick-Baez Mirror[11, 12], Wolter Mirror[11, 12], Capillary Tube[11, 12], Waveguide[12], Fresnel Zone Plate[11, 12], Bragg-Fresnel Lens[12], and Compound Refractive Lenses[13]. Since most of TIT light sources are not coherent light sources, we chose focusing system with grazing incidence reflection type. They are considered to be the most efficient system for incoherent radiation sources[11-13].

We have made a Wolter mirror[11]. The mirror has coaxial geometry and it usually used as a microscope in the VYV and soft X-ray region. When used in a reverse direction, the mirror is used as an X-ray focusing system.

A schematic of the Wolter mirror and its design parameters are shown in Fig.7. The mirror is made of Pyrex-glass and has a magnification factor of 1/20.

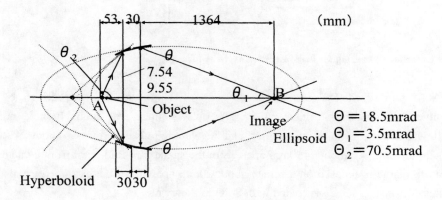

Fig.7 Schematics and design parameters of Wolter Mirror. Magnification/reduction factor is 1/20.

The Wolter mirror is set inside of the stainless pipe, and installed in the chamber.

Fig.8 Photograph of Wolter mirror and its setting.

5. PRELIMINARY EXPERIMENTS

Using above described Wolter mirror and small vacuum spark plasma, focusing of incoherent point radiation source is demonstrated experimentally. Schematics of experimental set up are illustrated in Fig.9.

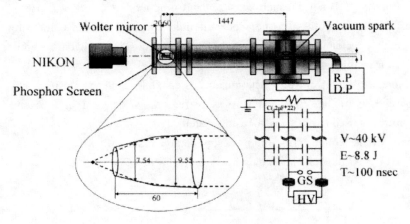

Fig.9 Experimental setting for demonstration of X-ray focusing.

At a pressure 1×10^{-4}Torr, a vacuum gap of 1mm distance was driven by a LC-inversion type, high-voltage pulser. The pulse voltage was about 40 keV, pulse width was about several micro seconds, and the electric pulse energy was about 9J. The Wolter Mirror is set about 1450mm apart from the spark gap. Radiation from a vacuum spark discharge is focused on a screen about 20mm apart from the exit aperture of the Wolter mirror. The screen is coated with Sodium Salicylate.

Typical results are shown in Fig.11. The phosphor screen image was taken by Minolta f 2.8 camera with Fuji ASA 1600 film. In Fig.11, faint shadow of supporting rods and supporting pipe are seen in the photo of several shots. In the figure the bright small spot at the center appears to correspond to the spark plasma. The diameter of the spot is 0.1~0.2mm. Considering the gap distance of the vacuum spark is about 1mm, and produced plasma would prevail around the area of at most be about 2mm, the bright spot size of 0.1~ 0.2 can be regarded as the plasma image of the vacuum spark with the reduction factor of 1/20.

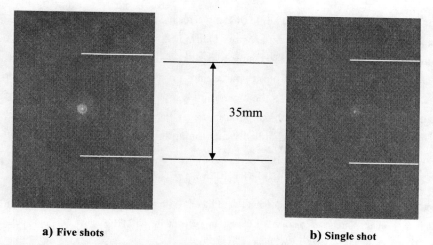

35mm

a) Five shots　　　　　　　　　　　　　**b) Single shot**

Fig.10 Photograph of a focal spot by Wolter Mirror :Minolta F2.8, FUJI ASA1600.

a) is photo of five shots and faint shadow of supporting rod is observed. b) is photo of single shot. Of the center bright spot, diameter is less than 0.1mm.

6.　CONCLUSION AND DISCUSSIONS

As a proof-of-principle experiment, the radiation from vacuum arc plasma was successfully focused less than mm by the Wolter mirror with the reduction factor of 1/20. In the present experiment, the driving energy of vacuum spark is about 9J. In order to investigate absorption process of VUV－X-ray radiation, we have to increase intensity of X-ray source. For this, we are planning to prepare a kJ pulse powered radiation source, e.g., tapered pinch plasma source, X-pinch plasma source.

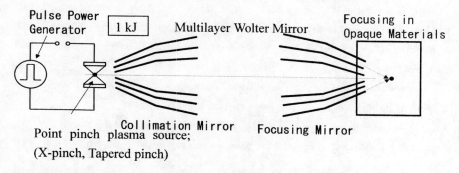

Fig.11 Schematic view of the kJ radiation source and Multilayer Wolter mirror in TIT

483

As for focusing system, the solid angle of the wolter mirror is about 5×10^{-7}.
In order to increase this solid angle we are now planning to fabricate multi-layer Wolter mirror schematically shown in Fig.11. Using two sets of the multilayer Wolter mirror, we can gather X-ray radiation from point source and produce pararell beam. Another multi-layer Wolter mirror would focus the beam onto a spot.

If we can control intense radiations at photon energy levels of more than keV region, they will be useful not only for mechanically manipulating the object but also radiation therapy.

REFERENCES

1. A.Kantrowits, "Propulsion to Orbit by Ground-Based Laser," Astronaut. 10, pp. 74, 1972

2. C.R.Phipps,Jr. et.al., "Impulse coupling to targets in vacuum by KrF, HF, and CO_2 single-pulse lasers," J.Appl.Phys. 64, pp.1083-1096, 1088

3. L.N.Myrabo and F.B.Mead, Jr., " Ground and Fligh Tests of a Laser Propelled Vehicle," AIAA98-1001, Aerospace Sciences Meeting & Exhibit, 36th, Jan12-15, 1998

4. C.R.Phipps and M.Michaelis, " LISP: Laser impulse space propulsion," Laser and Particle Beams 12, pp.23-54, 1994

5. T.Yabe, C.Phipps, K.Aoki, M.Yamaguchi, Y.Ogata, M.Shiho, G.Inoue, M.Onda, K.Horioka, I.Kajiwara, K.Yoshida, "Laser Driven Vehicle" SPIE Proceedings, Vol 4760, pp867-878 (2002)

6. J.J.Rocca, V.Shlyaptsev, F.G.Tomasel, O.D.Cortazar, D.Harshorn, J.L.A.Chilla,"Development of a Discharge Pumped Table-top Soft X-ray Laser", Phys. Rev. Letters, Vol.77, pp.1476-1479 (2002)

7. G.Niimi, Y.Hayashi, N.Sakamoto, M.Nakajima, A.Okino, K.Horioka, E.Hotta, "Development and Characterization of a Low Current Capillary Discharge for X-ray Laser Studies, IEEE Trans. Plasma Science, Vol.30, NO.2, pp.616-621 (2002)

8. M.Masnavi, T.Kikuchi, M.Nakajima, K.Horioka, "Influence of Opacity on gain Coefficients in Static and Fast Moving Neon-like Krypton Plasmas", J. Appl. Phys., Vol.92, No.7, pp.3480-3486 (2002)

9. M.A.Liberman, J.S.DeGroot, A.Toor, R.B.Spielman, "Physics of High-Density Z-pinch Plasmas", Springer (1998)

10. G.A.Mesyats, D.I.Proskurovsky, "Pulsed Electrical Discharge in Vacuum", Springer-Verlag, (1989)

11. E.F.Kaeble, Handbook of X-Rays (Mcgraw-Hill, New Yoek, 1969)

12. J.Kirz, X-ray Microscopy III (Springer-Verlag, Berlin, 1992), and ser.I,and II.

13. A.Snigirev, V.Kohn, I.Snigireva, and B.Lengeler: Nature(London)384, p.p.49-51(1996)

14. V.H.Wolter, Ann.Phys.10,p.p.94(1952)

15. S.Aoki et.al., Jpn.J.Appl.Phys. 31, Part 1, p.p.3477 (1992)

16. S.Aoki, T.Ogata, S.Sudo and T.Onuki: Jpn. J. Appl. Phys. Vol.31 p.p.3477-3480 (1992)

Experimental and Computational Investigation of Hypersonic Electric-Arc Airspikes

R.M. Bracken*, C.S. Hartley*, G. Mann*, L.N. Myrabo*,
H.T. Nagamatsu*, M.N. Shneider[¶], and Y.P. Raizer[§]

*Department of Mechanical, Aerospace, and Nuclear Engineering
Rensselaer Polytechnic Institute, Troy, New York 12180*

[¶]*Princeton University, Princeton NJ*

[§]*RAS Institute for Problems in Mechanics, Moscow, Russia*

Abstract. Drag reduction effects of an electric arc airspike in a hypersonic flow are currently being studied in the Rensselaer Polytechnic Institute 24-inch Hypersonic Shock Tunnel (RPI HST). In tandem these results are being modeled computationally, and compared to existing theory. The arc is driven by a high current lead-acid battery array, producing a maximum of 75-kilowatts into the self-sustaining electrical discharge. The test conditions were for Mach 10, 260 psia stagnation pressure, and 560 K stagnation temperature flow – a low enthalpy, "ideal gas" condition. Schlieren photographs are taken of the arc apparatus and downstream blunt body, with a variety of arc powers and source/body distances. Fast-response accelerometers are used to measure drag on the hanging blunt body. These tests are conducted with and without the arc to establish the most efficient placement and power of the airspike. The computational effort employs the Euler gasdynamic equations to represent a heat source in flow conditions and geometries identical to those tested in the RPI HST. The objective of the combined experimental/computational parametric study is to enhance understanding of the drag reduction features inherent to the airspike phenomenon.

INTRODUCTION

Among the most critical factors in the design of a hypersonic vehicle are aerodynamic drag and heating. To design for low aerodynamic drag would leave the body with sharp edges and a high fineness ratio. Heat transfer concerns would lead the design to a blunt body with a very low fineness ratio, which leads to a substantial increase in wave drag. It has been suggested by several authors [1-7] that both the drag and heating of a hypersonic Trans-Atmospheric Vehicle (TAV) could be greatly reduced by adding energy to the air ahead of the craft. This energy addition could be accomplished through the focusing of beamed electromagnetic radiation from a laser or microwave source. Figure 1 shows a diagram of this concept for focusing a powerful laser beam to support the Directed-Energy AirSpike (DEAS). Marsh, *et al.* [8], have experimentally demonstrated that the strong bow shock wave normally produced by a blunt body would become a more benign conical shock wave when an

CP664, Beamed Energy Propulsion: First International Symposium on Beamed Energy Propulsion,
edited by A. V. Pakhomov

electric arc torch is fired ahead of the body. Figure 2 is a schlieren photo from these Mach 10 tests.

Later, Toro *et al.* [10] conducted additional tests using the same apparatus, but fitting the blunt body model with piezoelectric pressure transducers and platinum thin-film heat transfer gauges. The surface and impact (taken at the model periphery) pressure measurements indicated a decrease in aerodynamic drag; the heat transfer measurements revealed a decrease in surface heat transfer. These experiments were conducted under Mach 10 ideal gas conditions with arc torch electric powers up to 127kW. Nevertheless, these impressive results are affected by the physical presence of the plasma torch itself, which behaves as a mechanical spike [11] and provides beneficial counter airflow [12] at the torch tip.

This combined experimental/computational study implements a "torchless" apparatus, designed to minimize flow disturbance and mitigate the two aforementioned properties of the torch. In initial testing, a Maxwell capacitor was used as the power supply, providing the high power necessary to induce and sustain an electric arc between two thin electrodes diametrically opposed and perpendicular to the hypersonic flow [13]. The current system employs a direct current car battery array power supply to alleviate measurement difficulties incurred by the transient nature of the capacitor discharge.

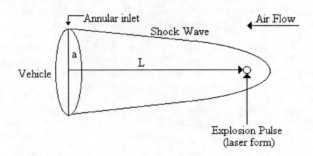

FIGURE 1. Laser supported directed-energy airspike.

FIGURE 2. Effect of the electric plasma torch (35kW) on the Mach 10 flow over a blunt body.

EXPERIMENTAL APPARATUS

The Electric Arc Simulated DEAS tests are conducted in RPI's Hypersonic Shock Tunnel. The facility is capable of producing test section Mach numbers ranging from 8 to 25, stagnation temperatures and pressures of 4,100 K and 1,500 psi, respectively, and useful test times on the order of 4 milliseconds. The high enthalpy condition can be achieved by operating the tunnel in the equilibrium interface condition [14] using argon as the gaseous piston [15,16] A more comprehensive description of the facility can be found elsewhere [17].

The RPI HST data acquisition capabilities consist of both electronic and optical measurement systems. Pressure and heat transfer measurements, made by PCB piezoelectric pressure transducers and platinum thin-film heat transfer gauges, as well as any other necessary electronic measurements, are recorded utilizing Tektronix VXI and 2520 Test Lab systems. These electronic data systems are integrated and controlled by a Labview™ program.

Optical data capabilities include both natural luminosity and single-pass laser schlieren photography. A borrowed Beckman & Whitley Inc. model 350 high-speed framing camera has increased luminosity capabilities from a single time-integrated photograph to up to 35,000 frames per second, for a maximum of 224 pictures per test. The camera is also adaptable to a streak configuration, allowing Schlieren picture rates from 1,000 to 30,000 frames per second, employing a repetitively pulsed Oxford 15 W copper vapor laser. All flow visualization photography uses the two 9 in diameter windows at opposite sides of the test section. Figure 3 is a schematic view of the hypersonic shock tunnel driven section end and dump tank, indicating the position of the aforementioned components.

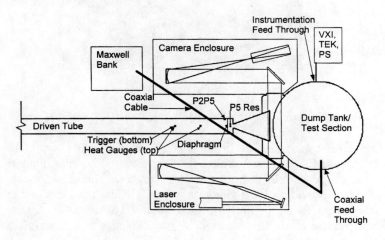

FIGURE 3. HST instrumentation layout.

Four electrical power supplies are commonly used in conjunction with experiments in the RPI HST. 1) A specially constructed 1000 V battery unit provides high voltage, low current discharges for applications requiring a low noise, moderate power direct current [13]. 2) For low voltage, high current needs, a Miller Model SRH 333 arc welder power supply can produce 70 VDC at 600 A. As previously mentioned, the welder was used by [8-10] in similar flow conditions, to produce an arc at the end of a physical spike (i.e. a plasma torch), instead of externally in the freestream, as it will used here. 3) Experiments requiring both high voltage and current utilize a donated Maxwell capacitor bank (Figure 4). This power supply can produce 10 kV with currents up to 100 kA, for short durations. The capacitor bank has been successfully used in the past for other shock tunnel experiments [18]. 4) Moderate voltage and high current applications can utilize a car battery bank. The DC battery bank can produce low noise power at over 100 kW.

The power supply chosen for a given experiment is connected to the apparatus in the test section employing a coaxial copper pipe-welding cable power line. A ceramic feed-through plate allows the coaxial power line to pass through the dump tank wall (Figure 4), while retaining vacuum and insulating the line from the steel wall.

FIGURE 4. Test section electric feed-through.

DEAS APPARATUS AND TEST PLAN

The principal apparatus used in the present experiment is a nylon ring supported electrode structure used to create an arc in a hypersonic flow. The nylon ring serves to insulate the copper terminals from the 24 in diameter aluminum nozzle exit of the HST, as well as providing some vibration isolation. This "torchless" arc configuration was fabricated in RPI's Central Machine Shop. To minimize disturbance of the hypersonic freestream, the thin electrodes where designed to span as much of the 24 in nozzle diameter as possible.

The test matrix consists primarily of a two-dimensional parametric study to quantify the effects of power and arc position on drag force. This includes several arc powers and upstream arc distances. The results of this study reveal the optimum arc placement and power for various desirable characteristics: e.g., a) minimum drag, b) maximum output-drag-power-reduction to input-power efficiency, as well as for favorable shock geometries.

To conduct these tests, an accelerometer was employed to measure drag, with the full gamut of optical data acquisition and pressure measurements. Drag reduction versus input power and arc positions have been listed to identify the major physical mechanisms involved in the airspike phenomena.

COMPUTATIONAL STUDY

The primary objectives of the computational efforts are to improve understanding of the physics of heat addition to a hypersonic flow and to gain a perspective of losses due to radiation and other phenomena. Figure 5 shows a preliminary result at Mach

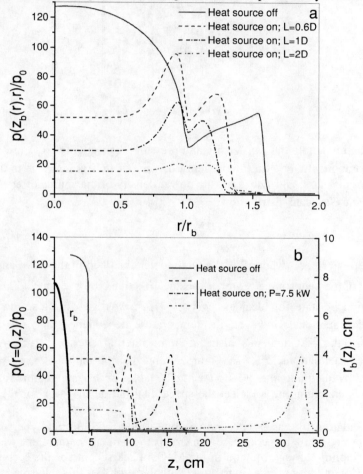

FIGURE 5. Stationary pressure distributions on (a) the blunt body surface and (b) along the central axis.

10 RPI HST low enthalpy conditions. Computations were conducted to simulate all experimental test conditions in order to gain a better understanding of the losses due to radiation, ablation, and plasma-aerodynamics.

Equations and Method of Calculation

The full set of Euler gasdynamic equations was solved in cylindrical coordinates,

$$\frac{\partial}{\partial t}U + \frac{\partial}{\partial r}R + \frac{\partial}{\partial z}Z = H, \tag{1}$$

$$U = \begin{Vmatrix} \rho \\ \rho w \\ \rho u \\ e \end{Vmatrix}, \quad R = \begin{Vmatrix} \rho w \\ \rho w^2 + p \\ \rho w u \\ (e+p)w \end{Vmatrix}, \quad Z = \begin{Vmatrix} \rho u \\ \rho u w \\ \rho u^2 + p \\ (e+p)u \end{Vmatrix}, \tag{2}$$

$$e = \rho\left[\varepsilon + (u^2 + w^2)/2\right], \tag{3}$$

where ρ and p denote the gas density and pressure, respectively; w, u are the r and z velocity components; e is the total energy of gas per unit volume, ε is the internal energy per unit mass; and $Q(r, z)$ is the power density of the source of energy release (W/cm^3). The Gaussian distribution

$$e = \rho\left[\varepsilon + (u^2 + w^2)/2\right], \tag{4}$$

is assumed for $Q(r, z)$, where P is the total power absorbed in the flow, and z_q is the coordinate of the heat source center, $r_{\it eff}$ is the effective radius of heat addition region. The equation of state for ideal gas $p = (\gamma - 1)\rho\varepsilon$, was used to close the system of gasdynamic equations. However, the temperature dependence of γ was included. Namely, for the instantaneous distribution of internal energy $\varepsilon(r, z) = p/(\gamma - 1)\rho$ derived in calculations, the gas temperature distribution $T(r, z) \equiv T(\varepsilon, \rho)$ was determined by quadratic interpolation using the tables of thermodynamic functions for hot air, thus enabling one to select the space distribution of $\gamma(T)$ for the next time step.

The calculation for cases with and without a blunt body in the hypersonic stream were performed until the complete time relaxation of all parameters was attained, using the rippling through method in the McCormack scheme of the second order of accuracy. Computations were done on rectangular grid. For this, physical coordinates (r, z) were transformed into $\overline{z}, \overline{r} \in [0,1]$. The flow velocity component normal to the blunt body surface was assumed to be zero.

Calculations were performed for the incident flow parameters corresponding to the RPI HST experimental conditions, namely: incident flow Mach number M=10.1; static pressure, 38.6 Pa and static temperature, 37.7 K. Calculations were continued until steady state conditions were achieved.

RESULTS

The results presented on Fig. 7 show very clearly how the oblique shock wave deflects the lines of flow far in front of the body, and how the pressure at the body surface is decreased in comparison with the case of no power addition Fig. 6. Without energy addition, the computation model predicts that the maximum pressure on the surface of this blunt body is about 130 p_0, and the bow shock is located 2.3 cm in front of the body, in close agreement with experiment as shown in Figure 6. For the case of the energy addition located in front of the body at a distance corresponding to 60% of the diameter of the body, $D = 2a$, a heating source term of 7.5 kW gives the best agreement between the experiment and computation, as illustrated in Figure 7-1,a and b. This suggests that once again roughly 27% of the arc power is deposited in the flow as thermal energy.

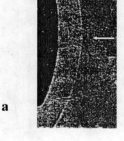

a

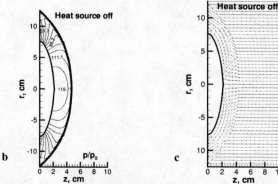

FIGURE 6. Bow shock wave in the case of no heat addition in the flow: a - experiment; b, c –numerical modeling: b - isobars; c – flow directions.

Unlike the previous case with energy addition to a free stream flow without the blunt body [19-20] the shock wave here is conical. However when the heating source is moved farther from the body, as L is increased, the shock wave becomes parabolic again (as shown in Fig. 8 and 9) and the drag on the body decreases. The results of these calculations show that the appearance of a reverse velocity region is intrinsic to the power case only (as clearly shown by comparing Figures 6,c and 7,c). "Stand off distance" increases drastically due to the essential lowering of the Mach number behind the heat source. The dependence on time of the surface pressure distribution after the heat source turning on is shown on Fig. 10.

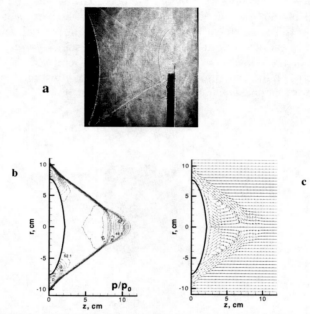

FIGURE 7. a) Schlieren for Mach 10, 27 kW arc; arc-body separation is 60% of blunt body diameter and results of numerical modeling: shock wave structure (isobars) (b), and flow directions (c) ; P=7.5 kW, L=0.6D.

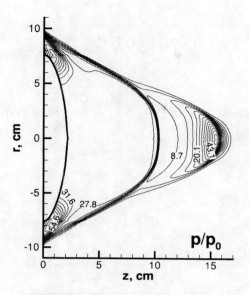

FIGURE 8. Isobars at *P*=7.5 kW and *L*=*D*.

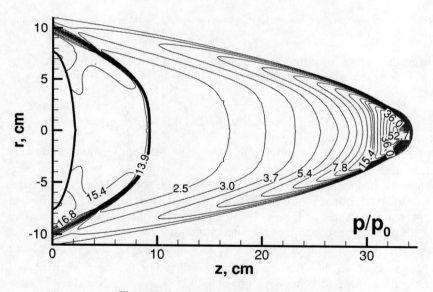

Figure 9. Isobars at *P*=7.5 kW and *L*=2*D*.

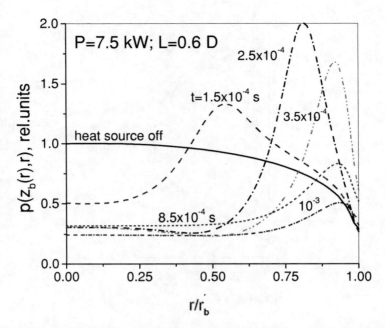

FIGURE 10. Pressure on the blunt body surface (in units of p_{max} at no heat addition case). Drag can be found by integrating p on the surface.

Using the results of numerical calculations it is possible to find the integral of pressure along the body surface that is drag force

$$F = \oint_S p_S \cos\alpha \cdot 2\pi(r/\cos\alpha)dr = 2\pi \int_0^{r_b} p_S r\,dr \qquad (5)$$

where p_s - pressure on the surface; α - the angle between the normal to the surface and free stream direction.

When this is done, the drag force is found to be 13.3 N ($L=2D$) as compared to 48.3 N when the heat source is moved closer to the body ($L=0.6D$) and 70.33N when there is no heat source. If this body were flying through the atmosphere, then to maintain the flight speed u_0 corresponding to the energy addition experiment, the total propulsive power expended would need to be 21 kW, according to $P_{total} = Fu_0 + P$ [6], where P is the power added to the flow by volume heating. For the case when $L=0.6D$, the corresponding value of propulsive power is 55.8 kW and when there is no energy addition, $P=0$, it is 84.1 kW. Note that considered heat source power 7.5 kW essentially exceeds the optimal for given blunt body and flow conditions.

Experimental tests were conducted under previously explained conditions employing the use of the direct current "battery box" array and the diametrically opposed electrodes creating an airspike in front of a free hanging blunt body, outfitted with a fast response PCB accelerometer. The preliminary results of these tests are presented in Table 1.

TABLE 1. Preliminary Drag Results for HST Airspike Tests (Mach 10 ideal gas conditions

Test	Distance: Body-to-arc (inches)	Arc Power (Kilo-Watts)	Drag on Body (Newtons)
#1	Power off	Power off	77.37
#2	6.875	59.75	21.99
#3	6.375	60.32	25.80
#4	Power off	Power off	83.01
#5	3.75	53.58	31.42
#6	7.00	39.70	24.78
#7	6.00	46.92	29.43
#8	5.875	19.34	70.11

Tests #1 and #4 were both conducted at "power-off" conditions to give a baseline for drag readings (i.e., normal shock wave on blunt body). The baseline case for Test #1 uses the same accelerometer as that used in Tests #2 and #3. Tests #4 through #8 employ a different accelerometer, for which the baseline case is given in Test #4. Both accelerometers were fast-response PCB type 352A10.

Due to the 9 inch Schlieren viewport, the maximum arc-to-body distance that can be observed is 1 vehicle diameter. While computational results predict the most efficient arc source distance to be well beyond this point, preliminary results show interesting trends. It is altogether possible that with greater separations between arc and body, the required arc input power might dramatically decrease. However, this increased separation may negatively impact the airbreathing inlet function for the Air-Spike.

It is interesting to compare the preliminary results of Tests #2 and #6. Test #2 shows a drag reduction of 71% from the baseline (i.e., normal shock) reading, while Test #6 indicates a 70% drag reduction. The two arc-to-body distances were within 1/8 of an inch of each other; however, Test #6 was conducted at 33% less arc power than that of Test #2. It is possible that the excessive arc power of Test #2 opens the airspike well beyond the vehicle perimeter, and diverts more air than is necessary. Computational findings reinforce this hypothesis.

Direct numerical modeling for a completely different body geometry and flow conditions has recently shown that an optimal heat source (position and power level) for drag reduction does indeed exist [21]. This will be the subject for the next phase of combined experimental/ numerical investigations on the directed-energy airspike.

ACKNOWLEDGMENTS

First and foremost the authors would like to thank Matt Filippelli and Jeff Alvarez for their help in carrying out these HST experiments. Special thanks are due to the RPI Central Machine shop for their expert machining skills and technical assistance. Finally the authors would like to thank the NASA Marshall Space Flight Center for their continuing support of this Air-Spike research.

REFERENCES

1. Myrabo, L.N., "Solar-Powered Global Air Transportation," AIAA Paper 78-689, April 1978.
2. Tidman, D.A., "Apparatus and Method for Facilitating Supersonic Motion of Bodies Through the Atmosphere," United States Patent Number 4,917,335, April 1990.
3. Myrabo, L.N., and Raizer, Yu. P., "Laser Induced Air Spike for Advanced Transatmospheric Vehicles," AIAA Paper 94-2451, June 1994.
4. Head, S.R., "Theoretical Analysis of the Hypersonic 'Air-Spike' Inlet," MS Thesis, Rensselaer Polytechnic Institute, August 1994.
5. Seo, J., "Analysis of Aerodynamic Properties Inside Airspike of Microwave Propelled Transatmospheric Vehicles," MS Thesis, Rensselaer Polytechnic Institute, August 1994.
6. Gurijanov, E.P., and Harsha, P.T. " AJAX: New Directions in Hypersonic Technology," AIAA Paper 96-4609, 1996.
7. Covault, G. " 'Global Presence' Objective Drives Hypersonic Research," *Aviation Week & Space Technology*, April 5, 1999, pp. 54-58.
8. Marsh, J.J., Myrabo, L.N., Messitt, D.G., and Nagamatsu, H.T., "Experimental Investigation of the Hypersonic "Air Spike" Inlet at Mach 10," AIAA Paper 96-0721, January 1996; see also, Kandebo, S.W., "Air spike could ease hypersonic flight problems," Aviation Week & Space Technology, V. 142, No.20, 15 May 1995, 66-67.
9. Toro, P.G.P., Nagamatsu, H.T., Minucci, M.A.S., and Myrabo, L.N, "Experimental Pressure Investigation of a "Directed-Energy Air Spike" Inlet at Mach 10," AIAA Paper 99-2843, July 1999.
10. Toro, P.G.P., Nagamatsu, H.T., Myrabo, L.N., and Minucci, M.A.S., "Experimental Heat Transfer Investigation of "Directed-Energy Air Spike" Inlet at Mach 10," AIAA Paper 99-2844, July 1999.
11. Bogdonoff, S. M., and Vas, I. E. "Exploratory studies of a Spiked Body for Hypersonic Flight," Gas Dynamics Laboratory, Princeton University, Mar. 1958.
12. Moraes Jr., P., and Ganzer, U. "The Flowfield of a Sonic Jet Exhausting against a Supersonic Flow," *VIII COBEM*, pp. 113- 115, Brazil, Dec. 1985.
13. Minucci, M.A.S., Bracken, R.M., Myrabo, L.N., Nagamatsu, H.T., and Shanahan, K.J., "Experimental Investigation of an Electric Arc 'Air-Spike' in Hypersonic Flow," AIAA Paper 00-0715, January 2000.
14. Minucci, M.A.S., and Nagamatsu, H.T., "Hypersonic Shock-Tunnel Testing at an Equilibrium Interface Condition of 4100 K," *Journal of Thermophysics and Heat Transfer*, **7** 251-260 (1994).
15. Nascimento, M.A.C., "Gaseous Piston Effect in Shock Tube/Tunnel When Operating in the Equilibrium Interface Condition," Ph.D. Thesis Dissertation, Instituto Tecnologico de Aeronautica – ITA, Sao Jose dos campos, Sao Paulo, Brazil, October 1998.
16. Nascimento, M.A.C., Minucci, M.A.S., Ramos, A.G., and Nagamatsu, H.T., "Numerical and Experimental Studies on the Hypersonic Gaseous Piston Shock Tunnel," *Proceedings of the 21st International Symposium on Shock Waves*, September 1997.
17. Minucci, M.A.S., "An Experimental Investigation of a 2-D Scramjet Inlet at Flow Mach Numbers of 8 to 25 and Stagnation Temperatures of 800 to 4.100 K, " Ph.D. Thesis Dissertation, Department of Mechanical Engineering, Aeronautical Engineering & Mechanics, Rensselaer Polytechnic Institute, Troy, New York, USA, May 1991.
18. Kerl, J.M., Myrabo, L.N., Nagamatsu, H.T., Minucci, M.A.S., and Meloney, E.D., "MHD Slipstream Accelerator Investigation in the RPI Hypersonic Shock Tunnel," AIAA Paper 99-2842, July 1999.
19. Bracken, R.M., Myrabo, L.N., Nagamatsu, H.T., Meloney, E.D., and Shneider, M.N., "Experimental Investigation of an Electric Arc Air-Spike With and Without Blunt Body in Hypersonic Flow," AIAA Paper 01-0796, January 2001.
20. Bracken, R.M., Myrabo, L.N., Nagamatsu, H.T., Meloney, E.D., and Shneider, M.N., "Experimental Investigation of an Electric Arc Air-Spike Air-Spike in Mach 10 Flow with Preliminary Drag Measurements," AIAA Paper 01-2734, June 2001.
21. Girgis, I.G., Shneider, M.N., Macheret, S.O., Brown, G.L., and Miles, R.G., "Creation of steering moments in supersonic flow by off-axis Plasma Heat Addition," AIAA Paper 2002-0129, January 2002.

Brazilian Activities On The Laser-Supported DEAS In Hypersonic Flow

M. A. S. Minucci, P. G. P. Toro, J. B. Chanes Jr., A. G. Ramos and A. L. Pereira

Laboratory of Aerothermodynamics and Hypersonics, IEAv-CTA
São José dos Campos - SP 12228-840 - BRAZIL
and

H. T. Nagamatsu and L. N. Myrabo

Department of Mechanical Engineering, Aeronautical Engineering, and Mechanics
Rensselaer Polytechnic Institute, Troy, NY 12180-3590 – USA

Abstract. The present paper presents recent (original) and previous experimental results on the Laser-Supported Directed Energy "Air Spike" – DEAS in hypersonic flow achieved by the Laboratory of Aerothermodynamics and Hypersonics – LAH, Brazil. A CO_2 TEA laser has been used in conjunction with the IEAv 0.3m Hypersonic Shock Tunnel - HST to demonstrate the Laser-Supported DEAS concept. A single laser pulse generated during the tunnel useful test time was focused through a NaCl lens ahead of an aluminum hemisphere-cylinder model fitted with a piezoelectric pressure transducer at the stagnation point. In the more recent experiments, the simple hemisphere-cylinder model was substituted by a Double Apollo Disc model fitted with seven piezoelectric pressure transducers and six platinum thin film heat transfer gauges. The objective being to corroborate the past results as well as to obtain additional pressure distribution information and new heat transfer data.

INTRODUCTION

It has been suggested by several authors[1-4] that aerodynamic drag and heating of a hypersonic Trans-Atmospheric Vehicle, TAV, could be greatly reduced by adding energy to the air ahead of it. Such energy addition could be accomplished by a plasma torch mounted at the nose of the TAV[5-7], by an electric breakdown ahead of the TAV[8, 9] or by focusing a powerful laser or microwave beam ahead of the TAV flight path, as it has been originally suggested by Myrabo and Raizer[2] in 1994.

Myrabo and Raizer called the effect of reducing aerodynamic drag and heating through the use of electromagnetic radiation by DEAS effect. A laser driven Trans-Atmospheric Vehicle - TAV, resembling two Apollo re-entry heat shields mounted back to back, was even suggested by the former author. The experimental TAV, which makes use of the Laser-Supported DEAS effect is depicted in Fig.1.

The first experimental confirmation of such effect came in 1996 when a model of the proposed TAV, fitted with an electric arc plasma torch, was tested in the Rensselaer

CP664, Beamed Energy Propulsion: First International Symposium on Beamed Energy Propulsion,
edited by A. V. Pakhomov

Polytechnic Institute - RPI - 0.6m Hypersonic Shock Tunnel as a part of electrical engineer Jack Marsh Master Thesis.

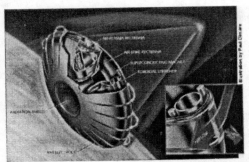

FIGURE 1. Conceptual Lightcraft TAV using the DEAS effect.

In these tests, the laser focus was represented by air plasma at the tip of the slender plasma torch mounted at the model centerline. It was observed that when the plasma torch was turned on at 35 kW, the conical shock wave, originating at the tip of the torch, would assume a parabolic shape indicating a change in the hypersonic, Mach 10, flow, due to the energy addition. Continuing Marsh's exploratory work, Paulo Toro, then a graduate student at RPI, extended the DEAS investigation by measuring both the surface pressure distribution and the surface heat transfer distribution for several plasma torch power levels. The results once more corroborated Myrabo & Raiser predictions but the presence of the torch itself would make it difficult to completely isolate the torch assembly beneficial effects from those of the energy addition.

To isolate the above mentioned effects and to more closely simulate the focusing of a laser beam or a microwave beam ahead of the model, the torch assembly had to be eliminated. To that end, the first author, during his Post Doctoral program at RPI, suggested that the energy addition could be performed by establishing an electric arc between two slender, 1.5mm dia., tungsten electrodes mounted at the exit plane of the hypersonic shock tunnel conical nozzle. In this way, the electrodes would be thin enough not to disturb the hypersonic flow and, at the same time, would eliminate the need to use the torch mounting. This experiment is still in progress[8, 9] but has already produced some interesting results.

The next natural step, which constituted the motivation for the exploratory work[10-12] and the recent investigation carried out at LAH, was to use a laser beam to drive the DEAS, as suggested by Myrabo & Raizer. In this situation, the DEAS in front of a vehicle is created by a shock wave propagating from a Laser-Supported Detonation (LSD) wave. The pressure at the wave front, being higher than atmospheric pressure, deflects the incident hypersonic airflow from the axial direction and forces it to flow over the air-spike to the periphery of the vehicle.

A propulsion system design of a transatmospheric vehicle using a DEAS inlet presents two important advantages: 1) it employs a detached conical (parabolic-shaped) shock wave to contain a rarefied "hot air pocket" which substantially reduces the flow Mach number impacting the vehicle forebody, thus decreasing the aerodynamic drag, and, most importantly, 2) it deflects the oncoming hypersonic air flow from the vehicle's path into an annular hypersonic inlet at the periphery of the vehicle where a MagnetoHydroDynamic - MHD engine could be located. The inlet air can either be subsequently accelerated by an MHD slipstream accelerator to produce thrust, or decelerated to extract onboard electric power.

EXPERIMENTAL APPARATUS

The Laser-Supported DEAS experiments were conducted at the Laboratory of Aerothermodynamics and Hypersonics, LAH, in Brazil. The IEAv 0.3m Hypersonic Shock Tunnel was used to produce high and low enthalpy hypersonic flow conditions. An excellent description of the facility and its capabilities can be found in Reference 13. In the high enthalpy runs, helium was used as the driver gas and the tunnel was operated in the equilibrium interface condition to produce a useful test time of roughly 500μs and reservoir conditions of 5,000 K, temperature, and 120 bar, pressure. In the low enthalpy case, air was used as the driver gas to produce a useful test time of 1.0 ms and reservoir conditions of 950 K, temperature, and 25 bar, pressure. The test section airflow Mach number was 6.2 in the high enthalpy tests and 7.8 in the low enthapy ones. The same conical, 15° half angle, 300mm exit diameter, nozzle with a throat diameter of 22.5mm was used in all cases. The different Mach numbers achieved are the result of the different reservoir conditions and real gas effects present in the tests.

One of the tunnel test section access windows had to be modified to accommodate the laser beam delivery system. This system consisted of a 50mm diameter NaCl lens with a focal distance of 180 mm mounted in a telescope. The telescope is free to move inside a support mounted to the test section window so that the focus can be adjusted to be in the nozzle centerline. Once positioned, the telescope is locked in position so that it does not move during the test. Due to geometrical constraints the telescope had to be installed 45° with respect to the nozzle centerline. This causes the lens to be damaged frequently and, sometimes, destroyed by high-speed particles/debris that reach the test section. In addition to that, due to the fact the air plasma, created in the focal point, tends to propagate towards the laser source[14] the energy addition region is not symmetrical with respect to the nozzle centerline. Figure 2 shows a drawing of the beam delivery system mounted to the test section access window.

FIGURE 2. Infrared laser beam delivery system mounted to the shock tunnel test section access window.

A <u>T</u>ransversely <u>E</u>xcited <u>A</u>tmospheric pressure, TEA, Carbon Dioxide Laser, designed and built by Watanuki *et al*[15], was used to drive the DEAS. Figure 3 shows the laser head, the beam delivery system and the hypersonic shock tunnel test section. The laser, in multimode operation, is capable of producing a single high energy, 7.5 J, short, 120 ns (FWHM), pulse at 10.6 μm. The output beam has a rectangular cross section, 34mm X 17mm. The experimental set up is shown in Figure 4.

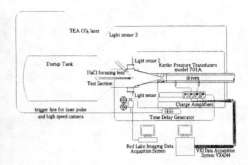

FIGURE 3. HST test section and the CO_2 TEA laser head.　　　**FIGURE 4.** Experimental set up.

The laser pulse was synchronized with the shock tunnel useful test time via a time delay generator triggered from a Kistler piezoelectric pressure transducer model 701A, located immediately upstream the nozzle entrance. Three Hamamatsu Ge photodiodes model B1720-02, as indicated in Fig. 4, were used as light sensors to monitor the generation of the laser pulse inside the laser head (sensor # 3), the production of the laser induced air ignition inside the test section (sensors # 1 & 2) and the natural air luminosity of the hypersonic/hypervelocity flow around the model (sensor # 2). Two additional Kistler pressure transducers model 701A, 0.5 m apart, located in the tunnel driven section, were used to time the incident shock wave.

In the previous experimental results the time-lapse type photographs of the luminous air flow around the model and of the laser-induced air ignition were taken by using a Nikon camera model N6006 with AF35-70mm f/3.3-f/4.5 Nikkor lenses and ISO 100 color film. In some selected tests, a Redlake high speed CCD camera model MotionScope PCI 8000S at 8,000 fps and a shutter speed of 1/24,000 s was used to investigate the generation and the extinction of the air ignition in the hypersonic flow. The high-speed camera was triggered simultaneously with the laser pulse and was set with a pre-trigger acquisition time of 0.5 s and a post-trigger acquisition time of 0.5 s, also. In the more recent investigation an ultra high-speed camera, Hadland IMACON 790 was also used for flow visualization.

All the data, with the exception of the flow visualization, were recorded using a Tektronix VX4244 16-channel 200kHz data acquisition system.

In the preliminary experiments, a very simple model consisting of an aluminum hemisphere-cylinder 55mm in diameter was mounted 60 mm downstream of the laser focal point. The model, Fig. 5, was fitted with a Piezoelectric pressure transducer so that the impact pressure downstream of the laser-driven air breakdown could be recorded. A centerline cylindrical channel, 4.5mm long and 2mm in diameter, connected the pressure transducer diaphragm to the model surface.

In order to obtain more detailed pressure information and heat transfer data a new model was designed and built. This model has the same geometry as that currently under investigation (electric arc – driven DEAS) by Rensselaer Polytechnic Institute – RPI researchers, i.e., double Apollo disc configuration. A schematic view of the model, with the instrumentation ports, is shown in Fig. 6 and a photograph of the actual model is shown in Fig. 7. Figure 8 depicts an artist view of the model installed in the tunnel test section.

FIGURE 5. Hemisphere-Cylinder model.

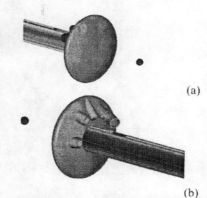

(a)

(b)

FIGURE 6. Front (a) and rear (b) views of the Double Apollo Disc model showing the position of the instrumentation. The red dot indicates the position of the laser focal point.

FIGURE 7. Photograph of the instrumented Double Apollo model

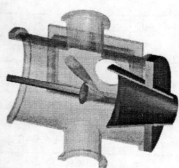

FIGURE 8. The Double Apollo Disc model in the IEAv HST test section.

The 100mm diameter model houses pressure transducers and platinum thin film heat transfer gauges, as indicated in Fig.6, and it is positioned 100mm downstream of the air ignition. For each pressure transducer there is a companion heat transfer gauge located at the same radial distance from the model centerline. In this way, simultaneous pressure and heat transfer data can be obtained at a given distance from the model axis of symmetry. The heat transfer gauges and their power supplies and amplifiers were manufactured and calibrated at LAH trough the use of a low-pressure shock tube. The same shock tube was also used to calibrate the pressure transducers. Figure 9 shows a detailed view of the heat transfer gauge.

FIGURE 9. Photograph of the 1.6 mm diameter platinum thin film heat transfer gauge.

EXPERIMENTAL RESULTS AND DISCUSSION

The nominal shock tunnel test conditions are presented in Table I. These conditions did not vary more than 5% from run to run. The laser operating conditions can be found in Table II.

Table I Shock Tunnel Test Conditions.

		High Enthalpy	Medium Enthalpy	Low Enthalpy
Reservoir	Pressure (bar)	120.0	173.0	25.0
	Temperature (K)	5,000.0	1,685.0	950.0
Conditions	Enthalpy MJ/kg)	9.0	1.9	1.0
	Pressure (mbar)	12.0	23.2	4.0
Free	Temperature (K)	1,000	158.4	77.0
Stream	Density (g/m^3)	4.0	51.1	17.0
Conditions	Mach Number	6.2	7.35	7.8

Table II CO_2 TEA laser operating conditions.

Energy per pulse (Joules)[*]	7.5
Pulse Duration (ns)	120
Gas Mixture	7%CO_2-54%N_2-39%He

• **average between the energy meter readings immediately before and after the test.**

Previous Results

From Table I it is quite evident that the static pressures present in the hypersonic flow at the nozzle exit were quite low. As a consequence, the authors were expecting some difficulty in inducing air ignition under these conditions[16], for the laser energy available shown in Table II. However, it was experimentally observed that, for the high enthalpy runs, laser-induced air breakdown would always take place in the low-pressure conditions existing in the hypersonic flow upstream of the model. At first, the present authors were lead to believe that the air ignition was being triggered by particles/debris contaminating the high-speed flow. Once the low enthalpy tests were performed, and the laser-induced air breakdown could not be obtained as easily as in the high enthalpy case, the particle contamination idea became uncertain. Since the same type of test gas, in the present case air, was used in both cases the high and low enthalpy tests, any contaminants that could trigger the air ignition were present in both scenarios. The authors then concluded that probably a combination of a high static temperature, 1,000 K, and non-equilibrium effects present in the flow could be responsible for the successful laser-induced air ignition at these low static pressures.

Figure 10 shows typical reservoir and the model impact pressure traces. The impact pressure trace shows a vibration noise, which could not be eliminated during the tests.

Figure 11 shows the same impact pressure trace depicted in Figure 10 as well as the output traces from light sensors # 1 and 3. Light sensor # 3 clearly indicates the moment the TEA laser fired by recording the luminosity coming out from the laser cavity electric discharge. On the other hand, the natural air luminosity of the hypervelocity flow around the model, as well as the luminosity of the laser-induced air breakdown, can be seen in the output trace from light sensor # 1.

Interestingly enough, the impact pressure trace seems to show the effect of the laser-induced air breakdown. Shortly after the air ignition is generated, a high frequency "ringing" indicates the impact of the detonation wave against the model front surface and a drop in pressure occurs. This drop in the impact pressure coincides with the duration of the air ignition and may indicate a decrease in the aerodynamic drag caused by the DEAS effect. This trend, however, is going to be corroborated by more recent tests to be discussed in the next section.

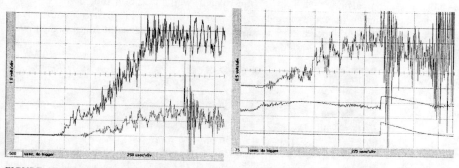

FIGURE 10. Typical high enthalpy reservoir (top) and impact (bottom) pressure traces.

FIGURE 11. Typical high enthalpy impact pressure (top), light sensor # 1 (middle) and light sensor # 3 (bottom) traces.

As stated previously, low enthalpy tests were also conducted with the hemisphere-cylinder model but no consistent laser induced air ignition was ever achieved. As a consequence, no change in the impact pressure was verified. Additional information on these tests can be found in Reference 12.

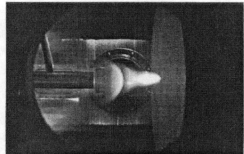

FIGURE 12. Open shutter photograph of the laser-induced air ignition in Mach 6.2 flow and of the strong normal and weak conical shock structures in front of the model.

Figure 12 shows a time-lapse photograph of the laser-induced air breakdown upstream of the model at Mach 6.2 flow conditions. Due to stray light it is possible to see internal details of the test section, the sting mount, the nozzle exit and even the infrared telescope mounting behind the sting. Since it is a time-lapse photograph, every luminous phenomenon that took place inside the test section was recorded onto the photographic film. Therefore, it is also possible to observe in Fig. 12 both the bow shock in front of the model and the conical flow structure upstream of it.

The conical flow structure seen in Fig.12 seems to agree with the impact pressure drop observed in Fig. 11 and also with the DEAS mechanism proposed by Myrabo and Raizer. As soon as the laser-induced air breakdown develops, the air is pushed by the LSD wave from the region immediately upstream of the model, and over the "Air Spike" to the periphery of the hemisphere generating the conical flow structure seen in Fig. 12. This conical flow structure creates a detached conical shock wave (parabolic-shaped) well ahead of hemisphere.

Additional information on the dynamics of the establishment of the DEAS came with the utilization of a Redlake high-speed camera. Figure 13 shows a sequence of frames taken every 125µs and an exposure time of 1/24,000 s.

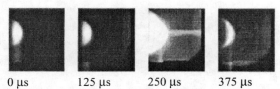

| 0 µs | 125 µs | 250 µs | 375 µs |

FIGURE 13. Time history of the generation and extinction of the laser-supported DEAS in high enthalpy flow.

From the sequence of frames depicted in Fig. 13, one can see that the bow shock is established over the hemisphere-cylinder model until the laser-induced air ignition creates the DEAS, at 250 µs, and the shock wave becomes conical. After that, when the air ignition is extinguished, at 375 µs, the bow shock structure is reestablished. This behavior agrees with the trend observed in the high enthalpy impact pressure trace, Fig. 11, in which the pressure decreases shortly after the establishment of the air breakdown and increases when ignition is extinguished.

Present Results

For the medium enthalpy conditions present in Table I, and the conditions show in Table II, a more instrumented model, Fig. 7, was tested. Surface pressure measurements can be found in Figure 14 (left). Such pressure measurements are non-dimensionalized by the impact pressure and the radial position of the pressure taps, r, by the model radius, R. As one can see, the surface pressure level over most of model front surface for DEAS-off is much higher than that given by DEAS-on. As a consequence, the laser-supported DEAS was able to generate a decrease in the impact pressure over the new model tested and, therefore, a considerable decrease in the aerodynamic drag. This result is in agreement to that obtained with the simple hemisphere-cylinder model discussed earlier in the present paper. On the other hand, a pressure increase was detected near the model periphery. Although the contribution of such pressure increase to the aerodynamic drag is minimal, it is an interesting behavior since, as stated previously, such configuration may be used as an annular air inlet. The same behavior was observed by Toro[6] (Fig.14 right) using a plasma torch and the same model geometry. Additional testing may reveal whether this pressure increase can be minimized by increasing the distance from the model front surface to the air ignition location.

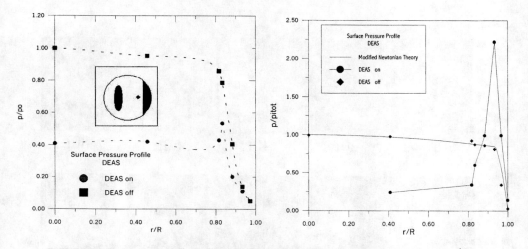

FIGURE 14. Surface pressure distribution over the Double Apollo Disc model (present) and Toro.[6]

Surface heat transfer measurements were conducted but unfortunately the Platinum thin film heat transfer sensors did not survive the test, and the data will not be presented at this time.

A more detailed history (than that depicted in Fig. 13) of the generation and the extinction of the Laser-Supported DEAS was obtained through the use of the ultra high speed Hadland IMACOM 790, Fig. 15. The camera was triggered by a pressure transducer upstream of the nozzle entrance as shown in Fig. 4, and operated at 100,000 frames per second. The exposure time of each frame was 2 μs resulting in a total of 10 frames each one taken every 18 μs. All 10 frames were recorded onto a single Polaroid 667 film. In order to facilitate the understanding of the sequence of events the film was digitized and each individual frame displayed in chronological order. In Fig. 15, the darkened areas seen in the two first frames (as well as in the two last frames) correspond to the edge of the camera recording phosphor screen. The positions of the model surface and the laser focal point are indicated in the inset in Fig. 15. In the two first frames only the luminosity of the bow shock wave over the model surface is seen. The third frame depicts the moment of the laser breakdown. In the next 4 frames one can see the resulting plasma being pushed against the model surface and the conical flow structure established ahead of it. Finally, in the last three frames it is visible the extinction of the DEAS and something that resembles re-establishment of the original bow shock wave.

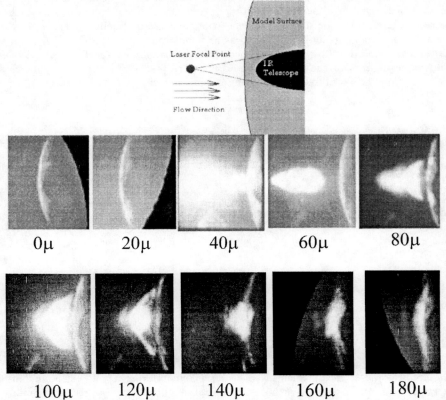

FIGURE 15. Time history of the generation and extinction of the laser-supported DEAS in medium enthalpy flow.

CONCLUSIONS

Preliminary experiments to demonstrate the laser supported DEAS concept in Mach 6.2 flow (real air) and Mach 7.35 flow (ideal air) were conducted in the IEAv 0.3 m Hypersonic Shock Tunnel. A CO_2 TEA laser was used to drive the air ignition upstream of a hemisphere-cylinder installed in the modified shock tunnel test section. It was observed that, for the high enthalpy reservoir conditions, the laser-induced air ignition could be established consistently in spite of the low static pressures present in the hypersonic/hypervelocity airflow. A piezoelectric pressure transducer, installed in the model, indicated a drop in the impact pressure that coincided with the duration of the luminosity generated by the laser-induced air breakdown. Time-lapse type photographs have shown a conical flow structure superimposed to the bow shock wave standing in front of the hemisphere- cylinder. The dynamics of the formation of the conical flow structure was revealed through the use of a high-speed camera. Low enthalpy, ideal air, runs were also performed but the laser-supported air ignition could not be established either consistently or completely.

In order to further investigate the laser supported DEAS and corroborate the promising results herein presented, a new model, fitted with seven pressure transducers and six platinum thin film heat transfer gauges was tested. The geometry of the model is the same as the one being tested by RPI researchers using an electric arc discharge to drive the DEAS. A drop in the surface pressure was observed on most of the model front surface when laser-supported DEAS was on. An increase in the surface pressure was also noticed near the model periphery but its contribution to the aerodynamic drag is negligible. The net result was a lower surface pressure over the model indicating a decrease in the net drag. Due to a premature failure of the heat transfer gauges, no conclusion about a reduction in the surface heat flux can be drawn yet. More experiments and flow visualization are needed to better understand the phenomenon.

ACKNOWLEDGMENTS

The authors would like to express their gratitude to Drs. Antônio C. Oliveira and Laurentino C. V. Neto for their invaluable support and advice.

The authors are also in debt to Messrs. André Jacobovitz and Lincon T. Tan, from LYNX Com. Imp. Ltda., for the technical support and graceful use of the Redlake high speed CCD camera.

The financial support received from Instituto Tecnológico de Aeronáutica-ITA is also acknowledged.

REFERENCES

1. Tidman, D. A., "Apparatus and Method for Facilitating Supersonic Motion of Bodies Through the Atmosphere," United States Patent Number 4,917,335, April 1990.
2. Myrabo, L.N., and Raizer, Yu. P. "Laser Induced Air Spike for Advanced Tansatmospheric Vehicles," AIAA Paper 94-2451, 25th AIAA Plasmadynamics and Lasers Conference, Colorado Springs, CO, June 20-23,1994.
3. Gurijanov, E.P., and Harsha, P.T., "AJAX: New Directions in Hypersonic Technology," AIAA Paper 96-4609, 1996.

4. Covault, G. "'Global Presence' Objective Drives Hypersonic Research," Aviation Week & Space Technology, April 5, 1999, pp. 54-58.
5. Marsh, J.J., Myrabo, L.N., Messitt, D.G., and Nagamatsu, H.T., "Experimental Investigation of the Hypersonic "Air Spike" Inlet at Mach 10,"AIAA Paper 96-0721, January 1996.
6. Toro, P.G.P., Nagamatsu, H.T., Minucci, M.A.S., and Myrabo, L.N., Experimental Pressure Investigation of a "Directed-Energy Air Spike" Inlet at Mach 10," AIAA Paper 99-2843, July 1999.
7. Toro, P.G.P., Nagamatsu, H.T., Myrabo, L.N., and Minucci, M.A.S., "Experimental Heat Transfer Investigation of a "Directed-Energy Air Spike" Inlet at Mach 10," AIAA Paper 99-2844, July 1999.
8. Minucci, M.A.S., Bracken, R.M., Myrabo, L.N., Nagamatsu, H.T., and Shanahan, K.J., "Experimental Investigation of an Electric Arc Simulated "Air Spike" in Hypersonic Flow," AIAA Paper 00-0715, 38th Aerospace Sciences Meeting & Exhibit, Reno, NV, January 10-13, 2000.
9. Bracken, R.M., Myrabo, L.N., Nagamatsu, H.T., Meloney, E.D., and Minucci, M.A.S., "Arc Simulated "Air-Spike" in the RPI Hypersonic Shock Tunnel at Mach 10 and Arc powers up to 11MW," 7th International Workshop on Shock Tube Technology, Long Island, NY, September 18-20, 2000.
10. Minucci, M.A.S., Toro, P.G.P., Chanes Jr, J.B., Ramos, A.G., Pereira, A.L., Nagamatsu, H.T. and Myrabo, L.N. "Experimental Investigation of a Laser-Supported Directed-Energy "Air Spike" in Hypersonic Flow – Preliminary Results," 7th International Workshop on Shock Tube Technology, Long Island, NY, September 18-20, 2000.
11. Minucci, M.A.S., Toro, P.G.P., Chanes Jr, J.B., Ramos, A.G., Pereira, A.L., Nagamatsu, H.T. and Myrabo, L.N. "Recent Developments on the Laser-Induced Air Spike in Hypersonic Flow," 23rd International Symposium on Shock Waves, Fort Worth, TX, July 23-27, 2001.
12. Minucci, M.A.S., Toro, P.G.P., Chanes Jr, J.B., Pereira, A.L., Nagamatsu, H.T. and Myrabo, L.N. "Investigation of a Laser-Supported Directed-Energy "Air Spike" in Mach 6.2 Air Flow – Preliminary Results," 39th AIAA Aerospace Sciences Meeting & Exhibit, Reno, NV, January 8-11, 2001.
13. Nascimento, M.A.C., "Gaseous Piston Effect in Shock Tube/Tunnel When Operating in the Equilibrium Interface Condition," Doctoral Thesis Dissertation, Instituto Tecnológico de Aeronáutica - ITA, São José dos Campos, SP, Brazil, October 1997.
14. Raizer, Y.P., Laser-Induced Discharge Phenomena, Studies In Soviet Science, Consultants Bureau, New York, 1977.
15. Watanuki, J.T., Oliva, J.L.S., Lobo, M.F.G., and Rodrigues, N.A. S., "Laser de CO_2 -Híbrido com Célula de Baixa Pressão Pulsada," Revista de Física Aplicada e Instrumentação, Vol. 3, No. 3, 1988, pp.207-216.(in Portuguese)

The Impact Imperative: Laser Ablation for Deflecting Asteroids, Meteoroids, and Comets from Impacting the Earth

Jonathan W. Campbell*, Claude Phipps**, Larry Smalley***, James Reilly****, and Dona Boccio*****

*Advanced Projects/FD02, National Space Science and Technology Center, NASA/MSFC, Huntsville, Alabama, 35812 jonathan.campbell@msfc.nasa.gov
**Photonics Associates 200A Ojo de la Vaca Road Santa Fe, NM 87505
***Department of Physics, University of Alabama, Huntsville
**** Northeast Science and Technology, East Sandwich, MA
*****Queensborough Community College of the City, University of New York, New York

ABSTRACT

Impacting at hypervelocity, an asteroid struck the Earth approximately 65 million years ago in the Yucatan Peninsula area. This triggered the extinction of almost 70% of the species of life on Earth including the dinosaurs. Other impacts prior to this one have caused even greater extinctions.

Preventing collisions with the Earth by hypervelocity asteroids, meteoroids, and comets is the most important immediate space challenge facing human civilization. This is the **Impact Imperative**.

We now believe that while there are about 2000 earth orbit crossing rocks greater than 1 kilometer in diameter, there may be as many as 200,000 or more objects in the 100 m size range. Can anything be done about this fundamental existence question facing our civilization? The answer is a resounding **yes**!

By using an intelligent combination of Earth and space based sensors coupled with an infra-structure of high-energy laser stations and other secondary mitigation options, we can deflect inbound asteroids, meteoroids, and comets and prevent them from striking the Earth. This can be accomplished by irradiating the surface of an inbound rock with sufficiently intense pulses so that ablation occurs. This ablation acts as a small rocket incrementally changing the shape of the rock's orbit around the Sun. One-kilometer size rocks can be moved sufficiently in about a month while smaller rocks may be moved in a shorter time span.

CP664, *Beamed Energy Propulsion: First International Symposium on Beamed Energy Propulsion*,
edited by A. V. Pakhomov
2003 American Institute of Physics 0-7354-0126-8

We recommend that space objectives be immediately reprioritized to start us moving quickly towards an infrastructure that will support a multiple option defense capability. While lasers should be the primary approach initially, all mitigation options depend on robust early warning, detection, and tracking resources to find objects sufficiently prior to Earth orbit passage in time to allow mitigation. Infrastructure options should include ground, LEO, GEO, Lunar, and libration point laser and sensor stations for providing early warning, tracking, and deflection. Other options should include space interceptors that will carry both laser and nuclear ablators for close range work. Response options must be developed to deal with the consequences of an impact should we move too slowly.

INTRODUCTION

Astronomical telescopes and deep space radar systems have verified the existence of a large number of near-Earth objects (NEOs), such as asteroids, meteoroids, and comets that potentially could destroy most life on Earth. An asteroid with a diameter of 1-10 km would strike the Earth with a power rivaling the strength of a multiple warhead attack with the most powerful hydrogen bombs known to man. Computational fluid dynamics studies have indicated that an ocean strike by an asteroid this size would create a gigantic tsunami that would flood and obliterate coastal regions. More significantly perhaps, this strike would eject a massive dust cloud rivaling the most powerful volcanic explosion, which could seriously affect climate on the scale of two to three years. It could alter our biosphere to the point that life as we know it would cease to exist.

As recent as five years ago, it was thought by the astronomical and astrophysics community that most of the known NEOs do not pose a near term threat, and therefore that these objects do not present any danger to the Earth and its biosphere. However, the relatively recent collision of the comet Shoemaker-Levy 9 with Jupiter and continuing discoveries of uncatalogued asteroids passing near Earth without any advanced warning have increased concerns. It is worthwhile to note that one striking feature of practically every celestial body in our solar system is the abundance of impact craters. [See *The Threat of Large Earth-Orbit Crossing Asteroids*, 103rd Congress, First Session, Hearing House Committee on Science, Space and Technology, Subcommittee on Space (Washington, DC: March 24, 1993), which discusses NASA and international research on detecting and deflecting asteroids before these hit the earth.]

Since collisions with asteroids, meteoroids, and/or comets have caused major havoc to the Earth's biosphere on several occasions in the geological past, one reality of our civilization's continued existence is that the Earth will experience another impact in the future. The idea presented here is to use lasers to defend against Earth impacting asteroids and comets.

BACKGROUND

Impacts from Near-Earth Objects (NEO's) are not "academic" problems. Direct impact by a NEO approximately 10 km in diameter will annihilate most biota because of the resulting firestorm and nuclear winter. Such objects have a kinetic energy release of order 30TT (teratons), create tidal waves [Hills, 1992] and earthquakes. The last such epoch-ending event occurred 65M years ago at the so-called "K/T boundary". The location of the impact is now known to be the Chicxulub site off the coast Yucatan [see Sharpton 1993].

A multiple body impactor of greater energies (Comet Shoemaker-Levy) struck Jupiter in 1994. Each body left a mark the size of Earth in its upper atmosphere. A more recent (and more likely) example is the Tunguska event of June 30, 1908, in which an object probably 110 m in diameter impacted with 10MT explosive equivalent, clear cutting 2150 km^2 of forest. It was probably a "snowball" NEO [BBC 2001]. NEO's include Earth-crossing Asteroids (ECA's), meteoroids and comets. Impacting NEO's cause damage via 6 mechanisms, whose relative importance depends on site, energy, diameter and path. Only three of these require the NEO to strike land.

For the 10-km-size "doomsday asteroids," Earth impact frequency is about one per 100My. However, impact probability is a strong function of asteroid diameter d, so that NEO impacts of the size that initiated the Tunguska event happen every few centuries. Where diameter d is in meters, NEO impact frequency (per year) is given by [see Shoemaker 1995]

$$N(d) = 80/d^x \text{ where } 2.5 < x < 3 \qquad (1)$$

Each month, about 30 of these small (40-80m) diameter objects pass through the Moon's orbit, offering excellent opportunities for diagnostics and experiments. Epoch-ending NEO's have also passed within fractions of an AU in the past decade. Small NEO's are the most likely threat in our lifetime [see Eq. 1]. However, small NEO's are extremely difficult to detect in time to take action. For example, assuming detection at visual magnitude $m_V=23$, an 80-m-diameter, 30 km/s "dirty snowball" NEO with albedo 0.025 will be 200 light-seconds distant (0.4AU) on detection and just 23 days from Earth impact.

Nuclear deflection has been suggested [Solem 1993]. In this approach, a multi-MT weapon is detonated in the vicinity of, but not adjacent to, the NEO. Orbit modification occurs through rapid ablation of the object as opposed to gradual ablation from the laser approach. Considering the additional time required to verify orbit, 23 days leaves inadequate time for launching any kind of nuclear-tipped conventional interceptor, transporting the payload to the NEO, and matching its speed (in the reverse direction) and detonating optimally.

In contrast, laser deflection offers instant response, agility, and low cost compared to the nuclear alternative. Lasers do not have to be transported to the target. Laser deflection is also attractive relative to putting nuclear weapons in orbit, a suggestion that may not be embraced by the general public. Laser deflection uses the thrust produced by a jet produced on the surface of the NEO by laser ablation [Phipps 1992-5, 1997-8].

LASER EARTH DEFENSE CONCEPT

Many schemes have been discussed for dealing with NEOs on collision courses with the earth. These include the use of nuclear weapons to fragment the NEO, or landing on them using various methods (propulsive, explosive, etc.) to steer the asteroid into a passing orbit.

Fragmentation may not be a viable solution because the center of mass of the cloud would continue on the original collision trajectory as the parent mass. This would result in multiple impact events similar to the Shoemaker-Levy 9 collision with Jupiter. Also, fragmentation may make subsequent orbit shaping more difficult.

Many issues and engineering solutions need to be addressed in order to land on a NEO and place nuclear devices or other trajectory altering systems there. Although the cost of any NEO protection system will likely be significant, any system requiring a deep-space rendezvous would also require sufficient warning of an impact to be implemented. Additionally, a failure of such a defense system may not allow for a second mitigation effort to be attempted before the object impacts the Earth.

A better system would be one that is "on station" and could be used routinely to shape asteroid orbits over long periods of time so that they do not pose a potential threat. The system should also be able to handle the wide range of materials and sizes that constitute the NEO population (current or yet to be discovered). Phased Array Laser Systems (PALS) could be developed and placed in space, either orbiting or lunar based. Space-based laser constellations (SBL) are presently under development and will be flown during the next decade. The feasibility for a PALS based system is discussed below.

Laboratory experiments using a 20 kW pulsed laser have shown that the impulse imparted to aluminum targets due to the ejected plasma cloud gives an average surface pressure $p = 6.5 \times 10^{-4}$ N/cm^2, or equivalently, an acceleration $\mathbf{a} = 1.25 \times 10^{-6}$ m/s^2.

Thus, with present technology, an array of laser beam directors can be aimed at an asteroid, meteoroid, or a comet, providing sufficient power to ablate its surface. It is simply a matter of putting in place a sufficient number of lasers to accomplish the mission.

To generate ablation thrust, the main requirement is that the minimum laser intensity

$$I_{min} = 24/\tau^{0.55} kW/cm^2 \tag{2}$$

be delivered the NEO surface, either during a pulse or continuously. A laser momentum-coupling coefficient (thrust to optical power ratio)

$$C_m = F/P = 50 \text{ N/MW} \tag{3}$$

can be assumed [Phipps 1997].

Deflecting a 1 km diameter iron asteroid, as we will see in the simulation results that follow will require a peak laser power of approximately 200 GW. Several alternate potential approaches are available to power the array including nuclear or electric generation and solar power arrays.

Let us assume that the asteroid is at infinity moving toward the Earth with a closing velocity v_0. The closest point of approach R_e is given by

$$R_e \cong R_E \left[1 + 2g \left(\frac{R_E}{v_0^2} \right) \right]^{\frac{1}{2}} \tag{4}$$

where R_E is the radius of the Earth, and g is the gravitation acceleration at the surface of the Earth. Clearly, for the large anticipated values of v_0, the Earth's gravitational pull will be insignificant in the encounter. There are two cases of interest:

- "Head-on" collision:
  $v_o = 40$ km/s $\longrightarrow$ R$_e$=1.04 R$_E$

- "Catch-up" collision:
  $v_o = 5$ km/s $\longrightarrow$ R$_e$=1.1 R$_E$

Hence, we may define a threshold for success for the two possible encounter scenarios. A minimum of 38.8 days of illuminating the target is necessary for the case of a head-on collision, and in most cases would take much less illumination time. The warning time of impending impact is of critical significance, which highlights the importance of deep space surveillance of NEOs in addition to long-term monitoring and orbital calculations. Early orbit shaping should be extraordinarily effective using a PALS. Also it is important that PALS be deployed at positions that are allow sufficient target illumination time to properly alter the trajectory of a confirmed impactor.

If the collision scenario depicted in Figure 1 was encountered. The PALS firing with a good aspect from L_5 and sufficient lead time (as shown in the figure,) would have 2-3 months to move the asteroid away from a collision path with the Earth. Only with a sufficiently capable detection system would there be adequate time in advance for the PALS to deflect the asteroid away from the Earth. This fact stresses the need for coupling with PALS an early warning system using optical and/or radar imaging techniques.

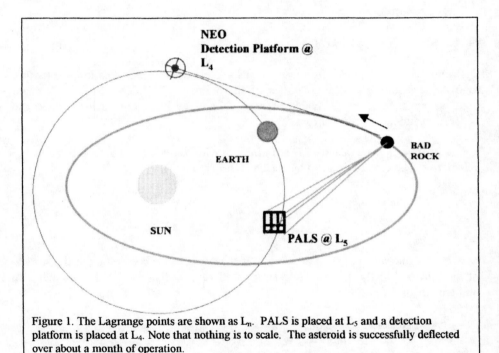

Figure 1. The Lagrange points are shown as L_n. PALS is placed at L_5 and a detection platform is placed at L_4. Note that nothing is to scale. The asteroid is successfully deflected over about a month of operation.

The ΔV of 5 km/s is an example of an impulse that yields a "miss distance." In this case, the simulation yields that the asteroid passes in front of the Earth by 1.25 Earth diameters. An approach requiring significantly less power for PALS would be a gradual shift in the orbit by a long duration, low intensity impulse. This lower energy impulse would reshape the orbit over a long time period, perhaps several orbits. Ideally, for the asteroidal orbit shown in Figures 1, 2 and 3, it might conceivable to move the asteroid into an orbit that removes any potential threat to the Earth. From a non-defensive standpoint, it is interesting to contemplate asteroid orbit modification for the purpose of scientific exploration and/or commercial exploitation (i.e., asteroid mining). This application of a PALS may be particularly feasible for small asteroids (less than 100 m) in orbits that are "easily" modified to a desired rendezvous location for processing.

CURRENT LASER TECHNOLOGIES

The US Air Force Airborne Laser (ABL) is a major weapon system development by the United States Air Force to provide an airborne, multi-megawatt laser system with a state-of-the-art atmospheric compensation system to destroy enemy theater ballistic missiles at long ranges [Lamberson 2002]. The Space Based Laser (SBL) program will use a high-energy laser to destroy boosting missiles in flight. The principal kill mechanism is to cause mechanical weakening of the booster skin, so that internal pressures will cause the missile to explode while it is still boosting [Riker 2002].

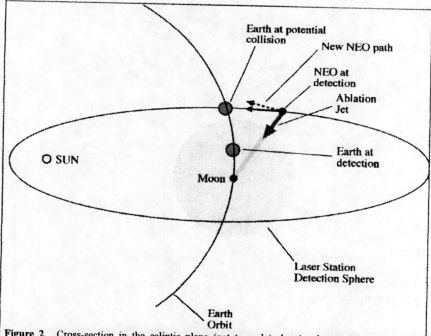

Figure 2. Cross-section in the ecliptic plane (not to scale) showing laser station. At least 5 stations on the Moon provide omnidirectional coverage. The NEO is handed off from one to the other during the interaction. Detection at 0.2AU (100Lt-s) gives at least 20 days action time. Electrical power for the lasers is beamed up from Earth.

Both are examples of very high power lasers which are available now, and which could be deployed for preliminary asteroid thruster tests without much further development.

CURRENT SENSOR TECHNOLOGIES

In general, acquisition of remote objects for observation and tracking is accomplished by the observation of either self-emitted or reflected optical energy, RF energy, acoustic energy or other quanta in comparison to some background level. In particular, only optical and radar sensors are usable to acquire targets at long range. The three approaches below are ones that currently appear promising given the ranges, object sizes and sensor characteristics involved. The first is microwave radar

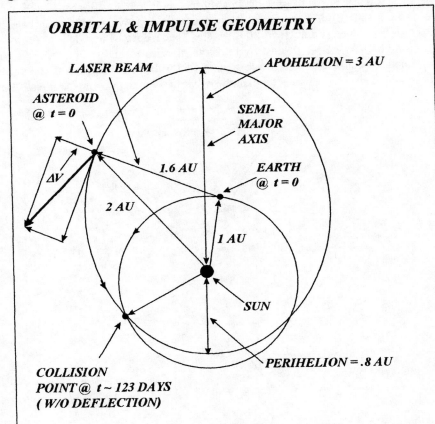

Figure 3. The ΔV is imparted to the asteroid by laser ablation over a period of time of about 40 days. As a result collision is averted and the asteroid will cross in front of the Earth.

with characteristics similar to the MIT/LL HAYSTACK, DoD PAVE PAWS or DEW Line radars, but with a very-much-higher-power electronically scanned beam (repeated linear two-dimensional scan or other acquisition strategy) for wide-angle search at long range.

The second is a passive optical system - an astronomical-class telescope perhaps with an angle-scanning capability along the lines suggested by MIT/LL in the NASA ORION study for a modified HAYSTACK-type, DoD PAVE PAWS type or DEW Line type radar. The illumination of the objects would be by sunlight. The size of the instantaneous Field of View of the system fixes the instantaneous spot size being viewed, while the angle-scanning capability determines the search Field of Regard. Ecliptic Plane as well as out-of-plane threat asteroid objects must be considered.

The third is an active illuminator laser-radar (LADAR) ranging system. Economy dictates that if this option were chosen, the transmitter would use the pusher laser as the energy source, but would use a de-focused beam to interrogate a large spot in space for the detection function. The beam would be then be narrowed to perform the ranging and tracking functions.

TECHNICAL OBJECTIVES

Define the laser and pointer-tracker (PT) system's characteristics including capabilities of the laser and PT system that a potential Phase II test planning might require for thrust and impulse-production applications. This definition must be the first objective accomplished, since it sets the technical environment for the tasks in the rest of the program.

Complete the conceptual design for a rare-gas laser system (0.3-1 micron wavelength) and solid-state system (0.3-1 micron wavelength) that satisfies the requirements of the potential application as defined above.
Identify, characterize, prioritize and select laser parameters including wavelengths in repped-pulse operation, specific wavelengths, and range of gain medium options proven reliable, as obtained from ongoing test programs and analyses.

Adapt laser designs including solid-state and sealed-off gas laser designs to be compatible with the empirically determined laser operation envelope into a preliminary design of the solid-state cooled laser and the sealed-off cooled rare-gas laser.

With the concept for a solid-state and a sealed-off waveguide-array rare gas laser in place at the end of Phase I, the logical continuation into Phase II would be first the testing of the chosen waveforms and wavelengths on appropriate materials and objects to validate impulse and thrust production. Those options that survive Phase I scrutiny will then be tested in Phase II, optimized to satisfy the requirements of the Phase II and Phase III demonstrations.

517

Compare sensor technology options. Geometry and sensor technology will be studied in combination to determine the best approach. Areas of investigation will include back-illuminated CCD's, crossed photon-counting delay lines and other novel options.

Compare location options.

      Moon - The Moon has strong advantages: providing a reaction mass for the station is critical. Disadvantages include wide temperature extremes.

      Libration Points – These offer advantages and should be considered as well.

      Earth – The most convenient location and least expensive superficially. Must overcome problems working through the atmosphere.

      Mars - Mars is interesting as an early-warning outpost.

      Rendezvous - Taking a smaller pusher laser to the target may be another option.

      Examine Energy-gain Options. Study creative options for providing substantial energy gain in the laser-NEO interaction. Two of these are: a) the billiard-ball option, in which a small NEO is deflected into the path of the larger one at distance sufficient for most of the resulting fragments to clear Earth and b) the scattering option, in which the orbit of a NEO which is substantially similar to Earth's orbit is modified using Earth's gravitational field.

The impact of cost sharing should be considered. Other applications can support the cost of a NEO-deflection laser system. These include capturing small asteroids and mining their rich rare-metal deposits [Blacic 1993] and deflecting Earth-orbiting space junk so that it burns up in the atmosphere [ORION concept: Phipps, et al. 1996; Campbell 1996].

SUMMARY, CONCLUSIONS, AND RECOMMENDATIONS

An elegant, cost effective, feasible laser technology approach has been identified - a global solution to solve a global problem. This solution is truly international in scope in that it solves the problem for everyone.

If a high energy, laser pulse of sufficient intensity strikes an asteroid, meteoroid, or comet in space; a micro-thin layer of material is ablated from its surface. This super hot vapor rapidly expands outward imparting a tiny amount of force to the object. Since current laser technology produces 10 to 100 pulses per second, the ablation interaction is rapidly repeated over and over again. This cumulative thrust acting on the object if applied at the appropriate point in the object's orbit is sufficient to deflect it from impacting the Earth.

In addition, the additional promise of orbit shaping capability for asteroids, meteoroids, and comets is that the orbit may be modified sufficiently to make it convenient for utilization such as mining or in situ materials utilization. One final note on statistics in an investment context: the probability of the Earth being struck by a hazardous asteroid in the near future is approximately a thousand times more likely than winning a recent Florida lottery.

We recommend a two-year program that will take these concepts to laboratory demonstration level as regards laser performance, laser-target interaction, detection and a lab-scale test of phased array performance.

We further recommend a follow-on program that will consist of an experimental program to prove the concepts at significant range, including detection of remote objects and pushing surrogate targets released by the Shuttle. This program will include a test in which an existing very high power laser (e.g., HELSTF, ABL,) is employed to illuminate and measurably push one of the 30 or so 40-m-size NEO's that pass through the Moon's orbit each month.

In general, we recommend that the World's space objectives be immediately reprioritized to start us moving quickly towards a multiple option defense capability – an integrated ground and space infrastructure. While lasers should be the primary approach, all mitigation options depend on robust early warning, detection, and tracking resources to find objects sufficiently prior to Earth orbit passage in time to allow mitigation. Infrastructure options should include ground, LEO, GEO, Lunar, and libration point laser and sensor stations for providing early warning, tracking, and deflection. Other options should include space interceptors that will carry both laser and nuclear ablators for close range work. Response options must be developed to deal with the consequences of an impact should we move too slowly.

Preventing collisions with the Earth by hypervelocity asteroids, meteoroids, and comets is the most important immediate problem facing human civilization. This is the **Impact Imperative**.

REFERENCES

1. BBC News Online, 30 October 2001
2. Center for Astronomical Adaptive Optics (1997) http://athene.as.arizona.edu:8000/caao/.
3. Starfire Optical Range (1997) http://wwwsor.plk.af.mil/.
4. Cleghorn, George et al (1995) *Orbital Debris: A Technical Assessment*, National Academy Press, Washington, DC, http://www.nas.edu/ccts/aseb/debris1.html.
5. Blacic, J., "Mining Near-Earth Objects for Resources to Benefit Earth", Los Alamos National Laboratory internal white paper (1993)
6. Campbell, J.W., *Project ORION: Orbital Debris Removal Using Ground-Based Sensors and Lasers*, NASA Marshall Spaceflight Center Technical Memorandum 108522(1996) http://infinity.msfc.nasa.gov/Public/orion/default.html.
7. Hills, J. G., "Fragmentation of Small Asteroids in the Atmosphere", Los Alamos National Laboratory report LA-UR-92-2321 (1992)

8. Kantrowitz, A. (1972) *Aeronaut. Astronaut.* **10**, 74

9. Lamberson, S., "The Airborne Laser", *Proc. SPIE High Power Laser Ablation IV* (2002) to appear

10. Phipps, C.R., "Dynamics of NEO Interception," *Report of the NASA Near-Earth-Object Interception Workshop*, John D. G. Rather, Chair, Report LA-12476-C, Los Alamos National Laboratory, Los Alamos NM (1992)

11. Phipps, C.R., "Astrodynamics of Interception," in *Report of the NASA Near-Earth-Object Interception Workshop*, John D. G. Rather, Chair (workshop summary), Report LA-12476-C, Los Alamos National Laboratory, Los Alamos, NM (1992)

12. Phipps, C.R., "Laser Deflection of NEO's," *Report of the NASA Near-Earth-Object Interception Workshop*, John D. G. Rather, Chair, Report LA-12476-C, Los Alamos National Laboratory, Los Alamos, NM (1992)

13. Phipps, C.R., "A laser concept for clearing space junk," in *AIP Conference Proceedings* **318**, Laser Interaction and Related Plasma Phenomena, 11th International Workshop, Monterey, CA October, 1993, George Miley, ed. American Institute of Physics, New York (1994) pp. 466-8

14. Phipps, C.R., "Lasers can play a rôle in planetary defense" in *Proc. Planetary Defense Workshop*, Report CONF-9505266, Lawrence Livermore National Laboratory, Livermore CA (1995)

15. Phipps, C.R., and Michaelis, M.M, "NEO-LISP: deflecting near-earth objects using high average power, repetitively pulsed lasers", *Inst. Phys. Conf. Ser. 140 section 9*, pp. 383-7, ICP Publishing, Bristol (1995)

16. Phipps, C.R., Friedman, H., Gavel, D., Murray, J., Albrecht, G., George, E.V., Ho, C., Priedhorsky, W., Michaelis M.M., and Reilly, J.P., "ORION: Clearing near-Earth space debris using a 20-kW, 530-nm, Earth-based, repetitively pulsed laser", *Laser and Particle Beams*, **14** (1996) pp. 1-44

17. Phipps, C.R., "Laser Deflection of Near-Earth Asteroids and Comet Nuclei", *Proc. International Conference on Lasers 96*, STS Press, McLean, VA (1997) pp. 580-7

18. Phipps, C.R., "Requirements for Laser Acquisition of NEO's", *Proc. International Conference on Lasers 97*, STS Press, McLean, VA (1998) pp. 928-34

19. Phipps, C.R., "Review of Direct-Drive Laser Space Propulsion Concepts", *AIP Conference Proceedings* **420**, Space Technology and Applications International Forum 1998, M. El-Genk, ed., American Institute of Physics, Woodbury, NY (1998) pp. 1073-80

20. Phipps, C.R., Reilly, J.P., and Campbell, J.W., "Optimum Parameters for Laser-launching Objects into Low Earth Orbit", J. Laser and Particle Beams, **18** no. 4 pp. 661-695 (2000)

21. Phipps, C.R. and Luke, J.R., "Diode Laser-driven Microthrusters: A New Departure for Micropropulsion", *AIAA Journal*, 40, no. 1, pp. 1-9 (2002)

22. Reilly, J.P., Phipps, C.R., and Campbell, J.W., "Comparison of Repetitive-pulse laser Approaches for Boosting Small Payloads into LEO," Proc. Santa Fe High Power Laser Ablation Conference III, SPIE **4065** (2000) pp. 946

23. Riker, J., "Space Based Laser Overview and Target Interactions," *Proc. SPIE High Power Laser Ablation IV* (2002) to appear

24. Sharpton, V.L., Nature, October 29, 1993; *Sky and Telescope*, July 1991, page 38; *Sky and Telescope*, January 1993, page 12.

25. Shoemaker, E. M. "The NEO Flux, Present and Past", *Proc. Lawrence Livermore National Laboratory Planetary Defense Workshop*, Report CONF-9505266, Lawrence Livermore National Laboratory, Livermore CA (1995)

26. Solem, J. C. "Nuclear Explosive Propelled Interceptor for Deflecting Comets and Asteroids on a Potentially Catastrophic Collision Course with Earth", *Report of the NASA Near-Earth-Object Interception Workshop, January 14-16, 1992, Los Alamos, NM, John D. G. Rather, Chair*, Los Alamos National Laboratory Report LA-12476-C (1993).

BEAM-PROPELLED MICROAIRCRAFT

Beam Driven Stratospheric Airship

Masahiko Onda

*Intelligent Systems Institute(ISI),
National Institute of Advanced Industrial Science and Technologies(AIST),
Namiki 1-2-1, Tsukuba Science City, Ibaraki Pref., 305-8564 JAPAN*

Abstract. Even though satellite, balloons and aircraft have served admirably as aerospace platforms for remote sensing and telecommunication, requirements for a new kind of platforms – an easily modifiable, sub-orbital platform – have been widely identified. The High-Altitude Long-Range Observational Platform(HALROP) was at first conceptualized as a solar power driven unmanned LTA (Lighter-Than-Air) vehicle or an airship to maintain a station-keeping position in the lower stratosphere for long-durations and to carry out missions such as high-resolution monitoring and high-speed informational relays. Nevertheless solar power is not available in winter seasons in the high-latitudinal regions. Therefore, alternative power sources are necessary and the candidates are surface-to-air transmission of microwave energy and high-power laser beams. The author introduces a wireless power transmission test by microwave carried in 1995 in Kobe, Japan, and then, discusses possibilities of using laser beam for powering such LTA platforms.

BACKGROUND

The development of stratospheric LTA platforms has begun to attract great attention of late in countries such as Japan, the USA, Europe and Korea[1]. This is because the stratosphere is now being considered as the new dimension for novel informational systems. It is expected that the development of such systems would contribute to the vitalization of the economies of the countries concerned. Stratospheric LTA platforms have to differ from conventional airships in four major ways; propulsive efficiency, power source, endurance, and operational altitude. LTA type stratospheric platforms are helium-filled re-deployable unmanned vehicles that can play identical roles to artificial satellites in such applications as remote sensing and telecommunications. Satellites, however, after their lifetime service, become, without exception, space debris, eventually falling down to the lower atmosphere and burning up, resulting in atmospheric contamination. Advantageous application areas have been identified for LTA type stratospheric platforms in contrast to fixed-wing aircraft as well as satellites.

The views outlined herein are solely those of the author and do not reflect any opinions or formal decisions taken by any governmental agencies or ministries.

CP664, Beamed Energy Propulsion: First International Symposium on Beamed Energy Propulsion,
edited by A. V. Pakhomov

HIGH ALTITUDE PLATFORM - HALROP

A feasibility study and scale model developments have been carried out on a high altitude (15-22 km) super-pressured helium-filled powered LTA platform as an ideal platform for environmental observation and telecommunication relay (FIGURE 1). It has a long service life and larger payload than that of a large artificial satellite. This LTA platform named HALROP (High Altitude Long Range Observational Platform) has a solar-powered electric propulsion system to maintain its position in the stratosphere against wind currents. The altitude where the platform resides is the least windy area in the lower stratosphere, at a height from which one can have a direct line of sight on the ground within and over a 1,000 km diameter range.

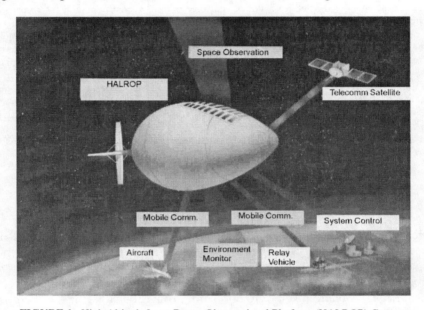

FIGURE 1. High Altitude Long Range Observational Platform (HALROP) Concept.

High altitude platforms already in use are listed together with the proposed stratospheric LTA platform in TABLE 1. The table shows comparisons among the various high-altitude platforms being used and the stratospheric LTA platforms. LTA stratospheric platforms appear outstandingly sustainable and environmentally friendly with wide variety and flexibility of mission capability.

TABLE 1. High Altitude Platforms and their Characteristics

	Tethered Balloon	Free Balloon	Fixed Wing Aircraft	Rotor Wing Aircraft	Artificial Satellite	Stratospheric LTA
Mission altitude (km)	~5	20~30	~15	~5	500~36,000	20
Mission durability	Long (Yr.)	Long (Yr.)	Short (hr.)	Short (hr.)	Long (Yr.)	Long (Yr.)
Weather dependency	Limited	Free	Limited	Limited	Free	Free
Payload (ton)	~3	~2	~50	~10	~5	~3
Mission costs	Low	Low	Medium	Medium	High	High
System upgrade	Easy	Easy	Easy	Easy	Difficult	Easy
System growth	Difficult	Easy	Easy	Easy	Difficult	Easy
Weather control	Impossible	Impossible	Impossible	Impossible	Impossible	Possible
Sensing resolution	Fine	Fine	Fine	Fine	Coarse	Fine
Deployment timing	Easy	Difficult	Easy	Easy	Difficult	Easy
Land coverage (km^2)	~100	~1	~200	~100	~100	~1200
Location intended	Fixed	Unable	Arbitrary	Arbitrary	Fixed	Arbitrary
Sustainability	Fair	Poor	Fair	Fair	Poor	Good

In this section, discussion is focused onto the major differences between artificial satellites and stratospheric LTA platforms. Large-scale artificial satellites need monstrous rockets to get them into orbit, but stratospheric LTA platforms have no such need. Obviously, rocket launches create a lot of problems; gas emission, space debris of used rocket body fragments, large off-limit areas designated around and far away from the launching site, etc. Even low-earth orbit artificial satellites cannot be repaired during their missions except by hiring the highly expensive Space Shuttles for repair missions. Satellites therefore need redundant systems to cover partial failures, and these boost up costs. Alternatively, if malfunctions or failures take place in satellites right after launch, they are left as they are.

On the contrary, stratospheric LTA platforms can be pulled down back to the earth, repaired and launched again. This features not only ensure less cost compared to satellites but also make the total system environmentally friendly. Stratospheric LTA platforms emit no pollution at all in their deployment stage. Stratospheric LTA platform mission altitudes are much lower than those of satellites and with their position keeping capability, their sensing resolutions are much higher, as stated in TABLE 1. LTA platforms can also move from place to place and can carry out any necessary observational missions.

LTA type stratospheric platforms allow for special missions owing to their physical characteristics which are different to satellites and high-altitude airplanes. An LTA vehicle can keep its position at mission altitude, while airplanes need a considerable broad air space to maintain air speed and thereby keep flying by making circles at mission altitude. For instance, most modern stratospheric-flying fixed-wing aircraft being developed need a 20-km diameter circle to maintain flying speed. These characteristics cause totally different effects both in the remote sensing area and the telecommunications area. LEO (Low- Earth Orbit) satellites are now being used for telecommunication relays, but due to the wireless propagation distant path and the nature of the high-speed moving platform, the relay system is made much more complex. In the field of telecommunications relays, an LTA will always keep a good wireless footprint on the ground. They are suitable particularly for high frequency telecommunications relays such as millimeter wave carriers. Airplanes on the other hand are unable to keep the same wireless footprint all the time, since they have to cover considerable airspace. Consequently, telecommunication relays by airplanes need high-gain automatic tracking antenna for all their customers' terrestrial equipment. This makes the total system cost much higher than telecommunications relays by unmanned position-keeping stratospheric airships with terrestrial antennae without tracking mechanisms.

Unlike satellites, the forte of a LTA platform is that it is situated in the atmosphere at high-altitude with long duration characteristics and mobility. This enables it to collect in situ data from the atmosphere. Among the most interesting scientific applications are polar observations, such as monitoring of the ozone level or ice-sheet movements and thickness. Apart from assessing the condition of the ozone layer, there have been other proposals e.g. an LTA for in situ repair of the ozone hole.

As for weather control, an LTA platform could be used in its position; for example, over specific clouds that yield large hail stones, to spray volcanic ash to render the hail grains smaller. In this way, the LTA platform can relieve the damage caused by hail. Similar methods could be applied to alleviate heavy rainfalls or heavy snow falls, or, in the future, to mitigate typhoon, hurricane, or tornado power to decrease damage.

FIGURE 2. Stratospheric Climbing Test (1999).

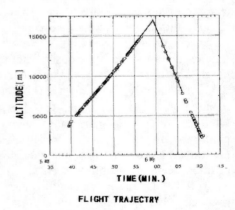

FLIGHT TRAJECTRY

FIGURE 3. Ascent and Descent of Launched Balloon

On the basis of the HALROP concept, flight tests to reach the stratosphere began in

1998. FIGURE 2 and 3 show a high-speed stratospheric climbing test carried in 1999. Possible regional areas should be identified where solar-powered LTA platforms can be used as an informational system since the daily harvestable solar energy depend on latitudinal location and the day of the year (FIGURE 4) [2].

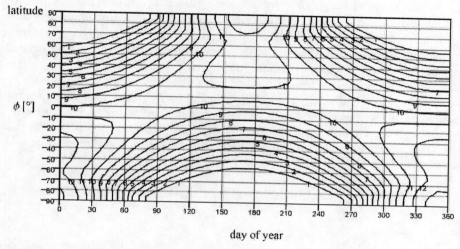

FIGURE 4. Yearly Global Variation of Daily Extra-terrestrial Irradiance [kWh/m²/day] by B.H.Kroeplin and M.A. Rehmet.[2]

If nighttime becomes longer, as in the polar regions in winter, or to cope with windy climates, the platform requires ground-based wireless power transmission systems for continuous thrust powering. LTA stratospheric platforms can be also driven by ground-to-air transmitted microwave power. This allows platforms to be designed into much smaller sizes than solar-powered platforms.

GENERAL DESIGN

In this section, generic matters are discussed regarding microwave transmission technology for energy beaming. In this method, a stratospheric LTA can be powered solely by ground-to-air transmission of microwave energy.

Design Considerations for Generic Stratospheric LTA Platforms

All power for propulsion and onboard systems for the stratospheric LTA must be supplied by microwave beaming. To accomplish this, two major components are required: a microwave rectifying antenna or rectenna and a ground-based microwave transmitter. For a generic stratospheric LTA aircraft, which maintains a station-keeping flight (within 2km diameter) at roughly 20 km altitude for indefinitely long periods of time, the ground based emitter (2.45GHz) must have a diameter between 100-170 meters to transmit power over a distance from 20 to 30km. Major design parameters are summarized in TABLE 2. In this case tracking accuracy, aircraft flight path, emitter diameter, and various other factors should be considered. All of the parameters should be kept to make infinite endurance, thereby, the aircraft can stay aloft indefinitely.

It is important to note that the microwave frequency chosen for this design is 2.45GHz. There are several issues associated with the selection of frequency. Signal attenuation between a transmitter and a receiver depends not only on microwave incident angle, but also on the atmospheric conditions (FIGURE 5). Consequently, such factors must always be taken into considerations to select the optimum frequency for surface-to-air transmission of microwave energy. In any case, a microwave emitter and receiver should be aligned on the same axis, which means if no matter how far the aircraft flies away from the zenith of the emitter, it ought to provide a 90 degree angle of incidence for the microwave beam. This implies that a good tracking control mechanisms are necessary for both of an emitter side and a receiver side.

TABLE 2. Basic Design Considerations of S-Band (2.45GHz) Microwave Transmission

Power Density	>500 W/ m^2
Vehicle Attitude	+/-3 degrees
Antennae Separation	20km +/-1km
Circling Radius of Aircraft	1km
Tracking Accuracy of Emitter	0.01 degrees
Diameter of Receiver	15m
Radius of Emitter	100m
Emitter Power	1MW

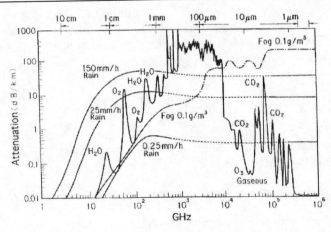

FIGURE 5. Microwave Power Attenuation vs. Frequency by M. Shiho.

ETHER Project

During the International Wireless Power Transmission (WPT '95) Conference held on October 16~19, 1995 in Kobe, Japan, a hovering test of a microwave-powered airship was demonstrated to the conference participants by the author's team at the Western Branch of Communication Research Laboratory, the ex-Japanese Ministry of Posts and Telecommunications. The test was named as ETHER(Energy Transmission to a High altitude long Endurance aiRship experiment) and was aimed at establishing two basic technologies – wireless power transmission and an environment-friendly aerial platform[3,4,5].

FIGURE 6. Completed ETHER-HALROP prior to Test

FIGURE 7. Hovering at 50m Altitude by Microwave Power

In the test, 10kW power was transmitted via microwave to a rectenna equipped to an airship hovering at 50 meters above the ground. The airship tried to keep hovering in the air solely by a power supplied via microwave.

High efficiency propulsion and high controllability are in general indispensable for an unmanned airship. For this low altitude flying test, however, only high controllability was emphasized in designing the configuration of the vehicle to cope with turbulent surface air conditions. In order to give the airship a high controllability and keep it hovering at a required position in the air, a pair of vectored thrusters were placed in the fore part of the hull. Beneath the airship hull, 60 panels with 20 rectenna elements on each panel (1200 elements in total) were placed to receive 10kW ground-emitted power.

The airship hull was made as a non-rigid fabric and double layered skin structure in order to minimize helium leakage, as well as to reduce the construction costs and to improve durability.

A pair of thruster motors and servomotors was driven only by microwave power supplied through aerial transmission from the ground and kept the airship in an aerial position.

The antenna for the power emission was a parabolic type which was three meter in diameter. The antenna was designed to generate a high power density at an altitude of 50~100 meters above the ground where the airship tried to keep its position.

Power level fluctuation was expected due to relative rotational deviation between transmission and receiving systems. To overcome this problem, partition of transmission power was planned by applying the dual polarization method to both of the transmitting and receiving systems. This dual polarization method also enabled to double the output power. Each of 5kW Magnetron was connected to two polarized oscillators respectively and the total output power amounted to 10kW.

FIGURE 8. Parabolic Antenna for Microwave Transmission with Hydro-pneumatic Tracking System

FIGURE 9. Power Sources (Dual units) and CRT Monitors for Tracking Telescopes

The rectenna to absorb the ground-emitted power was composed of a two polarized micro-strip antenna with a round shape printed on an aramid-honeycomb board and had two independent rectifying circuits to realize a lightweight structure as well as high output power. As the result, a very high direct current output was attained, namely 2.5W per one rectifying circuit or 5W per one rectenna element[6,7].

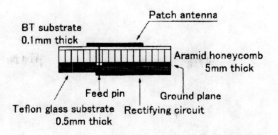

FIGURE 10. Sectional View of Rectenna Element

TABLE 3. Specifications of ETHER - HALROP

Hull Length	16.5m
Envelop Volume	340m³
Envelop Length	15.8m
Envelop Maximum Diameter	6.6m
Ballonet Volume	85m³
Total Buoyancy	approx. 300kg
Empennage	3 tail fins
Empennage Total Surface Area	13.5m²(4.5m²x3 units)
Propulsion Motor (Direct Drive)	2 units
Propulsion Motor Rated Power	2.5kW each
Propulsion Motor Rated Rotational Speed	3,000rpm
Microwave Frequency	2.45GHz
Rectenna Dimension	approx. 3mx3m
Rectenna Weight (Including Peripheral Systems)	approx. 30kg
Rectenna Expected Output Power	5kW

TABLE 3 shows overall specifications of the ETHER-HALROP test. FIGURE 11 shows on-board powering circuit, where excessive power absorbed by the rectenna should be dissipated as Joule heat. FIGURE 12 shows a power dissipater mounted beneath the aft-hull.

FIGURE 13 presents a hovering ETHER-HALROP, which shows its rectenna mounted at the bottom center of the hull.

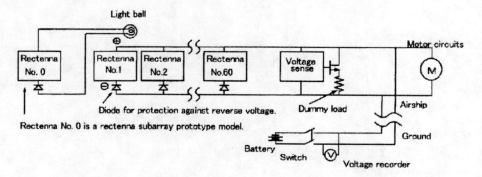

FIGURE 11. Block Diagram of Power Reception System and Thruster Driving Circuits

FIGURE 12. Excessive Power Dissipator

FIGURE 13. Bottom View of HALROP Airship with Rectenna

LASER POWERED LTA

Laser propulsion can be considered as a good candidate for the microwave beaming method. In this section, laser beaming to propel a stratospheric vehicle and associated technical issues are discussed.

Application to Stratospheric LTA

For an energy beaming to the stratosphere from the ground, laser beam will meet difficulties in penetrating the thick atmosphere, particularly in unfavorable weather conditions. In this case, in order to secure a laser propagation path through the atmosphere, optical fibers will be useful. FIGURE 14 shows a concept of using optical fibers, and then the laser is emit from a ground tethered LTA to a remotely located LTA in the stratosphere over a mission area to supply propulsive and mission executing power.

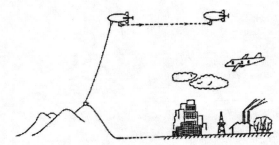

FIGURE 14. Laser Power Supply to Stratospheric LTA Platform

Laser Propelled Low Altitude Flying LTA

Basic experiments were conducted in aiming at realization of energy beaming for propulsion of a small balloon. Basic configuration of this tests are shown in FIGURE 15. A set-up of the test is shown in FIGURE 16. In this test, a small helium filled balloon in half a meter length was mounted on an air slider to make supporting friction minimal and its aft located aluminum target was hit by a series of successive laser pulses. A water tank kept continuous water supply to the target so that the balloon could slowly move forward at about a half of human walking pace. The mass of the balloon with its support is approximately 310g. Pulse energy was 0.6 Joule.

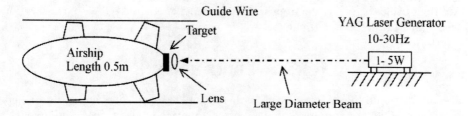

FIGURE 15. General Configuration of Demo Test

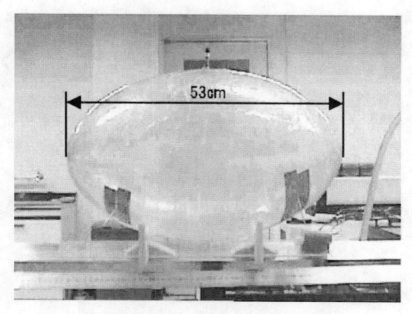

FIGURE 16. Balloon Propelled by Laser on Air Slider

CONCLUSION

Long-Endurance Station-keeping LTA Stratospheric Platforms have great potential for environmental protection, surveillance and telecommunication relay. Wireless power transmission technologies are inevitable to high-altitude long-endurance flying vehicles, especially in high latitudinal regions or windy flying conditions, where solar power utilization is disadvantageous. Microwave power transmission was demonstrated in 1995 in Kobe, Japan, to power an unmanned LTA vehicle.

Microwave power transmissions are useful to drive high-altitude long-endurance LTAs, but depending on frequencies due to moisture attenuation.

Laser power transmissions to high-altitude long-endurance LTAs will need an optical fiber cable traversing the troposphere. Namely, a tethered LTA is necessary for powering an LTA in mission flight.

ACKNOWLEDGMENTS

The author would like to express gratitude to Prof. Takashi Yabe who has provided to the author with an opportunity to give a presentation to the first ISBEP conference. The author also would like to thank Dr.Yoichi Ogata who has been helping in laser propulsion tests of miniature airship models.

REFERENCES

1. M.Onda, Stratospheric LTA Stationary Platform Development Activities in Japan, 3rd International Convention and Exhibition, 2000, Paper B-6, July 1-5, Friedlichshafen, Germany
2. M.A.Rehmet and B.H.Kroeplin, *Comparison between Airship and Aircraft*, Proc. of 3rd International Airship Convention and Exhibition, 2000, Paper A-23, CD-ROM
3. M.Onda, et al, *A Ground-to-Airship Microwave Transmission Test Plan for Stationary Aerial Platform*, AIAA-95-1603-CP, 1995
4. M.L.Ford & M.Onda, *Powering of High-Altitude LTA's by Surface-to-Air Microwave B Transmissions*, Proc. of Association for Unmanned Vehicle Systems International (AUVSI), Orland, FL, July 15-19, 1996
5. Y.Fujio, M.Onda, et al, *Wireless Power Receiving System for Microwave Propelled Airship Experiment*, Space Technol., Vol.17 No. 2, pp. 89-93, 1997 T
6. M.Onda, et al., A Stratospheric Stationary LTA Platform Concept and Ground-Vehicle Microwave Power Transmission, Proc. of 37th AIAA Aerospace Sciences Meeting and Exhibit, January 11-14, 1999, Reno, NV, USA M
7. M.Onda, *Electric Propulsion by Wireless Power Transmission*, Proc. of FOGL (Research Group for Airship Technology) Workshop, University of Stuttgart, Sept. 22-23.

Laser-Driven Micro-Ship and Micro-Turbine by Water-Powered Propulsion

Tomomasa Ohkubo*, Masashi Yamaguchi, Takashi Yabe, Keiichi Aoki, Hirokazu Oozono, Takehiro Oku, Kazumoto Taniguchi, and Masamichi Nakagawa

*Tokyo Institute of Technology, Dept. of Energy Science, 2-12-1 O-okayama, Meguro-ku, Tokyo 152-8552, Japan, Tokyo Institute of Technology,

Dept. of Mechanical Engineering and Science, 2-12-1 O-okayama, Meguro-ku, Tokyo 152-8552, Japan,

Abstract. In this paper, we report experimental demonstration of propulsion of 100-weight object with only 668mJ/5ns YAG laser. This is made possible by water overlay structure and the effort to reduce the friction by putting the object on water surface or using a levitating system which we call on "air-slider". Furthermore, several water supply systems provided the repetitive propulsion. In addition, we found the laser-driven micro-turbine would provide an interesting application area in driving micro-obstacle.

1. INTRODUCTION

We successfully demonstrated the flight of a paper airplane by experiments in a previous paper [1], in which we clarified the effectiveness of double-layered structures that consist of metal layer covered with transparent overlay. In particular, it was shown that the water overlay is more effective than solid overlay by both experiments and simulations [2]. This result has significant means in the sense that water can be repetitively used and is free from environmental problem, can be collected from the atmosphere during the flight and so on.

The major purpose of this paper is to find a way to drive an object as heavy as possible. For this purpose, we propose several methods of water supply for repetitive propulsion. In addition, we investigated the propulsion on the water surface and the levitation of airplane in order to avoid the friction that was dominated in the previous experiments.

2. LASER DRIVEN BUS

In the previous paper, we have succeeded to fly a micro airplane of 3.9 cm by water overlay on the aluminum foil and achieved 587[N.s/MJ], while the airplane did not move without water overlay. High-speed video camera shows that the unusual movement of the airplane during the acceleration on the launching rail. We believe that this is due to large friction in acceleration phase.

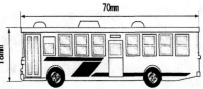

In addition, in our CIP-CUP method [3-7] based numerical simulation and experiment using pendulum and load cell demonstrated

FIGURE1. Size and shape of miniature bus.

much larger Cm if the laser spot size and energy are properly chosen [8]. Therefore, we hope that water overlay structure must have ability to drive much heavier object by choosing laser condition and reducing the friction.

As a first step to reduce the friction, we used 9.4g - weight miniature bus shown in Fig.1 expecting that the wheel might reduce the friction because of small contact area.

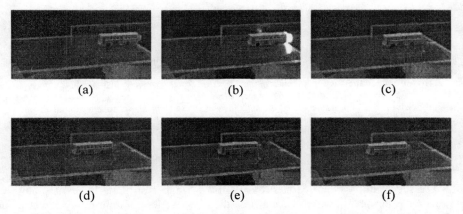

FIGURE 2 The miniature bus of 9.4g –weight with water overlay structure is driven by only 668mJ/5ns YAG laser . (a): initial position, (b): laser irradiation, (c)-(f): the bus moves.

We paste an Al foil and acrylic pipe filled with water at rear of the bus as a target irradiated by a 668mJ/5ns YAG laser. As shown in Fig. 2, the bus moved by 7.1cm after one pulse and the initial speed was 0.19 m/s and therefore Cm is 2673(Ns/MJ).

3. AIR SLIDER

Although we achieve larger Cm with miniature bus having wheels than that of airplane, Cm is still low. Since the friction is still large, we must find alternative way to reduce the friction.

For this purpose, we built a new system called "air-slider" which suspends objects using air-flow from air-compressor (1.57×10^{-3} m^3/s). Figure 3 shows the whole system. This system reduces the effect of friction by air-flow from the holes made in intervals of 1cm, and lifts up the object so that the object can move very smoothly.

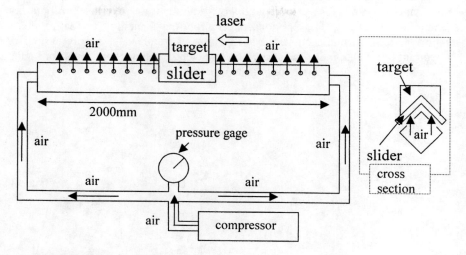

FIGURE 3 whole system of "air-slider".

Figure 4 shows the measured momentum of 10g-weight object irradiated by 640mJ/5ns YAG laser with this system and the data given in previous section. As shown here, we achieved Cm of over 4000.   Speed was 0.181 m/s.

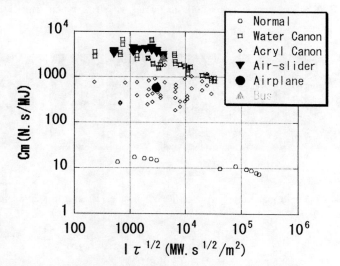

FIGURE 4 Experimental results with "air-slider" and previous experiments.

4. WATER SUPPLY SYSTEM

Up to this point, we have used only single pulse laser in experiments. It is of great importance to find a way to repetitively generate water-overlay configuration for repetitive propulsion. The previous paper proposed a method to realize this water supply system. In this section, we shall discuss several water supply systems. Most important point is to make the system be sufficiently compact and light for being used in weak laser energy, and not complicated structure.

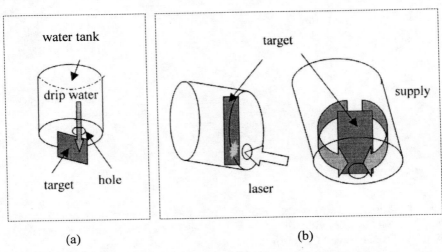

(a)   (b)

FIGURE 5 Water supply system. (a): simple supply system, (b): on-demand supply system

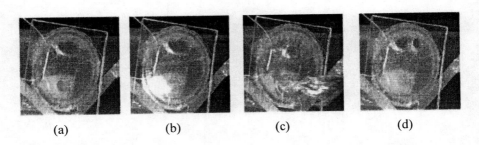

(a)   (b)   (c)   (d)

FIGURE 6 on-demand supply system. (a): initial state, (b): laser irradiation, (c): after irradiation, (d): return to the origin

4.1 Simple Supply System

At first, we built very simple water supply system shown in Fig. 5(a). This system only drips water to the target from a 0.5mm radius hole bottom of the water tank. This system can supply water at 0.42 cm^3/s.

FIGURE 7 Picture of the micro turbine

(a)   (b)

(c)   (d)

FIGURE 8 Micro turbine can supply water automatically and get power of rotation. (a): initial state, (b): laser irradiation, (c)-(d): rotating and supplying water for itself.

4.2 On-demand Supply System

Secondly, we built on-demand water supply system. The above-mentioned simple supply system is nothing more than pouring water, so we must build supply system which automatically replenishes water with target in accordance with the timing of laser irradiation (Fig. 5(b)). In this system, water is contained in a narrow space between two plates with a small hole in one side and water tank is on the back of the target. Figure 6 shows the actual movement of water supply in this system. The water is supplied from the back reservoir only when water around the target is ejected and removed by laser irradiation, thus the loss of water is small.

4.3 Micro-Turbine

Thirdly, we built another automatic water supply system. A basic idea of this system is to rotate the turbine dipped in water by the laser ablation. The shape and size of this turbine is shown in Fig. 7 and momentum of inertia is $9.9 \times 10\text{-}9\text{kg} \cdot \text{m2}$. This turbine is made of titanium because rigidity of aluminum is small so that aluminum turbine bends without rotating when the laser is irradiated.

If we dip about one thirds of this turbine into the water, this system can automatically replenishing water with the target by its rotation, and gains the powers of the rotation at the same time (Fig 8). Though this system can not directly drive objects , rotation power is usable for some other purpose. For example, rotate shaft with gear, paddling water directly, and so on.

5. CONTINUOUS DRIVE OF MICRO- SHIP

Although initial speed is not large enough, heavier object can be easily driven on water surface because of low friction and hence it is better to examines the water supply system with ship floating on wave surface.

5.1. Micro-ship using simple supply system

At first, we tried to move 131.2g-weight miniature ship (Fig. 9) including additional 100g weight and water tank of simple water supply system by 10Hz 640mJ/5ns YAG laser being unfocused whose spot radius is 31 mm.

Figure 10 shows the result of this experiment. We succeeded to move over 100g – weight object on water and average speed was 0.044m/s. In addition, we succeeded to move the same ship at 0.035m/s using only 275mJ laser. As shown here, the friction is very low and makes it possible to drive a heavy object even with small Cm or velocity on the water surface. Furthermore, the movement is very stable. Therefore the situation is suitable to test various water supply systems.

5.2. Micro-ship using on-demand supply system

Next, we tried to move the same ship (131.2g - weight) using on-demand supply system by 640mJ/5ns 10Hz YAG laser. The result is shown in Fig. 11. In these experiments, momentum does not increase in proportion to frequency or total energy in other word. This means Cm decreases with frequency. This is due to the time delay of water supply compared with laser repetition. High speed camera actually shows such delay.

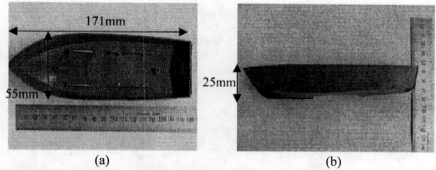

(a)   (b)

FIGURE 9 Size and shape of miniature ship. (a): top view, (b): side view

FIGURE 10 Propulsion of 131.2g – weight ship by 640mJ/5ns 10Hz YAG Laser.

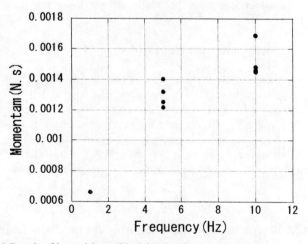

FIGURE 11 Result of laser driven ship with simple supply system

5.2. Micro-ship using micro-turbine

We tried to move the same ship using micro-turbine which was attached at the rear of ship , and the laser is 640mJ/5ns 10Hz YAG Laser. On the water surface, we do not need to care how to supply water because there is already water around. Figure 12 is a result. Average speed of this ship was 0.025 m/s. Thus, micro-turbine can supply water automatically but also can get rotation power to drive objects.

FIGURE 12 Propulsion of 131.2g – weight miniature ship using micro-turbine

6. CONTINUOUS PROPULSION ON "AIR-SLIDER"

As mentioned above, because of large virtual mass (water) attached to the ship on the water, it is too difficult to accurately measure the speed or Cm. However, there is a merit that we can sustain the heavy objects stably with low friction. In this section, we shall try to use "air-slider" system although the suspension is less stable compared with water surface.

6.1. Experiments using on-demand supply system on "air-slider"

We used on-demand supply system and drive about 50g - weight object including water tank and acrylic angle that fits to the slider. The laser is 10Hz 640mJ/5ns YAG laser and is unfocused with spot radius of 31 mm.

The result is shown in Fig.13. As shown here, we succeeded to accelerate object using air-slider and on-demand water supply system. As shown in the previous section, on-demand supply system has time delay for supplying water. Therefore water supply is not sufficiently fast to be operated at higher frequency, but 2Hz operation is the limit.

6.2 Propulsion of heavier objects on "air-slider"

In addition, we tried to move much heavier objects with this system. At this trial, finally we succeeded to move 298.7g (+ about 5g water) - weight object with these systems. The pictures are shown in Fig. 14. Acceleration is 0.002 m/s^2 and Cm is 93.3 N.s/MJ.

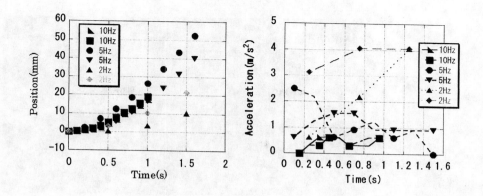

FIGURE 13 Multi pulse laser accelerates objects on "air-slider"

FIGURE 14 Even more than 300g – weight object can be driven using water-cannon target and air-slider by only 10Hz 640mJ YAG Laser.

7. APPLICATION OF MICRO-TURBINE

The above-mentioned micro-turbine can be used not only for water supply but getting rotation power. Because the structure is able to gain energy outside by laser, many applications are expected. This structure uses only metal and water as target, and we can deliver energy from a distant place far away.

As for one of the applications, we here propose NEBOT (Non Electric roBOT). This is the idea that drives the robot using only this turbine as a driving source. Such a micro-robot can be used at the accident of nuclear power reactor where a lot of neutron and gamma ray will be emitted and make electronic devices useless and normal robot can not reach such place.

In order to demonstrate the possibility of this turbine, we tried to lift up 1 yen coin (1g-weight) using this turbine with water by 10 Hz 640mJ/5ns YAG laser (Fig. 15).

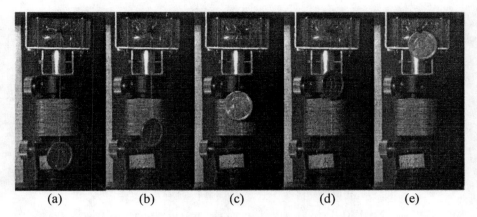

| (a) | (b) | (c) | (d) | (e) |

FIGURE 15 Lifting up 1-yen coin using micro turbine. (a): initial position, (b)-(e): laser irradiation and lifting up 1-yen coin.

8. CONCLUSION

We confirmed that objects much heavier than paper airplane can be propelled by only single pulse of 668mJ/5ns YAG laser using water overlay structure if we can reduce the friction. We also succeeded to make continuous propulsion and acceleration using water supply system. Especially we succeeded to drive 132g-weight object on water and about 300g-weight object on "air-slider" system by only 10Hz 668mJ/5ns YAG laser.

REFERENCES

1. T.Yabe et.al. Appl.Phys.Lett. 80, 4318(2002).
2. T.Yabe et.al. "Laser-Driven Vehicles – from Inner-Space to Outer-Space –" Proc.SPIE 4760 High-Power Laser Ablation IV, 867-878 April 2002,Taos, NM, USA
3. T.Yabe and T.Aoki, A Universal Solver for Hyperbolic Equations by Cubic-Polynomial Interpolation I. One-Dimensional Solver. *Comput.Phys.Commun.* **66**, 219 (1991)
4. T.Yabe, F.Xiao and T.Utsumi : Constrained Interpolation Profile Method for Multiphase Analysis. *J. Comput. Phys.* 169, 556 (2001)
5. T.Yabe and P.Y.Wang : Unified Numerical Procedure for Compressible and Incompressible Fluid. *J.Phys. Soc. Japan*, **60**, 2105 (1991).
6. F.Xiao, T.Yabe, N.Konma, A.Uchiyama, K.Akutsu and T.Ito : An Efficient Model for Driven Flow and Application to a Gas Circuit Breaker *Comput. Model. Simul. Eng.* **1**, 235 (1996).
7. H.Takewaki, A.Nishiguchi and T.Yabe, Cubic Interpolated Pseudoparticle Method (CIP) for Solving Hyperbolic Type Equations. *J.Comput.Phys.*, **61** , 261 (1985) .
8. K.Aoki et.al. "Numerical and experimental studies of laser propulsion toward micro-airplane" Proc.SPIE 4760 High-Power Laser Ablation IV, 918-928 April 2002,Taos, NM, USA
9. C.R.Phipps, D.B.Seibert II, R.Royse, G.King and J.W.Campbell, Very High Coupling Coefficients at Low Laser Fluence with a Structured Traget. *III International Symbosium on High Power Laser Ablation*, Santa Fe, 200 0, SPIE Vol.**4065**.

Control of Wing for Micro-Airplane with Smart Material and Laser

Itsuro Kajiwara[*], Hiroyasu Ishikawa[*], Shunsuke Furuya[*], Takashi Yabe[†] and Chiaki Nishidome[¶]

[*]Department of Mechanical and Aerospace Engineering, Tokyo Institute of Technology,
2-12-1 O-okayama, Meguro-ku, Tokyo 152-8552, JAPAN
[†]Department of Mechanical Science and Engineering, Tokyo Institute of Technology,
2-12-1 O-okayama, Meguro-ku, Tokyo 152-8552, JAPAN
[¶]Engineering Department, CATEC Inc., 5-16-3 Asakusabashi, Taito-ku, Tokyo 111-0053, JAPAN

Abstract. A laser tracking system and a smart wing are proposed and developed to control the flight of a laser-driven micro-airplane. The laser tracking system is composed of a light source, detector and optical system with a galvano-mirror control mechanism, and used for controlling the smart wing constituted by a wing, phototransistor, battery and smart material. The light source is installed on the ground and so this system is effective for miniaturization and lightening of the micro-airplane. Experiments are carried out to evaluate the performance of the laser tracking system and the smart wing.

INTRODUCTION

In the field of airplane technology, research for higher performance, miniaturization and lightening has been attracting the attention of the aircraft engineers. But the characteristic interference between the aerodynamics based on the aircraft shape and the control system can't be ignored when the miniaturization is going to be promoted. Micro-airplanes are potentially useful for surveillance and monitoring applications [1]-[2] and have drawn much recent interest. As the size of the micro-airplane decreases, stability suffers because angular accelerations tend to increase with inverse of length dimension. The stability is improved by controlling the wing of the micro-airplane. The control performance is highly influenced by the aerodynamics determined by the aircraft shape. Multidisciplinary optimization (MDO) of the micro-airplane for aerodynamic shape and control system design has been studied in order to achieve high performance and stability [3]-[4]. However, a new technology on the propulsion and the control of the micro-airplane has to be developed for realizing both miniaturization and lightening with the required stability.

The purpose of this paper is to develop a laser tracking system and a smart wing for realizing a laser-driven micro-airplane [5]. The laser propulsion [6]-[9] is significantly effective to achieve the miniaturization and lightening of the

CP664, Beamed Energy Propulsion: First International Symposium on Beamed Energy Propulsion,
edited by A. V. Pakhomov

micro-airplane. The laser-driven micro-airplane has been studied by Yabe et al. with a paper-craft airplane and a Yag laser, resulting in a successful glide of the airplane [10]-[12]. In the next stage of the laser-driven micro-airplane development, the laser tracking system and the smart wing are expected as the key technologies to achieve a continuous propulsion and flight control. The laser tracking system and the smart wing are developed with their control strategies in this study. The laser tracking system is composed of a He-Ne laser, detector and optical system with a galvano-mirror control mechanism, and used for controlling the smart wing constituted by a wing, phototransistor, battery and shape memory alloy (SMA). Experiments are carried out to evaluate the performance of the laser tracking system and the smart wing.

LASER-DRIVEN MICRO-AIRPLANE

The use of ablation induced by laser irradiation is significantly effective to obtain the propulsion for the micro-plane [10]-[12]. At the rear edge of the micro-airplane, thin aluminum film is pasted. And it is over-coated with acryl or water. When laser irradiates aluminum, it penetrates water or acryl and reaches aluminum film, then aluminum evaporates. The micro-airplane gains momentum as a reaction to this ablation. So no extra devices except laser are required for flying micro-airplane.

A paper-craft airplane of 39mm × 56mm × 15mm size and 0.2 g-weight is placed on the platform with a guiding groove and is irradiated by one pulse of a YAG laser. A 3.5mm × 3.5mm × 0.1 mm-thick thin aluminum film is pasted onto the rear edge of the airplane. Figure 1 shows the flight trajectories of the micro-airplane with water droplet overlay. In this case, the laser target is 3mm-diameter, 0.014g water droplet attached to the aluminum foil. The measured velocity of the airplane in the case where the flight path is normal to the camera is initially 1.4m/s. In Fig.1, the large bright spot on the left shows the laser irradiation zone.

It is required that the laser is continuously irradiated to the micro-airplane as shown in Fig.2 in order to obtain a continuous propulsion. The development of a laser tracking system is carried out to achieve this requirement.

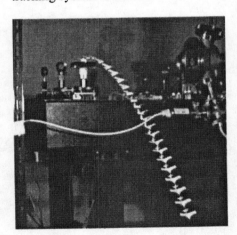

FIGURE 1. Flight trajectories of paper-craft micro-airplane.

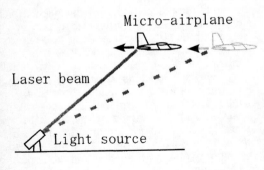

FIGURE 2. Conceptual scheme of laser tracking.

LASER TRACKING SYSTEM

The laser beam has to be tracked to the flight trajectories of the micro-airplane for obtaining the continuous propulsion. An optical system for the laser tracking is shown in Fig.3. A corner cube reflector is installed on the micro-airplane and reflects the laser to the light source. The direction error between the irradiated laser and the reflected laser is measured by a photodetector. The laser tracking is conducted by the galvano-mirror and case control. The case control functions a rough tracking control, and a fine tracking control is achieved by the galvano-mirror control. This laser is also used for controlling the smart wing described later.

Consider the H_2 control so as to reduce the direction error between the irradiated laser and the reflected laser. A control system is shown in Fig.4. u_g and u_c are the inputs to the galvano-mirror and the case controllers, respectively. z_1, z_{21} and z_{22} are the performance indices with respect to the direction error and the control inputs, respectively. The vector of the controlled variables is described as

$$z = \begin{Bmatrix} z_1 \\ z_{21} \\ z_{22} \end{Bmatrix} \tag{1}$$

where

$$z_1 = W_1 e, \quad z_{21} = W_{21} \begin{Bmatrix} u_{g1} \\ u_{g2} \end{Bmatrix}, \quad z_{22} = W_{22} \begin{Bmatrix} u_{c1} \\ u_{c2} \end{Bmatrix} \tag{2}$$

The H_2 norm of the transfer function matrix T_{zr} from the command signal r to the controlled variable is minimized by the H_2 control problem, resulting in the required performance on the laser tracking and the control energy. In case of a mixed H_2/ H_∞ control, the H_2 norm of the transfer function matrix T_{z1r} from r to z_1 is minimized under the constraint on the robust stability in which the H_∞ norm of the transfer function matrix T_{z2r} from r to z_2 is less than 1, where $z_2 = (z_{21}^T, z_{22}^T)^T$. The weight function W_1 is largely weighted in low frequency region, on the other hand, W_{21} and W_{22} are relatively large in high frequency region to satisfy the robust stability against a system error as shown in Fig.5. The designed controller $K(s)$ is described as a state-space form:

$$\begin{aligned} \dot{q}_c &= A_c q_c + B_c e \\ u &= C_c q_c + D_c e \end{aligned} \tag{3}$$

where e is the feedback output from the photodetector.

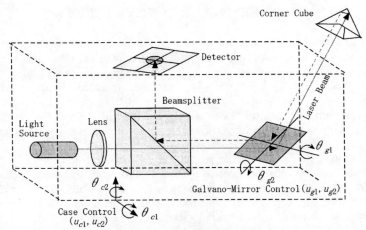

FIGURE 3. Constitution of laser tracking system.

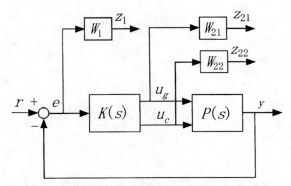

FIGURE 4. Block diagram of control system.

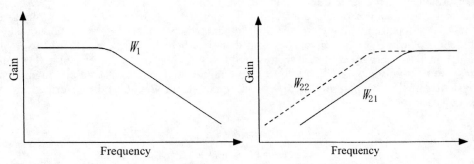

(a) Weight for controlled response   (b) Weight for control input

FIGURE 5. Frequency weight functions.

SMART WING

Typical actuators such as servomotor are not available for the micro-plane because of its large scale and weight. A smart structure is a key technology to control the deformation of a small and light wing [13]. A mechanism of a smart wing is proposed and developed to control the flight of the micro-airplane. This mechanism of the smart wing is composed of the flexible wing, SMA, battery, phototransistor and optical system as shown in Fig.6. The SMA is installed in the wing and the electrical circuit is constituted by the SMA, battery and phototransistor. The phototransistor has a switch function and turns on when a laser is irradiated. The SMA shrinks due to a temperature rise when a voltage is applied and extends to the origin when the voltage is released. The lightening of the smart wing is realized by using a small and light battery. The smart wing is also available to improve an aerodynamic instability caused by air flow with low Reynolds number.

In this system, the irradiated laser is separated into the incident radiations to the reflector and the phototransistor by the beamsplitter. The reflected laser from the reflector goes back to the light source and is used for the laser tracking. On the other hand, the voltage applied to the SMA is switched by another irradiated laser into the phototransistor. As a result, the shape of the wing can be changed by the SMA deformation. The mechanism of the wing shape control is shown in Fig.7. The flight control is achieved by changing the brightness of the laser irradiated to the phototransistor.

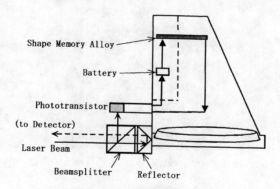

FIGURE 6. Constitution of smart Wing.

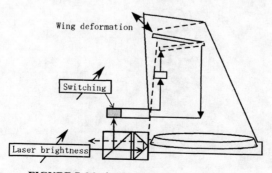

FIGURE 7. Mechanism of wing shape control.

SYSTEM DEVELOPMENT

The developed laser tracking system is shown in Fig.8. The system is composed of the He-Ne laser, condenser, beamsplitter, PSD of a photodetector and galvano-mirror. The direction of the laser beam is controlled by the galvano-mirror control. The control mechanism of the galvano-mirror is shown in Fig.9. The mirror is moved to the lateral and the longitudinal directions by two DC motors for each direction. The irradiation range of the laser with this galvano-mirror is ± 10 degrees in both lateral and longitudinal directions. The errors of the lateral and the longitudinal directions between the irradiated laser from the light source and the reflected laser from the flying object are detected by the PSD. Accordingly, this system is two inputs and two outputs system in the lateral and the longitudinal directions. The specifications of the He-Ne laser of the light source, the PSD of the photodetector and the DC motor used in the galvano-mirror control are shown in Table 1.

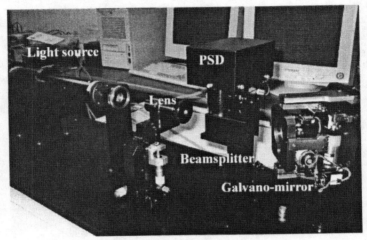

FIGURE 8. Composition of developed laser tracking system.

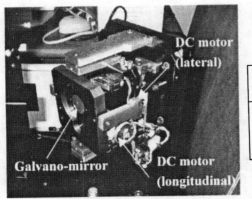

FIGURE 9. Composition of galvano-mirror.

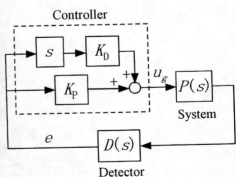

FIGURE 10. Control system of laser tracking.

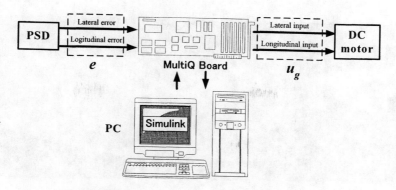

FIGURE 11. Constitution of digital control system

TABLE 1. Specifications of He-Ne laser, PSD and DC motor in laser tracking system.

He-Ne laser		PSD		DC motor	
CW output power (mW)	17.0	Active area (mm)	12×12	Rated output (W)	0.8
Beam diameter (mm)	0.96	Resistance length (mm)	14×14	Rated volt (V)	12
Beam divergence (mrad)	0.84	Interelectrode resistance Vb=0.1V (kΩ)	10	Rated torque (Nm)	0.029
Vertical mode (MHz)	257	Spectral response range (nm)	320-1060	Rated speed (rpm)	47
Operation current (mA)	7.0	Structure	Pin-cushion type	Weight (g)	30
Polarization ratio	500:1	Package	Ceramic	Reduction ratio	1:150

The control system of the laser tracking system is shown in Fig.10. In Fig.10, e is the error detected by the PSD and u_g is the control input to the DC motors, and both have two elements in the lateral and the longitudinal directions. The controller is composed of the feedback of the error e and its derivative $\dot{e}$ with the gains K_P and K_D, respectively, and executed by the digital control system. The constitution of the digital control system is shown in Fig.11.

Figure 12 shows the fundamental mechanism of the smart wing which is developed to evaluate the characteristics of the wing movement by the SMA and the laser. This mechanism is composed of the SMA, coil springs, power supply, phototransistor and plate of 47mm×72mm×1.5mm size as a wing corresponding to Fig.6. The bio-metal-fiber of Ti-Ni based SMA is employed as actuator in this system. The diameter of the bio-metal-fiber is 200 μ m and the fiber is tensioned by the coil springs. The specifications of the bio-metal-fiber and the phototransistor are shown in Table 2. When a voltage is applied to the bio-metal-fiber, it is axially shortened by the rise of the temperature, and when the voltage is released, it extends to its original shape. The deformation of the wing is changed by controlling the irradiated laser to the phototransistor and switching the voltage applied to the fiber.

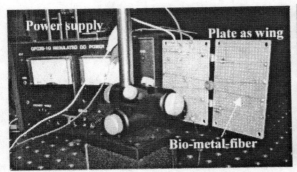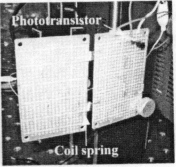

(a) Front view   (b) Back view

FIGURE 12. Fundamental composition of smart wing

TABLE 2. Specifications of bio-metal-fiber and phototransistor in smart wing system

Bio-Metal-Fiber				Phototransistor				
	Min.	Typ.	Max.			Min.	Typ.	Max.
Diameter (μ m)	190	200	210	Dark current[*1] (μ A)		—	0.01	0.2
Linear density (mg/m)	—	200	—	Light current[*2] (μ A)		100	—	—
Electric resistance (Ω/m)	—	40	—	Collector-Emitter Saturation Voltage[*3] (V)		—	0.25	0.4
Min. current to move (mA)	100	—	300	Switching time[*4] (μ s)	Rise time	—	2	—
Standard current (mA)	—	800	—		Fall time	—	2	—
Wattage to move (W/m)	—	15	—	Peak sensitivity wavelength (nm)		—	800	—
Reaction frequency (Hz)	—	0.7	—	Half value angle (degree)		—	± 10	—
Max. actual load (N)	—	4	—	[*1] V_{CE}=30V, E=0mW/cm^2 [*2] V_{CE} =3V, E=0.1mW/cm^2 [*3] I_C =30 μ A, E=0.1mW/cm^2 [*4] V_{CC} =5V, I_C =10 μ A				
Max. actual strain (%)	4.0	5.0	—					
Operation life (Cycle)	10^5	10^7	—					

EXPERIMENTAL RESULT

First, the characteristics between the input voltage to the DC motor and the output voltage from the PSD are examined to evaluate the open-loop property of the laser tracking system. The laser is irradiated to the reflector which is fixed at a point on the height of the light source. The rectangular input with 1V is applied to the lateral DC motor for 0.04 seconds. At this time, the galvano-mirror rotates to the lateral direction so that the light spot of the reflected laser from the reflector moves almost from the left end to the right end on the mirror as shown in Fig.13. Figure 14 shows the output signal of the lateral error detected by the PSD. In Fig.14, the solid line is the measured response and the broken line is the identified one when the open-loop system is assumed to be a third order system with respect to s. The identified open-loop transfer function becomes

$$P(s)D(s) = \frac{1}{s} \cdot \frac{3.19 \times 10^6}{s^2 + 190s + 6.25 \times 10^4} \tag{4}$$

Next, the control system is constituted by giving the gains K_P and K_D in Fig.10 as

$$K_P = 7$$
$$K_D = 0.063 \tag{5}$$

The laser tracking ability is evaluated by a simple experiment shown in Fig.15 in which the reflector is installed at the tip of the pendulum and a swing motion is caused by releasing the reflector from a position with an angle θ. When $\theta = 20°$, $L = 780$ mm, and the distance between the galvano-mirror and the reflector is $D = 3400$ mm, the trajectories of the reflector and the laser are shown in Fig.16. The natural period of this pendulum is 1.8 seconds. It is observed from Fig.16 that the laser is exactly traced to the movement of the reflector. The measured outputs from the PSD are shown in Fig.17 in which Fig.17 (a) and (b) show the lateral and the longitudinal errors, respectively. It is also confirmed from Fig.17 that the maximum output voltage is smaller than that in Fig.14 and so a successful tracking of the laser beam is achieved.

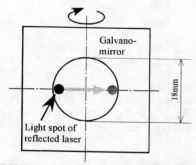

FIGURE 13. Region of light spot on mirror.

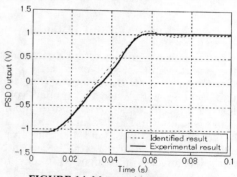

FIGURE 14. Measured lateral error by PSD.

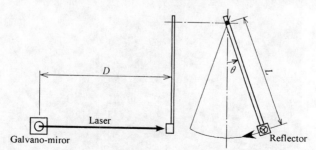

FIGURE 15. Movement of reflector for evaluating laser tracking.

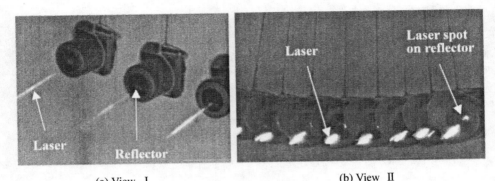

(a) View I　　　　　　　　　　(b) View II

FIGURE 16. Trajectories of reflector and laser

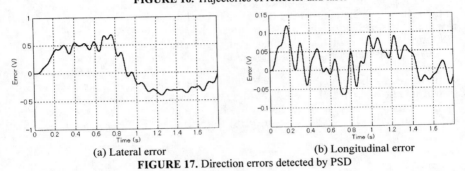

(a) Lateral error　　　　　　　(b) Longitudinal error

FIGURE 17. Direction errors detected by PSD

Next, the characteristics of the smart wing are examined. The measurement system of the smart wing is shown in Fig.18 in which the laser is irradiated to the phototransistor and the deformation at the corner of the wing is measured by the laser displacement sensor. When the applied voltage with the power supply is 1.9V and switched by the irradiated laser and the phototransistor, the measured deformation of the wing is shown in Fig.19. It is observed from Fig.19 that the switching mechanism with the irradiated laser and the phototransistor is normally functioned and the effective deformation of the wing is caused by the bio-metal-fiber. It is also expected that the mechanism of this smart wing functions well in an actual micro-airplane.

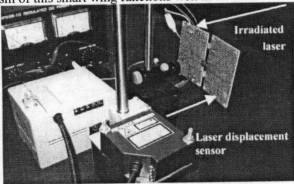

FIGURE 18. Measurement system of smart wing

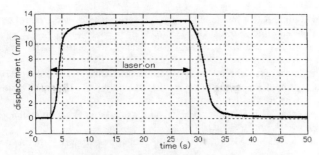

FIGURE 19. Measured deformation of smart wing

From these results, it has been verified that the laser tracking system and the smart wing mechanism are successfully developed and have the required functions. In the future works of this study, it is necessary for realizing an actual flight of the micro-airplane to synthesize the laser with high power, the laser tracking system and the smart wing. It is also required that the case control will be installed in the present laser tracking system in order to extend the laser irradiation range and the smart wing will be controlled by the irradiated laser to realize an advanced flight of the micro-airplane.

CONCLUSIONS

The laser tracking system and the smart wing mechanism for the laser-driven micro-airplane have been developed in this study. The characteristics and the performance of the developed laser tracking system composed of the He-Ne laser, condenser, beamsplitter, PSD and galvano-mirror have been evaluated by the experiment. It has been verified that the laser is traced to the moving reflector with the required accuracy in this system. The mechanism of the smart wing has been developed with the bio-metal-fiber, coil springs, power supply, phototransistor and plate as a wing. The switching function of the electrical circuit with the phototransistor and the laser has been confirmed by the experiment, resulting in the effective deformation of the wing. The developed system and mechanism can be expected to be available in an actual micro-airplane. A future work is synthesizing the laser with high power, the laser tracking system and the smart wing developed in this study to achieve an advanced flight of the micro-airplane.

REFERENCES

[1] D. C. Jenn, R. L. Vitale, G. M. Lee and T. B. Bibson, Microwave Powered Micro-RPV, *Proceedings of 1st International Conference on Emerging Technology for Micro Air Vehicles*, 1997, pp. 74-86.
[2] S. J. Morris, Design and Flight Test Results for Micro-Sized Fixed-Wing and VTOL Air craft, *Proceedings of 1st International Conference on Emerging Technology for Micro Air Vehicles*, 1997, pp. 117-131.
[3] Kajiwara, I. and Haftka, R. T., Simultaneous Optimum Design of Shape and Control System for Micro Air Vehicles, *Proceedings of AIAA, 40th SDM Conference*, 1999, pp. 1612-1621.

[4] Kajiwara, I. and Haftka, R. T., Integrated Design of Aerodynamics and Control System for Micro Air Vehicles, *JSME International Journal*, **Vol. 43, No. 3**, 2000, pp. 684-690.

[5] Ishikawa, H., Kajiwara, I. and Yabe, T., Design of Shape and Control System for Micro-Airplane Development, *SICE Annual Conference 2002*, 2002, CD-ROM(No.MM06-2).

[6] Kantrowitz, A., Propulsion to Orbit by Ground-Based Laser, *Astronaut. Aeronaut.*, **10**, 1972, p. 74.

[7] Phipps, C. R., et al., Impulse coupling to targets in vacuum by KrF, HF, and CO_2 single-pulse lasers, *J. Appl. Phys.*, **64**, 1998, pp. 1083-1096.

[8] Myrabo, L. N. and Mead, F. B., Ground and Flight Tests of a Laser Propelled Vehicle, AIAA98-1001, *36th Aerospace Sciences Meeting & Exhibit*, 1998.

[9] Phipps, C. R. and Michaelis, M. M., LISP: Laser Impulse space propulsion, *Laser and Particle Beams*, **12**, 1994, pp. 23-54.

[10] Yabe, T., Phipps, C., Aoki, K., Yamaguchi, M., Nakagawa, R., Mine, H., Ogata, Y., Baasandash, C., Nakagawa, M., Fujiwara, E., Yoshida, K. and Kajiwara, I., Proposal and Demonstration of Laser-Driven Micro-Airplane, *J. Plasma Fusion Res.*, **Vol.77, No.12**, 2001, pp. 1177-1179.

[11] Yabe, T., Phipps, C., Yamaguchi, M., Nakagawa, R., Aoki, K., Mine, H., Ogata, Y., Baasandash, C., Nakagawa, M., Fujiwara, E., Yoshida, K., Nishiguchi, A. and Kajiwara, I., Microairplane propelled by laser driven exotic target, *Applied Physics Letters*, **Vol. 80, No. 23**, 2002, pp. 4318-4320.

[12] Yabe, T., Phipps, C. R., Aoki, K., Yamaguchi, M., Ogata, Y., Shiho, M., Inoue, G., Onda, M., Horioka, K., Kajiwara, I. and Yoshida, K., Laser-driven vehicles: from inner space to outer space, High-Power Laser Ablation IV, *Proceedings of SPIE*, **Vol. 4760**, 2002, pp. 867-878.

[13] Kajiwara, I. and Uehara, M., Design of Shape and Control System for Smart Structures with Piezoelectric Films, *Proceedings of AIAA, 42nd SDM Conference*, 2001, CD-ROM(No.2001-1555).

Laser-Driven Water-Powered Propulsion and Air Curtain for Vacuum Insulation

Masashi Yamaguchi ,Ryou Nakagawa , Takashi Yabe ,
Choijil Baasandash, Keiichi Aoki, Tomomasa Ohkubo, Masashi Sakata,
Youichi Ogata and Masamichi Nakagawa

Tokyo Institute of Technology, Dept. of Mechanical Engineering and Science
2-12-1 O-okayama, Meguro-ku, Tokyo 152-8552, Japan

Abstract. In previous papers, we reported the successful flight of paper-airplane-about 5 cm-size [1]. Since then, we try to find out a new possibility of the laser propulsion that uses overlay structure especially using water leading to high efficiency [2]. In order to this concept in high altitude, we need to insulate the water layer from vacuum, otherwise water will be evaporated or freeze in vacuum condition. For this purpose, we here propose an "air curtain". We have done simulation and experiments on this concept.

Furthermore, we examined in detail the various features of water exotic target by pendulum and semi-conductor load cell. The coupling coefficient and specific impulse are discussed both by simulation and experiments.

1.INTRODUCTION

Kantrowitz proposed the concept of using laser ablation as a propelling way for space vehicles in 1972 [3]. In November 1997, 140mm-diameter 60g-weight rocket has reached at 15.25m altitude[4-6]. Unlike chemical propulsion that is common now, power source is located on the ground or in space and heating of target material by laser causes the reactive force and we use it to propel objects. This promising way of propulsion will open a new field of applications.

As an application of this concept being propelled with minimum payload such as communication machine, we succeeded in propulsion of a paper airplane[1], in which we demonstrated effectiveness of the double-layered structure that consists of metal layer covered with transparent overlay. This result has significant means in the sense that water can be repetitively used and is free from environmental problem, can be collected from the atmosphere during the flight and so on.

In the previous paper, we proposed laser propulsion of a large airplane in stratosphere. Because the same as micro-airplane, since it's difficult to realize the combustion engine for airplanes flying at high altitude and there exist environmental problems like destruction of ozone layer at stratosphere when the airplane exhausts NOx there, the use of water-powered laser propulsion can be a promising alternative.

In this paper, we shall discuss various features of the target with water overlay and its possibility by experiments and simulations. For this purpose, we examined the system of repetitive for the practical application.

CP664, *Beamed Energy Propulsion: First International Symposium on Beamed Energy Propulsion,*
edited by A. V. Pakhomov
© 2003 American Institute of Physics 0-7354-0126-8/03/$20.00

Although it may work, the water is vaporized or freezed at high altitude. Thus we propose an air curtain to insulate water reservoir from vacuum.

2.TOWARD STRATOSPHERIC FLIGHT

Stratosphere

Stratosphere is an atmospheric region at an altitude of about 11~50km. The density and the pressure of this region is less than 1/10 of surface of earth, and temperature is -60~-70 degrees centigrade. There is no cloud, so always sunny. Using this feature, the idea to communicate by using flight vehicle in the stratosphere instead of the satellite is proposed.

The difficulty here is the realization of the engine. Due to low density, it's difficult to propel by the propeller. The efficiency of jet engine is low, thus a lot of fuel is necessary to be loaded. There exists more important and serious problem. That is the existence of the ozone layer. About 90% of the ozone of the entire globe exists in stratosphere. Using a reciprocal engine or a jet engine that burns fossil fuel, NO_x or CO of exhaust gas indirectly contributes to the destruction of ozone layer by catalysis. While the influence of chloro-fluorocarbon appears after several decades, this influence is direct and thus a serious damage can be expected.

Therefore the application of laser propulsion has big advantages such as no need to load fuel, no harmful emission (emission of only water) and the possibility of getting water from atmosphere.

Flight vehicle driven by laser

A micro-plane is useful to observe a small space like a crater. However, the flight of micro-airplane needs realization of a small and powerful engine. We applied laser propulsion for micro-plane and tried to propel paper airplane by laser. Figure1 shows trajectories of paper airplane with a 590mJ/5ns YAG laser. The airplane flies in a curved trajectory. The size of airplane is 38mm×30mm×5mm and the weight is 0.1g. The target structure is a 3.5mm × 3.5mm × 0.1mm aluminum over-coated with 0.6mm-thick acrylic. The coupling coefficient of $237[N \cdot s / MJ]$ was recorded as a result of measuring initial velocity that was 1.4 m/s with a high-speed camera. Figure 2 is the trajectory of another type of airplane. The size of airplane is 39mm×56mm×15mm and the weight is 0.2g and about 3mm-diameter water droplet was attached to the target aluminum. C_m of the airplane with water is about 2.5 times as large as that of with acrylic. In contrast, with single-layered target the paper airplane did not fly and this shows the importance of double-layered target in practical application.

As shown in Fig.3, we distinguish two types of targets one is double layered target that consists of a transparent material and laser absorbing material and the single-layered target. Hereafter, we call them "Transparent Cannon (TC)" and

" Single Layer (SL)", respectively.   We evaluate the efficiency of laser propulsion by the momentum coupling coefficient C_m.   This C_m defined by

$$C_m = \frac{m\Delta u}{W} \quad [N \cdot s / MJ] \tag{1}$$

where target momentum $m\Delta u$ and incident laser pulse energy W.

In the TC target, the ablation spreads in a space between laser absorbing material and transparent material.   Thus pressure becomes high leading to several orders of magnitude larger C_m in comparison with SL like simulation results[2].   This increase is due to the Cannon-ball effect suggested by Yabe [9], Winterberg [8] and Azechi et al. [7], independently and introduced to laser propulsion by Fabbro [10].

FIGURE 1.   The flight trajectories of paper airplane with acrylic.

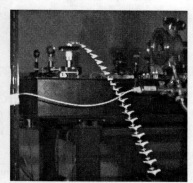

FIGURE 2.   The flight trajectories of paper airplane with water.

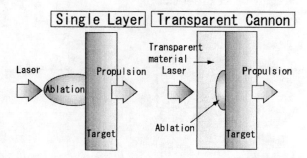

FIGURE 3.   The structure of Single Layered (SL) and Transparent Cannon   (TC) targets.

The characteristics of water-cannon target

For practical application, repetitive supply of transparent material is necessary for repetitive propulsion of airplane.   As a transparent material, liquid is easier to be handled for such purpose.   In addition, we found the effectiveness of water in the previous paper[1].

The experiments were performed by pendulum, and the maximum angle was measured with high-speed camera. This angle is used to calculate momentum of the target. Laser intensity is defined by

$$I = W /(\pi r_0^2 \cdot \sqrt{\pi}\tau/2) \qquad [\text{MW/m}^2] \qquad (2)$$

Laser radius r_0 was measured by knife-edge method with a power meter Model AN/2 made by OPHIR Optronics LTD. [11].

In Fig.4, the experimental result using same quantity of water and acrylic for transparent material is shown which we call Water Cannon (WC) and Acrylic Cannon (AC) hereafter. The C_m of water is about 4 times as large as that of acrylic. Figure 5 shows the result of the experiments using $60\,\mu l$ water and $10\text{mm} \times 10\text{mm} \times 2\text{mm}$ size of acrylic. In this example, C_m both of water and acrylic is same. The mass of water is 0.06g, and that of acrylic is 0.22g. It is interesting to note that only the water of about 1/4 of acrylic in mass is sufficient to obtain the same C_m.

We performed a simulation with 0.1mm-thick water overlay and acrylic overlay. Numerical simulation was done by hydrodynamic code PARCIPHAL based on the CIP-CUP (Cubic-Interpolated Pseudo-particle Combined Unified Procedure)[12-16]. Laser pulse τ is 5ns and wavelength is 1064 nm. Figure 6 shows a time development of C_m given by simulation Both Fig.7(a) and Fig.7(b) are the results at $3\mu s$. Laser intensity is 1.43×10^8 MW/m^2. The C_m of water is larger than that of acrylic and increases gradually. This increase is due to fluidity of water, and a part of water around the laser spot is blown away in Fig.7(a). In contrast, all of acrylic layer moves except for plastic deformation. We speculate that the loss due to this plastic deformation is the cause of the difference of C_m.

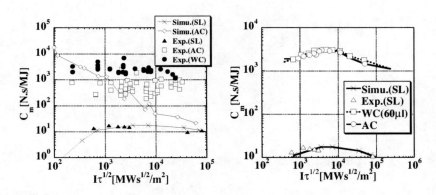

FIGURE 4. The C_m of water cannon (WC) and acrylic cannon (AC) overlays. **FIGURE 5.** The C_m of AC and WC.

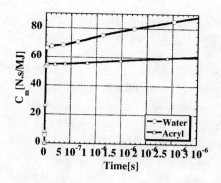

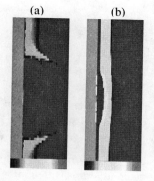

FIGURE 6. Time development of C_m (simulation result)

FIGURE 7. Density contour by simulation for (a) water overlay and (b) acrylic overlay.

Influence of material properties of TC

In this section, we try to find optimal conditions for the efficient propulsion with TC. Using acrylic for transparent material (AC), we measured the velocity of acrylic under the constant laser energy of 410mJ with the spot radius being kept constant at 661 μm but the size of acrylic was changed. In Fig.8, the horizontal axis is specific impulse (I_{SP}) defined by

$$I_{SP} = v/g \quad [s] \tag{3}$$

As I_{SP} increases (mass decreases), C_m decreases if the hydrodynamic efficiency (=Kinetic energy/Laser energy) is constant because of energy conservation (hydrodynamic efficiency=vCm/2). I_{SP} is a parameter shows the performance of the propelling system. As I_{SP} is higher, flying vehicle can get thrust by smaller amount of propellant. This parameter is important for saving propellant in a rocket because of the limit of the payload of propellant. The airplane we propose, however, can get propellant that is water from the atmosphere. Therefore we don't have to care of saving propellant. We should rather determine the optimal laser intensity or the mass of target material matching to the airplane than increase of I_{SP}.

Figure 9 is a relation of an acrylic mass and the ratio of kinetic energy (KE) of the pendulum and the acrylic given by experiments. As the acrylic mass increases, share of KE to pendulum increases. This can be explained as follows. The conservation of momentum is defined by equation (4). Thus, the ratio of KE is defined by equation (5).

$$M_1 V_1 = M_2 V_2 \tag{4}$$

$$\frac{E_1}{E_2} = \frac{M_1 V_1^2}{M_2 V_2^2} = \frac{M_2}{M_1} \tag{5}$$

Where M is the total mass, V the velocity and suffix 1 indicates pendulum and 2 acrylic in this case. This is the cannon-ball theory by Winterberg[8] ,and Yabe[9] for more accurate description. Since the mass of pendulum doesn't change in our experiment, the ratio of KE must be linear according to this theory. Actually, the result of experiment shows this dependency but slightly different from Eq.(5) like $E_1 / E_2 = 0.81 M_2 / M_1$.

Figure 10 shows the result of experiments in which the amount of water in nozzle on the pendulum and the diameter of the nozzle were changed. The peak of C_m at $90 \mu l$ is 3 times larger than that of $10 \mu l$. The increase of water contributes to the increase of C_m. As the amount of water increased, we observed some of water remained in the nozzle for low intensity lasers. In this region of $60 \mu l$ or $90 \mu l$, the curved line drastically falls. The most important point is that all of the peaks of each line are located at almost same intensity. We can use this point as an optimal point.

At $10 \mu l$, the peak of C_m with 3mm.-nozzle is 1.5 times larger that of 6mm.-nozzle, and the peak moves to high intensity region. The horizontal axis in Fig.11 is the ratio of the laser radius to the nozzle radius. Both 3mm and 6mm have a peak at the ratio of 0.5 and the shape of the curved line is almost the same. Experiments were performed under the condition that a laser radius is less than nozzle radius and laser energy is constant at 670mJ. Thus there exists the influence of the change of laser intensity, but we think the effect of the nozzle is the cause of the movement of the peak. In the past research, the nozzle has an effect on rectification of the ablation. About Fig.11, this effect is up to about 2 times of a laser radius. Therefore we think there are two ways to increase the efficiency for water cannon (WC). It's necessary to perform further experiments to search optimal amount of water and nozzle radius.

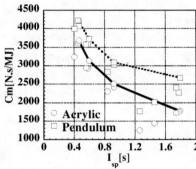

FIGURE 8. The graph of C_m vs. I_{sp}

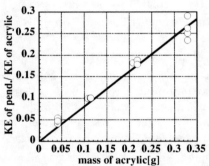

FIGURE 9. The ratio of KE of pendulum-KE of acrylic

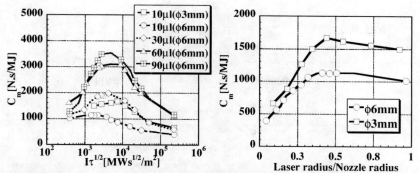

FIGURE 10. C_m graph with the changes of water quantity and nozzle radius.

FIGURE 11. C_m vs. the ratio of the laser radius and the nozzle radius.

Repetitive propulsion with water supply

In previous paper [2], we proposed the structure for repetitive water supply. This structure uses the effect of surface tension for sustaining water reservoir. For practical use, the airplane is used mainly near the earth, and gravitation is not negligible and is larger than surface tension. Thus we here propose alternative concept by using the effect of gravity positively. Figure 12 shows the experimental configuration, water supply system is placed between two wires and its movement is restricted to one-dimension. The water of tank drops down and fills up the nozzle. It is easy to estimate the time for the water filling up the nozzle. The velocity of water at the outlet is calculated by equation (6).

$$v = \sqrt{2gh} \qquad (6)$$

Considering nozzle volume ($\phi6mm$, 1mm-thick), water drops about 5Hz.

In Fig.13, the equipment was propelled at the rate of 0.08 m/s, with 5Hz of irradiation without focusing laser beam. Laser energy was 516mJ, and laser radius was $2896\,\mu m$. The mass of the supply system is 1.6g, and including water, it is 2.6g. Since the laser is not focused and therefore laser can reach the target with the same spot size after many shots, we expect that faster or heavier objects can be propelled by repetitive propulsion in an optimal condition.

In addition, the major problem is the friction at the point of gliding. In reducing this friction, we designed the air slider shown in Fig.14, using one edge of the square rod, then propelling object is restricted to one-dimensional movement. Since the exposure was done in the same interval in Fig.14, acceleration is clearly observed by the distance of object between each exposure time. Using this slider for takeoff, we are able to make the flight of heavy airplane possible.

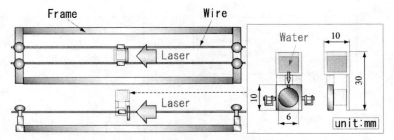

FIGURE 12. The system of the experiment of repetitive propulsion.

FIGURE 13. The tracks of water supply equipment.

FIGURE 14. The tracks of water supply equipment on the air slider..

3.AIR-CURTAIN

If we apply this water supply system to the stratospheric flight, one problem arises owing to the freezing of water. When we reduce the pressure, the water at first begins to evaporate and finally freezes because of latent heat. The experiments show that thrust was not sufficient when this ice was used for the propellant material, because large absorption occurs in ice including bubbles. Although we can use this ice as an ablator instead of metal and resolve environmental problem due to ablation of metal, we are not able to use it as WC. Therefore, the freeze of water by pressure or temperature decrease is not desirable and needs to be solved.

As a way to solve this problem, we propose air curtain. The system of air curtain is shown in Fig.15. The purpose of air curtain is to keep the pressure of the cell surrounded by the pipe and the plate by using airflow. Only the hole that is open to the cell is exposed to the atmosphere.

Simulation result

We performed simulations by the hydrodynamic code PARCIPHAL as used in the laser propulsion. The simulation was done in two dimensions with the Cartesian but non-equally spaced grid. The arrangement of grids is shown in Fig.16. The area of simulation is $4.16cm \times 3.88cm$. The deep gray region is the wall of the cell made of solid and the light gray region is air of 1atm, 274K. The white region is air of 0.0118atm, 226.509K (the condition of 30km altitude). The air in narrow space between two walls (nozzle) flows over the hole opened to the cell. This flow velocity was changed, and we investigated the influence on the pressure of cell (inner pressure).

Figure17 is the contour of the pressure for three different air-flow velocity, 0, 170, 340m/s. The contour range is $10^4 \sim 10^6 [Pa]$. After $10\mu s$ that is about time air past on the hole at 170m/s, each result shows a little difference. After $100\mu s$, obvious difference observed. Mainly there are two directional pressure changes at 170m/s around the hole. One is vertical (nozzle axis) direction and another is horizontal (hole axis) direction. Horizontal component is due to leak from the cell to the atmosphere. The velocity and the pressure at the center of hole are shown in Table 1. Those velocities are 24.7m/s and 100m/s at air-flow velocity of 340m/s and 170m/s, respectively. Therefore, air curtain can not prevent the leak and can not keep inner pressure at 170m/s at this time. After $1ms$, at 0m/s, 170m/s, the pressure change in horizontal direction decreases. After $10ms$, constant flow builds up and inner and outer pressures are balanced by the effect of air curtain even at 170m/s.

Figure 18 is the time development of inner pressure at five different flow velocities. Atmospheric pressure is almost constant at about 0.01atm. At 340m/s, inner pressure is kept to a constant value 0.778atm. As the flow velocity decreases, inner pressure approaches the atmospheric pressure. In order to use water in liquid phase, the pressure must be higher than that of triple point of water(0.06atm). Therefore, more than 34 m/s is needed according to this result.

The simulation result in which the inlet air density is doubled is shown in Fig.19. The flow velocity is 170m/s. In case the density is doubled, inner pressure is 0.511atm at 10ms, while inner pressure in previous density is 0.328atm. Thus, the effect of air curtain is due to mass flow rate not only volumetric flow rate. Therefore, at the same velocity, the use of dense gas is effective to keep inner pressure.

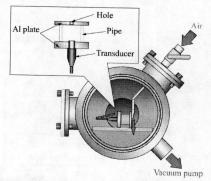

FIGURE 15. The experimental apparatus.

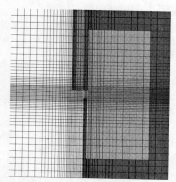

FIGURE 16. The arrangement of grids.

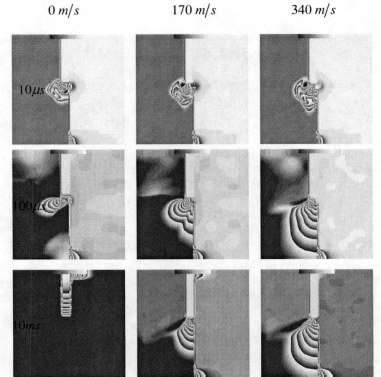

FIGURE 17. Simulation result of the contour of the pressure

Table1. The velocity and the pressure at the center of hall at $100\mu s$			
Flow velocity	0m/s	170m/s	340m/s
Velocity [m/s]	278	100	24.7
Pressure [Pa]	4.81×10^5	8.61×10^5	9.84×10^5

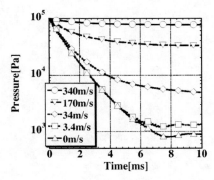

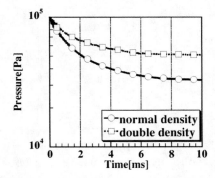

FIGURE 18. Time development of the pressure at 5 kinds of velocity.

FIGURE 19. Time development of the pressure with changing nozzle density at170m/s

Experimental results

The experimental apparatus for air curtain is shown in Fig.15. One side of the pipe is open to the atmosphere in the chamber and the another side is attached to the semi-conductor gauge pressure transducer (diffusion types) DD102A TOYODA MACHINE WORKS, LTD.. This transducer with one side of the diffusion type gauge opened to the atmosphere can measure a gauge pressure, and the maximum measurement pressure range is 30kPa. Its precision of non-linearity is $\pm 0.3\%$, and hysteresis is $\pm 0.02\%$. The pressure in the chamber was measured by Pirani gauge, series 345 MKS INSTRUMENTS. The air supplied by compressor flows on the hole of the cell. The air in the chamber is exhausted by a vacuum pump. We measured the differential pressure between atmosphere and inside of the cell by changing the nozzle shape, the hole diameter and the pipe length. We use two types of nozzle shown in Fig.20. One is relatively straight type with narrower nozzle(Type1), and another has an expansion on the flow path(Type2). Figure 21 is the time development of differential pressure with Type1 and Type 2 nozzles. Since the atmospheric pressure was 3.4×10^3 Pa, Type1 and Type2 nozzles sustained 121% and 5% of the atmospheric pressure, respectively.

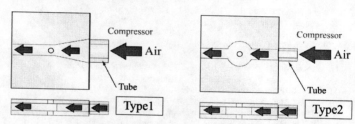

FIGURE 20. The shape of the nozzle

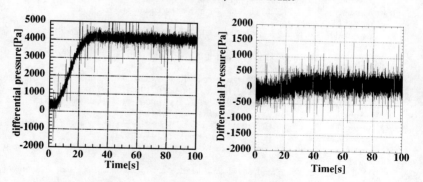

FIGURE 21. Time development of the differential pressure with Type1(left) and Type2(right) nozzle

4.CONCLUSIONS

To realize stratospheric flight, we investigated the characteristics of water cannon (WC) target. The amount of water and the nozzle diameter contribute to the increase of C_m but there exists the optimal laser intensity. For practical application, we demonstrated repetitive propulsion. Water supply equipment of 2.6g was propelled at 0.08m/s with 0.5J laser.

Finally, we performed the simulation and experiments of air curtain in order to prevent the freeze of water at high altitude. Both the volumetric flow rate and the mass flow rate from the nozzle influences on the effect to sustain the pressure of the cell from the result of simulations. The experimental results show that nozzle structure influences to sustain the pressure. Especially, on the condition of low flow rate, the nozzle with the contraction on flow path can sustain inner pressure higher than the expansion nozzle.

REFERENCES

[1]T.Yabe et.al., Micro-airplane Propelled by Laser-Driven Exotic Target   Appl.Phys.Lett. 80, 4318(2002).
[2]T.Yabe et.al., Laser-Driven Vehicles - from Inner-Space to Outer-Space Proc.SPIE 4760 High-Power Laser Ablation IV, 21-26 April 2002,Taos, NM, USA
[3]A.Kantrowitz " Propulsion to Orbit by Ground-Based Laser" Astronaut.Aeronaut.10,74.1972
[4]Franklin B.Mead,Jr and L.N.Myrabo, "Flight Experiments and Evolutionnary Deveroment of a Laser Propelled, Trans-Atmospheric Vehicle", STAIF-98   Congress,Albuquerque(NM),USA,JAN.25-29,1998
[5]L.N.Myrabo and F.B.Mead,Jr., "Ground and Flight Tests of a Laser Propelled Vehicle", AIAA98-1001,Aerospace Sciences Meeting & Exhibit,36th,Jan12-15,1998
[6]Franklin B. Mead, Jr and Leik N. Myrabo,"Flight and Ground Tests of a Laser-Boosted Vehicle", AIAA 98-3735, AIAA/ASME/SAE/ASSEE Joint Propulsion Conference &Exhibit,36th July13-15,1998
[7]H.Azechi, N.Miyanaga, S.Sakabe, T.Yamanaka and C.Yamanaka, Model for Cannonball-Like Acceleration of Laser-Irradiated Targets Jpn.J.Appl.Phys.20, L477(1981).
[8]F.Winterberg, Recoil Free Implosion of Large-Aspect Ratio Thermonuclear Microexplosion, Lettere al Nuovo Cimento 16, 216 (1976).
[9]T.Yabe and K.Niu, Numerical Analysis on Implosion of Laser-Driven Target Plasma.   J.Phys.Soc.Japan, 40 , 863(1976).
[10]F.J.Fabbro, P.Ballard, D.Devaux and J.Virmont, Physical Study of Laser-Produced Plasma in Confined Geometry, J.Appl.Phys. 68, 775(1990).
[11]J.M.Khosrofian, B.A.Garetz, "Measurement of a Gaussian laser beam diameter through the direct inversion of knife-edge data, Appl. Opt. Vol.22, No.21, 3406-3410 (1983)
[12]H.Takewaki, A.Nishiguchi and T.Yabe, Cubic Interpolated Pseudoparticle Method (CIP) for Solving Hyperbolic Type Equations. J.Comput.Phys., 61 , 261 (1985) .
[13]T.Yabe and T.Aoki, A Universal Solver for Hyperbolic Equations by Cubic-Polynomial Interpolation I. One-Dimensional Solver. Comput.Phys.Commun. 66, 219 (1991)
[14]T.Yabe, F.Xiao and T.Utsumi, Constrained Interpolation Profile Method for Multiphase Analysis.   J. Comput. Phys. 169,   556 (2001)
[15]T.Yabe and P.Y.Wang, Unified Numerical Procedure for Compressible and Incompressible Fluid.   J.Phys. Soc. Japan, 60, 2105 (1991).
[16]F.Xiao, T.Yabe, N.Konma, A.Uchiyama, K.Akutsu and T.Ito, An Efficient Model for Driven Flow and Application to a Gas Circuit Breaker Comput. Model. Simul. Eng. 1, 235 (1996).

BEAM GENERATION, PROPAGATION, AND RECEPTION

Laser Spot Size Control In Space

H.E. Bennett

Bennett Optical Research Inc.
Ridgecrest, California 93555

Abstract. Beaming of laser power to satellites in space may involve sending laser beams over tremendous distances. For example, sending a laser beam from the earth to satellites in geosynchronous orbit, 35,900 km above the earth's equator, may involve a trip of about 40,000 km. In the far field at these distances the size of the laser spot is determined entirely by the coherent diameter of the beam director primary mirror from which the laser beam came. It is independent of the mirror's focal length. Even at shorter distances the attempt to control the beam diameter on the solar cells of the spacecraft by focusing the primary mirror may involve attempting to set the mirror to accuracies that are shorter than the telescopes' focal range, $4\lambda f^2$, where λ is the wavelength and f the f/number. For an f/4 system at 0.6 μm wavelength, for example, the focal range is 38 μm. Setting the focus more closely than that accomplishes nothing. However if the spot size is not controlled accurately enough, it may become too small and the laser beam may ruin the solar panels of the space elevator, orbital transfer vehicle, or other satellite being irradiated. This problem can be averted by controlling the effective diameter of the beam director system primary mirror rather than trying to control the focus. By making the primary mirror an adaptive optic and continuously adjusting the size of the segment diameters over which correlation exists, the laser beam from each individual mirror segment can be maintained at the correct diameter on the solar panel at all distances less than the maximum distance for the particular primary mirror diameter.

INTRODUCTION

If laser energy is to be transferred to space to power a satellite such as an Orbital Transfer Vehicle, the diameter of the laser spot size that powers the solar cells must remain constant. The laser beam carries the energy so that the satellite does not need to carry the chemically energetic material required for conventional propellants. If an ion propulsion system is powered in this way, the increase in power per unit weight of propellant is increased by about a factor of ten. Large ion engines, too large to be powered efficiently by anything else except nuclear power, can be powered by the laser. This approach furnishes an attractive alternative to sending rockets carrying radioactive material into orbit.

CP664, *Beamed Energy Propulsion: First International Symposium on Beamed Energy Propulsion*,
edited by A. V. Pakhomov
© 2003 American Institute of Physics 0-7354-0126-8/03/$20.00

Spot Size

The spot size of a laser beam in space depends on the focal length and diameter of the projecting telescope, often called a beam projector, and on the lack of aberrations in the laser beam introduced by the laser, the optics or the atmosphere. If the aberrations are negligible then the beam diameter in the near field is determined by the diameter of the telescope entrance or exit pupil and on the beam focus. If the beam is in the far field beyond the beam waist, then the beam spread, which is determined only by diffraction, (which is a function only of the diameter of the mirror) and does not depend on the mirror focal length. An ideal laser beam is illustrated in Figure 1. At distances closer to the mirror than the beam waist, changing the focus affects the beam diameter. At distances greater than the beam waist, the beam divergence is determined only by diffraction, which depends on mirror diameter alone, and is not affected by the focal length or geometrical optics point of focus of the mirror. Beyond the waist the minimum beam diameter is given[1] by $\alpha = 2.44\ \lambda/D$ where α is the full angle of the diffraction-limited beam, λ is the wavelength of the light and D is the diameter of the entrance pupil of the telescope.

Mirror Defined Beam Diameter

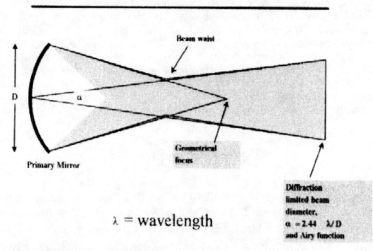

FIGURE 1. Diffraction limited laser beam. In the near field beam size is determined by both the geometrical focus and the exit pupil diameter D. In the far field it is determined only by D.

Additional spreading of the beam may be caused by atmospheric distortion. Figure 2 shows a laser beam incident on a satellite in low earth orbit.[2] As the laser wavefront travels through the atmosphere different areas of the wavefront encounter turbulence caused by air at different temperatures. As it passes through this air the wavefront becomes distorted and cannot be focused to an image, as seen for the uncompensated image on the left of Figure 2.

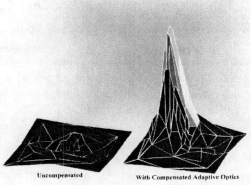

FIGURE 2. The cross-sectional intensity of a transmitted laser beam striking a satellite (LACE) in low earth orbit.

Astronomers know this phenomenon well, since it causes blurring of stellar objects. Even the best, most perfect telescopes, if uncompensated, will produce a somewhat fuzzy picture, since the aberrations are caused by the atmosphere, not the telescope itself. Under good conditions approximately 96% of this atmospheric aberration may be removed[3] by use of adaptive optics. The clean Airy pattern seen on the right side of Figure 2 then results. Using adaptive optics astronomers can form a sharp image of stellar phenomena. Both in the case of light coming into a telescope and light going out of the telescope the surface of the adaptive optic mirror is distorted to match the distorted or to be distorted wavefront, as seen in Figure 3. Reflection of the wavefront has been of an incoming beam by the mirror corrects its distortion, allowing it to be focused to form a sharp image. Similarly the wavefront distortion of a laser spot projected through the atmosphere can be corrected by introducing the negative of the wavefront distortion it will develop in passing through the atmosphere. The atmosphere then corrects the aberrated wavefront and produces an unaberrated wavefront which can be focused to a desired spot size in space. This corrected spot can be directed to fall on a solar panel and create electricity to power a satellite or a space vehicle. This paper describes how to focus the adaptive optic telescope to produce the desired spot size in space and how to configure the receiving solar panel to accept it and transform it into electricity without overheating the solar panels.

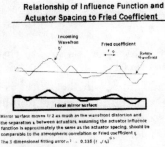

FIGURE 3. Principle of adaptive optics. Disturbed wavefront reflects from a matching mirror, giving plane reflected wavefront.

Constant Laser Spot Size Independent of Distance

Since the solar panel is a fixed size, the light beam should be configured to be the same diameter independent of the distance of the solar panel from the telescope. A simple example of a calculation to achieve this outcome is given below. The separation of the solar panel from the telescope may vary from a few hundred kilometers to as much as 40,000 kilometers. Since the ultimate possible minimum size of the laser spot is fixed through diffraction of light by the diameter of the telescope primary mirror, the minimum telescope primary mirror diameter is determined by the maximum distance at which the satellite may be from the telescope. As seen in Figure 4, this diameter is given by a simple analysis. If α is the full angle of the laser beam divergence, which is determined for an initially distortion-free laser beam by the diameter D of the primary mirror of the beam projector telescope, and λ is the wavelength of the laser light, then $\alpha = 2.44\ \lambda/D$. As an example, assume a solar panel of 10 meters diameter. Require that the laser spot size be 6 meters in diameter to allow for reflection of the laser light from the cone-shaped receiver to be described and also for pointing and tracking errors, errors in beam diameter compensation, or other aberrations not entirely removed from the laser beam when projected to the satellite. The minimum diameter of the beam projector primary mirror is D = 2.44 $\lambda/\ \alpha$ where $\alpha = 6/40,000,000 = 1.5 \times 10^{-7}$ radians. If $\lambda = 0.84$ micrometers (the wavelength of peak sensitivity of either silicon or gallium arsenide solar cells) then D = 2.44 x 0.84 x $10^{-6}/1.5 \times 10^{-7}$ = 14 meters in diameter.

**TECHNIQUE FOR MAINTAINING CONSTANT DIAM
LASER BEAM ON SOLAR CELL**

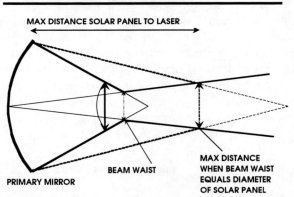

FIGURE 4. Schematic of a laser beam adjusted to produce a constant diameter spot on the satellite solar panel from LEO to GEO.

The largest monolithic mirror in the world is 8 meters in diameter. It is so massive that in all probability no larger monolithic mirror will ever be built. By using one meter hexagonal mirror segments phased together, McDonald Observatory in Texas has mounted in the Hobby-Eberly telescope the largest telescope mirror yet made. It is of segmented mirror design, phased, 11 meters in diameter, and is thought to be the wave for the future in astronomy. Using the same size one meter segments, a 15 meter telescope, rather than a 14 meter telescope just discussed, naturally results. As an example, then, let us assume that the primary mirror of a potential laser beam projector is of segmented construction, is 15 meters in diameter, and has an f number of 2. The f number is defined as the ratio of the diameter of the entrance pupil of the telescope, taken here as the diameter of the primary mirror to its focal length, which is then 22.5m. How can we use the 15 meter diameter $f/2$ mirror telescope to power a solar array? The telescope must maintain a constant beam diameter over distances that differ by as much as a factor of 100. The solution to this problem is illustrated in Figure 4. The maximum distance at which the laser beam can have a 6 meter diameter, the assumed laser spot size, is determined by the size of the beam waist. The beam waist is determined by the telescope mirror divergence α, which depends only on the diameter of the mirror and the wavelength. Beyond the 6 meter beam waist in the example discussed, the focal length of the mirror has no effect on the laser spot size. At shorter distances the focus can be shifted to maintain the 6 meter diameter spot in our example. At a distance labeled x on Figure 4 the spot size is to be 6 meters. The focus is then at a distance q, where from similar triangles $q = Dx/(D - 6) = 1.667x$ in our example and is independent of f number. Given x we can calculate q from this equation. From the thin lens equation $1/p + 1/q = 1/f$ where f is the focal length of the mirror, p is the effective object distance of the source, and q is the image distance. Let $p = f + \delta p$. Then $\delta p \approx f^2/q$. Given q and knowing the mirror focal length f we can calculate the distance away from the focus of the laser beam that would provide a spot size of 6 meters. This analysis can be performed at any distance less than or equal to the maximum distance of 40,000 km in our example. There is a caveat for longer distances. The focal range[3] of a lens or mirror is the distance along the mirror axis over which, according to diffraction theory, a change in focus is undetectable. The focal range of the primary mirror is $\delta p' = 4\lambda f^2$ where f is the f-number of the mirror. If $\delta p < \delta p'$, diffraction makes the setting indistinguishable from the focal length setting, which focuses the beam at infinity. If the f-number of the primary mirror is $f/2$, then at a wavelength of 0.84 μm the focal range is 13.4 μm. The focal length of the mirror is then 22.5 m in this example. At a distance of 500 km in near earth orbit the value of δp is 608 μm or 0.6 mm. At a distance of 40,000 km the value of δp is 13.5 μm, which is equal to the focal range of the mirror. At all distances up to 40,000 km the mirror can then be focused to put the 6 meter diameter focused spot on the solar panel at the satellite distance. To put a 6 meter diameter spot farther than 40,000 km the mirror diameter would have to be larger than 15 meters.

Beam Projector Design

A generic form of the beam projector is shown in Figure 5. The laser output is at the bottom of the figure. The laser may be a free electron laser with an out put power of 200-1000 kW, as described in "Ignition Feedback Regenerative Free Electron Laser (FEL) Amplifier, "Bennett Optical Research Patent No.

6,285,690 B1. Although this application is not limited to the above laser, sometimes abbreviated as the "IFRA" laser, free electron lasers should have the most aberration-free output beam of any laser. In a fel the lasing action occurs in ultrahigh vacuum and there is no material medium to introduce beam distortion. In correcting for atmospheric distortion using adaptive optics it is desirable to have an aberration-free laser beam to input into the adaptive optics. The IFRA laser output beam initially has a diameter of a few mm, and while in vacuum it expands through diffraction to a meter in diameter. It then passes through a fluoride glass window[4], which has the property that it is resistant to developing optical distortion caused by laser heating. Once it exits the window the beam cannot be brought to a focus in the beam train or air breakdown will result.

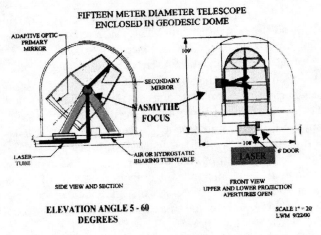

FIFTEEN METER DIAMETER TELESCOPE
ENCLOSED IN GEODESIC DOME

FIGURE 5. Laser beam projector with 15m primary adaptive optic mirror. Laser is introduced into the beam projector utilizing a mirror reflection of the nearly parallel laser beam at what would normally be the Nasmythe focus.

The nearly parallel laser output beam strikes the secondary mirror, which is convex, and is reflected to the primary mirror and then to the satellite or orbital transfer vehicle (OTV) in space. The focal point for the primary is behind the secondary mirror and is thus virtual, which prevents air breakdown. The minute changes in focus required as the satellite moves farther out in space are achieved by moving the secondary mirror. To achieve both reproducibility and accuracy a three axis precision linear stage such as the CTS25 motorized linear stage controller manufactured by Newport Corporation can be used. These double-row linear ball bearing positioning stages have a bi-directional repeatability of 0.2 micrometer and an on-axis accuracy of 2 micrometers. An incremental linear encoder reads the position of each stage to 0.1 micrometer. The controllers are driven by a model MM4006 Controller Driver and can carry a load of 100 lbs. The infinity setting for the telescope, which forms the starting point for the controller, can be determined interferometrically using a parallel plate interferometer. If measurements are taken at wavelengths other than 0.84 µm, care must be taken to correct for the difference in the phase change that occurs at the mirror surface when the wavelength is different from 0.84 μm. Such a correction is particularly important when there are multilayer protective coatings on the mirrors.

Solar Panel Receiver

Changes in focus are very important as the satellite moves farther away in space. Unless the focus is set accurately, the intensity of the laser beam on the solar panel may be sufficient to damage the solar panel. AECAble, an experienced supplier of space studies and components for the National Aeronautical and Space Administration (NASA) performed a study in year 2000 under Contract No. A5753220, "Photovoltaic Array Design for Power Beaming to GEO Spacecraft" for Bennett Optical Research (BOR). The conclusion reached was that it is feasible to power a satellite in orbit from a ground-based laser. Their starting assumptions for power transmitted to the satellite, although derived independently, are similar to those assumed earlier by BOR, and are given in Figure 6. AECAble Corporation has been working since 1995 on an ultralight large solar panel, and has developed an ingenious flexible system they call an "Ultraflex" Flight Wing. It was used on the Mars 01-Lander and was to supply power to the instruments to be used on Mars. System performance achieved in tests was over 100 W/kg of wing mass, twice as high a power density as any competing solar array. It contains accordion-type gores which unfold much as does a Japanese fan. The completed circle is held together with an ingenious latching device and slightly cupped for added strength. Figure 7 shows the Flight Wing. In order to track the satellite and get it to provide feedback to the pointing-tracking system, BOR had planned to use a diode laser mounted on the solar cell and feedback sensors to relay the actual shape and positioning of the beam on the solar panel.

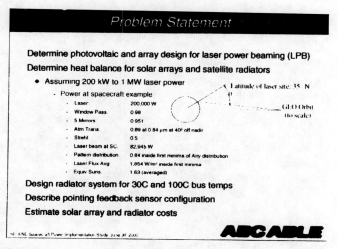

FIGURE 6. Initial assumptions for the laser output and losses as the beam is transmitted to the satellite and absorbed by the solar panel. The number of suns for thermal heating calculations is 1.63, but for electrical equivalent calculations it is 7.0.

577

FIGURE 7. Ultraflex Flight Wing developed by AECAble. It folds into a long package with a diameter that can be carried on a rocket. For laser use the panels will flex upward to form a cone.

AECAble pointed out that the spar lengths of the Ultraflex system could be extended to perform this function, as shown in Figure 8. The IFRA free electron laser has a macropulse spacing that is 1/6th as long as the 10-100 ns minority carrier lifetime of the gallium arsenide, so that cell response should be similar to that for continuous wave monochromatic illumination.

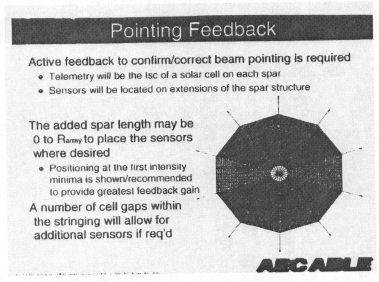

FIGURE 8. Spar extension on Ultraflex solar panel on which to mount laser diodes for tracking. Sensor would also be mounted along the spar to relay beam position on panel.

Pointing at and tracking the solar array would be greatly enhanced if the array carried a diode laser operating at a wavelength of 0.84 μm and similar to the one shown in Figure 9 as a beacon for the ground telescope. The position of the laser beam on the array could be relayed to earth by sensors on each spar in the ultraflex array. By extending the spars as shown in Figure 8 the effective size of the array for pointing purposes could be made much larger than diameter of the array.

1 WATT LASER DIODE

820 to 840 nm Devices

Specifications:

Laser Output: 1w **Beam divergence**: <10deg (slow axis)
Emitter area: 100μm x 1μm   <35deg (fast axis)
Spectral width: <2nm **Power**: 2V, 1350 ma,. 2.7 watts
Temperature coefficient: 0.28nm/°C **Operating temperature**: -20°C to +30°C
Life time: 200,000 hours (23 years)

FIGURE 9. Typical laser diode to mount on end of spars for tracking solar panel. The wavelength of 0.84 μm is near the wavelength of maximum sensitivity for the solar cells and also at an atmospheric window.

Rejection of heat generated in the solar cells by the laser is a challenge. Approximately 52% of the incident laser light is changed into electricity, and thus only about 1/4th of the incident energy goes into laser heating, as compared to 2/3rds of the incident energy going into heating when the same cells are exposed to solar energy. Thermal back-loading of heat to the main spacecraft can also be a problem, but AECAble concluded that that could be corrected by appropriate design. They pointed out that above laser power levels of 0.137 W/cm^2 the relative change in efficiency to be expected in gallium arsenide solar cells would be –0.2% per °C change in cell temperature. For gallium arsenide the cell survival maximum temperature is 380°C. Various heat rejection schemes have been considered. They include beam wavering, achieved through using alternate stringing techniques to prevent current limiting and consequent power loss; solar cell type; distribution; packing; improving thermal substrate conductance; backside fins mounted on the solar arrays for increased radiative transfer; and heat pipes. They were evaluated in part using the program AECAble developed for thermal heating of solar panels. None of these techniques was effective in significantly reducing the increase in heating of the solar panels.

There are differences between the solar irradiation of solar panels and laser irradiation. First, in laser irradiation about three times as much of the laser energy incident on solar panels is converted into electricity than if solar energy is used. There is thus less energy going into thermal heating per watt of incident power. Second, the sun's energy intensity is uniformly distributed over space objects with a solar constant of 1.4 kW/m,^{2} whereas laser energy is sharply peaked in intensity. One way to reduce localized heating of the solar cells by a peaked laser beam is to tilt the cells so that the laser beam strikes them at a large angle of incidence. The intensity I received per unit area of solar cell is then reduced by the cosine of the angle of incidence. The intensity of laser light on the cell is given by $I = -I_o \cos\theta$ where θ is the angle of incidence of light onto the solar cell and I_o the incident laser intensity for normal incidence. If the Airy function is simulated in the center of the Ultraflex wing, transforming the planar Ultraflex Flight Wing in Figure 7 into a quasi-cone, the power levels incident on all of the solar cells is approximately equal. The string lengths then all have the same number of solar cells, and the power level on the cell at the most intense point on the Gaussian beam is reduced by a factor of five. The solar cells are mounted on the cone as seen in Figure 10. The theoretical curve is given by the solid line in the figure.

LINEARIZING SOLAR CELL OUTPUT

Shape of a solar cell covered structure for GaAs (assumed antireflection coated) which would reduce intensity on all solar cells to 20% of Airy maximum out to 20% intensity point. Reduces solar heating by 5 times and allows constant string lengths.

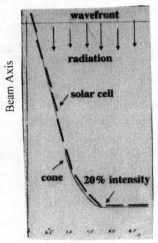

FIGURE 10. Cone shape of solar panel suggested to approximately equalize solar cell exposure to the laser beam.

There will be some light reflected from the inclined solar cells. AECAble suggested that the solar cell array could be extended to absorb this reflected light, thus approximately equalizing the absorbed flux on these cells and on the tilted cells. Their calculations suggest that a laser as large as one megawatt could be usefully employed for powering a solar array 10 meter in diameter and containing a gallium arsenide solar array if thermal considerations were attended to.

CONCLUSIONS

It is technically possible to power the next generation of satellites at distances ranging from low earth orbit to GEO using ground-based lasers, achievable beam directors supplied with adaptive optics, existing fine positioning systems, space mounted laser diode beacons and conventional gallium arsenide solar cells. Applications include powering ion engines for ComSats, orbital tugs, space-based radar, worldwide surveillance systems, and the Space Based Laser. Typical performance using a ready-to-build 200 kW free electron laser design with components which, in nearly all cases, have been demonstrated experimentally, are compared to current state of the art solar arrays below. The following data is reproduced from the AECAble Corporation Report.

Predicted system performance	33 kW	600W/kg	$350/watt	1400 W/m^2
SOA solar arrays	13 kW	60W/kg	$1500/watt	140 W/m^2
Benefit from Laser Power Beaming	2.5X	10 X W/kW	0.2X $/watt	10X W/m^2

REFERENCES

1. Dr. Sadegh Siahatgar, Naval Sea Systems Command, 1338 Isaac Hull Ave. SE, Washington Naval Shipyard, D.C. 20376, Phone (202) 781-3565, e-mail: siahatgar@navsea.navy.mil (private communication).
2. Dr. John D.G. Rather, "Project SELENE, Space Laser Electric Energy, National Aeronautics an Space Administration (1994).
3. Dr. Robert Q. Fugate, "Ground-based Laser Energy Projection," AFRL Technology Horizons, September 2001, pp. 12-14.
4. Max Born and Emil Wolf, **Principles of Optics,** 7th (expanded) ed. Cambridge, England, 1999), pp. 461-463.
5. A.E. Conrady, **Applied Optics and Optical Design,** Part II (Dover Publications, Inc., New York (1960)), p. 627.

Ground-Based Adaptive Optic Transfer Mirrors For Space Applications: I. Design and Materials

H.E. Bennett*, J.J. Shaffer*, R.C. Romeo+ and P.C. Chen+

*Bennett Optical Research Inc, Ridgecrest, California 93555
+Composite Mirror Applications Inc., Tucson, Arizona 85710

Abstract. Projectors 15 meters or more in diameter and of excellent quality will be needed to successfully transmit laser beams from the earth to satellites in space. They can be made up of adaptive optic mirror segments one or two meters in diameter connected to make a phase-continuous mirror surface. Since they must be used in poor as well as good seeing conditions, the faceplate influence function radius must be short and the actuators closely enough spaced to use with Fried coefficients of 2 to 4 cm. Under a Phase II NASA program we are developing a lightweight prototype adaptive optic mirror segment meeting these requirements. It uses a graphite impregnated composite cyanate ester faceplate. Design and materials considerations will be discussed..

INTRODUCTION

The largest astronomical mirrors in the world at present are 8 – 10 meters in diameter. To transmit laser beams successfully to geosynchronous orbit to power orbital transfer vehicles and other satellites larger diameter mirrors are needed. In the far field the diameter of a laser beam transmitted into space is determined only by the diameter of the primary mirror of the beam director. The half angle α of the projected laser beam from a round mirror of diameter D and wavelength λ is[1]

$$\alpha = 1.22 \; \lambda/D \qquad\qquad (1).$$

To make α smaller it is necessary to make D bigger. If it is bigger the telescope resolution is improved, assuming the limiting factor is the telescope, the light gathering power is larger and the size of the laser spot can be controlled to a value of a few meters over a longer distance. To reach from low earth orbit to geosynchronous orbit with a 6 meter constant diameter spot size the mirror diameter must be nearly double what the largest mirror diameters are today[2]. Larger mirrors is the direction that the telescope community is heading. A paradigm shift in mirror design is required to reach this goal.

CP664, *Beamed Energy Propulsion: First International Symposium on Beamed Energy Propulsion*,
edited by A. V. Pakhomov
© 2003 American Institute of Physics 0-7354-0126-8/03/$20.00

To maximize the efficiency of a conventional silicon or gallium arsenide solar cell the wavelength should be ~0.84 μm. Power generation efficiencies of over 50% can be obtained from gallium arsenide and over 40% from silicon cells using approximately this wavelength.[3] Fortunately an excellent atmospheric window is centered at about 0.84 μm, so only about 10% of the light is absorbed in passing through the atmosphere into space.[4] If the solar panel is to be a reasonable size, say 10 meters in diameter, the primary mirror diameter of the ground-based beam projector telescope should be at least 15 meters in diameter[2]. Astronomers are planning to build mirrors of 35, 50 and 100 meters in diameter[5]. Gravity sag and cost make mirrors of these diameters infeasible if the mirror is of unitary construction. A new technology is needed.

The paradigm shift will be to use segmented mirrors a meter or two in diameter which are phased together to form a very large, optically coherent mirrors. The ability to phase segments to form a coherent mirror has been demonstrated at the Keck Telescopes in Hawaii. The novel Hobby-Eberly Telescope in Texas, seen in Fig. 1, illustrates the concept. Phasing the mirror edges of this segmented telescope was accomplished recently under the leadership of Marshall Space Flight Center and Blueline Engineering[6].

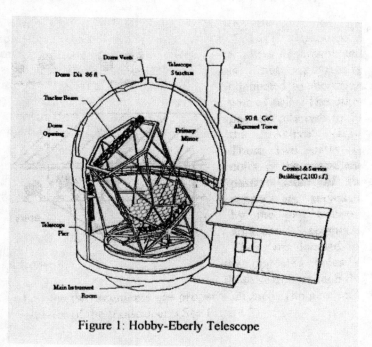

Figure 1: Hobby-Eberly Telescope

FIGURE 1. Hobby-Eberly Telescope with 11m segmented mirror, McDonald Observatory, TX.

One of the major problems with conventional designs is the need to make the one piece primary mirror thick enough so that it will hold an optical figure. That problem can be eliminated by the segmented mirror approach, since the mirror is now a two dimensional, not a three dimensional optic. The rule of thumb is that to maintain the optical figure of a glass mirror it should have a thickness which is 1/6th of its diameter. For a one meter diameter mirror segment the thickness needs be 17 cm (6 ½") or more independent of the diameter of the entire mirror.

Segments are relatively portable, mirror blanks are much less expensive to manufacture than the unitary mirror blanks, and the segments are much easier to remove for coating or partial replacement. The 11 meter diameter Hobby Eberly telescope was designed to provide light to a spectrograph, not to provide high resolution. Its one meter diameter Zerodur mirror segments are only 5 cm (2") thick, and thus are not designed to hold a good optical figure. They are still quite heavy to lift and are not adaptive. There are problems with the rigidity of the mirror mount and other factors. However the Hobby-Eberly telescope concept points the way to the segmented high-resolution mirrors of the future.

Glass Vs Composite Mirror Faceplates

Fig. 2 shows a 15 meter diameter mirror and a 35 meter diameter mirror made up of one meter segments. If the mirrors are to be ground-based, the telescope must incorporate adaptive optics in order to send an undistorted laser beam through the atmosphere. There are good arguments for making that adaptive optic system the primary mirror: (1) the alternative, a much smaller adaptive optic mirror farther down in the optical train, may have difficulty handling the laser power if the telescope is used as a beam projector. (2) The Fried coefficient, measured in cm, drops down into the low single digits in bad weather even at high altitude observatories, and averages in the single digits for most lower altitude observatories. The separation between actuators and thus the influence function of the faceplate should be lower than the Fried coefficient. One can, with difficulty, obtain faceplates that are thin enough so that their influence functions are as low as 2 cm. However if the secondary or tertiary mirrors are made adaptive, the resulting demagnification of the beam makes the demagnified Fried coefficient too small for glass or composite faceplates to meet the desired criteria and microsystems are the only alternative. Bennett Optical Research Inc. and Composite Mirror Applications Inc. are therefore working to develop adaptive optics to be used in primary or possibly secondary mirror systems.

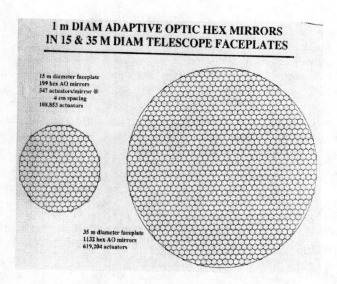

FIGURE 2. A 15 meter diameter and a 35 meter diameter mirror made up of one meter diameter hexagonal segments. There are 199 segments and 108,853 actuators spaced 4 cm apart in the 15 meter mirror and 1132 segments with 619,204 actuators in the 35 meter mirror.

The mirrors being developed consist of transfer mirror faceplate segments of graphite-filled cyanate ester composite material. The work is being carried out under a Phase II Contract with NASA[7]. This composite material has a lateral expansion coefficient of ~5 x 10[-8], which is similar to that of Zerodur and ULE low expansion glass[8]. Low expansion glass has been used for many years to maintain a good optical figure in telescopes under varying temperature conditions. Advantages of the cyanate ester material besides low expansion coefficient are that it is not hygroscopic[9], does not swell in a humid environment[9], and is exceedingly tough.[9] Unlike glass it can readily be made into large sheets a few tenths of a mm thick for adaptive optic mirror faceplates with short influence functions. It is lightweight, about 1/3rd the density of Zerodur[9], and can be produced with a superpolish under 1 nm rms[9]. Scattered light from these mirrors is thus held to a level as much as a factor of ten lower than conventional telescope mirrors. The mirror, mirror mount and telescope can all be made of materials with similar expansion coefficients. In this way the distortion which often results in a change in the mirror's optical figure with change in temperature can be minimized.

Young's Modulus

Young's Modulus for silica-based glasses[10] is about 10.4 x 10[6] psi. A glass microscope slide 1.65 mm thick, 20 mm wide and 63 mm long was clamped on one end and weights were placed on the other end so that the distance between the clamp and weight was 5.5 cm. A Federal model 432 millionths indicator was used beneath the weight to measure the deflection and determine Young's modulus using the equation

$$y = - (4F/Eb)(L/d)^3 \qquad\qquad (2)$$

where y is the deflection, F the force, b the width, d its thickness, L the effective length and E Young's Modulus. A series of measurements were made for different end masses and a straight line obtained when the deflection was plotted against the weight. A composite graphite impregnated cyanate ester copy of the glass slide was made and the measurements repeated. Young's modulus for the composite was measured to be 13.4×10^6 psi, slightly more than the glass. ULE fused silica is similar to Pyrex and about 74% that of Cer-Vit and Zerodur.[11]

Glass Faceplates

It becomes increasingly difficult to obtain the optimum influence function with a glass faceplate as the number of actuators per unit area becomes large. Thin glass sheets are very apt to fracture. Nevertheless most if not all facesheets planned for adaptive optic mirrors at the present time appear to be made of glass. An example is the adaptive secondary optic for the Multiple Mirror Telescope being developed by Media Lario and others in Italy [12]. The glass facesheet is 64.2 cm in diameter and 1.6 mm thick. An even more difficult effort is the deformable mirror being made for the Multiple Mirror Telescope at the Stewart Observatory at the University of Arizona. Its faceplate is 51 cm in diameter and 0.4 mm thick. The 3000 actuators will be spaced at an interactuator distance of 0.8 cm[13]. Our calculations agree with the assumption that each actuator influence function extends only to the distance between actuators for this remarkably difficult to make glass faceplate.

Composite Faceplates

The cyanate ester composite does not fracture as glass does and is extremely tough. It can be produced in large sheets in thin layers with thicknesses as low as 0.2 - 0.3 mm by replication. Work done under the NASA-funded Gossamer program[14] demonstrated these thicknesses. For a 0.6 mm thickness faceplate the actuator spacings would be 2 cm, making it possible to use such a faceplate on a secondary mirror at a high altitude observatory with a significant beam demagnification. There are some complications as the thickness becomes still less because the number of plies used for very thin layers is small and warping can occur. This problem can probably be lessened with more work.

Faceplates measuring 0.6 to 0.8 mm in thickness require less care to produce than thinner faceplates. By casting faceplates on a superpolished glass mandrel we have obtained 0.6 mm thick faceplates under 0.8 nm rms in microroughness. Since from Eq (5) the deflection of the plate decreases as the cube of its thickness, for interactuator distances of 4 cm the faceplates should be about 0.8 mm thick. The first prototype mirror we made, seen in Fig. 3, was 23 cm in diameter. It had a 0.8 mm thick glass faceplate. Larger diameters become increasingly difficult. The second prototype, a composite optic, had a diameter of 30.5 cm and a thickness of 0.76 mm. The influence function of this mirror was 5 cm and the actuator spacing 4 cm. Although it was low scatter, the optical figure was not as good as we wished. The faceplate was convex. The problem was that the mandrel used had a larger expansion coefficient than the composite faceplate. A second, better faceplate has just been cast on a Zerodur mandrel by Composite Mirror Applications Inc.

FIGURE 3. First prototype adaptive optic mirror assembled by BOR using cyanate Ester backplate made by CMA.

Actuator spacing was 1.22 kg with aerial density of 16.8 kg/m^2. With 2 cm actuator spacing it is predicted to be 2.44 kg with an aerial density of 23.2 kg/m^2. Aerial densities of 3.2 kg/m^2 have been achieved for a lightweighted composite mirror, which is one of the lowest values yet achieved for high quality mirror optics[9]. This mirror was not adaptive. However mirrors of 90 cm in diameter have been made using this technology.

Some of the key characteristics of this composites materiel are (1) expansion coefficients comparable to those of Zerodur and ULE, the lowest expansion coefficient materials known. Values of 0.05 ppm/ ^{0}K have been reported for the graphite filled composite[8]. The material has a low thermal conductivity[9]. It cannot be used[9] above its glass transition temperature, 200 – 260°C. Its density is low[9], 1.61 g/cm^3. It is only 73% that of ULE (ultralow expansion quartz)(which is 2.20 g/cm,^{3}) and 64% that of Zerodur (which is 2.53 g/cm^3).

Mirror Applications

Fig. 4 reproduces some data taken from Mount Haleakula on the Island of Maui in 1990 of the intensity of a laser beam after traveling through the atmosphere with and without use of an adaptive optic. With the use of an adaptive optic the observed intensity is an Airy function as it should be if the atmosphere did not strongly affect the beam. Without the use of an adaptive optic the laser intensity is greatly reduced. This data was taken using a laser operating in the visible region of the spectrum. Approximately 96% of the turbulence induced atmospheric distortion can be removed in this region using adaptive optics.[15] Sharp pictures can be obtained from the ground as well as from space, as seen in Fig. 5. Adaptive optics has clearly opened a new chapter in the field of astronomy. It is often not recognized that the requirements on preserving laser beam intensity require better correction than do imaging optics. An image is said to be diffraction limited in resolution, based on the quarter wave Rayleigh criterion,[16] when the Strehl ratio is 80% of the theoretical maximum it could have. At this point 20% of the peak intensity which the beam could have had in transmission has been lost. For a LIDAR system to operate at peak intensity when transiting tens of kilometers through the atmosphere, for example, adaptive optics is needed.

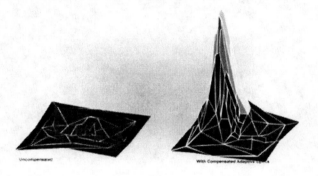

FIGURE 4. The cross-sectional intensity of a transmitted laser beam.

Figure 5. Resolution can be increased dramatically through use of adaptive optics.

A statistical measure involving the phase error is a coefficient r_0, often called the Fried (freed) coefficient after Professor Fried, the Professor at the Naval Postgraduate School in Monterey California who first suggested it. As the Fried coefficient becomes smaller the distortion becomes greater. For *zonal compensation* involving actuators pushing against the flexible faceplate of a mirror and affecting both displacement and local tip-tilt of the faceplate elements, the phase fitting error σ_F^2 of the adaptive optic mirror to the wavefront, is given as[17]

$$\sigma_F^2 \sim 35(r_s/r_o)^{5/3} \tag{3}$$

where r_s is the distance between actuators. If more actuators are used, for example let the actuator separation be decreased by a factor of 2, then the phase fitting error is decreased by a factor of about 3. Equation (3) implicitly assumes that the influence function of the faceplate, defined as the elementary deformation of the faceplate surface produced by one actuator, all other actuators acting only as springs, extends approximately to the next actuator. If the faceplate is not sufficiently flexible, so the influence function extends over many actuator separations, the mirror faceplate cannot distort to correct for peaks in the phase with spatial separations of the order of r_o and Eq. (3) does not represent the situation. As r_o becomes shorter the separation between actuators should not become significantly less than the faceplate influence function. If the faceplate is not sufficiently flexible, the fitting error can become unacceptably large.

The other important factor in adaptive optic operation is the time delay τ between the time the wavefront error is sensed and the time the actuator has moved to correct it. This time delay arises both from the sensing circuit τ_c and the time constant of the actuator τ_a. The relationship between the mean squared phase error σ_τ^2 and the delay time τ is again a 5/3rds power dependence, and is given by[15]

$$\sigma_\tau^2 \quad = (\tau / \tau_o)^{5/3} = 6.88(\tau v/r_o)^{5/3} \tag{4}$$

where the Greenwood time delay τ_o, the delay which results in a phase error of one radian, is given by $\tau_o = 0.314 \, r_o / v$. Here v is the modulus of the average propagation velocity through a turbulence layer. Two delay times contribute to τ, the time constant of the feedback control circuit τ_c and the time constant τ_a, representing the response of the actuator itself. In practice it is desirable that $\tau_c + \tau_a <$ one millisec.

The peak intensity through a large telescope degrades exponentially with the variance of the wavefront distortion. The intensity of the diffracted image of a point source may be represented[18] by the Strehl function R, where σ^2 is the mean square deformation (equivalent to wavefront fitting error or error variance) of the incoming wavefront of wavelength λ. Then

$$R = e^{-(2\pi/\lambda)^2 \sigma^2} \tag{5}.$$

The threshold goal for an aberrated image of point source relative to an aberration free point source image is often taken as a phase difference of one radian, since images with less than one radian of phase difference are relatively acceptable, whereas those with phase differences of more than one radian are undesirable. The Strehl ratio R_{min} corresponding to the threshold phase difference is $R_{min} = 1/e = 0.37$. The contribution of adaptive optics to the value of R is determined by two transfer functions, the telescope transfer function, and the atmospheric transfer function. In the far field the telescope transfer function, which is a maximum when R = 1 but which is said to be diffraction limited for image quality, according to the Rayleigh criterion if R = 0.80, is a function only of the diameter of the telescope entrance pupil relative to the wavelength. The atmospheric transfer function represents the effect of the atmosphere on the beam intensity. It has two parts, one related to the aberrations introduced into the outgoing, initially plane wave by the atmosphere and represented by σ_F, and one related to the time delay of the actuator and the control circuit and represented by σ_τ, as discussed above.

Assuming that the atmospheric rather than the telescope transfer function is the limiting parameter, that the faceplate influence function is optimal, and that $\sigma_\tau \ll \sigma_F$, the Strehl ratio becomes mainly a function of the number of actuators for a mirror of diameter D at a wavelength λ. It may be written as[19]

$$R \sim e^{-(2\pi/\lambda)^2 (0.335 r_s/r_o)^{5/3}} = e^{-(2\pi/\lambda)^2 (N_o/N)^{5/6}} \qquad (6).$$

Eq. (6) gives the theoretical Strehl ratio that should be observed as a function of the ratio of the actuator spacing to the Fried coefficient. The minimum number of actuators which theoretically should give acceptable performance and the improvement in performance to be expected by increasing the number of actuators can also be obtained from these equations, assuming that the mirror performance is limited by the atmosphere. Roddier[20] has compared the theoretical predictions above to experimental results for a number of large adaptive optic mirrors. He finds good agreement except that as the number of actuators increases their efficiency decreases. If the influence function of the faceplate remained approximately constant for the different mirrors, this would be the expected result.

CONCLUSIONS

Adaptive optics has opened up new opportunities both in astronomy and in beaming of laser power to satellites or orbital transfer vehicles operating with highly efficient, electrically powered ion engines. Telescopes with larger primary mirrors are needed to capitalize on these advances. A paradigm shift is required to make larger telescopes. Single primary mirrors are too heavy, with too much mass, to be practical in diameters above about 8 meters. Computer-controlled, segmented large mirrors will now be the future large telescope mirrors, and diameters of 35m, 50m and even 100 meters are being planned. The primary or secondary mirror of these giants is the ideal mirror to make adaptive. Glass faceplates of these sizes offer many challenges. However low expansion, lightweight and relatively inexpensive graphite-fiber filled cyanate ester mirrors furnish an attractive alternative. These mirrors are being developed by Bennett Optical Research in collaboration with Composite Mirror Applications, and appear promising.

REFERENCES

1. Jenkins, F.A. and White, H.E., **Fundamentals of Optics**, 2nd ed., (McGraw Hill, New York, 1950) p. 295.
2. Bennett, H.E., "Laser Spot Size Control in Space," First International Symposium on Beamed Energy Propulsion (sponsored by NASA Marshal Space Flight Center, Huntsville, AL and the University of Alabama in Huntsville Nov. 5-7, 2002) to be published by the American Institute of Physics (in press).
3. Murphy, David M., "Final Report on Photovoltaic Array Design for Power Beaming to GEO Spacecraft," (AEC-Able Engineering Co., Contract No.A5753220 for Bennett Optical Research, Inc. July 5, 2000) p. 9.
4. ibid. p. 8.
5. Future Giant Telescopes, (SPIE Conference 4840 in Astronomical Telescopes and Instrumentation Symposium, Hilton Waikoloa Village Hotel, Waikoloa, Hawaii, USA, August 2002) (in press).
6. Booth, J.A. et.al., "Development of the Segment Alignment Maintenance System (SAMS) for the Hobby-Eberly Telescope," Proc. SPIE v. 4003, International Symposium on Astronomical Telescopes and Instrumentation 2000.
7. NASA Contract # NAS8-02008
8. Chen, Peter C. et. al., "Advances in very lightweight composite mirror technology",*Opt. Eng.* **39**, 2320-2329 (2000).
9. Chen, Peter C. et. al., "Progress in very lightweight optics using graphite fiber composite materials", *Opt. Eng.* **37**, 666-676 (1998).
10. Shand, E.B., **Glass Engineering Handbook** (McGraw Hill, New York, 1958) p. 37.
11. **"The Infrared Handbook"**, 3rd Ed., William L. Wolfe and George J. Zissis, eds. (Environmental Research Institute of Michigan, 1985), p. 7-84.
12. Brusa, G. et. al., The adaptive secondary mirror for the 6.5 m conversion of the Multiple Mirror Telescope: first laboratory testing results," *Conference on Adaptive Optics Systems and Technology*, SPIE **3762**, Denver, 1999.
13. Langlois, M.P. et. al., "High order adaptive optics system with a high density spherical membrane deformable mirror", *Conference on Adaptive Optics Systems and Technology, SPIE* **3762**, Denver 1999.
14. Chen, Peter C. (private communication).
15. Fugate, Robert Q., "Ground-based laser energy projection," AFRL Technology Horizons, September, 2001, pp. 12-14.
16. Born and Wolf, **Principles of Optics** *7th ed.*, (Cambridge Univ. Press, New York, 1999) p. 522.
17. Roddier, F., "Imaging through the atmosphere", in **Adaptive Optics in Astronomy,** F. Roddier ed., (Cambridge Univ. Press, New York, 1999) p. 13.

18. ibid, p. 15.
19. ibid, p. 19.
20. Roddier, Francois, "Maximum Gain and Efficiency of Adaptive Optics Systems", *Astron. Soc. Pacific* **110,** 837-840 (1998).

Ground-Based Adaptive Optic Transfer Mirrors For Space Applications: II. Composite Prototype Mirror

H. E. Bennett[*], J.J. Shaffer*, R.C. Romeo+ and P.C. Chen+

Bennett Optical Research Inc., Ridgecrest, California 93555
+Composite Mirror Applications Inc., Tucson, Arizona 85710

Abstract. Beam Projectors 15 meters or more in diameter and of excellent quality will be needed to successfully transmit laser beams from the earth to satellites in space. They can be made up of adaptive optic mirror segments one or two meters in diameter connected to make a phase-continuous mirror surface. In a companion paper (titled I. Design and Materials) the concept and applications of mirror faceplates were discussed. In this paper the construction and performance of a prototype graphite impregnated composite cyanate ester adaptive optic mirror is evaluated.

INTRODUCTION

The potential and actual advantages of graphite impregnated composite cyanate ester adaptive optic mirrors were discussed in part one[1]. They are both lightweight and durable. Fig. 1 shows a 0.9 meter diameter composite faceplate mirror being held up by two small girls instead of the crane required for a solid glass mirror. The lateral expansion coefficient of the composite material is comparable to that of the lowest expansion coefficient materials known, ULE quartz and Zerodur[2]. Unlike epoxy mirrors composite mirrors exhibit almost no swelling in high humidity or under water immersion conditions[2]. They are very tough and do not fracture like glass. Mirror faceplates can be made which cover large areas and have thicknesses allowing the actuator spacing to exceed sea level requirements and still be comparable to the faceplate influence function[1]. The entire mirror and mirror mount can be built of very similar materials. The mirrors exhibit very little outgassing in vacuum conditions[3]. They can have superpolished surfaces and may scatter only a tenth as much light as typical astronomical mirrors[1].

CP664, *Beamed Energy Propulsion: First International Symposium on Beamed Energy Propulsion*,
edited by A. V. Pakhomov

CMA/GSFC RECENT ACHIEVEMENTS

- Sisters Hannah (l) and Sarah (r) Keyes hold an uncoated 0.9m f/1.2 composite mirror
- Wt. 9 lbs
- Areal density 6 kg/m²
- Smooth surface, clear absence of bond lines
- No data on optical figure of the convex mandrel

FIGURE 1. Lightweight 0.9 meter graphite filled cyanate ester mirror weighs only 9 pounds.

AO Mirror With 2-4 cm Actuator Spacing

The 30 cm diameter prototype AO mirror shown in Fig. 2 was designed to demonstrate that a lightweight mirror with a high performance, high-density actuator mirror faceplate with actuator spacings comparable to the influence function of the faceplate and with millisec response times can be built economically. It is initially equipped with 163 actuators and has an actuator separation and faceplate influence function of 4 cm. It is designed to have as an alternate design an actuator separation and influence function of 2 cm. The goal is to have the faceplate at rest have a relatively good figure, and be low scatter, with microroughness of 1 nm rms or less. A view of the mirror disassembled is shown in Fig. 2. In the figure the mirror surface is down. It is made entirely of graphite filled cyanate ester composite. The mirror material has a low thermal conductivity[2]. It cannot be used[3] above 200-250° C, its transition temperature. After fabrication the mirror surface was coated with aluminum. A close-up of the faceplate is seen in Fig. 3. Next the actuator structure is assembled by fitting it into the black mirror cell as indicated in Fig. 4. The actuators, holes for which are seen in the actuator structure, are glued to the back of the face sheet. Separation between actuators in this sheet is 4 cm, so the inter-actuator distance, which matches the influence function of the face-sheet[4], is equal to or less than the average Strehl coefficient even at sea level. Two cm spacings are made possible by using multiple actuator structures, as seen in Fig. 5. The v-shaped cover seen on the mirror cell is a push-on cover for the face-sheet.

FIGURE 2. Components of the prototype cyanate ester AO mirror fabricated by CMA using a BOR mandrel.

FIGURE 3. Faceplate surface is aluminized and has a lip so the coated mirror if inverted will not touch the surface of the shell.

FIGURE 4. Faceplate and initial actuator support structure are fitted to mirror shell.

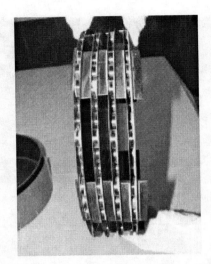

FIGURE 5. Edge view of actuator mount allowing 2 cm spacing of actuators.

Optical Figure

The face-sheet for this mirror was cast onto a 1/10th wave superpolished Pyrex optical flat. Heating is used in the process, and the Pyrex material shrank upon cooling, which contributed to giving the faceplate a convex optical figure. A similar effect is seen in a bimetallic strip. Pyrex has an expansion coefficient[5] of 3.2 ppm/°C and the graphite impregnated cyanate ester has an expansion coefficient[2,3] of 0.05 ppm/°C. The overall spherical shape of the faceplate was initially deduced from interferometric measurements made over small areas of the mirror. Over a very small area, a ¾ cm diameter spot size, straight fringes are obtained, as seen in Fig. 6. The lack of structure on these fringes suggests that the surface is very smooth, a result confirmed by the Wyko Interferometer measurements discussed later. When the spot size diameter is increased to 3 cm, nearly circular rings are obtained, as seen in Fig. 7, indicating a nearly spherical surface. Using the Saggital formula,

$$R = D^2 /8s \qquad (1)$$

where R is the radius of the spherical surface, D the diameter of the spot examined, and s the saggital depth of the surface over diameter D, the radius of the sphere can be calculated. There are 10 and 11 fringes going up and down vertically along the semi-major axis and 18 fringes in both directions along the semi-minor axis. Averaging their number and remembering that there are two fringes per wavelength, the average radius of curvature at the center of the quasi-spherical faceplate is calculated to be 25 meters. Repeating the process about an inch from the edge of the faceplate one obtains the fringe pattern seen in Fig. 8. The circular fringe patterns extend to near the edge of the mirror. There is a circular zone about 4 cm from the edge, shown by the saddle point in Fig. 8, which is approximately concentric with the mirror and indicates the presence of a zone. Our laboratory is completing an interferometer designed to measure the true optical figures of very long focal-length mirrors[6]. With it we will be able to obtain further interferometric details of samples such as this one.

FIGURE 6. Straight line fringes from ¾ cm diameter spot size show lack of structure, suggest small microroughness.

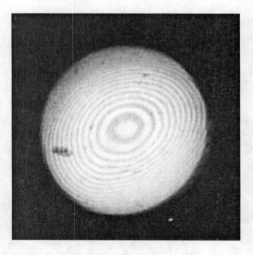

FIGURE 7. circular fringe pattern indicates long radius spherical faceplate.

FIGURE 8. Saddle point indicated by the X in the fringes near the edge of the mirror indicates a zone, possibly related to the bonding of the mirror to the mounting cup.

In another approach, the convexity is seen in Fig. 9 by placing a straight rod onto the mirror surface. At the center of the mirror the rod is in contact with the mirror surface, while at the edges of the mirror there is a separation of the rod and the mirror surface. A plot of the separation is seen in the graph above the photograph. It was measured by using a Federal millionths indicator gauge[7] and contacting the mirror surface at equally spaced points. The convexity is quite smooth, and reaches a maximum departure from the flat onto which it was cast of about 0.9 mm. The effective radius of curvature of the entire faceplate can be determined from this data using Eq. (1). The result is a radius of 13 meters, slightly over half of the 25 meters determined interferometrically for the mirror center.

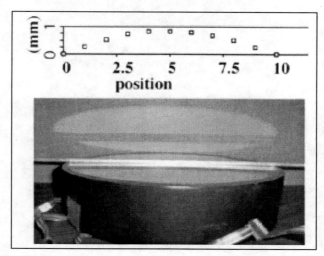

FIGURE 9. Direct measurement of the convexity of the faceplate at equally spaced positions using a millionths of an inch indicator gauge.

FIGURE 10. Reflected beam of light from the sun is measured to evaluate mirror.

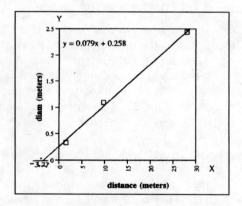

FIGURE 11. Linear relation between spot size and distance is found, gives radius as 3.27m.

Another type of measurement is illustrated in Fig. 10. When the mirror is exposed to the parallel rays of the sun a beam is formed on a wall. It is uniformly round, and exhibits a very bright ring of sunlight around the disc, possibly a reflection of the focusing action of the zone around the mirror periphery. The beam grows in diameter as the distance to the wall upon which it falls becomes larger. The diameter of the beam was measured as a function of the distance to the wall and is linerally dependent on distance, as shown in Fig. 11. The virtual source of the radiation from the convex mirror can be determined by extending the line to y = 0. Because parallel light striking the mirror is reflected through twice the angle formed by the intersection of the mirror radii behind the convex mirror surface, the position of the virtual source determined by the light beam is half of the radius of curvature of the mirror.

Measurements made on the newly received mirror in July 2002 (in the heat of the summer, when the temperature was close to 90°F) gave a virtual source value of 3.3 meters, and thus a mirror radius value of 6.6 meters, two times lower than the direct measurement technique described above. When the mirror was remeasured using the reflection technique in December, 2002 (when the temperature was about 50°F) the radius had increased by a factor of 2.2 and was now 14.4m, in relatively good agreement with the 13 meter value obtained from direct measurements. We do not understand the results of these two sunlight measurements. One possibility is that the temperature dependence of the cement used to bond the faceplate to the mirror shell may be the culprit. The higher the temperature the more convex the faceplate should then become. Another possibility is that either the cement or the cyanate ester itself may require a stabilization period to reach its final dimensions. Our second prototype mirror will use a different cement that does not shrink appreciably when it dries. Results from this experiment may help to solve the puzzle. Another experiment is to vary the temperature of the present mirror and observe the results. This problem is still under investigation.

Actuator Performance

Actuator performance is monitored[8] using a non-contact Michelson interferometer. A double-trace oscilloscope tracks the interference fringes on the lower trace, which is in phase with the upper trace that shows the drive voltage. This technique is a sensitive method for detecting both the displacement amplitude and the phase lag if any. The fringes in Fig. 12 were taken on an earlier prototype glass-faceplate mirror, fabricated by BOR, which was 23 cm in diameter. The face was one mm in thickness and the wavelength 0.6328 μm. Two fringes are equal to one wavelength of displacement.

The fringes were recorded on a digital camera. For low displacements, a few wavelengths of light, there was no hysteresis, as seen by the coincidence of the minimum of the drive voltage and the inflection point on the fringes in Fig. 12. As the displacement became larger and the voltage increases to 100 V peak to peak, hysteresis begins to appear. When displacement approaches 5 μm the hysteresis becomes as large as 16%. As the frequency becomes even higher but the displacement drops, the hysteresis again disappears.

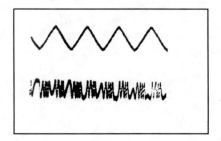

FIGURE 12. Top – triangular voltage wave applied to actuators. Bottom – interferometer trace, which shows no hysteresis in this case.

The displacement of the actuators depends on the voltage applied and on the force required to displace the membrane mirror. An actuator will lift a 2 kg weight. The stiffer and thicker the membrane the more force is required per μm of faceplate travel. Tests made on the unloaded actuators showed the displacement was linear with applied voltage up to a voltage of 250 V. The displacement was 175 μm at that point. It increased more slowly with increasing voltage after that. The response of the actuators was linear but virtually independent of frequency. It rose about 4% in going from a frequency of 1 Hz to a frequency of 1 kHz, as seen in Fig. 13.

ACTUATOR FREQUENCY RESPONSE

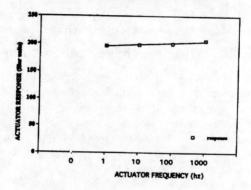

FIGURE 13. Actuator can be nearly insensitive to applied frequency.

Faceplate Resonances

Resonances in the faceplate can also occur for key frequencies, as seen in Fig. 14. This data was taken on the glass prototype mirror above. The thickness of this faceplate was one mm. The theoretical resonant frequency of a mirror is[9]

$$\nu = \frac{10.21\,h}{2\,\pi\,r^2}\sqrt{\frac{E}{12\,\rho\,(1-\sigma^2)}} \tag{2}$$

where h is the faceplate thickness, r the distance to the nearest clamped point (i.e. actuator), E is Young's modulus, ρ the mass density of the faceplate, and σ Poisson's ratio. The calculated fundamental frequency for this glass faceplate is 1.88 kHz, so the observed resonance is a second harmonic. The resonance for a composite should be similar except that the density of the composite is only 1.61[2], only 73% that of glass and Young's modulus is 1.34 x 10[6] psi, 1.29% higher than glass[1]. Both these factors will tend to push up the resonance to higher frequencies for the composite. If it goes up to ≥ 1 kHz, the resonance problem can largely be neglected. If not, the actuators can be electrically clamped at the resonant frequency, which will minimize the effect of the resonance.

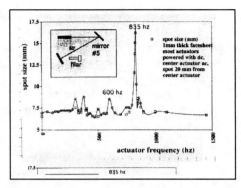

FIGURE 14. Second harmonic resonance of mirror faceplate. Measurement setup in upper left.

Surface Microroughness And Surface Ripple

Using conventional fresh-feed polishing techniques, glass surfaces, including telescope mirrors, are expected to have surface roughnesses of 20–25 A rms[10]. Since the scattered light level is proportional to the square of the rms roughness[11], a 2.5 nm rms mirror will scatter nearly ten times as much light as a 0.8 nm surface. Fig. 15 gives the result of roughness measurements made on this prototype sample. It is clearly in the superpolish range. The ability of a telescope to resolve a weak source in the neighborhood of a strong source depends heavily on the amount of scattered light generated by the optical surfaces in the telescope. In considering whether a graphite impregnated composite cyanate ester faceplate will make a good telescope mirror material, the scattering level achieved is important. A small 5 cm diameter coupon of this composite material was previously measured and had a microroughness of between 0.6 and 0.9 nm rms[3]. The question remained as to whether low scattered light can be achieved on a larger mirror. It clearly depends critically on the smoothness of the mandrel used to cast the ester material on and possibly on its temperature dependence. Three 30 cm Pyrex blanks were superpolished at Bennett Optical to be used in this test. They were tested against each other and the optical figure tested out at about 0.1 wave in the visible. One of them was used to form this faceplate. Pyrex was used, in part, to see if under production conditions the difference in expansion coefficient affects the rms roughness. It clearly does not. It probably does, however, affect the optical figure.

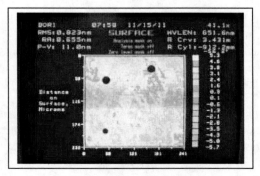

FIGURE 15. Microroughness at center of mirror is only 0.8nm, showing faceplate is superpolished.

Fig. 16 shows the result of a Wyco Interferometer[12] analysis of the transfer optic. The range of correlation lengths extended from ~ 0.6 µm wavelength to 241 µm, somewhat longer than the usual 100 µm upper limit. The wavelength was 651.6 nm and the roughness value obtained by BOR at China Lake was 0.8 nm, well within the superpolishing range. A Wyco TOPO 3-D instrument was used for the measurement. Three highly localized asperities were removed in the analysis and are shown by the black circles. They are clearly not endemic to the surface and were probably dust particles.

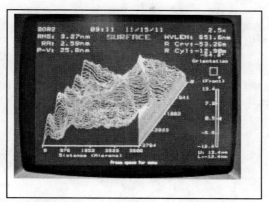

FIGURE 16. Ripple over this mirror area is ~3 nm rms.

The analysis was then extended to a 3.906 x 3.764 mm area to evaluate the surface ripple. Results at two spots were: 3.27 nm rms (seen in Fig. 16) and 43.8 nm rms (seen in Fig. 17). The order of magnitude increase resulted because we had discovered a saddle point on the surface as seen in Fig. 17. The microroughness is small but this large ripple artifact is an unexpected phenomena. Ripple is seldom reported when testing mirror surfaces.

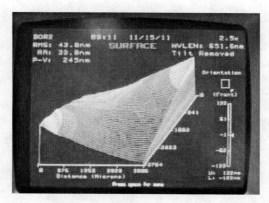

FIGURE 17. Ripple over this mirror area, which is of similar size to Fig. 16, is over 40 nm because of a saddle point.

Magnitude of Atmospheric Aberrations at Various Geographical Locations

There have been many measurements of the optical coherence length r_o of incoming light from space and they often disagree significantly in value. Fortunately there is a group at the Naval Postgraduate School in Monterey, California which has specialized in experimental measurements of the Fried Coefficient at different astronomical sites for over 20 years. It is headed by Professor Donald Walters, who has achieved a reputation for reliable and consistent results over both a long period of time and under various experimental circumstances. His r_o data is used in this analysis. It was all taken using a telescope with a 0.35 meter diameter primary mirror[13] and is for starlight measured over some period within the time from just after sunset to just before dawn. A series of high altitude balloon measurements made at Lawrence Livermore National Laboratory[14] are shown appended to the Walters et. al. data and are illustrated in Fig. 18. Agreement of the two sets of data in areas of altitude overlap is quite good.

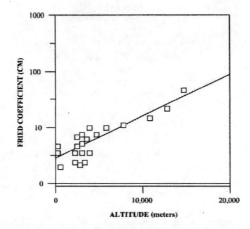

FIGURE 18. Fried coefficient over a range of sites plus balloon results.

The images were focused on a one dimensional array and exposure times were typically between ½ and 1 sec. Longer exposures increased susceptibility to wind and vibration induced image motion. Shorter exposures could suppress atmospheric low frequency tilt components and reduce signal to noise ratio. Eight of the locations measured had additional instrumentation including acoustic sounders, and microthermal probes either mounted on towers or carried aloft on meteorological balloons to obtain corroborating data. Measurements were made at about 750 nm just at the long wavelength edge of the visible spectrum. They were then scaled to give zenith values of r_o at a wavelength of 500 nm. A comparison with data from the U.S. Naval Observatory at Flagstaff, Arizona showed excellent correlation up to coherence lengths of 20 cm or so. Most of the measurements were made at astronomical sites chosen with the expectation that the Fried coefficient would be longer than at average sites. A few of the sites were not potential astronomical sites and were chosen to investigate possible differences at "average" locations, particularly those near sea level. Some of these sites were surprisingly good.

A more detailed view of the terrestrial data is given in Fig. 19. There are considerable differences between the sites. One of the most striking is Mt. Wilson, which is in the mountains overlooking Los Angeles. Its mean covariance length is 12.9 cm, nearly 4 cm larger than the curve fit at that altitude. The standard deviation of all measurements made at Mt. Wilson is only 0.17 cm, far too small to explain the discrepancy. The average standard deviation for all measured sites during astronomical observing times is only 0.15 cm, as seen on lower right of the figure. Each site is remarkably consistent, and the differences between sites are clearly not just a matter of altitude.

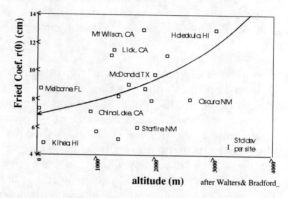

FIGURE 19. Closer view of scatter of Fried coefficient at level trial sites. The variance per site indicated at lower right.

It is important to remember that the reported data are not "average" data for locations at the various altitudes. Instead they are taken at some of the best sites available using optimum conditions. For example, no data are reported for periods just before climatic changes such as an advancing cold front. Test measurements made under approaching cold front conditions[15] led to r_o values in the 4–7 cm range at Mt. Wilson, which had average values under normal test conditions of 12.9 cm. Also no data were reported during the daytime when solar heating will lead to reduced r_o values. Data taken uniformly during 24 hour periods during average conditions in average places would surely reduce the r_o values to some extent. To provide good correction of atmospheric aberrations down to sea level the actuator separation should be no larger than the Fried coefficient. Fig. 19 suggests that the faceplate actuator separation should thus be equal or less than 4 cm. On the basis of the incomplete data we have on "average conditions" for 24 hour operation, a choice of 2 cm might be more prudent.

The advantages of using the primary mirror as the adaptive optic mirror instead of the secondary mirror should be mentioned. The number of actuators needed for atmospheric correction is the same whether the primary or secondary is used. However the demagnification in size of area between the primary and the secondary is reflected in the additional crowding of the already closely spaced actuators. Performance will be best if the influence function of the faceplate is of the same order as the actuator separation. The thickness of the mirror faceplate for the primary mirror needed to make this objective feasible has been demonstrated for composite mirrors. If the secondary is chosen as the adaptive optic element, an even thinner faceplate is required.

CONCLUSIONS

Graphite fiber reinforced cyanate ester has nearly as low an expansion coefficient as Zerodur or ULE. By processing the surface properly a 20 – 50 μm layer of pure resin can be formed on the surface which is not penetrated by roughness-inducing graphite fibers. By replicating a thin sheet of the cyanate material on a superpolished mandrel a surface can be generated which is very smooth, 5 – 10 A rms, and produces a mirror surface which scatters as little as one tenth the light scattered by conventional astronomical optics.

These cyanate mirrors we call "Transfer Optics" to distinguish them from the lower quality but much better known "Replica Optics" the Optics Community is familiar with. This paper describes the results to date of a program to utilize this cyanate ester technology together with some very compact, inexpensive, piezoelectric actuators to produce adaptive optic mirrors meeting the goals outlined above. It appears to us that by using this approach adaptive optic mirrors in diameters up to 1 to 2 meters can be made both achievable and affordable. They will be useable even at sea level to remove a large fraction of the distortion introduced by the atmosphere and should reduce time lost even at high altitude observatories when bad seeing days occur. We also expect them to make daytime observations feasible.

It is difficult to use commercially available actuators in such applications. Actuators depending on the linear expansion of piezoelectric elements can exert a large force, but they have limited throw per element and when multiple elements are required it makes them rather expensive. Many have a long time constant, so they cannot keep up with atmospheric changes very well. BOR actuators are not limited by these problems.

Based on the above discussion certain desirable features are achievable in the design of a zonal type adaptive optic mirror. The mirror can have actuators separated by 2 – 4 cm and the faceplate can have an influence function of a similar length. The actuators can and should have a response time of the order of one millisec or less. CMA has demonstrated that the diameter of a composite mirror can be at least 1 - 2 meters in diameter. It is difficult to make large diameter glass faceplates thin enough to meet adaptive optic requirements and not have them fracture. In order to have an adequate actuator influence function in either glass or composite, the faceplate thickness should be 0.6 to 0.8 mm thick, which corresponds to actuator separations of 2 cm to 4 cm. A cyanate ester composite mirror is fracture resistant, lightweight, can be low scatter, and can be produced in sizes of 1-2 meters. Achieving optical figure control is the remaining problem, but we expect to achieve it, which will make composite cyanate ester mirrors an excellent choice for use in segmented large mirror applications.

ACKNOWLEDGEMENTS

The authors would like to thank Dr. Donald Decker, Mr. Randy Dewees and Mr. Al Slomba of the Naval Air Warfare Center Weapons Division, China Lake, California for help with optical evaluation of the adaptive optic mirrors.

REFERENCES

1. Bennett, Shaffer, Romeo and Chen, "Ground-based adaptive optic transfer mirrors for space applications: "I. Design and Materials" (this volume).
2. Chen, Peter C. et. al., "Progress in very lightweight optics using graphite fiber composite materials," *Opt. Eng.* **37**, 666-676 (1998).
3. Chen, Peter C. et. al., "Advances in very lightweight composite mirror technology",*Opt. Eng.* **39**, 2320-2329 (2000).
4. Roddier, F., "Imaging through the atmosphere", in **Adaptive Optics in Astronomy,** F. Roddier ed., (Cambridge Univ. Press, New York, 1999) pp. 13, 15, 19.
5. **"The Infrared Handbook"**, 3rd Ed., William L. Wolfe and George J. Zissis, eds. (Environmental Research Institute of Michigan, 1985), p. 7-84.
6. Bennett, H.E. and J.J. Shaffer, "Test facility for long-focal-length mirrors", in *Laser-Induced Damage in Optical Materials: 1992*, Bennett, Chase, Guenther and Soileau eds. (SPIE v. 1848, 1992), pp. 117-124.
7. Model 2400, manufactured by Federal Products Corporation, a subsidiary of Esterline Company, Providence, Rhode Island, USA.
8. Patent applied for.
9. Tyson, Robert K., **Principles of Adaptive Optics,** *2nd ed.*, Academic Press, Boston (1998), p. 216.
10. Elson, J.M., H.E. Bennett and J.M. Bennett, "Scattering from optical surfaces" in **Applied Optics and Optical Engineering,** R. Shannon & J. Wyant eds., (Academic Press, New York 1979), p. 202.
11. Bennett, H. E. and J.O. Porteus, "Relation between surface roughness and specular reflectance at normal incidence", J. Opt. Soc. Am. **51,** 123-129 (1961).
12. Wyco Topo 3-D, Wyco Inc., Tucson, Arizona, USA.
13. Walters, Donald L. and William Bradford, "Measurements of r_o and θ_o: Two Decades and 18 Sites," *Appl. Opt. 36,* 7876-7886 (1997).
14. Chocol, C.J. and M.J. Newman, in *Energy and Technology Review,* (Lawrence Livermore National Laboratory, March, 1993), pp. 12-18.
15. Roddier, Francois, "Maximum Gain and Efficiency of Adaptive Optics Systems", *Astron. Soc. Pacific* **110,** 837-840 (1998).

<u>Note:</u> CMA is Composite Mirror Applications, Inc.
      BOR is Bennett Optical Research Inc.

SPACE-BORNE SOLAR LASER FOR POWER-BEAMING APPLICATIONS

Ja H. Lee and Bagher M. Tabibi

Research Center for Optical Physics, Hampton University
Hampton, VA 23668-0199

Abstract. A brief overview of the concept of space power beaming systems with solar-pumped lasers as the in-space energy source is presented. For the past two decades under auspicious of NASA-Langley Research Center we have pursued a solar-pumped laser development program and searched its space applications including laser-beamed propulsion. Among various laser media evaluated for suitability the iodine photodissociation laser was found to be the best qualified and could be scaled up to a lightweight space borne system. The results of the proof-of-the concept experiments will be provided.

INTRODUCTION

In space, sunlight is one of the few natural resources. However the technology for converting sunlight to electric or thermal power and the system mass and the scale required for accomplishing that conversion are primary constraints that limit flexibility of most of the space missions. The emerging concept of power beaming, which may provide higher power density level, increased flexibility, and improved economics to space missions, offers the potential for an important evolution in the primary space power source. Among various power-beaming methods, such as microwave beaming, concentrated solar power beaming, laser power beaming has advantages from the fundamental characteristics of the laser i.e. coherence and its short wavelength that result in extremely small beam divergence. Space-borne laser power beaming offers additional advantages over the earth-based laser power beaming. The advantages of space-borne solar lasers are due to that the pumping source, solar energy, is free, non-local, abundant and pollution-free. When high-power levels (>1kW) are required for missions beyond near earth orbits, a space-based laser operating as the central power plant could become as valuable[1] as a terrestrial power plant for electric or thermal machineries on earth. The first solar-pumped laser was successfully demonstrated in 1963 immediately after the invention of the laser itself. However, until recently, solar-pumped lasers attracted little attention. The past two decades NASA-Langley has pursued the high-power solar-laser development as part of the space power beaming program.[1-11] We revisit the program briefly before the discussion of space-borne laser application to beamed energy propulsion.

CP664, Beamed Energy Propulsion: First International Symposium on Beamed Energy Propulsion,
edited by A. V. Pakhomov

SOLAR-PUMPED LASERS

Direct solar-pumped lasers with high average powers pumped with the existing solar furnaces have been reported from Japan (H. Arasi, 18-Watts, Nd:YAG, 1984) and Israel (M. Weksler et al./, 100-Watts, Nd:Cr:GSGG, 1987). A recent high-power accomplishment in the US is the excitation of a 48-Watt CW iodine photodissociation laser using a solar AMO simulator at NASA Langley Research Center. The NASA program has evaluated theoretically and experimentally various laser systems with laser media including IBr, alkyliodides such as C2F5I, C3F7I, C4F9I and C6F13I in vapor phase, organic dyes and Nd in the solid/liquid phases, Nd^+ in various solid hosts such as YAG, YLF, GSGG, and alexandrite. In addition, the technology for an electrically powered (i.e. indirectly solar pumped) multiple laser-diode amplifier array was also studied.[6] This study found that with a 5-m diameter receiver at 10,000 km away from the 6-m laser transmitter can achieve a power collection efficiency of 80%. The laser power transmitter, when successfully developed to a wireless central power plant in space, could remotely provide the required thermal and/or electric power to advanced space stations on near earth orbit, lunar and Mars bases, manned long-range rovers, and planetary-mission spacesrafts.[9] In the following section we review the iodine photodissociation laser development in particular.

SOLAR-PUMPED IODINE PHOTODISSOCIATION LASERS

Direct solar-pumped laser systems to be operated in space environment demand special requirements on the system and the laser medium. The major requirements are: (1) the laser medium should be in a gas or liquid phase to be amenable to cooling and recharging for high-power and continuous operation, (2) high temperature operation should be possible at high laser efficiency to reduce cooling radiator size, (3) chemical reversibility for laser medium renewal in space to alleviate frequent refueling, and (4) efficient use must be made of the solar spectrum to reach laser-threshold excitation power below the solar concentration limit.

Among the various laser media evaluated, the iodine photodissociation laser (IPL) with perfluoroalkyliodes was found to be the best qualified with respect to the above-mentioned criteria. The work on the IPL at NASA-Langley was aimed to prove and validate the concept and to establish database needed for scaling up the output power of the IPL system. The laboratory laser system used for the purpose was consisted of a 40-kW (optical) solar-simulating arc lamp, an elliptical reflector/concentrator, a laser medium circulator loop, and a laser cavity. The laser cavity had a set of annular quartz tubes for laser medium and cooling water circulation. The IPL is pumped by the near UV (220 – 320nm) band of the solar spectrum to produce near IR (1.315 micrometer) laser output. The progressive experimental results are shown in Table 1.

In addition the IPL could easily be scaled up by forming a multi-module assembly as reported in the conceptual design for a 50-kW module in 1997.[10] This design work was based on a master oscillator-power amplifier (MOPA) system and has concluded that the 50-kW direct solar-pumped IPL system can be a simple and achievable concept and the components of the system are readily available. The launch weight estimate for the design system was 10,725 kg that gives the specific laser power of 4.7 W/kg. The technology of cooling the system in space was also elaborated in this report

Table 1. Performance of experimental iodine photodissociation lasers

Year	Iodide	Solar concentration	Laser power		Reference
1981	n-C3F7I	10,000	4	Watts	Lee and Weaver (1981)
1986	n-C3F7I	1300	10		Lee, Lee and Weaver (1986)
1989	t-C4F9I	1000	14		Lee et al. (1989
1990	n-C3F7I	1000	20		From Lab note
1992	i-C3F7I	1300	30		Tabibi et al. (1992)
1993	t-C4F9I	1000	48		Tabibi et al. (1993)

SOLAR LASER APPLICATION TO BEAMED ENERGY PROPULSION

Microwave and laser beam energy propulsion is already well conceived and a few flight-tests are reported. The lightcraft and lightsails program[12] is an example. However the microwave and laser transmitters used are earth-based and tests are limited yet to near the earth surface. The proposed concept, if realized, will revolutionize the space mission planning and spacecraft design because the central power plant, microwave or laser transmitter, will be space borne and readily available on demand from multiple spacecrafts. After the placement of a space-borne power plant the mission-assigned spacecraft, unlike current ones, need not to carry heavy fuel for energy. This will reduce the major payload currently taken up by the fuel and fuel tanks or large solar panels. In particular, the electric propulsion system has the power supply as the major penalty for the system efficiency. This penalty could be significantly reduced by the new concept. Also the high power density of laser beam makes it possible to reduce by more than an order of magnitude the area of the photovoltaic receiver panel required by the spacecraft for generating the same electric power that otherwise obtains by directly converting the solar energy.

The solar-pumped IPL can be Q-switched for producing a train of high-power laser pulses. This mode of operation is suitable for pulsed plasma propulsion system. The Q-switching of IPL is easily accomplished as reported in Ref.11.

SUMMARY AND CONCLUSION

A review of space power beaming effort of past two decades at NASA-Langley and Hampton University is made and the concept of direct solar-pumped laser beam generation in space is presented. In particular, atomic iodine photodissociation lasers with perfluoroiodides and a large-scale laser diode amplifier array are recommended for establishing a central power station in space for reducing the power generating payload from mission assigned spacecrafts.

ACKNOWLEDGMENTS

This work is supported in part by NASA Grant NCC-251 offered to the Research Center for Optical Physics of Hampton University.

REFERENCES

1. Lee, J. H., and Weaver, W. R., A Solar Simulator-Pumped Atomic Iodine Laser. Appl. Phys. Lett. 39, 137 (1981).
2. Lee, J. H., Wilson, J. W., Enderson, T., Humes, D. H., Weaver, W. R., and Tabibi, B. M., Threshold Kinetic Processes of t-C$_4$F$_9$I. Optics Communications, 53, 367 (1985).
3. Lee, J. H., Weaver, W. R., Tabibi, B. M., Perfluorobutyl Iodides as Gain Media for a Solar-Pumped Laser Amplifier. Opt. Comm., 67, 435 (1988).
4. Lee, J. H., Kim, K. C., and Kim, K. H., Threshold Pump Power of a Solar-Pumped Dye Laser. Appl. Phys. Lett. 53, 2021 (1988).
5. Lee, J. H., Tabibi, B. M., Humes, D. H., and Weaver, W. R., Perfluoro-n-Hexyl Iodide as Gain Media for High Power, Continuous Solar-Pumped Lasers. Optics Communications, 74, 380 (1990).
6. Kwon, J. H., Lee, J. H., and Williams, M. D., Far-field Pattern of a Coherently Combined Beam From Large- scale Laser Diode Arrays. Journal of Applied Physics, 69, 1177 (1991).
7. Hwang, I. H., and Lee, Ja. H., Efficiency and Threshold Pump Intensity of CW Solar-pumped Solid-state Lasers. IEEE J. of Quantun Electronics, E 27, 2129 (1991).
8. Tabibi, B. M., Terrell, C. H., Lee, J. H., and Miner, G. A., CW Iodine Laser Performance of t-C$_4$F$_9$I Under Closely Simulated Air-Mass-Zero Solar Pumping. Optics Communications, 109, 86 (1994).
9. De Young, R. J., Williams, M. D., Walker, G. H., Schuster, G. L., and Lee, J. H., A Lunar Rover Powered by an Orbiting Laser Diode Array. Space Power, 10, 103 (1991).
10. Choi, S.H., Lee, J.H., Meador, W.E., and Conway, E.J., A 50-kW Module Power Station of Directly Solar-Pumped Iodine Laser. J. Solar Energy Engineering, 119, 304 (1997).
11. Lee, J. H., Tabibi, B. M., Humes, D. H., and Weaver, W. R., High-Power Continuously Solar Pumped and Q-Switched Iodine Laser. Paper WP-8. 3rd International Laser Science Conference ILS-III, Atlantic City, NJ, Nov. 1-5, 1987. AIP Conf. Proc. 172, 109 (1988).
12. Wang, S., Cheng, Y.-S., Liu, J., Myrabo, L.N., and Mead Jr., F.B. "Performance Modeling of an Experimental Laser Propelled Lightcraft," AIAA 2000-2347, 31st AIAA Plasmadynamics and Lasers Conference, Denver, CO, 19-22 June 2000.

611

High-Energy Pulse-Repetitive CO$_2$-laser for Lightcraft Experiments

Anatoly V. Rodin, Valery G. Naumov, Anatoly F. Nastoyashchii and
Vladimir M. Shashkov

Troitsk Institute for Innovation and Fusion Research
Troitsk, Moscow Region, Russia 142190
Phone: (095) 334-5162, Fax: (095) 334-5776, E-mail: rodin@triniti.ru

Abstract. Result of experimental investigations of e-beam sustained high-repetition rate CO$_2$-laser are presented. This type of laser sources is known to have unique capability of high average power beam production (in our case it is in the range from several hundred Watts up to one hundred kiloWatts). Typically, pulse duration is shoter than the pause between successive pulses by the factor of a few thousand. Thus, the pulse pick power is significant higher than the average output power.

The general characteristics of this laser are given. The possibility of this laser used for acceleration of various targets in the technology "Lightcraft" is discussed.

INTRODUCTION

The e-beam-sustained discharge pumped lasers are actively investigated for the last 25 years. High-energy pulse-repetitive CO$_2$-lasers have been developed and experimentally tested. The investigations have been performed to develop various perspective areas of application of these lasers. They may be rather efficient in the technology "Lightcraft" too. This technology is based on laser ablation of materials. As it was reported on the conference Lasers'98 (Tucson, Arizona, USA) in the paper of Prof. Leik N.Myrabo et al "Flight and ground tests of laser-boosted vehicle" in 1998 in USA the experiments were performed on an acceleration of 30 g Al target to the height of 60 m by radiation of pulsed CO$_2$-laser with average power of 10 kW.

This paper gives characteristics of high repetition rate CO$_2$ lasers with power up to several tens kilowatts, that can be used for accelerating of more heavier targets.

HIGH POWER ATMOSPHERIC PRESSURE CO$_2$-LASER

In this paragraph the high-energy atmospheric pressure e-beam sustained CO2-laser is described. The results of some research works that have been carried out using this set-up are given also [1-4].

In a fig. 1 the scheme of this laser is shown. It is the laser with an open gas cycle. The gases CO$_2$, nitrogen and helium are stored in high-pressure vessels. In a special preliminary chamber the gas mixture of a given composition is prepared. It is usually 1:6:3. Here the homogeneous gas flow is formed. Further this mixture is blown

CP664, Beamed Energy Propulsion: First International Symposium on Beamed Energy Propulsion,
edited by A. V. Pakhomov
© 2003 American Institute of Physics 0-7354-0126-8/03/$20.00

through the discharge chamber and is exhausted into the atmosphere. The laser operates in a repetitively pulsed mode.

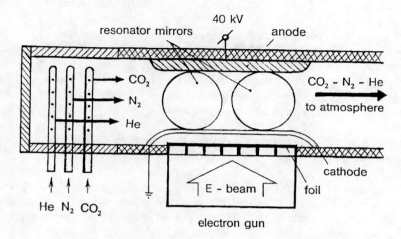

FIGURE 1. The scheme of high-power atmospheric pressure CO_2 laser

In the discharge chamber between the anode and cathode the non self-sustained discharge is supported. The electron beam is formed in the electron gun. As we said, this laser operates at atmospheric pressure. Therefore in the discharge chamber there is only anode and cathode. The discharge chamber has no walls. The resonator of the laser is arranged on the discharge chamber. The laser does not have special-purpose output window, through which the laser beam is usually injected into the atmosphere.

Some parameters of the laser are shown in the table 1. The length of a discharge zone along an optical axis of the laser is 80 cm, discharge gap is 8 cm. Gas flow rate is 6 kg/s. Usually we used unstable resonator with four mirrors of the rectangular shape. This resonator included a maximum possible volume of active medium and produced maximum laser output.

TABLE 1. Characteristics of atmospheric pressure repetitively-pulsed e-beam CO_2 laser

Parameter	Gas mixture	
	CO_2-N_2-He 1 : 6 : 3	CO_2-N_2-H_2O 1 : 9 : 0.06
Output radiation power, kW	100	80
Electrooptical efficiency, %	13	10
Radiation divergence, 10^{-3} rad	0.5 - 1	0.5 - 1
Pulse duration, μs	20 - 100	20 - 60
Pulse repetition rate, Hz	200	200
Energy per pulse, J	500	400

In such laser scheme there is a problem of geometric adjustment of a discharge volume with a resonator volume caused by scattering of electrons by a foil of an e -

gun output window. It was shown in our measurements, that if the relation of the dimension of e - gun window along a gas flow to an anode-cathode gap a is 1:1, in a resonator volume is located only 43 % of energy input in discharge. If this relation is 2:1 it makes 54 %.

Some methods of compensation of this effect were considered that allow increasing a part of energy located in a resonator volume. First of all we tried to use a strong magnetic field to change trajectory of electrons in a beam and to concentrate a majority of a beam in a resonator volume. This idea was realised by the installation of strong magnets on the outer side of the anode. It was obtained, that the part of the discharge energy concentrated in a resonator volume can be increased by 10 - 20 %.

Further an attempt was undertaken to use processes in discharge plasma for the same purpose. The physical idea is as follows. If the loss of electrons in plasma is controlled by a recombination, the concentration of electrons depends on the ionization rate q as the square root of q. If the loss of electrons is controlled by an attachment, the concentration of electrons varies in direct proportion to q. As a result at presence of an attachment the concentration of electrons decreases apart from an axis of a beam much faster. In this case the central part of discharge should contain more energy. This idea was realized by adding some oxygen to a mixture $CO_2:N_2:He$ = 1:6:3. Certainly oxygen addition results in decreasing of a total discharge current according to the increase of loss of electrons due to attachment. So there is the optimum concentration of oxygen. In experiments it was found to be about 0.5 %, that has allowed to increase efficiency of the laser approximately by 10 - 15 %.

One more way of localization of discharge in a resonator volume is the use of resistive anode [2]. Resistive anode was made as flat anode divided on strips, between which there was a dielectric. The individual ballast resistors have been connected to every metal strip. The profile of ballast resistors was chosen so that the potential of the anode smoothly decreased from the center to an edge. As a result we can localized more energy in the central region of the discharge. At e-gun window width of 16 cm the part of input energy in the resonator volume was increased from 54 up to 78 %.

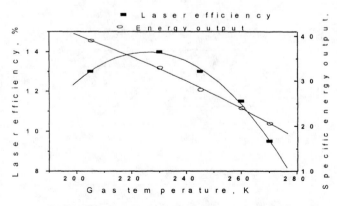

FIGURE 2. Characteristics of pulsed CO_2 laser at low temperature

It is known that such characteristics of CO_2 laser as efficiency and output energy can be improved by means of decreasing initial gas temperature. This is illustrated by fig. 2 in which the initial gas temperature dependencies of specific energy output and laser efficiency are shown. The special heat exchanger was used for cooling the gas down to required temperature before entering the discharge chamber. The reduction of temperature from 280 K down to 220 K results in increase of specific energy output and also of the laser efficiency.

These data are obtained at specific energy input equals to 0.2 J/cm^3 amaga. At the increase of specific energy input up to 0.3 J/cm^3 amaga specific energy output increases up to 60 J/litter and the laser efficiency up to 17% at temperature 220 K can be obtained. If the initial gas temperature was reduced from 200 K down to 160 K a decrease of laser efficiency was observed. It is because of the voltage of the discharge power supply system was limited to the value of 40 kV and at temperatures below 220 K the E/N value was less than optimum one. At temperature lower then 160 K laser generation was absent. In this case CO_2 is frozen in the heat exchanger.

THE DIVERGENCE OF LASER BEAM AND ITS VARIATION WITH DIFFERENT PARAMETERS

One of the principal field of our activity is directed to improving radiation quality of high-power e-beam sustained discharge pumped CO_2-lasers. We have carried out a number of experimental investigations to find out how output beam divergence depends on pulse duration, pulse repetition rate, flow velocity, gas mixture composition, method of flow arrangement. Besides, near- and far-field energy distributions have been obtained at various values of pulse duration and flow velocity. Spectrum of laser radiation has been investigated as well. These experimental data enables us to gain enough clear knowledge about spatial and temporal structure of laser radiation.

The flow velocity of gas mixture, consisting of carbon dioxide, nitrogen and helium could vary from 10 to 250 m/s, electron gun current density on passing through the foil - from 0.1 to 3 mA/cm^2; the laser pulse duration - from 10 to 200 μs. The unstable confocal resonator was used in these experiments; the active medium length along a resonator axis was about 2 m.

At the first stage of the experimental study of medium influence on the radiation divergence the laser pulse duration was less than 10 μs at 5kW/cm^3 input power per volume unit. In this case the optical inhomogenities of active medium caused by the laser beam (so called laser induced medium perturbations) should be negligible. Measurements of the divergence have been made by focal spot method. Near- and far-field energy distributions were determined using the infra-red photography technique. The intensity level of $0.2 \cdot I_{max}$, where I_{max} is maximum intensity in focal spot, was chosen as the criterion defining output beam divergence Θ.

Results of measurements Θ at various gas velocities plotted in Fig. 3 have shown that at flow velocities lower than 100 m/s the beam divergence is approximately twice as large as diffraction limit. The beam divergence increases by nearly 5 times in the range of flow velocities from 100 to 250 m/s.

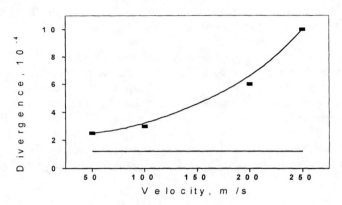

FIGURE 3. The single pulse laser beam divergence as a function of gas flow velocity.

In addition, the velocity dependence of the fraction of the output energy approaching a remote target has been obtained. These results show that this value decreases by a factor 2 with an increase of flow velocity from 100 to 250 m/s that is likely to demonstrate the enhancement of radiation scattering due to gas density fluctuations. Besides, near-field energy distributions at 1-meter distance from resonator mirror have been obtained at various gas velocities. These distributions appear either as an array of stripes extended along gas flow direction or has a spotted structure with typical dimension of spots ~ 5mm. The modulation depth of radiation intensity is close to 1. It should be noted that there is a correlation between the near- and far- field energy distributions. On the basis of present experimental data it may be concluded that one of cause being responsible for such near- and, correspondingly, far-field radiation structure is related to optical inhomogenities generated by initial gas flow.

As it is well known the other cause of deterioration of laser beam quality is a growth of a small-scale optical inhomogenities arising from, so-called, thermal action of radiation upon CO_2-laser active medium. Data of measurements (table 2) show that extending of pulse duration from 10 to 40 μs results in increase of laser beam divergence in 2.5-4 times depending on gas mixture composition. In these experiments the input power per volume unit was constant, about 5 kW/cm^3, and flow velocities were less than 100 m/s. When this CO_2-laser operated at pulse repetition frequency about 100 Hz the output beam divergence of individual pulses in this case differ little from single pulse operation one and observed in experiments a total (in time of operation) increase of divergence is associated with fluctuations of laser beam direction from pulse to pulse.

From the above it might be assumed that the study of optical inhomogeneities both inherent in initial gas flow and arising during the laser generation is the important task. We have carried out such investigations using Twiman-Green interferometer,

Talbot interferometer, thermo-anemometer method which enables us to obtain the information about the large- and small-scale medium perturbations, fluctuations of speed and temperature, to determine characteristic spatial scale of such fluctuations and to estimate their influence on the laser beam divergence.

However the most useful method, in our opinion, is the determination of small-scale inhomogeneities of turbulent gas flow on the basis of the analysis of average far-field energy distribution of probe Cu-laser radiation on passing through medium to be investigated. In this case Cu-laser radiation of 20 ns pulse duration and initial near diffraction limit divergence pass through gas flow and then focuses with use of 10-meter focal length optical system. The mathematical procedure has been elaborated to determine a correlation function of dielectric permittivity. The knowledge of this function enables with some degree of certainty to evaluate contribution of small-scale optical inhomogeneities in laser beam divergence. Moreover, the use of this method in combination with the others, for example, with anemometry makes possible to determine the correlation length of fluctuation gas density not only across propagation probe laser beam but along it as well.

TABLE 2. Divergence of radiation of atmospheric pressure repetitively-pulsed e-beam CO_2 laser .

Diffraction limit of level $0.2\ I_{max}$	$1.2 \times 10^{-4}\ (\theta_{dif})$
Initial gas flow ($\tau_p < 10\ \mu s$)	
a) V = 50 m/s	$2.5 \times 10^{-4}\ (2\ \theta_{dif})$
b) V = 100 m/s	$3.0 \times 10^{-4}\ (2.5\ \theta_{dif})$
c) V = 200 m/s	$6.0 \times 10^{-4}\ (5\ \theta_{dif})$
d) V = 250 m/s	$1.0 \times 10^{-3}\ (8\ \theta_{dif})$
Single pulse	
$\tau_p = 40\ \mu s$, V = 50 m/s	$(6 \div 8) \times 10^{-4}\ (5 \div 7\ \theta_{dif})$
Repetitively-pulsed mode	
$\tau_p = 40\ \mu s$, V = 50 m/s	
a) F = 50 Hz	$(7 \div 8) \times 10^{-4}\ (6 \div 7\ \theta_{dif})$
b) F = 100 Hz	$(7 \div 9) \times 10^{-4}\ (6 \div 7\ \theta_{dif})$

SCOPE OF EXPERIMENTS ON THE BASIS OF PULSE PERIODIC LASERS FOR PHOTOABLATION.

Theoretical background of problem and possibilities of experimetal testing applicable to acceleration of solid-state targets (pellets) were discussed in [5]. The main parameter in this problem is a recoil impulse, which is transmitting to accelerated target at the action of laser radiation. Recoil impulse depends upon the choice of pellet material, power density of laser radiation on the surface and geometrical factor (recoil impulse arising under univariate dilatation of evaporant is larger than under three-dimentional dilatation).

For recoil measurements the authors [5] proposed to apply the ballistic pendulum; the equation (1) for calculation of recoil impulse $\Delta m \cdot V_0$ by the angle of deflection of accelerated mass α was also presented there:

$$\Delta m\, V_0 = (m - \Delta m)\, \sqrt{gl} \,/\, \alpha, \tag{1}$$

where Δm is the value of lost pendulum mass;
$\quad\quad\quad V_0$ - velocity of vapours flowing out of the target;
$\quad\quad\quad m$ – pendulum mass;
$\quad\quad\quad l$ - length of the pendulum thread;
$\quad\quad\quad g$ - gravitational acceleration

In TRINITI there are pulse periodical CO_2-laser facilities with following parameters:

TABLE 3

Mean power, kW	Pulse duration, μs	Pulse repetition, Hz	Divergence, rad	Operation interval, s
30-50	150-200	Up to 100	$\leq 3 \times 10^{-4}$	Up to 300
Up to 100	20-100	Up to 400	$\leq 10^{-3}$	Up to 10

Experiments on irradiating plane targets on the pendulum with the length of 120 cm were carried out with the first installation [3] to measure recoil momentum. The target was placed 30 m far from the laser, the peak pulse power density on the target was $\leq 10^6$ W/cm^2, pulse repetition rate was 100 Hz, pulse duration was 150 μs. In the experiments the deviation amplitude of the pendulum have irradiated by pulse series was measured.

The results of experiments are presented in table 4.

TABLE 4

Target material	Target mass, g	Irradiation interval, s	Deviation amplitude, cm	Recoil momentum, g cm/s
Steel	650	10	2	3.8×10^3
Aluminum	235	5	3,5	2.4×10^3
Acrylic plastic	195	1	9	5×10^3

The last column of the table shows values of recoil momentum calculated according to equation (1), which the target obtained for pulse-periodical CO_2-laser with average power 30 kW.

Included laser beam divergence and lens distance to target L=30m it is possible to measure the value of focus spot ~1 cm and correspondingly the power density of radiation on the target ~ 1MW/cm^2, which is sufficient for realization of low-threshold optical breakdown near surface and intensive evaporation of the surface. Under the estimation the surface temperature for various materials ranged from 2000 to 4000 degrees.

The calculated values of recoil momentum were not large and therefore the irradiated target obtained rather low velocity ~ 6-25 cm/s.

The estimation shows that for the pellet mass 20 g from acrylic plastic applying laser radiation for the facility 2 (average power 100 kW) the pellet velocity about 1km/s in vacuum can be achieved for 100 s laser action.

Thus, the measured experimental values of recoil momentum show the possibility of pellet acceleration up to high velocities by laser pulse at the condition of solving the problem of laser beam propagation through atmosphere.

In the following experiments of pellet acceleration we are planning to carry out the measurements of evaporated mass and temperature of irradiated surface. This allows to make more accurate analysis of possibility of pellet acceleration by laser radiation and to find out the optimal regimes of acceleration.

CONCLUSION

Characteristics of high repetition rate CO_2 lasers with output power up to 100 kW are given. These lasers can be successfully used in experiments on accelerating of targets due to laser ablation.

REFERENCES

1. V.G.Vostrikov, V.G.Naumov, L.V.Shachkin. "On the effect of the specific pump power on the efficiency of the atmospheric-pressure electroionization CO_2-laser", *Sov. J. Quantum Electronics*, **9**, n.2, pp.413-415, 1982.
2. V.G.Vostrikov, A.I.Loboiko, et al. "Localization of non-self-sustained discharge in given volume with the help of resistive anode". *Sov. J. Tech. Phys.*, **57**, n.2, pp.268-272, 1987.
3. V.G.Naumov, A.V.Rodin, "Physics of low temperature plasma and CO_2-laser with average power up to 50 kW for industrial application", *Proc. Of Int. Conf. On Laser '94*, pp. 171-175, Quebec, Canada, 1994.
4. B.V.Bunkin, V.A.Glukhikh, G.Sh. Manukyan, V.D. Pis'mennyi, A.V.Rodin, V.V.Valuev, "High-power Fast-Flow CO_2-lasers", *Proc. Of XVII Int. Conf. "Laser Optics '93"*, SPIE, v. 2095, pp. 29-37, St. Petersburg, 1993.
5. A.F.Nastoyashchii, V.G.Naumov, A.V.Rodin, V.M.Shashkov. "Macroparticle acceleration in the regime of laser-driven rocket fraction and possibility to to conduct experiments using CO_2-laser". *Proc.of High-Power Laser Ablation III, Proc. of SPIE, v.4065, pp. 812-817, Santa Fe, 2000*

Nonlinear Optical Techniques of Laser Beam Control for Laser Propulsion Applications

Vladimir E. Sherstobitov, Aleksey A. Leshchev, and Leonid N. Soms

Institute for Laser Physics, 12 Birzhevaya line, 199034, St. Petersburg, Russia

Abstract. A brief overview of phase conjugation techniques of laser beam control in CO_2 and solid-state lasers in application to the task of laser propulsion of vehicles is presented. It is shown that nonlinear-optics methods of controlling the laser beam wavefront can be used as a very promising tool in solving the problem of laser power delivery to laser jet engine vehicles.

INTRODUCTION

The problem of using laser beams for generation of reactive thrust and propulsion of various objects has been discussed in the literature already for over more than 30 years. The interest in this area of research has increased substantially in the recent decade, which should be assigned to the increasing power of lasers, both currently available and in the design stage, as well as to the advent of projects aimed at solution of such problems of practical significance as launching nanosatellites into a geocentric orbit [1], development of a space system for transferring satellites from low geocentric to a geostationary orbit [2], construction of microairplanes for the purposes of environmental monitoring [3], etc.

In all these applications, attaining the necessary radiation intensity at the receiving collector of a vehicle with a laser jet engine (LJE) requires a high average beam power (from a few kW to 1 MW and more) and an extremely small angular divergence (10^{-6}-10^{-7} rad). It has to be added that the accuracy of beam pointing at the receiving collector and of tracking a moving vehicle should be an order of magnitude higher and reach in some cases 10^{-8} rad.

These parameters cannot be attained by traditional means. A promising way to solve this problem lies in application of nonlinear optics methods of laser beam distortion compensation, which are based primarily on the phase conjugation (PC) effect in nonlinear media.

The general scheme of PC-based distortion compensation, which involves double beam transit through the optical train producing distortions and an intermediate phase conjugation operation, is well known. Phase conjugation can be based on the use of Stimulated Brillouin Scattering (SBS), as well as of Four-Wave Mixing (FWM) in conventional nonlinear media and in media with a Brillouin-type nonlinearity (the so-called Brillouin-Enhanced FWM (BEFWM)). In some cases, application of dynamic

CP664, Beamed Energy Propulsion: First International Symposium on Beamed Energy Propulsion,
edited by A. V. Pakhomov

holography turns out to be useful in the correction of distortions. The present communication gives a brief overview of the studies made in the recent decade in this area at the Institute of Laser Physics, St. Petersburg, Russia (see also [4]).

GENERAL CONCEPT OF PC APPLICATION IN LASER PROPULSION

The large variety of presently available wavefront conjugation means offers a possibility of carrying out dynamic compensation of wavefront distortions for pulsed and repetitively-pulsed lasers within a broad range of wavelengths (from the UV to 10.6 μm inclusive), and of pulse durations (from the ns scale to cw).

Two slightly different conceptual schemes of a PC-based solution of this problem for the task of laser propulsion are presented in Fig. 1. In this scheme, a corner reflector mounted on the receiving collector of the LJE vehicle reflects part of the radiation delivered to the vehicle backward to form a probe beam, which accumulates the distortions in propagating along the optical path (the atmosphere, telescope, laser amplifier) and is fed into the PC mirror. The distortions acquired in the forward transit will be compensated by the PC mirror in the backward pass through the amplifier, telescope, and the atmosphere, because the difference between the distortions in the forward and backward transits through the atmosphere, i.e., during the time from the moment the probe beam enters the atmosphere to its exit from it (~20 μs) may be considered insignificant. As a result the cross section of the beam propagating outside the atmosphere toward the corner reflector will be close to $\lambda L_p/D_T$, in the collector plane, where λ – is the wavelength, L_p is the distance from the telescope to the vehicle,

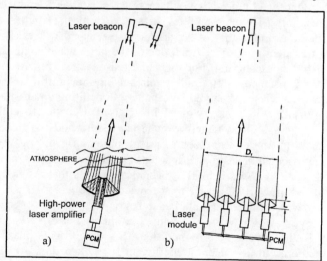

FIGURE 1. Phase conjugation concepts of laser beam delivery to a vehicle with LJE onboard a) for a single telescope transmitter and b) for a multi-channel telescope array.

and D_T is the diameter of the transmitting telescope (or telescope array). Obviously enough, because of the laser and the corner reflector moving with respect to one another, and of the velocity of light being finite, the phase conjugated beam may miss the vehicle collector. However, if the PC operation includes tilting the beam by an angle corresponding to the angular displacement of the vehicle with respect to the telescope axis during the time taken by light to travel from the vehicle to the transmitting telescope (the lead angle), and if within this angular displacement the distortions introduced by the atmosphere may be considered the same (i.e., if the lead angle is smaller than the isoplanatism angle), then the distortions the beam accumulates in the propagation train will be corrected, as before, when the beam travels back. The laser energy will be efficiently delivered to the lead point, which the vehicle collector will reach at the arrival of the beam, i.e., the vehicle will be tracked with the accuracy determined only by that of entering the lead angle. To achieve the distortion compensation and beam self-pointing, the field-of-view (FOV) of the system consisting of the telescope, the amplifier, and the PC mirror must exceed the lead angle, as well as the angles of deflection of the rays generated by wavefront distortions in the propagation train. It also appears obvious that the corner reflector size, the transmitting telescope diameter and the amplifier gain should be matched so as to provide an efficient extraction of energy stored in the amplifier.

PC COMPENSATION FOR DISTORTIONS IN CO_2 LASERS AND THE BEAM PROPAGATION TRAIN

High-power repetitively-pulsed CO_2 lasers have a considerable application potential for launching LJE nanosatellites in space from the Earth's surface [1], because their radiation, unlike that of solid-state lasers, is fairly weakly distorted in transit through a turbulent atmosphere and, due to the large wavelength, is practically not accompanied by backscattering. For the same reason, however, reaching a small beam divergence, for instance, of 10^{-6} rad, requires using transmitting telescopes 10 m or more in size. Such a telescope should be capable of operating under substantial heat loads and component vibrations, which are caused by the repetitively-pulsed operational regime of a high average-power laser. If one adds to this the thermal blooming of the beam in the telescope optical train, as well as the wavefront distortions the beam accumulates in propagating through the turbulent atmosphere and the wake the LJE vehicle leaves behind, it becomes obvious that efficient delivery of radiation from a high-power CO_2 laser to the receiving collector of a flying vehicle cannot be achieved without the potential of adaptive optics.

More than 15 years ago it was shown in experiments performed at the ILP (see review [5]) that FWM in isotopic ally modified $^{34}SF_6$ can provide phase conjugation of CO_2 laser radiation with a reflection coefficient R_{pcm} exceeding 100% and a high conjugation fidelity (Fig.2).

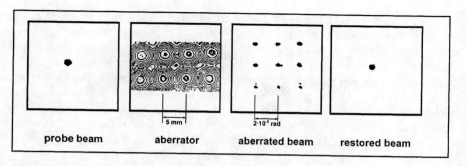

probe beam   aberrator   aberrated beam   restored beam

FIGURE 2. An example of PC correction of a CO_2 laser beam after two transits through the same aberrator.

To explore the possibility of achieving nonlinear-optics compensation of distortions in the train of a high-power CO_2 laser, engineers and physicists of the ILP, in cooperation with other Russian companies, used in 1990 a high-power E-beam-sustained gas-flow CO_2 laser producing an output beam 200 mm in diameter to perform large-scale studies of a double-pass amplifier with a PC mirror (MOPA-PCM) [6] (Fig. 3). The laser operated at a pulse repetition rate of up to 100 Hz with pulses 15 μs long. The three-stage amplifier included a relay optics system to provide a large FOV. The PC gas-flow cell with $^{34}SF_6$, pumped by two beams produced by the master oscillator, provided a phase-conjugate beam with an average power of up to 100 W. The gain per pass was $\sim 10^3$ for the phase conjugate output of 1.2 kJ. The maximum average power attained in the experiments was as high as 40 kW, with the beam divergence at the 0.8 energy level only twice the diffraction limit. As far as we know, these figures are the highest of the results reached thus far in similar experiments anywhere in the world.

The scheme in Fig. 1a, is presented mainly to illustrate the concept of PC-based beam self-pointing and distortion compensation. In actual fact, the scheme presented in

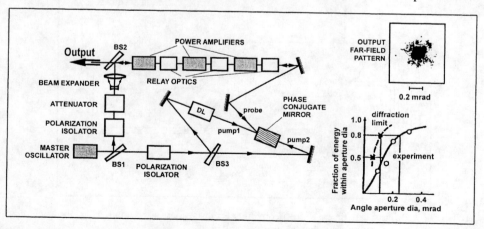

FIGURE 3. Block-diagram of the experiment on PC correction for distortions in a high-power CO_2 laser amplifier (see text).

Fig. 1b, in which one telescope is replaced by an array of smaller telescopes, is much more convenient from a practical point of view. This scheme allows straightforward laser power up-scaling by adding channels. Phase conjugation provides their coherent summation with correction of mutual angular and phase mismatch among the individual channels and compensation of the phase distortions occurring in each train. A multi-component telescope may have a substantially smaller length than a single telescope of the same diameter, which reduces considerably the mass of the telescope and facilitates the problem of its angular displacement when tracking an LJE vehicle. This cuts also markedly the cost of the telescope as well; indeed, it is dominated by that of its primary, which grows rapidly with the increase in the mirror size.

To determine the technical constraints associated with construction of a system depicted in Fig. 1b, we used a repetitively pulsed CO_2 laser in model experiments on PC correction of distortions in a telescopic system consisting of four telescopes of a square cross section (Fig. 4), each including a primary mirror $H/2 = 12$ cm in size, so that the total beam size at the exit from the transmitting telescope was 24 x 24 cm for a telescope length of 30 cm and magnification $M = 5$. In the course of the experiment [7], the beam from a point source, which was fixed on a mount that could be translated in the transverse direction and imitated a distant moving laser beacon, was intercepted by a telescope array and directed on an SF_6-based FWM PC mirror, as shown in the Figure. The spot of the beam delivered back to the point source was recorded by a Pyrocam-1 pyroelectric matrix and displayed on a Spiricon LBA monitor. By properly tilting one of the PC mirror pumps, one could turn the output beam to imitate introducing a lead angle in the course of tracking. The angular velocity of point source displacement was $3 \cdot 10^{-5}$ rad/s.

As shown by the experiments, when a PC mirror was replaced by a conventional one, the radiation pattern in the point source plane was a number of shapeless spots close in intensity to the pyroelectric matrix noise level (Fig. 5a). This radiation

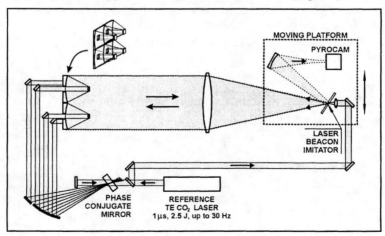

FIGURE 4. CO_2 laser experiment on PC correction for distortions in the multi-channel transmitting telescope and beam position control in the beacon plane.

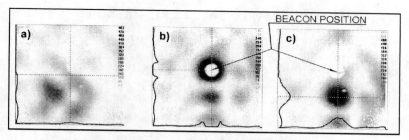

FIGURE 5. Far-field patterns of the beam in the beacon plane
  a) without PCM (PCM is replaced by a conventional mirror)
  b) with PCM (collinear pump waves)
  c) with PCM (one pump wave is tilted).

distribution pattern is due to the poor optical quality of the telescope mirrors, as well as to uncontrollable tilting of their axes and mutual beam phase mismatch (optical path difference between the individual channels). When the PC mirror was introduced, the beam acquired a distinct core $5 \cdot 10^{-5}$ rad wide, which was only 10% in excess of the diffraction-limited width of a beam with H = 24 cm (Fig. 5b). This implied efficient correction of distortions in the propagation train and phase matching of the radiation produced in individual channels. The radiation spot had the shape of a ring around the point source aperture (the bright disc in Fig. 5b), which is an evidence of the precise hitting the point source.

As anticipated (see [8]), when one of the pump waves was tilted by an angle, the phase conjugate beam turned through the same angle (see Fig. 5c). In our experiment, the maximum tilt was determined by the PC mirror parameters to be $\pm 13\lambda/H$. This angle agrees in order of magnitude with the lead angle, which has to be introduced into a CO_2 laser beam at the output of a telescope two meters in size when delivering power to an LJE vehicle moving in a low geocentric orbit.

As the point source moved within an angular range $\pm 13\lambda/H$, it was tracked with a subdiffraction-limited accuracy (the rms deflection was $\sim 0.12\lambda/H$ for an angular displacement velocity of $\sim 3 \cdot 10^{-5}$ rad/s.

Thus, we have shown that application of PC can provide compensation of distortions in the optical train of a CO_2 amplifier with a beam expander, including a multi-component beam shaping system consisting of several telescopes. Estimates show that, for instance, in the case of a telescope made up of four single square telescopes, 1 x 1 m in size, and a corner reflector ~ 10 cm in diameter, which is mounted on an LJE vehicle collector having a diameter of 1 m, the CO_2 amplifier saturated gain sufficient to launch the vehicle to an altitude of 200 km should be close to 10^3. In the above experiments performed with the double-pass CO_2 amplifier and the PC mirror [6], the gain was $\sim 10^3$ for a beam pulse energy of ~ 1.2 kJ. Assuming 1 kJ output at a 100-Hz repetition rate and a coupling coefficient $C_m = 400 \dfrac{N \cdot s.}{MJ}$, a realistic figure, the engine thrust T attainable with such a laser is ~ 40 N.

SOLID-STATE LASER EXPERIMENTS WITH

A HIGH-SENSITIVE PCM

Straightforward evaluation shows that, other conditions being equal (i.e., for the same corner reflector dimensions, the same telescope size, and equal atmospheric transmission coefficients), the gain required from the amplifier, is proportional to $\lambda^2 L_P^2 R_{PCM}^{1/2}$. Thus, for the same gain, the distance L_P at which PC compensation of distortions is efficient is proportional to $\lambda^{-1} R_{PCM}^{1/4}$. Whence it follows that when one goes over from the 10-micron wavelength region to $\lambda \sim 1$ μm, at which BEFWM-based PC can be attained with a reflection coefficient $R_{PCM} \approx 10^5 \text{-} 10^6$ [9], the gain in energy transport distance may be increased by about a factor 300, in other words, the region of applicability of the PC correction method may be extended to include missions involving vehicle transfer to a geostationary orbit.

A key problem encountered in development of such concepts is the design of BEFWM-based PC mirrors capable of operating within a large FOV $\sim$ 100 DL, where DL stands for the diffraction-limited beam divergence. In the recent years, experiments supporting the possibility of constructing such PC mirrors have been performed at the ILP. A simplified diagram of the BEFWM used in the experiments, as well as the four-wave mixing scheme employed are displayed in Fig. 6. The four-wave interaction of a probe beam of frequency ω_s and a pump of frequency ω_1 took place in a BEFWM amplifier cell filled by a nonlinear medium $TiCl_4$, where the above beams produced a

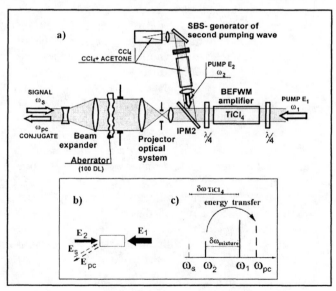

FIGURE 6. The simplified optical arrangement of BEFWM PCM used in experiments(a), four wave mixing geometry (b), and spectrum of interacting waves (c).

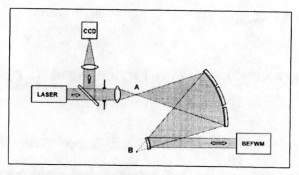

FIGURE 7. Demo experiment on double-pass PC correction for distortions in a segmented primary mirror telescope via BEFWM PCM.

refractive index grating moving with a hypersonic velocity. The second pump was formed due to SBS in a cell containing a mixture of CCl_4 with acetone and had a frequency ω_2, which was Stokes-shifted relative to ω_1 (see Fig. 6). This beam scattered from the moving hypersonic grating to generate a conjugate beam of frequency ω_{PC}. The conditions of PC mirror operation at a large angle of view (up to 100 DL) were simulated by placing an aberrator at the entrance to the PC mirror. The results of experiments showed the possibility of achieving reflection coefficients in excess of $R_{PCM} \approx 10^5$ for FOV~100 DL. The sensitivity of the PC mirror reached in these conditions was 10^{-10}-10^{-9} J.

This BEFWM was used in demo experiments on double-transit PC distortion compensation for a telescope with a segmented primary mirror 400 mm in diameter (Fig. 7). As seen from the interferogram (see Fig. 8), the mirror segments were of a fairly poor optical quality and were tilted with respect to one another by various angles. No special measures were undertaken to eliminate segment piston shifts in the direction perpendicular to the mirror surface. A comparison of the focal spot of the original laser beam at point A (Fig. 8, left) with the spot obtained after reflection from

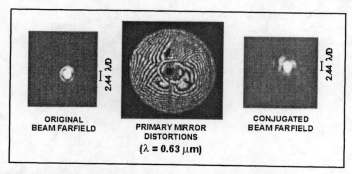

FIGURE 8. The interferogram of the primary mirror (middle) and the focal spots at point A (see Fig. 7) for the original laser beam (left) and for a PC restored beam (right).

627

the segmented primary mirror, phase conjugation, and repeated reflection from the same mirror (Fig.8, right) shows that the original beam quality can be restored with a practically diffraction-limited accuracy by using a high-sensitive BEFWM PCM.

FROZEN PC IN A CONCEPT OF A HIGH-POWER MOPA-PCM SYSTEM

BEFWM PC mirrors for distortion correction in solid-state lasers can be employed to advantage only in the ns-range of pulse durations. At the same time, literature dealing with application of lasers to LJE development discusses the possibility of using quasi-cw radiation, as well as the ps-range of pulse length. To broaden the pulse length range within which BEFWM PC mirrors are capable of realizing wavefront conjugation of ultra-weak signals with their simultaneous amplification in reflection, the concept of the so-called "frozen" PC was developed at the ILP (Fig. 9) [10].

The method consists essentially in that following phase conjugation by a BEFWM PC mirror and simultaneous amplification by 10^5-10^6 times, the ns-range pulses arriving with a certain period from the laser beacon are used to record a dynamic hologram in a holographic corrector (HC), which represents actually a liquid-crystal optically-addressed spatial light modulator (LC OA SLM). Between the record pulses, the hologram is read out by a beam of another laser having the required pulse length, and the diffracted radiation is fed into the amplifier. As a result, a phase-conjugate beam of desired power, with pulse duration variable from a few picoseconds to quasi-cw, can be obtained at the laser amplifier output.

A conceptual scheme of a 300-kW quasi-cw laser operating in combination with a laser beacon as a double-pass amplifier with a PC mirror, which implements the frozen-PC concept, is presented in Fig. 10 [7].

The BEFWM mirror pumps that should be coherent with the radiation arriving from the laser beacon and the reference beam required to write a dynamic hologram are generated in an additional receiving channel with a small telescope, in which the frequency of auxiliary master oscillators MO1 and MO2 is locked by heterodyne techniques to that of the laser beacon radiation. It is assumed that a signal of energy

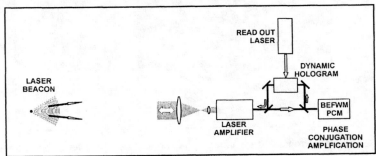

FIGURE 9. "Frozen" PC concept based on BEFWM PCM and a holographic corrector (HC).

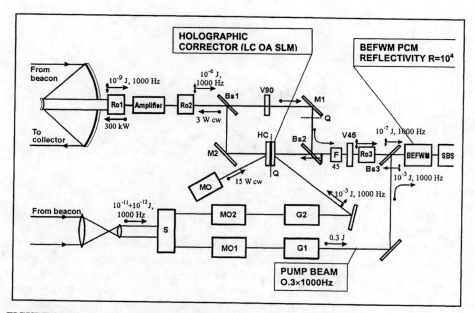

FIGURE 10. The concept of a laser-transmitter with PC correction for distortions via "frozen" PC.

10^{-11}-10^{-12} J arrives from the laser beacon at a repetition rate of 1 kHz to the input of the additional channel, and a signal of energy 10^{-9} J is fed into the input of the main telescope channel. The dynamic hologram produced in the holographic corrector HC is read out by means of a cw reference laser MO, ~15 W in power. As a result, the amplifier input should receive a quasi-cw signal ~3 W in power, with the wavefront phase conjugate to that of the laser beacon radiation. After amplification with a gain ~10^5, the 300-kW output beam, in which the distortions present in the propagation train are already compensated, is directed onto the collector of the vehicle. The required lead angle is introduced by tilting properly the beam of the master oscillator MO. To increase the FOV of the PC mirror amplifier train, relay optics RO1, RO2, and RO3 are included in the train.

PC CORRECTION WITH A LOCAL BEACON

Thus far, we have been considering the case of a ground-based transmitting laser. In the applications which would require placing a laser in space or upper atmosphere layers (onboard a plane or airship), where the effect of turbulence on the efficiency of beam transport delivery to the LJE vehicle collector may be neglected, the part played by the PC mirror reduces to correcting the distortions inside the transmitting system. One can now reject mounting on the receiving collector a laser beacon radiating a probe signal of a fairly high power at the operating wavelength and replace it by any appropriate low-power laser beacon mounted on the collector and a reference laser located inside the transmitting system and serving to produce pump waves and the

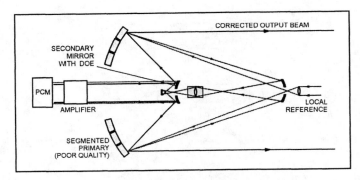

FIGURE 11. Segmented beam-forming telescope TENOCOM with PC correction for distortions.

probe beam. In this case, problems associated with the coherence of the probe beam and of the PC mirror pump waves no longer arise, and the stringent requirements on the amplifier gain are lifted, because the reference laser power is not lost, as in the preceding case, on the way from the beacon to the transmitting telescope. However, to compensate the distortions caused by vibration of the telescope primary mirror and the aberrations introduced by its components, one will have to develop special arrangements for injecting reference laser radiation into the optical train of the system.

During the recent decade, people at the ILP have designed and studied experimentally various versions of such arrangements, which provide correction of the distortions forming in both the amplifier and the beam expander [11-14]. Figure 11a displays one of these versions termed TENOCOM [13,14]. In this arrangement, the primary telescope mirror can be composed of individual segments, not necessarily phase matched, each of which can be light-weight and of a quality two orders of magnitude poorer than required for a traditional telescope. In this arrangement, despite mutual tilts of tens DL and piston shifts of the segments up to a hundred of wavelengths, the telescope produces a practically diffraction-limited radiation beam (Fig. 12). This is achieved by using at the secondary mirror of the telescope an annular diffraction structure, which enables double transit of the reference laser radiation through each point at the surface of the segmented primary and distortion compensation through phase conjugation in between the reflections. By properly

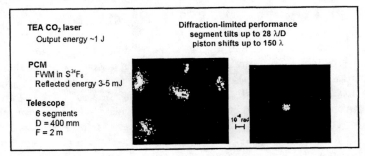

FIGURE 12. The far-field patterns of the beam at TENOCOM output without (left) and with (right) PCM

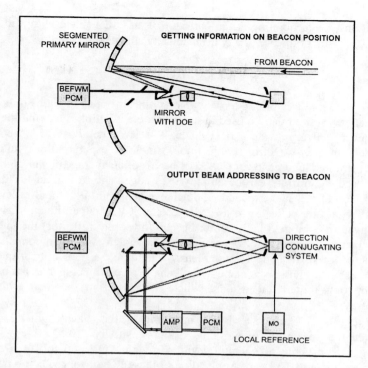

FIGURE 13. A possible PC arrangement for precision addressing of a high-power beam to a low-power beacon.

varying the reference laser beam tilt with respect to the telescope axis, one can control the direction of the exiting output beam within the FOV of the telescope-amplifier-PC mirror system and, in particular, introduce the lead angle into the output beam.

Figure 13 shows a possible PC arrangement [7] for precision addressing of a high-power beam to the collector of the vehicle. In this arrangement, a high-sensitivity BEFWM phase-conjugating mirror is employed to correct the distortions forming in the system in viewing a low-power laser beacon and determining its position in the telescope reference frame (Fig. 13, top). The distortions in the high-power beam accumulating as it passes through the optical train within the system (including the distortions in the amplifier and beam expander) are compensated by means of a conventional high-power PC mirror (Fig. 13, bottom). To match the focal spot produced by the reference laser (MO) beam with the image of the low-power laser beacon or, if required, to introduce a lead angle into the reference laser beam, one uses a direction conjugating system (see Fig. 13), which is based on an OA SLM controlled laser and operates by "write in-read out" algorithm [15].

The transmitting system provided by a beam expander arrangement illustrated by Fig. 13 is expected to be only weakly sensitive to vibrations, misalignments, and thermal deformations of components which do not drive the system out of its FOV. In these conditions, beam pointing at the collector of an LJE vehicle within the FOV is all-optical (i.e., without the use of any mechanically adjustable components).

CONCLUSION

We have shown in this brief overview that nonlinear-optics methods of controlling the laser beam wavefront can be used as a very promising tool in solving the problem of laser power delivery to an LJE vehicle. We have demonstrated the currently available techniques of wavefront conjugation, which are capable of providing a close to diffraction-limited divergence of CO_2 laser beams of an average power of 40 kW.

FWM-based PC mirrors can be employed to advantage to correct the distortions arising in a CO_2 laser beam expander and achieve self-pointing at and efficient beam energy delivery to the collector of an LJE vehicle. It was experimentally shown that phase conjugation combined with a laser beacon (or corner reflector) mounted on the vehicle permits one to use for power transport to a vehicle multi-component telescope arrays made up of small optical components, which have a poor optical quality and, hence, low cost. The spot in the plane of the vehicle receiving collector can be brought to the diffraction-limited dimensions, which correspond to the total width of the telescope array.

Experiments with solid-state lasers operating at the wavelength ~1 μm have demonstrated the possibility of building a high-sensitivity (10^{-9}--10^{-10} J) BEFWM-based PC mirror capable of operating within a wide FOV (up to 100 DL) with a reflection coefficient of 10^5--10^6. Application of these PC mirrors in combination with high-power micron-wavelength range lasers will make it possible to increase the power transport distance to LJE vehicles to tens of thousands of km.

In the cases where the transmission train is outside the atmosphere, or its influence may be neglected, PC compensation arrangements with an onboard reference laser appear to be advantageous. The TENOCOM developed at the ILP is one of such arrangements. Experimental data have been presented, which support the high potential of such arrangements for power delivery to space-borne LJE vehicles.

ACKNOWLEDGMENTS

Some results presented in the paper have been obtained in the scope of the ISTC Project #929. Authors are grateful to Kunihisa Eguchi and Hiiro Yugami, as well as to Takeshi Sakamoto, and Yurii Malakhov, for their collaboration and support.

REFERENCES

1. Myrabo, L.N., and Mead, F.B., Jr. "Ground and Flight Tests of a Laser Propelled Vehicle", AIAA 98-1001, Aerospace Sciences Meeting & Exhibit, 36th, Jan 12-15, 1998, pp.1-10.
2. Niino, M., et al., "Study and Research Results of Advanced Space Propulsion Investigation Committee (ASPIC), 48th International Astronautical Congress, 1997.

3. Yabe, T., Phipps, C., et al., "Laser-Driven Vehicles-from Inner-Space to Outer-Space, HPLA 2002, pp. 21-26, April 2002, Taos, N-M, USA, SPIE 4760.

4. Mak, A.A., Sherstobitov, V.E., Kuprenyuk, V.I., Leshchev, A.A., Soms, L.N., J. of Optical Technology, **65**, N 12, 52-61, 1998, (in Russian).

5. Sherstobitov, V.E., "Phase Conjugation of CO_2 Laser Radiation", International Journal of Nonlinear Optical Physics, **2**, N 3, pp. 465-482,1993

6. Sherstobitov, V.E., Ageichik, A.A., Bulaev, V.D., et al, "Phase conjugation in a high-power E-beam sustained CO_2 laser", SPIE 1841, Nonlinear Processes in Solids, pp 135-145, 1991.

7. "Laser Beam Control by Means of Nonlinear and Coherent Optics Techniques", Final Technical Report on ISTC Project #929, Institute for Laser Physics, St.Petersburg, 2001.

8. Zel'dovich, B.Ya, Pilipetskiy, N.E., Shkunov, V.V., "Wavefront reversal", Nauka, Moscow, 1985, (in Russian).

9. Andreev, N.F., Bespalov, V.I., Kiselev, A.M., Matveev, A.Z., Pasmanik, G.A., Shilov, A.A., Sov. Phys. JETP Lett., **32**, 625, 1981.

10. Berenberg, V.A., Leshchev, A.A., Soms, L.N., Vasil'ev, M.V., Venediktov V.Yu., "Adaptive system of laser energy transport with OA LC SLM correction element", XIII Int. Symp. on Gas-Flow and Chemical Lasers, SPIE 4184, pp. 465-468, 2001.

11. Vasil'ev, M.V., Venediktov, V.Yu., Leshchev, A.A., Pasmanik, G.A., Sidorovich, V.G., Izv. Akad. Nauk SSSR, ser. Phys., **55**, 260-266, 1991 (in Russian).

12. Andreeev, R.B., Volosov, V.D., Irtuganov, V.M., Kalinin, V.P., Kononov, V.V., Sherstobitov, V.E., Kvantovaya elektronika, **18**, 762-765, 1991 (in Russian).

13. Vasil'ev, M.V., Venediktov, V.Yu., Leshchev, A.A., Semenov, P.M., Kvantovaya elektronika, **20**, 317-318, 1993 (in Russian).

14. Ageichik, A.A., Kotyaev, O.G., Leshchev, A.A., Rezunkov Yu.A., et al., "Experimental study on phase conjugation correction of distortions imposed by the telescope elements", in. V.E. Sherstobitov (ed.), Laser Optics'95: Phase Conjugation and Adaptive Optics, SPIE 2771, pp 136-139, 1996.

15. Kornev, A.F., Pokrovsky, V.P., Soms L.N., and Stupnikov, V.K., J. of Optical Technology, **61**, N 1, 10-25, 1994 (in Russian).

100 MW 1.6-μm Pr^{+3}:LaCl$_3$ Propulsion Laser Pumped by a Nuclear-Pumped He/Ar/Xe Laser

Frederick P. Boody

Ion Light Technologies GmbH
Lessingstrasse 2c, 93077 Bad Abbach, Germany

Abstract. A 20 kHz, 100-MW average power, pulsed 1.6-μm Pr^{+3}:LaCl$_3$ up-conversion laser for space propulsion, pumped at high intensity by a *continuous* 212-MW, 2.026-μm He/Ar/Xe nuclear-pumped laser, is proposed. 1.6 μm falls within the 1.72 μm $> \lambda >$ 1.53 μm atmospheric transmission window and is sufficiently distant from 1.4 μm to be relatively eye safe. The high pulse rate minimizes dribble and thus maximizes specific impulse. The nuclear-pumped laser is also a steady-state nuclear reactor. The Pr^{+3}:LaCl$_3$ solid-state laser combines the many small beams produced by the reactor-laser with high beam quality. It consists of an array of face-pumped thin disks at Brewster's angle, face-cooled by high-pressure turbulent helium. The inhomogeneities introduced by the disk array are corrected by using a double pass geometry with a phase-conjugate mirror, as suggested by Magda. The fraction of the pump energy extracted from the Pr^{+3}:LaCl$_3$ laser is 47% at a pump intensity of 2×10^5 W/cm^2. The He/Ar/Xe laser is 0.4% efficient, giving an overall efficiency of 0.2%, equivalent to ~0.8% if electrically pumped.

INTRODUCTION

Recent publications [1-3] have revived interest in laser propulsion by demonstrating significantly higher propulsive force per watt of laser power due to ablative interaction than was the case for laser-plasma interaction. For a given specific impulse, propulsive force per unit laser power is determined by energy conversion efficiency. Pakhomov, et al. [1] have achieved nearly 90% conversion efficiency with 100 ps pulses.

With long-pulse, lower intensity laser radiation, most radiation absorbed is absorbed by the plasma, resulting in low coupling coefficient, C_m (force / laser power) due to plasma reflection, most of the absorbed energy heating plasma rather than ablating mass (favoring specific impulse, I_{sp}, over C_m), and the nondirectionality of the interaction. For short laser pulses of adequate intensity, the laser pulse ends before significant plasma is formed, most energy is coupled into solid material and the reaction is strictly ablative, with ions emitted with a highly directional cosine distribution. For this ablative laser propulsion (ALP), C_m is near the theoretical maximum, $C_m = 2/v_i$, where v_i is the ion velocity of the ablated material. For ALP, Pakhomov, et al. [1] have reported efficiency of 87% for Al, resulting in $C_m = 4.5 \times 10^{-5}$ N/W at $I_{sp} = 4 \times 10^3$ s.

Most recently, water covered ablators have produced very high $C_m = 5 \times 10^{-3}$ N/W [2] and simulation predicts $C_m = 0.1$ N/W [3]. However, since energy must be conserved, I_{sp} is lowered and necessary propellant mass is raised in the same proportion (for the same end velocity), if the efficiency is also near 100% (otherwise the specific impulse would be lowered even more than proportionately).

CP664, *Beamed Energy Propulsion: First International Symposium on Beamed Energy Propulsion,*
edited by A. V. Pakhomov
© 2003 American Institute of Physics 0-7354-0126-8/03/$20.00

The plasma created by each laser pulse lasts about 10 μs and propagates away from the interaction point at 2-10×10^4 m/s. To avoid interaction with the plasma from the previous pulse, the period between laser pulses must be >10 μs, plus time to allow outgoing plasma to clear the interaction area, another 10 to 40 μs. Thus, maximum pulse rate should be ≤ 20-50 kHz, depending on clearance speed of the interaction volume. For continuous pumping, since upper laser level (ULL) lifetimes are of order 200 μs or less, the time between pulses should not be more than about 50 μs, giving a minimum pulse rate of about 20 kHz, the same as the maximum pulse rate.

For pulsed pumping of pulse bursts or low repetition rate individual pulses, acceleration limits on payload and dribble must be accounted for. To take off from the earth's surface, average acceleration must be greater than earth's gravity. Taking 10 g's as the maximum allowable acceleration for people, the duty cycle would have to be $\geq 10\%$. For pulsed nuclear reactors, the thermal neutron pulse length is about 10 ms and the maximum pulse rate is about 1 Hz, giving a maximum duty cycle of about 1% – much too low – thus continuous pumping is required. Even without acceleration limits, duty cycle is important for the minimization of dribble – the evaporation of hot bulk propellant outside the ablation volume between laser pulses: High duty cycle = low time for dribble. Short high intensity pulses also help minimize dribble by minimizing the time for heat transfer to the bulk propellant. Cold bulk propellant does not dribble.

For propulsion to earth orbit and other long-range propulsion applications, an additional requirement comes from the need for average acceleration high enough to achieve orbital velocity within the range of the laser. While several estimates of optimum values have been published, a rough estimate can be made from basic Newtonian physics, giving a range of about 2500 km and a launch time of about 700 s.

Finally, the laser wavelength should be one with high atmospheric transmission that is eye-safe for scattered or reflected laser light ($\lambda > 1.4$ μm). The best window for atmospheric transmission at wavelengths $\lambda > 1.4$ μm is 1.72 μm $> \lambda > 1.53$ μm.

NUCLEAR-PUMPED LASERS

Nuclear-pumped lasers (NPLs) directly convert ion energy from nuclear reactions into coherent light [4]. One high-grade form of energy is converted directly into another, bypassing the thermalization, generation, and transmission required with electrically powered lasers. Since the delivered thermal efficiency of electric power systems (generation and transmission) in only ~25%, NPLs are four times more efficient overall for the same laser efficiency. Perhaps more important for very large lasers that would require more power than many electrical systems can produce, a large NPL puts no demands on the electrical grid. This enables siting the laser in remote locations distant from the commercial power grid and enhances reliability for an application where power interruption is not acceptable.

On the negative side, the only suitable pumping source for large scale direct NPLs is fission fragments (ff) [4] from ^{235}U or ^{235}UO$_2$ coatings on the walls of circular or rectangular cross-section channels, Figure 1. The short range of heavy ions in gases restricts the transverse dimensions of each laser channel to a few centimeters or less, requiring very many small-diameter beams, with low beam quality, for high power operation. Combining the many beams with good beam quality is difficult.

The even shorter range of ff in solids means that even for coatings ~2 μm thick, ~70% [5] of the fission fragment energy is deposited in the coating or substrate and does not make it to the laser gas. The fission fragments are also only 87% of the fission energy, so that ≤26% of the fission energy is available for pumping the laser.

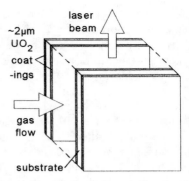

FIGURE 1. Coating Nuclear-Pumped Laser Geometry.

Large temperature gradients are need to transport the other 74% of the fission energy, as heat, across the film at the interface of the laser gas with the coating surface. The non-directionality of the fission ions results in more energy being deposited in the gas near the coating surface (sine of angle to normal weighting), although this is somewhat offset by the gas density gradient resulting from the heat removal from the surface. The net effect is strong density gradients [6], which lower beam quality, due to thermal lensing, and limit the fraction of the cross-sectional area from which a laser beam can be extracted [7]. These beam quality problems can be made manageable, for applications that do not require high beam quality, by use of corrective optics [7].

A further problem is transient radiation-induced absorption in silica-based optics, particularly windows, exposed to high radiation [8]. This absorption is so high in the UV and blue as to rule out lasing. Thus, direct nuclear-pumped lasers seem to be limited to long wavelength visible and preferably IR wavelengths.

Both beam quality and beam combination problems can be solved by using the nuclear pumped photons to pump another laser medium, which has previously been termed the dual-medial approach to NPLs [9]. This approach has been proposed for fluorescence pumping of gas [10] and solid-state [8] lasers. For large laser systems however, the fluorescence approach does not work because of radiation-induced absorption or the large volume required for large D/L light pipes.

Thus, the best approach to propulsion NPLs seems to be a coating NPL pumping a second laser that combines the many small beams into one large high quality beam. Based on beam quality considerations, solid-state lasers appear to be the best choice.

This paper combines approaches presented at the recent Conference on Nuclear-Pumped Lasers, NPL-2002, in Snezhinsk, Russia. It continues the idea of a continuously pumped, high pulse rate, short-pulse solid-state laser pumped by nuclear-pumped photons presented by this author [11], but adopts the use of a Pr^{+3}:$LaCl_3$ solid-state laser pumped by a 2.026 μm He-Xe-Ar NPL as presented by Magda [12].

REACTOR-LASER

He/Ar/Xe at 2.026 μm is the NPL of choice for steady-state operation for two reasons: (1) it lases at a pumping power of ~50 W/cm³ [5], tied for the lowest of any high energy NPL, and (2) He has the best heat transfer properties of any NPL gas. He is probably the only buffer gas suitable for steady-state laser operation. He/Ar/Xe has the additional advantage, as pointed out by Magda, that its wavelength is very near the absorption maximum of the Pr^{+3}:$LaCl_3$ solid-state laser.

Modeling [13] predicts ~3% He/Ar/Xe power efficiency at room temperature, with weak He/Ar ratio dependence and strong temperature dependence. Experimental measurements, with temperature uncoupled from pumping power or pumping time, show He/Ar/Xe efficiency drops by a factor of about 2 as temperature increases from 300 K to 600 K [14]. Experiments without gas flow, where temperature is a function of time, show optimum pressure and mixture of 3 atm and 800:200:1 of He/Ar/Xe (T not given). In these experiments, at 2 atm the efficiency increased strongly with pumping power up to about 50 W/cm^3, then rose more slowly for pumping power up to 80 W/cm^3, and then dropped as temperature rose further [5]. For a broad range of He fraction, efficiency can be expected to decrease from 2.8% at 300 K to 1.4% at 600 K.

Only 165 MeV of 190 MeV per fission is charged particle energy, 30% of which escapes to the gas [5], where <90% is absorbed. This ~23% efficiency combined with ~2% He/Ar/Xe laser efficiency means that ~99.5% of the nuclear energy must be removed as heat. For pumping powers of order 50 W/cm^3, ~220 W/cm^3 of heat must be removed. The amount of heat and thus ΔT is proportional to channel length. Because laser efficiency decreases with temperature, it also decreases with channel length. Thus, from an efficiency point of view, the flow channel should be as short as possible. As nuclear reactors, however, steady-state NPLs must also be critical. Criticality requires a minimum reactor volume-to-surface-area, giving a minimum volume and thus minimum channel length. The question, then, is whether the minimum channel length for criticality is shorter than the maximum length for acceptable efficiency.

In steady state, no heat can be stored in structural materials for later removal; the flowing laser gas must continuously remove all heat generated. For 74 W/cm^3 average pumping power and 23% efficiency, this is 320 W/cm^3. Because the volume heat capacity of rare gases, 8.4×10^{-4} J/cm^3-K-atm at 300K, is low, gas temperature rises rapidly. Average gas temperature at the channel outlet is the product of the rate of T rise (the ratio of heat load to heat capacity) and the time required to reach the end of the channel (channel length divided by flow velocity). Flow velocity is limited to less than the sound speed, $v_s = (\gamma R T / M)^{1/2}$, where for rare gases $\gamma = 5/3$, $R = 8.314$ J/mole-K, and M is the gas mixture molar mass in kg/mole. For $T_{out} = 600$K and 9:1 He/Ar gas mixture, $v_s = 1046$ m/s. For $T_{in} = 300$ K, v_{in} must be ≤ 523 m/s to remain subsonic at the outlet. For $v_{in} = 500$ m/s, $p = 4$ atm, and $L_{chan} = 160$ cm, $\Delta T = (qL)/(p\rho c_p v_{in}) = (320 \text{ W/cm}^3)$ $(1.6 \text{ m})/(4 \text{ atm})(8.4 \times 10^{-4} \text{ J/cm}^3\text{-K-atm})(500 \text{ m/s}) = 300$ K, giving $T_{out} = 600$K as assumed.

The 77% of the heat not deposited in the gas directly must be transferred to it by a temperature gradient from the channel wall to the bulk gas. Even with turbulent flow and high He pressure, large $\Delta T_{trans} = T_{wall} - T_{bulk}$ is required. ΔT_{trans} is calculated from $\Delta T_{trans} = q/hA$ where the heat transfer coefficient h is given by [15]

$$ h = 0.036 \frac{k}{D} \left(\frac{Dv\rho}{\mu} \right)^{0.8} \left(\frac{c_p \mu}{k} \right)^{0.33} \left(\frac{\mu}{\mu_s} \right)^{0.14}, \tag{1} $$

where k is thermal conductivity, D is effective channel diameter (~ 2 x channel width), v is bulk gas velocity, ρ is mean gas mass density, c_p is gas specific heat, and μ_s and μ are coating surface and bulk gas viscosities, respectively. h varies strongly (0.8-power) with gas velocity and density – from ~0.5 W/cm^2-K at 400 m/s to ~0.7 W/cm^2-K at 600 m/s, for rare gases at 4 atm, but only by ~4% for temperatures from 300 K to 600 K (combined effect of thermal conductivity and viscosity) and 13% when D varies by a factor of 2. Since velocity and pressure dependence are similar to those axially (exponent 0.8 vs. 1.0), optimizing these for axial heat transfer also optimizes them for transverse heat transfer. For 74 W/cm^3 pumping power, 320 W/cm^3 heat load, and 11 mm wide channel, 135 W/cm^2 must be transferred. For v_{in}=500 m/s and p_{in}=4 atm, h=~0.6 W/cm^2, giving ΔT_{trans}=225 K.

For highly turbulent flow the relative temperature profile, $(T_{wall}-T)/(T_{wall}-T_0)$ where T_0 is the centerline temperature, depends only on relative position and

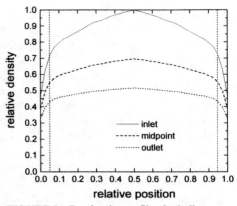

FIGURE 2. Gas density profiles for bulk temperatures of 300, 450, and 600 K and ΔT=300 K.

the Prandtl number [16], which is ~0.7 for all gases. From this universal temperature profile, a density profile can be calculated from T_{wall} and T_0, Figure 2.

Five factors govern pumping power deposition in the laser gas. Coating thickness and neutron flux determine the ion source characteristics while laser gas composition and density and channel width determine the ion absorption characteristics. Coating thickness is chosen for maximum efficiency to minimize cooling problems associated with steady-state operation. However, for steady-state operation, the neutron flux and thus the ion flux entering the gas are limited only by thermal considerations. Since gas density and He fraction are chosen as high as possible consistent with laser efficiency, for cooling reasons, ion absorption is determined by varying the channel width.

Ion energy deposition is an exponential of form $E = E_0(1-s/R_0)^n$ [17] where s is the distance traveled by an ion with initial energy E_0 and E_0-dependent range R_0. But since ion emission is isotropic, it is more likely at an angle to the wall than perpendicular to it, increasing energy deposition near the wall. Rather than writing a complicated computer code to solve this equation for varying angle and temperature as in [6], energy deposition profiles were determined by inserting the gas density profiles of Figure 2 into the SRIM-2000 computer code [18]. Figure 3 shows the results for 11-mm channel width, 9:1 He/Ar, and densities corresponding to 4 atm at 300 K, 450 K, and 600 K. Similar profiles would result for 13 mm of 95:5 He/Ar or 15 mm of 99:1 He/Ar. The 300 K inlet profile results in 89% absorption of the ion energy in the flow

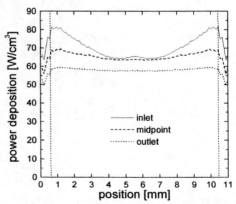

FIGURE 3. Ion power deposition profiles in 9:1 He/Ar for 1.1-cm wide channel pumped from each side by 46 W/cm^2 of fission fragments. Pressure: 4 atm; Temperature: 300 K inlet, 600 K outlet. Dotted vertical lines are 90% masking boundaries.

channel. The reduction in gas density as T rises to 600 K at the outlet reduces the absorbed fraction to 84% at the channel midpoint and 73% at the outlet. For 46 W/cm² energy flux into the channel from each side, average pump powers over the central 90% of the channel are 71, 66, and 58 W/cm³ at inlet, midpoint, and outlet. A neutron flux of 2.2×10^{15} cm⁻²s⁻¹ is required for a fission fragment energy flux into the laser gas of 46 W/cm², from 2-μm thick fully enriched UO_2 coatings.

Thermal reactors minimally consist of fissile material ($^{235}UO_2$), moderator (Be), and coolant (He/Ar/Xe). Be is the only moderator (and structural/substrate) material considered because it has the shortest thermalization length of any solid [19], 1/4 that of next-best graphite, and is the stiffest element per unit weight, enabling a 2-mm thick substrate. UO_2 thickness and the channel width determine ^{235}U density, while Be thickness and channel width determine moderator density. Since the UO_2 coating is much thinner than the neutron absorption length and the Be thickness is much less than the fast neutron thermalization length, the reactor can be treated as a homogeneous Be/^{235}U mixture, a textbook case [19]. For 2-mm Be and 2-μm UO_2, the moderator to fuel ratio is 2.77×10^3 and the infinite multiplication factor is 1.98 [19].

For a rectangular solid core with sides $1.5L$, $1.5L$, and L and the parameters above, the allowed leakage can be found from textbooks [19] to require $L \geq 55$ cm. For this L, neutrons released in the volume just balance those leaking from the surface:

$$(\rho V/A) = \rho L^3 / 6 L^2 = \rho L / 6 = \text{critical value} \qquad (2)$$

If ρ decreases, L must increase to maintain reactivity. With 2-mm Be and a 11-mm channel, ρ is 15% that of a solid reactor. Thus, L must be 6.5 times larger or 356 cm. L can be reduced by using a reflector/moderator that replaces fuel without affecting criticality [19]. For a Be moderated/reflected reactor, 20 cm of Be replaces 15 cm of fuel [19]. This "reflector savings" is similarly multiplied by the density ratio so 130 cm of reflector "saves" 98 cm of fuel and L is reduced to 161 cm. Reflectors also dramatically increase flux at the edge of the fueled region, making thermal flux relatively flat, important for flat pumping power density in the laser. L could be reduced still further by introducing the gas flow into the middle of the core, which halves L to 98 cm.

Averaging pumping power and laser efficiency over channel length and width, the extracted power (central 90% = 1 cm of width) is 1.4 W/cm³ and net efficiency is 0.4%, equivalent to an electrically pumped laser efficiency of 1.6%. Each 240 cm x 160 cm x 1 cm channel thus produces 54 kW and the 186 channels in one minimum size critical reactor-laser produce 10 MW. 210 MW requires 21 modules for pumping the solid-state laser. Stacking the modules three high would save the moderator between them. Cylindrical lenses between modules would correct thermal lensing due to the density profile, enabling the entire masked area to be extracted. Depending on the correction needs, the channel laser length can be increased, also saving Be. The complete laser-reactor including reflector would be 19.4-27.2 m long, 7.6 m high and 4.2 m wide.

The cost is estimated to be equal to that of a commercial power reactor. Fuel is a minor part of power reactor operating costs. 53 GW$_{th}$ is equal to ~17 1-GW$_e$ power reactors, but capital costs would be those for a single gas-cooled power reactor because power density would be ~10x higher, making sizes equivalent. The lack of electrical generation hardware would offset optical materials costs and 50-100 M$ for Be.

SOLID-STATE LASER

Pr^{+3}:LaCl$_3$ has demonstrated lasing at 1.6 μm and 5.2 μm [20] as well as 7.2 μm [21] on transitions to the $^3H_4$, $^3H_6$, and $^3F_2$ levels, respectively, from the $^3F_3$ level which was up-conversion pumped by a pulsed 2.02-μm Tm:YAG laser. The Pr^3 energy level scheme [20] is shown schematically in Figure 4. The reactions and their rate constants [22] are given in Table 1. With strong pulsed pumping (such as used for the demon-

strated lasers), the ground level is emptied and, due to the long lifetimes of all of the excited levels ($^3F_3$ 180 μs, $^3H_6$ 5 ms, $^3H_5$ 33 ms), it refills only slowly. Interpreting the results of [20] and [21] based on the level structure, the 1.6-μm laser worked because it was pumped strongly enough. The 5.2-μm laser oscillated only after the higher gain 1.6-μm laser reduced the inversion of the $^3F_3$ level compared to ground state to zero. Then it worked well only when the temperature was low enough (150 K for optimal efficiency) that the upper levels of the $^3H_6$ manifold were not thermally pop-

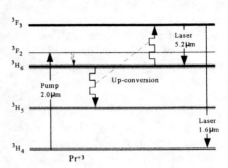

FIGURE 4. Energy level diagram of Pr^{+3}:LaCl$_3$ showing pumping and lasing lines. [11]

ulated because, based on the difference in lifetimes, the total population of the N2 manifold should be more than an order of magnitude greater than that of the $^3F_3$ level. The 7.2-μm laser worked only after pumping ended and the population of the $^3F_2$ level thermally relaxed to the $^3H_6$ manifold

The energy efficiency of the 1.6-μm laser produced by this scheme, assuming unity quantity efficiency, is 62%. However there are several competing processes, including a phonon emitting transition from $^3F_3$ to $^3H_6$ as well as the usual radiative transitions between the levels [22]. The up-conversion and cross-relaxation coefficients for most of the transitions have been determined [22]. The transitions and their coefficients are

included in Table 1, which presents the input to an OLCHEM rate equation integrator model [23]. Notation used is: N3 - $^3F_3$, N2 - $^3F_2$ + $^3H_6$, N1 - $^3H_5$, N0 - $^3H_4$.

The model was first run for several milliseconds to establish approximate steady-state populations for levels N1 and N2. The N0 and N3 steady-state populations were then summed and divided by two and the population of both levels set equal to this value, the assumed result of a lasing pulse. The

TABLE 1. Pr^{+3}:LaCl$_3$ transitions and rate constants.

Transition						Rate Constant
1. N0	I	→	N2	I	abs	3.0000E-17
2. N0	N3	→	N1	N2		9.6000E-17
3. N1	N1	→	N0	N2		5.0000E-18
4. N1	N3	→	N2	N2		1.0000E-17
5. N2	N0	→	N1	N1		1.1000E-16
6. N2	N1	→	N0	N3		4.0000E-17
7. N2	N2	→	N1	N3		1.1000E-17
8. N3		→	N0	16mu		2.3000E+03
9. N3		→	N1	24mu		5.3000E+02
10. N3		→	N2	52mu		1.2000E+02
11. N3		→	N2	phonon		2.6000E+03
12. N2		→	N1	47mu		2.6000E+01
13. N2		→	N0	24mu		1.7400E+02
14. N1		→	N0	48mu		3.0000E+01

medium was then pumped for 50 μs and the resulting populations of N0 and N3 were again summed and divided by two and the population of both levels set equal to this value (again assumed as result of lasing pulse). The new populations of N1 and N2 as well as the averaged populations of N3 and N0 were then used as input to a new run for 50 μs. This was continued until equilibrium was reached, until the populations of N1 and N2 and the averaged populations of N0 and N3 were the same at the end of the 50 μs pumping period as they were at the beginning. This procedure was repeated for various constant pump intensities and for intensities that increased linearly from an initial value, for example from 0.5×10^4 to $1.5\, 10^4\, W/cm^2$.

The extractable absorbed energy efficiency was calculated from the number of extractable N3 states produced ([N3-N0]/2) times the photon energy (0.77 eV for 1.6 μm) divided by the energy absorbed. It is highly dependent on average pump intensity, increasing from 0% for $I=2 \times 10^4\, W/cm^2$ to 55% for $I=2 \times 10^5\, W/cm^2$, as shown in Table 2. Energy efficiency is higher than photon (or quantum) efficiency because laser photon energy is higher than pump photon energy: 55% vs. 44% at $I=2 \times 10^5\, W/cm$.

The ground state population is greatly depleted by the high intensities required for high efficiency, to ~1.3% of its unpumped level of $7 \times 10^{19}\, cm^{-3}$ before a laser pulse and ~4% afterwards (at the ends). The electrons are stored mostly in the $^3H_6$ manifold (65%), due to the strong pumping, and the $^3H_5$ manifold (28%), due to its long lifetime (33 ms). Because of the dramatic jump in ground state population after each pulse (a factor of 2 in the middle and 3 at the ends) the absorption length is dramatically reduced, concentrating absorption at the ends, and the pump intensity varies strongly with time, even with pumping from both crystal ends, as shown in Figure 6.

For high efficiency, the laser should be long enough that nearly all the pump energy is absorbed. If this reduces the average intensity near the center too much however, the decrease in energy efficiency will more than compensate for the improved absorption. A time and position dependent model is required to determine if a crystal length and pump intensity combination exists which will give both high absorption and high efficiency over the entire crystal length. Such a model was constructed in Mathcad [24] using the parameters in Table 1. Figure 5 shows the resulting intensity profiles just before and just after a pulse for 50 cm of $Pr^{+3}:LaCl_3$ with input $I=2 \times 10^5\, W/cm^2$ at each end. By summing the transmissions for each time increment, the average absorption over the 50 μs pump period was calculated to be 94.5% for the given parameters. Figure 6 shows the variation of 2-way pump intensity with time at 5 cm intervals. The variation can be treated as linear without significant introduction of error and used in OLCHEM to calculate the extractable energy efficiency. For $2 \times 10^5\, W/cm^2$, 50 cm, and 50 μs, extractable absorbed energy efficiency varies from 49% at the middle to 55% at

TABLE 2. Efficiency and Heat Deposition vs. pump intensity.

parameter	2.026 μm pump intensity [$10^4\, W/cm^2$]			
	0.4	0.7	1.0	2.0
energy efficiency	0.30	0.43	0.485	0.54
fraction of maximum efficiency	0.54	0.77	0.86	0.95
pumping power density [kW/cm^3]	5.3	6.6	7.3	8.1
heat deposition [kW/cm^3]	1.47	1.68	1.71	1.81
fraction of pump thermalized	0.276	0.248	0.235	0.223

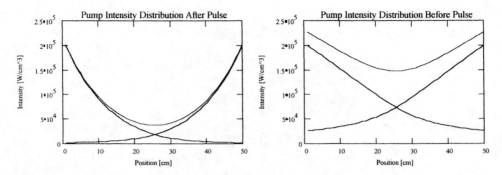

FIGURE 5. Pump intensity just before and just after a laser pulse. The exponential curves are the distribution produced by irradiation from one end and the U-shaped curve is the combined 2-way intensity.

the edge, giving an average efficiency of 52%. The average extractable pump energy efficiency, the product of absorption efficiency (94.5%) and extractable absorbed energy efficiency (52%), is 49%. For $2\,10^5$ W/cm^2 pump intensity on each end of the laser and a 50 μs pump period, the stored fluence is 9.8 J/cm^2, which is ~200 J/liter for a 50 cm length. Assuming a stimulated emission coefficient of $\sigma = 3\times10^{-20}$ cm^2 [25], $g_o l = 4.74$ and $\exp(g_o l) = 115$.

For 2×10^5 W/cm^2, 50 cm, and an average absorption of 0.945, average power density is 7.6 kW/cm^3, varying from 7.1 to 8.2 kW/cm^3, despite the high pump intensity. Although the laser's average extractable absorbed energy efficiency is 52%, of the remaining 48%, 24% is emitted as fluorescence and only 24% is thermalized. Since all of the fluorescence lines (except N3-N2, 0.2%) are inverted, the crystal does not reabsorb the fluorescence, so only the 24% or 1.6 to 1.9 kW/cm^3 must be removed as heat.

Contrary to the impression given by the fact that the 5.2-μm laser needed to be cooled below room temperature to be efficient [20], as shown by these calculations, there is no need to cool the 1.6-μm laser below room temperature. The 5.2-μm laser relied on cooling to depopulate the upper level of the $^3H_6$ manifold in order to achieve

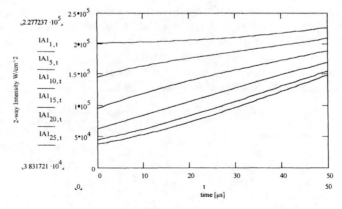

FIGURE 6. 2-way pump intensity as a function of time after laser pulse at 5-cm intervals. The progressive bleaching of the center of the laser can be seen.

642

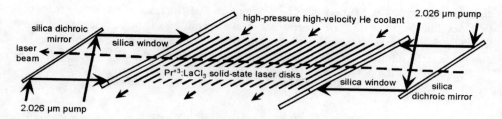

FIGURE 7. Pr^{+3}:LaCl$_3$ laser amplifier formed by 1000 0.5-mm thick disks face pumped by high intensity 2.026 μm radiation and cooled by high pressure, turbulent He. Disks are at Brewster's angle.

an inversion. With the 1.6-μm laser, continuous pumping and the long life of the excited levels depopulate the ground state. However, heavy-duty cooling of the crystals is required to remove the significant continuous heat deposition and to prevent excessive temperatures that would change the physical properties of the crystal, and through excessive stress the optical properties of the crystal.

The only geometry that allows sufficient cooling for steady-state heat deposition of up to 1.88 kW/cm^3 is face-pumped disks with turbulent, high-pressure He flowing between them, as shown in Figure 7. Thermal stress considerations determine the maximum thickness of the disk. Using equations in [26], one can derive that $t < (12 R / Q)^{1/2}$ where R is the thermal shock parameter and t is the disk thickness. Only one R value has been published for LaCl$_3$ [27] and it is probably too low, because LaCl$_3$ is quite hygroscopic and discoloration from absorption of water was reported. However, using this value, 0.5 W/cm, as a conservative estimate gives $t < 0.56$ mm for $Q = 1880$ W/cm^3. The ΔT between the disk face and its centerline is given by $\Delta T = t^2 Q / 8 k$ [26]. Taking the disks to be 0.5 mm thick, similar to Si wafers for integrated circuits, $\Delta T = 12$ K for 1880 W/cm^3 and $k = 0.05$ W/cm-K.

Since Pr^{+3}:LaCl$_3$'s properties are not affected by the gas pressure, as they were for the nuclear pumped laser, a higher He density of 12 atm can be assumed. If the spacing between disks is 1 cm, nearly that of the reactor fuel plates, and the velocity is 500 m/s, then the heat transfer coefficient will be 2.4 times ($3^{0.8}$) higher than for the reactor cooling calculation for 500 m/s or $h = 1.4$ W/cm^2-K. The heat deposition per cm^2 of surface area, ($t = 0.5$ mm), is 47 W. thus $\Delta T = q / hA = 34$ K. Such a ΔT will result in a modest gas density gradient, but because it is in the direction of the laser propagation, rather than transverse to it as in the nuclear-pumped laser case, it will have no lensing effect on beam propagation. The total ΔT is 46 K.

In disk geometry, the important temperature variation is that across the face of the disk: $\Delta T = (q L / (v \rho c_p)) = \Delta T = (94 \text{ W/cm}^3 \times L)/(5 \times 10^4 \text{ cm/s} \times 2.1 \times 10^{-3} \text{ g/cm}^3 \times 5.2 \text{ J/g-K})$, where L is the transverse dimension of the disk. $\Delta T = 0.17 L$ for the parameters given above. For a disk area of 360 cm^2 (see below), L is 19 cm and ΔT would be 3.3 K.

The extractable photons stored in the ULL can be extracted efficiently only by a strongly saturating flux of photons. It is proposed to use the arrangement shown in Figure 8, which is a version of that proposed by Magda [12] that has been modified to reflect the differences between his concept and this one. Spatial and temporal control elements such as spatial filters, Faraday rotators, and Pockels cells have been omitted. The key to this arrangement is the use of double-passed amplifiers with a phase-conjugated mirror (PCM) between the first passes and the second passes. During the first pass the beam senses the aberrations in the amplifiers, which are conjugated by the PCM and compensated during the second pass. Another advantage of this arrangement is that the amplifiers function as both preamps and power amps.

The fluence at the output of each stage in the amplifier chain is given by [26]

$$E_{out} = E_s \ln\left\{1 + \left[\exp\left(\frac{E_{in}}{E_s}\right) - 1\right] \exp(g_0 l)\right\} \tag{3}$$

where $E_s = h c / \lambda 2\sigma = 2.07\,\text{J/cm}^2$ is the saturation fluence and $\exp(g_0 l) = 115$ as discussed above. The extracted fluence is $E_{ext} = E_{out} - E_{in}$ and the extraction efficiency is $\eta_{ext} = E_{ext}/9.8\,\text{J/cm}^2$, where $9.8\,\text{J/cm}^2$ is the stored fluence determined earlier. Thus, the extracted fluence and the extraction efficiency are dependent on the input fluence. Another way of putting this is that for each extraction efficiency there is an associated output fluence – the higher the extraction efficiency desired, the higher the output fluence must be. Put yet another way, the extraction efficiency is limited by the maximum fluence associated with surface damage constraints.

If, for example, an extraction efficiency of 96% is desired, extracted fluence must be $9.4\,\text{J/cm}^2$. Assuming an input fluence of say $3\,\text{J/cm}^2$, $E_{out} = E_{ext} + E_{in} = 12.4\,\text{J/cm}^2$ but Equation (2) gives $E_{out} = 12.3\,\text{J/cm}^2$. Increasing E_{in} to $3.6\,\text{J/cm}^2$ gives $E_{out} = 13.0\,\text{J/cm}^2$ from both equations. Because the PCM is only about 80% reflective, it slightly reduces the extraction efficiency for this case to 95%. The extraction efficiency is 85% for a maximum fluence of $10\,\text{J/cm}^2$, including the effect of the PCM. An electrically pumped oscillator with a pulse energy of 1 mJ (average power of 20 W) or less is sufficient to drive the laser amplifiers because of the double-passed configuration and the high gain of the first two amplifiers for the first pass.

Because the disks are extremely thin (required for steady-state heat transfer), have *ultra*polished surfaces (required to minimize scatter from the very many thin disks), and will never be exposed to the atmosphere or touched by human hands (required by hydroscopicity of $LaCl_3$ and due to He cooling) they should have very high surface damage thresholds and the extraction efficiency should be $\geq 96\%$. The nearness of the 1.6-μm laser wavelength to the absorption minimum of fused silica means that absorption by and damage to fused silica optical components such as windows, mirrors, and polarizers should be minimal, also supporting operation at very high fluences. Overall laser efficiency for 96% extraction of the photons that were stored with 49% efficien-

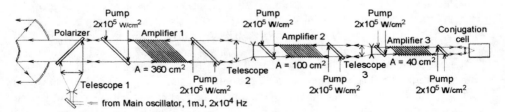

FIGURE 8. Double-pass amplifier chain with phase-conjugate mirror for efficiently extracting stored photons from Pr^{+3}:$LaCl_3$ laser amplifiers. Spatial and temporal control elements such as spatial filters, Faraday rotators, and Pockels cells are omitted. Modified from Magda [12] for steady-state operation.

cy is 47%. For 100 MW output power, 212 MW of pumping power would be required.

REFERENCES

1. Pakhomov, A. V., Gregory, D. A., and Thompson, M. S., *AIAA Journal* **40**, 947-952 (2002).
2. Phipps, C. R., Seibert, D. B., Royse, R. W., King, G. and Campbell, J. W., *Proc. SPIE* **4065**, 931-938 (2000)
3. Yabe, T., Phipps, C., Yamaguchi, M., et al., *Appl. Phys. Letters* **80**, 4318-4320 (2001).
4. Boody, F. P., Prelas, M. A., Anderson, J. H., Nagalingam, S. J. S., and Miley, G. H., "Progress in Nuclear-Pumped Lasers" in *Radiation Energy Conversion in Space*, edited by K.W. Billman, New York: AIAA, 1978, pp 379-410.
5. Magda, E.P., Bochkov, A. V., Lukin, A. V., Magda, L. E., and Pogrebov, I. S., "Optimal Characteristics of Nuclear-Pumped Gas Lasers" in *Proceedings of the 3rd International Conference on Nuclear Pumped Lasers*, Snezhinsk, Russia, 16-20 Sept. 2002.
6. Montierth, L.M., Neuman, W.A., Nigg, D.W., and Merrill, B.J., *J. Appl. Phys.* **69**, 6676-6788 (1991).
7. Neuman, W. A., and Fincke, J. R., *J. Appl. Phys.* **69**, 6689-6798 (1991).
8. Boody, F. P., and Prelas, M. A., "Very High Average Power Solid-State Lasers Pumped by Remotely-Located Nuclear-Driven Fluorescers" in *OSA Proceedings on Advanced Solid-State Lasers*, edited by H. P. Jenssen and G. Dubé, Washington, D.C.: Optical Society of America, 1991, pp. 192-199.
9. Boody, F. P., "The Dual-Media Approach to Nuclear-Pumped Lasers" in *Proceedings of the 1981 IEEE International Conference on Plasma Science*, Santa Fe NM, May 1981, Paper 6B2, 81CH1640-2 NPS
10. Boody, F. P., and Prelas M. A., "Photolytic Dual-Media Nuclear Pumping of Excimer Lasers" in *Excimer Lasers - 83*, edited by C.K. Rhodes, H. Egger, and H. Pummer, AIP Conference Proceedings 100, New York: American Institute of Physics, 1983, pp. 349-354.
11. Boody, F. P., "Nuclear-Driven Fluorescence-Pumped Lasers" in *Proceedings of the 3rd International Conference on Nuclear Pumped Lasers*, Snezhinsk, Russia, 16-20 Sept. 2002.
12. Magda, E.P., "Laser-Reactor Driver for Launching Useful Loads into Low Earth Orbit" in *Proceedings of the 3rd International Conference on Nuclear Pumped Lasers*, Snezhinsk, Russia, 16-20 Sept. 2002.
13. Alford, W. J., Hays, G. N., Ohwa, M., and Kushner, M. J., *J. Appl. Phys.* **69**, 1843-1848 (1991).
14. Tomizawa, H., Wieser, J., Ulrich, A., in *Proceedings of the 3rd International Conference on Nuclear Pumped Lasers* s, Snezhinsk, Russia, 16-20 Sept. 2002.
15. Sieder, E. N., and Tate, G. E., *Ind. Eng. Chem.* **28**, 1429 (1936). See also [19].
16. Martinelli, R. C., *Trans. Am. Soc. Mech. Engrs.* **69**, 947 (1949). See also [19].
17. Ziegler, J. F., Biersack, J. P., and Littmark, U., *The Stopping and Range of Ions in Solids*, New York: Pergamon, 1985.
18. Ziegler, J. F., and Biersack, J. P., *SRIM (Stopping and Range of Ions in Matter)*, Version 2000.40 from www.SRIM.org, Technical reference is [17].
19. Glasstone, S., and Sesonske, A., *Nuclear Reactor Engineering*, Princeton: D. Van Nostrand Company, Inc., 1963, Chpt. 4.
20. Bowman, S. R., Ganem, J., Feldman, B. J., and Kueny, A. W., *IEEE J. Quantum Electron.* **30**, 2925-2928 (1994).
21. Bowman, S. R., Shaw, L. B., Feldman, B. J., and Ganem, J., *IEEE J. Quantum Electron.* **32**, 646-649 (1996).
22. Kirkpatrick, S. M., Bowman, S.R., Shaw, B. J., and Ganem, J., *J. Appl. Phys.* **82**, 2759-2765 (1997).
23. *OLCHEM Chemical-Rate-Equation Integrator*, SRI International (1986) http://www-mpl.sri.com/software/olchem/olchem.html,
24. *Mathcad 7: Users Guide*, Cambridge: MathSoft, Inc., 1997.
25. Choi, Y. G., Kim, K. H., Park, H. J., and Heo, J., *Appl. Phys. Letters* **78**, 1249-1251 (2001).
26. Koechner, W., *Solid-State Laser Engineering*, Berlin: Springer Verlag, 1992.
27. Payne, S. A., Beach, R.J., Bibeau, C., et al., *IEEE J. Select. Top. Quantum Electron.* **3**, 71-81 (1997).

BEAMED ENERGY PROPULSION:
SYSTEM ANALYSIS

Powering Ion-Engine Equipped Orbital Transfer Vehicles With A Ground-Based Free Electron Laser

H.E. Bennett

Bennett Optical Research Inc.
Ridgecrest, California 93555

Abstract. Ion engines are a promising way to provide fuel-efficient continuous thrust propulsion for orbital transfer vehicles in the low earth to geosynchronous orbit range. However the power required for such an application is 50-100 kW, too much for reasonable-sized solar panels. Adequate power can be provided by a ground-based 200 kW to 1 megawatt free electron laser. Using rf photoinjectors and bare copper cathodes output powers of 200-250 kW are anticipated. New photoinjector technology should raise the output power to one megawatt, the design value for the laser. The beam projector makes the laser beam eyesafe in the atmosphere and the entire system is based on proven technology.

INTRODUCTION

Ion thrusters for space vehicles are receiving increasing attention as relatively low cost, efficient and long lived substitutes for vehicles propelled by chemical rockets. A typical two stage chemical rocket to geosynchronous orbit (GEO) such as United Technologies Orbus[1] has as a first stage the Orbus 21 with an average thrust in vacuum of 196 kN, a specific impulse of 303 sec and a burn time of 154 sec. The second stage Orbus 6/6E has an average thrust of 80.95 kN, a specific impulse of 303 sec and a thrust lasting 103 sec. These thrust values are orders of magnitude above those envisioned for ion thrusters. However the exhaust velocity of typical chemical propellants is about 2 km/sec and never above 3.5 km/sec (i.e. an Isp of 200 to 350 sec)[2]. As a result, the mass of propellant of a chemical rocket must be over an order of magnitude larger than that of an ion engine. Electrostatic ion propulsion systems have demonstrated[2] Isp's of 10,000 sec (29 to 50 times that of chemical propellants). The advantage of chemical propellants is that the thrust comes over a short time period, which allows the spacecraft to overcome the gravity well resulting from the earth. After being initially launched from earth the satellite is quickly carried up and established in an orbit around the earth. No ion propulsion system develops anything like the thrust of a large chemical rocket, so the ability to establish a satellite into

CP664, *Beamed Energy Propulsion: First International Symposium on Beamed Energy Propulsion*,
edited by A. V. Pakhomov

orbit around the earth resides with chemical rockets or other approaches. Once the orbit is established, however, the need for high thrust is often reduced and the efficiency of the propulsion system in utilizing its propellant becomes very desirable. Ion engines allow a tradeoff to be made between high thrust and high efficiency rockets.

If the chemical rocket is used to send the payload into geosynchronous orbit, typically 99% of the initial weight of the liftoff mass is rocket and only 1% payload. As a result the cost per kg to send a payload to GEO using chemical rockets is high. The fuel efficiency of Deep Space 1, seen in Figure 1, was ten times that of a typical rocket.[3] This deep space satellite used only an ion engine. Its thrust was only one ten thousandth that of a typical chemical propulsion system, but since the thrust from Deep Space 1 was continuous, the satellite built up a speed of 3,600 meters/sec in a one year period. It logged more than 14,000 hours of operation in its first voyage, has a predicted engine lifetime of 20,000 hours,[2] and uses only 1-3 mg/sec or ~100 g of xenon per day[2] of fuel.

FIGURE 1. First ion engine powered deep space satellite. It uses only 100 g of propellant per day and has a projected lifetime of 2.3 years.

Energy Requirements for Electric Propulsion

A plot of the energy required to send a payload from the earth to GEO is shown in Figure 2. Rockets or other means may be used to overcome the deepest part of the gravity well and establish the satellite in orbit. Once there the ion engine may carry the payload to its ultimate destination using a spiral course rather the elliptical orbit usually used by chemical propulsion, but with greatly increased propellant efficiency and thus lower cost because of the higher Isp. It may be 29 to 50 times that of chemical propellants.

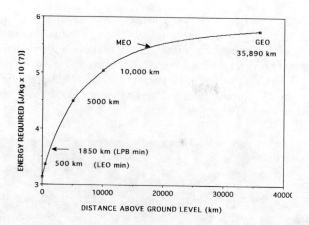

FIGURE 2. Energy in joules per kg required to lift an OTV from Earth to GEO

Much of the present effort on electric propulsion has been focused on increasing the thrust, while maintaining the specific impulse and thus the efficiency at approximately the same levels as achieved for the earlier systems, about 0.06N/kW. Charging of exit surfaces limits the available thrust of electrothermal thrusters, Resistojets and electrostatic or ion propulsion systems. Typical ion thrusters now range in thrust from 10^{-3} to 10^{-1} Newtons with efficiencies of 0.03 to 0.06 N/kW.

Interest is now focusing on Hall-effect and magnetoplasmadynamic thrusters, which do not have as severe limitations in the charge-buildup areas as electrostatic thrusters. Hall effect thrusters have achieved[2] thrusts of 0.4 N with 6 kW of power and an efficiency of 0.06N/kW. They are expected to go much higher in thrust and may be the most promising of the different approaches. Magnetoplasmadynamic thrusters have achieved[2] thrusts of 12.5 N (thrust densities of 400N/m^2) for input powers of 200 kW and an efficiency of 0.0625N/kW. The effort to achieve larger thrusts in electrical propulsion is continuing. A program initiated in late 2000 at the NASA Glen Research Center is designed to develop a 50 kW Hall thruster[4]. If the efficiency stays the same, the thrust would then be 3.3N and the electrical power needed to operate the satellite will be >50 kW. Where is that space power to come from?

The largest commercial satellites, at present, generate power levels from their solar powered-solar cells of less than 15 kW. Most are under 10 kW. The Space Station mounts a solar array of 8 solar panels built by the United States, each 32.6 meters in length (about 1/3rd of a football field) and 11.6 meters in width. It contains space qualified silicon solar cells and is shown in Figure 3. Jane's Space Directory[5] reports that each solar panel generates 23 kW, less than half of the power needed by the Hall Thruster. These solar panels are extremely expensive and because of their size they had to be assembled in space. The blue circle represents the 10 meter solar panel to be mounted on the orbital transfer vehicle. Using a 200 kW fel the power generated by this panel is 1.8 times that generated by one of the 33 foot long space station panels. Using a megawatt laser it is 8.7 times as much power. Its area is 4.8 times smaller. It has been suggested[2] that the only real possibility for powering large ion engines is nuclear power. One engineer, who is currently working on developing a non-nuclear 50 kW Hall Thruster, told the author that he hoped he was not just "playing in a sandbox", since it was unclear whether an alternate practical, non-nuclear power source can be developed. It can be.

FIGURE 3. Space station silicon cell solar panel. Length is 32.6 meters, width 11.6 meters. The circle represents the solar panel to be mounted on the OTV, which using the fel will generate more power than the space station panel.

Ground-Based Energy Source for Satellites

Fortunately a source of ground-based energy for satellites is already largely designed and is waiting to be built[6,7]. The space part of the system consists of a solar panel with conventional gallium arsenide cells mounted on a receiver 1/5th of the area of a single Space Station Solar Panel and delivering almost twice as much power. The power comes from a ground-based, ready-to-build free electron laser. The laser components have largely already been tested and are now in operation at the Stanford Linear Accelerator Complex (SLAC). Dr. Barletta and his group at the Lawrence Berkeley National Laboratory (LBNL) would be expected to take the lead in constructing the new laser, with help from personnel at Lawrence Livermore National Laboratory (LLNL), SLAC, and Bennett Optical Research (BOR). We would welcome help from the rest of the U.S. fel community including those at Lockheed Grumann, Thomas Jefferson National Laboratory, and the group at the University of Hawaii headed by Dr. John Madey, inventor of the free electron laser. LBNL designed most of the components for the accelerator train now being used at the Stanford Linear Accelerator Complex and seen in Figure 4. These components, which are of a very advanced design, have been in use now for over two years without major problems. They handle beam currents of as much as 0.7 amperes, well over an order of magnitude higher than other systems. The electron current is key to the power output. LBNL personnel designed and were ready to build a novel free electron laser under contract to BOR, which holds one patent and has a patent pending on the laser design. The United States Government has a royalty-free license on the patent. The laser was designed to initially develop an output beam of 200 kW in power expandable to one megawatt after an improved injector system is developed. Recent materials advances lead BOR to believe that the difficulties in developing improved injectors have now been greatly diminished. A schematic diagram of the laser and transmission system is given in Figure 5. The transmission system includes lightweight, relatively inexpensive graphite fiber impregnated cyanate ester superpolished mirrors. The primary mirror will be an adaptive optic. A small prototype of this mirror was demonstrated at the Beamed Energy Propulsion Symposium.[8] The time frame for putting the first laser and its associated transmission system together and making it operate at the 200 kW level is 4-5 years, approximately the time allotted to develop the 50 kW Hall thruster.

FIGURE 4. Stanford Linear Accelerator line used as a prototype for the ignition feedback regenerative fel proposed for powering an OTV.

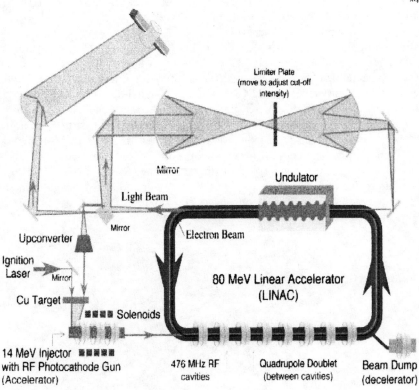

Limiter Plate
(move to adjust cut-off
intensity)

Mirror

Undulator

Light Beam

Electron Beam

Mirror

Upconverter

Ignition
Laser   Mirror

80 MeV Linear Accelerator
(LINAC)

Cu Target

Solenoids

14 MeV Injector
with RF Photocathode Gun
(Accelerator)

476 MHz RF
cavities

Quadrupole Doublet
(between cavities)

Beam Dump
(decelerator)

FIGURE 5. 200-1000 kW output regenerative free electron laser with no resonator and a non-radioactive beam dump. Output goes to a large beam director that sends it to the orbital transfer vehicle.

Beam Director

The requirements for the beam director necessary to maintain a constant diameter laser beam for propelling an orbital transfer vehicle in space are discussed in a companion paper.[9] In order to maintain a 6 meter diameter beam to geosynchronous orbit (GEO) the beam director must have a 15-17 meter diameter primary mirror. It would be segmented and somewhat larger than the segmented Keck telescope now in operation. A one watt diode laser mounted on the orbital transfer vehicle (OTV) would aid in maintaining the laser beam on the OTV solar panel as illustrated in Figure 6. Sensors on the panel would indicate beam position in real time. A lead-ahead angle would be programmed in or a companion satellite used to correct for the lag caused by the velocity of light, which is important even when the satellite is in GEO and appears stationary, as pointed out by Zeiders.[10]

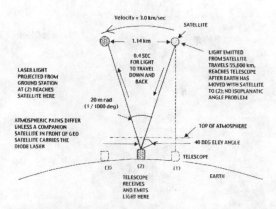

FIGURE 6. Pointing and tracking procedure for the laser beam.

Laser Beam Is Eye Safe

Because of the large diameter of the beam director mirror, which must be 15 –17 meters in diameter[9], the laser intensity in the atmosphere is lower than that of the sun and far below the visual damage threshold for pulsed lasers. Figure 7 shows the transmission and absorption of near infrared radiation by the eye. The laser wavelength is within the transmission region of the eye. However there is no safety problem when birds or airplanes fly through the laser beam. The rf laser has a pulse width of 3.3 picosec and a pulse separation of 80 ns. The retinal damage threshold per pulse of laser energy[11] is 10^{-5} J/cm^2, the eye lens damage threshold/pulse[11] is >4 x 10^{-3} J/cm^2. A 200 kW laser projected through a 15 meter diameter beam projector, assuming a 1% loss per window surface and 95% total reflectance loss from the 5 mirrors, has a total exit power of 186 kW and an average power of one joule/sec/cm^2 exiting the mirror. This energy is in the form of pulses 3.3 psec in width separated by 80 nsec. The average energy per pulse is then 3 x 10^{-13} J/cm^2, eight orders of magnitude below the retinal damage threshold and ten orders of magnitude below the eye lens retinal damage threshold.

655

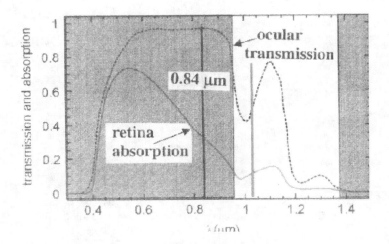

FIGURE 7. Retinal absorption and ocular transmission in the visible and near infrared region of the spectrum. The ocular and retina laser damage threshold for the laser beam projector are many orders of magnitude higher than the laser beam intensity.

Electrical Power Generated

When the laser reaches the satellite the laser spot size will be 6 meters in diameter. Assuming[12] the Strehl ratio is 0.5, the atmospheric transmission is 0.89 at a wavelength of 0.84μm, and 84% of the laser beam is inside the first Airy minimum, the laser energy from a 200 kW laser striking the solar cell that is inside the first Airy maximum will be 83 kW. The electrical conversion efficiency for gallium arsenide irradiated by the laser light at a wavelength of 0.84μm[12] is 52%. The solar irradiation outside the atmosphere is shown by the smooth dark line and after it has passed through the atmosphere (airmass 2) by the jagged dark line. An excellent atmospheric window occurs at 0.84 μm, right at the peak sensitivity of the gallium arsenide cell. As a result, about 3 times as much energy is produced by the laser irradiation as by an equivalent intensity solar irradiation. The electrical energy generated by the solar panel will then be 42 kW, which is greater than the electrical energy generated by the 33 meter long Space Station Solar Panel seen in Figure 3. By increasing the laser power it will be possible to raise the electrical power generated by the laser to as much as 200 kW.

Propulsion in Low Earth Orbit

Curvature of the earth limits the length of time a satellite is in view of any one laser, particularly at low elevations, as seen in Figure 9. Initially there will be only one laser, and solar energy will furnish power when the satellite is not visible to the laser. Double sided solar cells may thus be desirable. In low earth orbit (LEO) an OTV in equatorial orbit spends as much as 1/3rd of its time in the shadow of the earth. It is then not receiving power from the sun. Even a single laser can supply power to the satellite for part of the time to help correct for the loss of power caused by this earth shadowing. At about the level where the Van Allen belt starts the laser will begin to supply most of the power for up to one-third of the time per laser. If double sided solar cells are used, the sun will continue to supply power also and act as a useful backup in case there is a cloud or other problem with the laser, now the main source of energy. The satellite will carry a 10 meter diameter solar panel and the laser beam will be a constant 6 meters[9] in diameter all the way from low earth orbit to geosynchronous orbit.

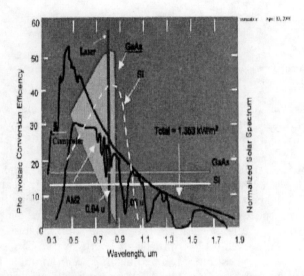

Figure 8. Special characteristics of solar cells. The lower jagged curve is the transmission spectrum of the sun through an atmospheric path length of air mass two (i.e., twice as much air as is found when pointing at the zenith). A laser beam transmitted through the atmospheric window at a wavelength of 0.84 mm, indicated by the heavy line, is close to the maximum sensitivity of both silicon and gallium arsenide solar cells. At this wavelength, both Si and GaAs cells produce nearly three times more electrical power per unit of laser light flux than they do when irradiated by the sun.

LONGITUDE COVERAGE vs. HEIGHT OF OTV

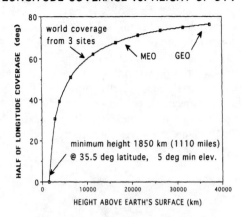

Figure 9. Longitudinal coverage of the Earth increases rapidly above 10,000 km and 3 laser sites can cover the Earth.

The laser output wavelength will be set at the peak efficiency wavelength for the solar cells. Fortunately that peak occurs at an atmospheric window, as seen in Figure 8. For laser irradiation the gallium arsenide cells have an efficiency of over 50% instead of the 17% obtained when irradiated by the sun[12]. The resulting 300% improvement in output power will be more than doubled by the increased output density of the laser beam as compared to the sun. Adequate heat loss in the solar array is crucial to keep the parts of the satellite from overheating in space. This loss can be achieved successfully for laser power levels up to one megawatt. The fact that more than half of the incident laser light is transformed into electricity reduces the amount of heating. Only 17% of the sun's energy is transformed into electricity. Most of the rest of it goes into unwanted heating. With multiple lasers, irradiation can be continuous from the Van Allen Belt outward, and electrical power levels of 50 kW to 100 kW or more to drive a Hall thruster or a magnetoplasmadynamic thruster appears feasible.

Laser Sites

Initially there will only be one laser site to supplement solar power for the OTV. That site should be near Ridgecrest, California, seen in Figure 10. It is in the rain-shadow of the High Sierra Mountains and has 260 clear days per year. This desert community of 25,000 people is home to the Naval Air Weapons Station (NAWC) and the Naval Air Warfare Center Weapons Division, largest landbased Research, Development and Test Facility in the United States Navy. The Navy Facilities are here in large part because of the clarity of the air. Visibilities of 100 miles are normal. The area has the highest solar insolation of any spot in the United States and the value of the Fried Coefficient r_0 averages over 7 cm. It often reaches into the double digits. Several feasible sites around Ridgecrest have elevations over 1200 m (4,000') but the best one is on NAWC. Over it is one of the largest restricted airspace areas in the United States. Finally, the community is full of highly technical, very tolerant people who take sonic booms and other technical demonstrations as a matter of course.

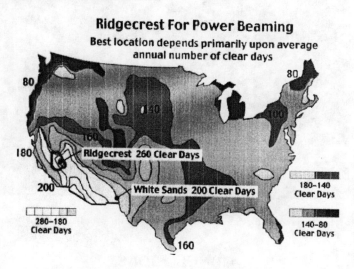

FIGURE 10. Clear days in various parts of the United States. The circled area around Ridgecrest is Restricted Flight Area 2506. Ridgecrest is in the rain shadow of the High Sierras.

A second laser site could be located in the Hawaiian Islands, possibly on 3055m (10,023') Mount Haleakula on the island of Maui. It is high enough to be above the clouds most of the time and houses a large telescope run by the Air Force. A third laser site could be in the Australian Desert. The coverage afforded by these sites is shown in Figure 11. A fourth site could be in Peru and additional sites in Morocco or in the Sahara mountains and in the Near East or India. Alternately sea sites could be used. The NASA sponsored Space Elevator Program is planning to mount this fel on a large, quasi-stationary barge similar to the one built by The Boeing Company, which has a rocket launching platform. It is situated in the ocean on the equator west of the Galapagoes Islands as shown in Figure 11. The fel is the most efficient and perhaps the only way of powering the climber ascending the nanotube cable to space. Without use of an adaptive optic system the laser beam would become dispersed in passing through the atmosphere. However the adaptive optic system being developed under NASA contract and reported in reference 10 has a faceplate influence function and actuator separation which makes it functional down to r_0 values of 2 cm or less. Therefore it should operate well even at sea level.

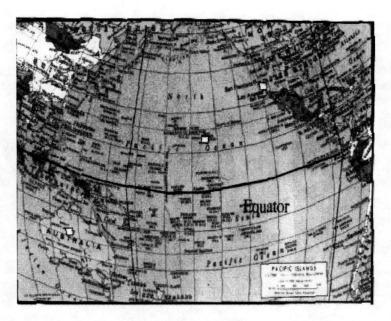

FIGURE 11. The white boxes indicate suggested laser sites to cover the Pacific Ocean. Additional sites are discussed in the text.

CONCLUSIONS

Ion thrusters offer great promise as economical, continuous thrust, reliable sources of propulsion, and would be much more promising if there were a way to use them without relying on nuclear sources of electrical energy. The electrical power requirements for large ion engines are too large to be met in any economical way by conventional solar panel designs. However they can be met by using novel photoelectric cell panel designs and powering the panel from the ground using a 200-1,000 kW free electron laser.

The design for such a laser exists and the operation of its most critical component, the accelerator train, has been successfully operating at SLAC for over two years. The operation of nearly all of the other laser components has also been demonstrated experimentally also. A means for maintaining a constant diameter laser beam from LEO to GEO has been described and a by-product of this technique is that the laser is eye-safe in the earth's atmosphere. Finally the location of possible laser sites has been discussed. The development under a currently-funded NASA program[13] of an adaptive optic that will correct atmospheric distortion of the transmitted laser beam even when the laser is at sea level gives additional flexibility in choosing the optimum laser sites.

ACKNOWLEDGEMENTS

The author would like to acknowledge the encouragement he received from Dr. John Rather, former head of New Concepts at NASA Headquarters, to work on the problem of powering orbital transfer vehicles. Dr. Glenn Zeiders, a NASA Consultant and Mr. Sandy Montgomery from Marshall Space Flight Center have also contributed materially to this program.

REFERENCES

1. "Vehicle Launch Propulsion" in *Jane's Space Directory, 2000-2001*,16th ed, David Baker ed. (Sentinel House, Coulsdon, UK, 2000) pp. 310-311.
2. Lerner, Eric J., "Plasma Propulsion in Space," *The Industrial Physicist*, October 2000, pp.16-19.
3. Herman, Albert L., "Advanced Spacecraft Propulsion Takes Flight," Launchspace Publications, 1997.
4. Jankovsky, Robert S., Glen Research Center, NASA (private communication).
5. "Human Space Flight" in *Jane's Space Directory, 2000-2001*,16th ed, David Baker ed. (Sentinel House, Coulsdon, UK, 2000) p. 527.
6. Zholents, A.A.; R. Rimmer, O. Walter, W. Wan, and M. Zolotorev, "FEL design for power beaming" in *Free Electron Challenges II*, H.E. Bennett & D.H. Dowell, eds., Proc. SPIE vol. 3614, 72-85 (1999).
7. Bennett, H.E., "FEL powering of satellites: a technically and economically viable program" op. cit, pp. 1514-167 (1999).
8. Bennett, H.E., J.J. Shaffer, R.C. Romeo and P.C. Chen, "Prototype Ground-Based Adaptive Optic Transfer Mirrors for Space Applications," First International Symposium on Beamed Energy Propulsion," Huntsville, Alabama, November 5-7, 2002 (in press).
9. Bennett, H.E., "Laser Spot Size Control in Space," First International Symposium On Beamed Energy Propulsion, op. cit. (in press).
10. Zeiders, G.W. Jr., The Sirius Group (private communication).
11. Winburn, D.C. *Practical Laser Safety*, (Marcel Dekker, New York 1985) pp. 20-21.
12. Murphy, David M., *Final Report on Photovoltaic Array Design for Power Beaming to GEO Spacecraft*, AECAble Engineering Company, Inc., Contract No. A5753220 with Bennett Optical Research Inc., July 2000, p. 7-10.
13. NASA Phase II Contract Number NAS8-02008, awarded to Bennett Optical Research Inc. on 1-23-02.

Vehicle And System Concepts For Laser Orbital Maneuvering And Interplanetary Propulsion

Jordin T. Kare

Kare Technical Consulting, 222 Canyon Lakes Pl., San Ramon CA 94583 jtkare@attglobal.net

ABSTRACT

In-space laser ablative propulsion using beamed power (as opposed to on-board lasers) may be superior to all alternatives, except possibly nuclear propulsion, for rapid, efficient transport within the Earth-Moon system and, eventually, the inner Solar system. The keys to this concept are large, lightweight transmitting optics and even larger, lighter vehicle-based collectors. We present scaling relationships for cislunar and interplanetary laser propulsion, and discuss some options for these components. In particular, large (10 - 50 meter) diffractive optics and holographic concentrators appear to be enabling technologies that can be demonstrated within a few years and deployed operationally in the 2010's.

INTRODUCTION

Using laser beamed energy for orbital maneuvering propulsion has been considered for many years [1]. The basic concept for laser orbital maneuvering is shown in Figure 1: a ground- or space-based laser transmits power, possibly via one or more relay points, to a spacecraft in orbit. In pulsed ablation propulsion, the laser is pulsed, and each pulse ablates a thin layer of an inert solid (or possibly liquid) propellant; the hot gas expands against the propellant surface to produce thrust without requiring a nozzle. By choosing the propellant and pulse characteristics, the specific impulse can be varied between a few hundred and several thousand seconds [2], and potentially higher. The achievable thruster efficiency η_{thr} (exhaust kinetic power / received beam power) is not well understood, but is likely to range between 20% and 50%

Other forms of laser propulsion have been considered for orbit transfer, including CW plasma propulsion [3], and laser-electric propulsion; in particular, the laser-electric orbital transfer vehicle (OTV) concept has been analyzed in some detail [4].

Relative to CW plasma propulsion, pulsed ablative propulsion offers superior propellant characteristics (dense, storable solid or liquid vs. cryogenic hydrogen) and

CP664, *Beamed Energy Propulsion: First International Symposium on Beamed Energy Propulsion,*
edited by A. V. Pakhomov
© 2003 American Institute of Physics 0-7354-0126-8/03/$20.00

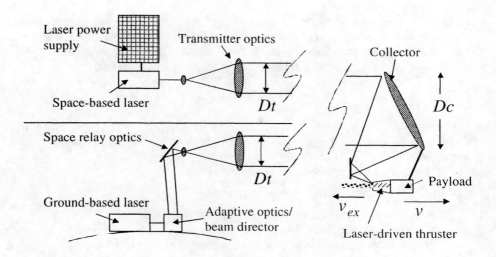

FIGURE 1. Laser propulsion orbital maneuvering system concept

higher specific impulse. Relative to laser-electric propulsion, pulsed ablative propulsion offers:

- Similar overall efficiency (e.g., 25% η_{thr} vs. ~50% $\eta_{photovoltaic}$ x 50% $\eta_{electric\_thr}$)
- Much wider range of usable laser wavelengths
- Lower propulsion system mass and cost

In particular, the cost of megawatt quantities of photovoltaics and electric thrusters probably limits laser-electric systems to reusable OTV's, while pulsed ablative propulsion can directly replace ordinary single-use chemical stages.

Pulsed ablative propulsion does require a laser with high peak power and low pulse duty cycle, but a variety of laser technologies can meet these requirements.

SYSTEM CONCEPT

Laser

For the purposes of this paper we assume a ground-based laser. For the foreseeable future, a ground-based laser will be substantially less expensive than a space-based laser, as well as easier to maintain.

The preferred laser wavelength depends on available laser and optics technology and on whether the system is "dual use" -- providing both propulsion and electric power -- or propulsive only. We assume a nominal wavelength of 1 μm, available from free-electron lasers (FELs) or diode-pumped solid-state lasers and suitable for dual use. If a suitable laser is available (e.g., an FEL) a slightly longer wavelength, nominally 1.5 μm, would be "eye safe" and therefore have much reduced safety concerns with respect to specular reflections (glints). (Target spacecraft could be designed to minimize or eliminate glints, but glints from accidental illumination of other satellites or debris would remain an issue for beams crossing low Earth orbit.) Other system characteristics discussed below can be scaled appropriately for the longer wavelength without significantly changing the system concept.

Relay Location

The aperture which actually sends power to a vehicle can be in various locations: on the ground, suspended from an aerostat, or in low, medium, or geosynchronous Earth orbit (LEO, MEO, or GEO).

Of these options, a GEO relay has multiple advantages:

- (Nearly) stationary relative to ground laser site

    - Ground aperture does not need a pointable mount
    - Relay collector does not need to point (assuming separate collector/transmitter optics, as opposed to a simple reflector)
    - Minimizes adaptive optics complexity; eliminates "pointahead" (effect of target motion during light round trip time).
    - Fixed beam path minimizes interference issues for aircraft and spacecraft

- Good access to all Earth orbits

The issue of interference with aircraft and spacecraft has not generally been appreciated, but is critical for the acceptability of a beamed-power system, due to the widespread use of sensitive optical detectors (in all wavebands) looking toward the Earth. A fixed beam at a known location can be incorporated into spacecraft operations plans; when spacecraft must pass close to the beam, the time of passage can be predicted and either the laser or the sensor(s) shut down. A beam which sweeps over any significant angle, even on a known schedule (e.g., to connect to a LEO or MEO relay) greatly complicates this scheduling, and creates risks for many more satellites. A high-power uplink beam which sweeps unpredictably (e.g., to track individual mission vehicles) is almost certainly unacceptable.

A ground-based telescope has extremely limited access to low Earth orbits and limited access even to high orbits (circa 25%, e.g., $\sim\pi$ steradians out of 4π). A LEO relay, used with a ground-based laser, has both limited access to the laser and limited access to most orbits (very limited for LEO, $\sim$50% for high orbits) and is thus the worst of all possible worlds, unless a large constellation of LEO relays is available.

Aerostat-carried relay mirrors (analyzed by Kare [5]) offer essentially horizon-to-horizon operation, similar to a LEO relay, but remain fixed over the laser. They may be an interesting alternative to a GEO relay, especially for development and demonstration of laser propulsion.

MEO relays offer coverage similar to GEO relays, with smaller optics, at the expense of requiring a tracking beam projector on the ground. Relays in Molniya orbits could be preferable for certain missions.

Relay Optics

Prior concepts for laser orbital maneuvering systems have been limited by the anticipated cost and feasibility of space-based optics. For both the ground-to-relay and relay-to-vehicle links, the minimum aperture sizes are set by diffraction:

$$D_t \, D_c = f_{opt} \, R \, \lambda. \tag{1}$$

For f_{opt}=2.44, this formula defines the range at which the first zero of an Airy pattern just fills Dc, assuming a uniform transmitter illumination. In this case, 84% of the transmitted power is collected. However, smaller values of f_{opt} are acceptable, especially if the transmitter aperture illumination is tapered (i.e., maximum in the center and decreasing toward the edge). Deliberately tapered illumination is common in RF engineering but less so in optics, although it has been suggested for laser power beaming (e.g., [6]). We assume $f_{opt} = 2$, which provides ~95% transmission efficiency with tapered source illumination.

For $\lambda = 1$ μm and R ~ 40,000 km, $D_t \, D_c > 80$ m^2, so typical concepts required a 10 meter ground telescope as well as an 8- to 10-meter relay mirror. 10-meter space optics have been regarded as at or beyond the state of the art, and certainly expensive; the 6.5-meter James Webb Space Telescope, optimized for infrared operation, will cost roughly $825 million [7]. Lightweight, low-cost alternatives such as inflatable membrane optics have not demonstrated anything approaching diffraction-limited optical performance at large sizes. Longer ranges (e.g. GEO to GEO) or longer wavelengths appeared unfeasible in the near term.

Recently, however, at least one near-term technology for large space-based optics has made significant progress: long-focal-length diffractive optics. The Eyeglass project [8, 9] is aimed at producing 25-meter-diameter space optics with diffraction-limited visible-wavelength resolution. The project has demonstrated a 5-meter f/100 diffractive optic fabricated on panels of ~0.5 mm thick glass, with an areal density of 1-2 kg/m^2. Glass films as thin as 50 um have been investigated as substrates, yielding a projected areal density of 0.1 - 0.2 kg/m^2.

The Eyeglass design takes advantage of the relaxed tolerances associated with high f/number, long focal length optics; the nominal 25 m f/100 Eyeglass primary has a 2.5 km focal length. An Eyeglass telescope therefore consists of a free-flying primary, kept flat and in tension by slow rotation (or possibly by a supporting ring), and a separate free-flying optical system at the focus of the primary. Eyeglass is capable of broad wavelength coverage by using corrective optics at the focus of the main aperture to compensate for the chromatic aberration of the diffractive primary lens, but such correction would not be required for a monochromatic laser.

Assuming 1 kg/m^2 for deployed Eyeglass apertures, a 25-meter aperture would mass only 491 kg; a reasonable estimate for a complete telescope is 1000 kg.

Transmitting a 1 μm beam from the ground to a 25 meter aperture at GEO requires a ground telescope diameter of only 3.2 m (for f_{opt} ~2.0). Quite modest telescopes

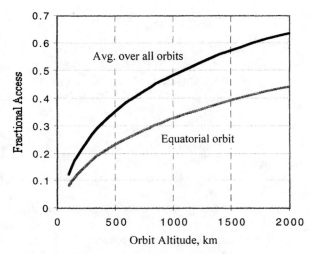

FIGURE 2. Line of sight access to LEO (fraction of entire LEO orbit) from a GEO relay, for a minimum beam altitude of 30 km

are therefore sufficient to transmit power to GEO, and wavelengths as long as ~2.5 um are usable with current (up to ~10 m) ground telescope technology.

Similarly, transmission from GEO to LEO, using a 25 meter transmitting aperture, requires a collector aperture of only a few meters. Since the spacecraft collector is a light bucket, and does not need anything like diffraction-limited quality, it can be implemented as a simple rigid panel or dish, and still fit easily in a launch shroud. Somewhat larger deployable collectors may be desirable to reduce beam pointing requirements or beam flux, or to allow use of longer wavelengths.

LEO Access and Multiple Relays

A single geosynchronous relay can supply power to spacecraft in slightly more than 50% of LEO space; the remainder being blocked by the Earth. Therefore, for general-purpose use, a laser power system should have at least two, and preferably three, GEO relays.

GEO to GEO relaying requires a range of up to 80,000 km, depending on the Earth-centric angle between relays. However, using 25-meter apertures at both ends of the link, diffraction losses can be kept small ($f_{opt} > 4$) for wavelengths as long as approximately 2 μm.

Even with an eye-safe laser wavelength, it is quite possible that pointing a high-flux, high power laser at the Earth will be determined to be an unacceptable risk. In fact, the risk could be made extremely low by various straightforward measures, such as requiring a feedback link from the target spacecraft to the relay, or establishing minimum beam slew rates and "keep out" zones around populated areas. However, it

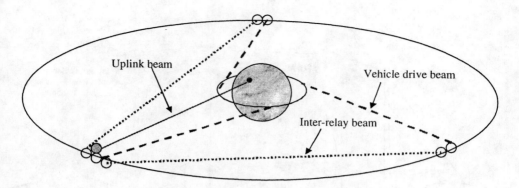

FIGURE 3. Laser orbital maneuvering system concept with one primary and two secondary geosynchronous-orbit relay stations

is not clear that these will be sufficient for public and regulatory acceptance of a high-power beam aimed at the ground.

If pointing the beam at the earth is not allowed, a GEO relay can still access a useful fraction of LEO space. Figure 2 shows the fraction of LEO space that is accessible from a GEO relay assuming a minimum beam altitude of 30 km. With three GEO relays, approximately 50% of a 300 km altitude orbit is accesible; with 6 relays, nearly all LEO space is accessible without pointing a beam at the Earth.

System Configuration

Figure 3 shows a conceptual overall system design for laser propulsion in near-Earth space (LEO/MEO/GEO and transfer orbits). Figure 4 shows a highly conceptual design for the primary relay site in GEO.

Comparatively small optics are used to transfer beams between the collector focal point and the transmitter focal point at each relay. These optics will presumably be sized by their power handling requirement, since the ranges involved are a few kilometers at most.

Separate apertures are used for each transmitted beam at the primary relay site to avoid having to rapidly re-point a large aperture. With this configuration, switching the beam from a vehicle to a secondary relay, or one secondary relay to another, can be done essentially instantaneously.

To avoid coupling loads into the main apertures, the focal point optics are not attached to the main optics. The focal point optics can be kept in position either propulsively (since the differential accelerations are of order 10^{-6} m/s^2, or 0.1 m/s per day) or, as shown in the figure, by using a counterbalancing mass on a tether, so that the centers of gravity of the main aperture and the tether assembly are co-orbiting.

For the output beam, which must track a moving vehicle, the focal point optics must be able to maneuver over an appropriate area in the focal plane. For tracking vehicles in LEO the relevant angles are +/-0.2 radians (8000 km) and the

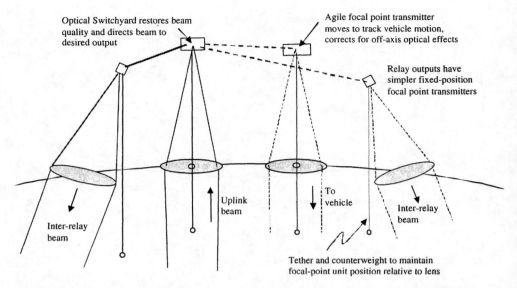

Optical Switchyard restores beam quality and directs beam to desired output

Agile focal point transmitter moves to track vehicle motion, corrects for off-axis optical effects

Relay outputs have simpler fixed-position focal point transmitters

Uplink beam

To vehicle

Inter-relay beam

Inter-relay beam

Tether and counterweight to maintain focal-point unit position relative to lens

FIGURE 4. Concept for primary GEO laser beam relay using multiple Eyeglass long-focal-length diffractive optics

corresponding motion range (for a 25-meter f/100 system) is +/-500 meters, over approximately 90 minutes. This motion may be implemented propulsively or by using a reaction structure, or by some combination.

Because of the high f/number of the Eyeglass apertures, beam aberrations will be small even for large off-axis beam angles, and can be corrected in the focal-point optics. A detailed optical design will be needed to determine how large an off-axis angle can be accommodated before the aperture itself needs to be rotated. If off-axis angles of ~30 degrees can be accommodated, it may be possible to use a common aperture for receiving and retransmitting the beam.

ORBIT TRANSFER SYSTEM PERFORMANCE

Figure 5 shows the payload mass (in kilograms per megawatt-day of laser energy) which can be transferred from LEO (300 km orbit) to GEO as a function of thruster Isp. (Note that this plot includes a nominal factor of 0.25 for thruster and transmission efficiency.)

Two curves are shown in Figure 5, corresponding to Hohmann transfer and spiral orbits. Spiral or5bits are associated with continuous thrust; the more efficient Hohmann transfer requires that impulse be delivered near perigee until the orbit apogee reaches geosynchronous altitude, and then near apogee until a circular orbit is achieved. In fact, laser propulsion transfers may optimize at some intermediate trajectory. For example, low orbits may be raised to ~1000 km circular orbits to increase laser access time before beginning the main perigee "burns" to raise the orbit apogee to geosynchronous altitude. The required laser energy (MW-days) for this

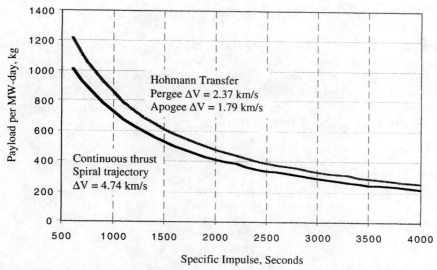

FIGURE 5. Mass transfer rate from 300 km orbit to geosynchronous orbit vs. Isp. Energy (MW-day) is laser output energy, assuming 25% thruster and transmission efficiency.

option would be larger than for a direct transfer from 300 km orbit, but the calendar time would be reduced because more time per orbit could be spent accelerating.

LUNAR AND INTERPLANETARY MISSIONS

A 25 meter transmitter can beam power to ranges well beyond GEO to GEO links, provided the collector on the vehicle is sufficiently large. We assume a collector technology similar to the Eyeglass telescope technology: large area diffractive optics, which can be held taut by a simple frame. However, for low-quality collector optics, either reflective or relatively fast (low f/number) transmissive diffraction gratings are reasonable; a transmissive grating has the advantage of greater tolerance to surface deformation (e.g. under acceleration) than a reflector. Curved reflectors, particularly inflatable reflectors, are also plausible, but may not scale well to very large sizes.

A diffractive concentrator made of 10-mil (0.25 mm) plastic film would have a mass of 0.25 - 0.5 kg/m^2; using 2 mil plastic would reduce this to of order 0.1 kg/m^2. 10-micron (0.4 mil) film would have an areal mass of ~0.01 kg/m^2, and still lighter concentrators should be possible with advanced solar sail-type materials, but much below 0.1 kg/m^2 support structures may dominate the collector mass. For a range of 400,000 km (approximately the maximum GEO-Moon distance) and $f_{opt} = 2$, a 32-meter collector is required; at 0.1 kg/m^2 such a collector would mass approximately 80 kg. Thus, transfers from LEO to Lunar orbit would be feasible with overall performance similar to that for LEO to GEO transfers.

Earth-Escape Missions

For missions beyond the Moon, a substantial fraction of the acceleration time will necessarily occur after Earth escape velocity is achieved. The range over which the spacecraft can accelerate is therefore proportional to the size of the collector, and the duration of the acceleration, and thus the energy delivered, increases with D_c.

For acceleration from rest in the absence of gravity, there is a direct relationship between range and velocity (derived in [10]):

$$R = \frac{m_f v_f^3}{2P_{exh}} \left(\beta^{-3} \right) \left(e^\beta - \beta - 1 \right) = \frac{m_f v_f^3}{2P_{exh}} f(\beta) \tag{2}$$

where $\beta = v_f / v_{exh}$, v_f is the mission final velocity (velocity at the end of acceleration, in the laser's frame) and

$$P_{exh} = (-dm/dt)\, v_{exh}^2 = P_{laser}\, \eta_{trans}\, \eta_{thr} \tag{3}$$

Figure 6 illustrates the relative variation of range (given by $f(\beta)$), burn duration, and mass ratio with beta. Note that very low values of beta (i.e., $v_{exh} \gg v_f$) yield very slow acceleration and therefore very long range.

Combining eq.s 2 and 3 with the diffraction range limit (eq. 1) yields

$$m_f = \left(\frac{D_t D_c}{f_{opt}\, \lambda} \right) \left(\frac{2P_{laser}\, \eta_{thr}\, \eta_{trans}}{v_f^3\, f(\beta)} \right) \tag{4}$$

Assuming a fixed fraction f_c of m_f is allocated for a collector (and its support structure) with a mean areal density σ:

$$f_c\, m_f = \frac{\pi}{4}\sigma D_c^2 \quad \text{or} \quad D_c = \sqrt{\frac{4f_c\, m_f}{\pi\sigma}} \tag{5}$$

Defining a "payload" mass (vehicle dry mass excluding collector) $m_{pay} = (1 - f_c) m_f$, yields a direct calculation of the mass that can be accelerated from rest to v_f by a given laser power:

$$m_{pay} = (1 - f_c) \left(\frac{4f_c}{\pi\sigma} \right) \left(\frac{D_t}{f_{opt}\lambda} \right)^2 \left(\frac{2P_{laser}\, \eta_{thr}\, \eta_{trans}}{f(\beta)} \right)^2 v_f^{-6} \tag{6}$$

Two notable features of this equation are the extremely rapid variation in m_{pay} with v_f, and the fact that m_{pay} varies not with P_{laser} but with P_{laser}^2. The latter is

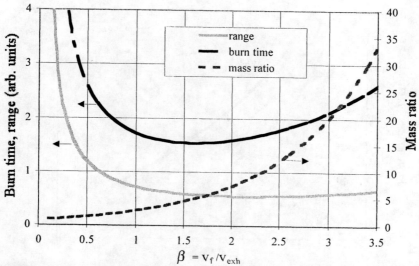

FIGURE 6. Variation in relative laser range, acceleration time ("burn time"), and mass ratio with β

qualitatively understandable, in that increasing the laser power allows an increase in the overall spacecraft mass, and therefore in the collector size.

The capabilities of a nominal 10 MW beamed power system (with $\eta_{trans} \times \eta_{thr} = 0.25$ and $f_c = 0.2$) are plotted in Figure 7. $f_c = 0.2$ yields a below-optimum final payload mass (by a factor of 0.64) but is probably more practical than an optimum collector mass fraction of 50%

The remarkably fast variation in m_{pay} with v_f leads to a highly counterintuitive result, which is that a it may be desirable to use a modest ΔV chemical propulsion stage after the <u>end</u> of a high-I_{sp} laser-boosted acceleration phase. There is a final mass advantage to doing so for $v_f < 3\ v_{exh\_boost}$ (where v_{exh_boost} is the chemical-stage exhaust velocity) if the laser collector is carried through the chemical boost, and up to $v_f = 6\ v_{exh\_boost}$ if the laser collector is dropped. Of course, using such a chemical boost imposes a penalty in mission mass ratio.

The actual value of m_f for a modest-velocity interplanetary launch will be somewhat different, due to the effect of Earth's gravity, and the ability to continue accelerating past the nominal diffraction-limited range, but will have similar scaling.

The collectors implied by Eq. 5 can be very large: for fixed collector mass fraction and areal density, the collector diameter simply scales with the $m_f^{1/2}$, and can thus be shown on the same axis as m_{pay} in Figure 7. Even larger collectors would be associated with lower collector areal densities. It can be argued that very large collectors (bigger than ~100 m diameter, for a 10 MW laser system) would provide more power if used as solar collectors for solar-electric or solar-thermal propulsion. Broadband solar collectors, however, will be significantly heavier than narrowband

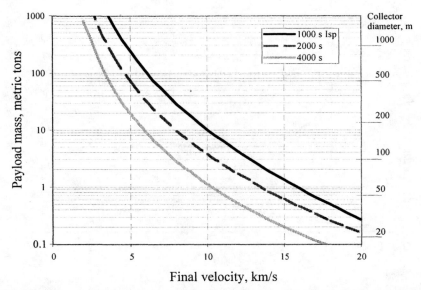

FIGURE 7. Payload mass (vehicle mass excluding collector) vs. final velocity for laser propulsion launch of interplanetary missions. Collector diameter indicated at right, assuming collector mass = 25% of vehice final mass and areal density of 0.1 kg/m^2.

diffractive laser collectors, especially at very large scales, and solar-electric systems will have a large cost and mass penalty for the power system and thruster.

The ranges involved are also large -- over a million km, for a 100 m collector -- but not sufficiently long to provide substantial thrust at Mars. However, reasonable extrapolations of the technology could provide useful propulsion at interplanetary ranges. For example, D_t = 100 m and D_c = 1000 m gives f_{opt} = 2 at 50 million km.

For manned Mars missions and other missions involving return to Earth, the laser system can provide deceleration into Earth orbit at high specific impulse. Given the mass ratios for round-trip missions, minimizing the propellant needed for deceleration has a very high leverage for reducing overall mission mass.

CONCLUSIONS

Assuming 25 meter-class diffractive optics can be deployed in space at reasonable cost, a highly useful multipurpose beamed-power architecture for in-space transportation can be build with readily foreseeable technology. Using pulsed ablative propulsion at 25% efficiency and a 10-MW laser, payloads of several thousand kg can be transferred from low orbit to geosynchronous orbit in 1-2 days of laser operating time and a few days of calendar time. The fastest transfers will be made with modest (for pulsed ablation) Isp's of order 1000 s. The recommended system configuration uses a ground-based laser and a primary relay station in geosynchronous orbit, plus two or more secondary relay stations also in GEO. By using separate apertures for the direct-to-spacecraft beam and for transmission to each secondary relay, beams can be rapidly switched without steering large apertures. Similarly, using long-focal-length diffractive apertures allows steering beams over moderate angles without moving the main aperture.

Finally, the same system, used with plausible large-area collectors, can launch very substantial masses into interplanetary trajectories -- up to several hundred tons at several km/s. However, the mass launched decreases rapidly with final velocity, so that, although fast missions can be done with reasonable mass ratios, they will require lasers significantly larger than 10 MW.

REFERENCES

1. Myrabo, L. N., "Power-Beaming Technology for Laser Propulsion, " in Orbit Raising and Maneuvering Propulsion: Research Status and Needs, L. H. Caveny, ed., Progress in Aeronautics and Astronautics V. 89, AIAA, New York, 1984, pp. 3-29.
2. Pakhomov, A. V., Thompson, M. S., and Gregory, D. A., "Ablative Laser Propulsion Efficiency," AIAA 2002-2157, presented at 33rd AIAA Plasmadynamics and Lasers Conference, Maui, HI, USA, 2002.
3. E.g., Kemp, N. H. and Legner, H. H., "Steady (Continuous Wave) Laser Propulsion: Research Areas," in Orbit Raising and Maneuvering Propulsion: Research Status and Needs, L. H. Caveny, ed., Progress in Aeronautics and Astronautics V. 89, AIAA, New York, 1984, pp. 109-128.
4. Montgomery, E. E. IV, "Beamed Energy Transportation Initiatives at MSFC", OE/LASE Symposium, Free Electron Laser Challenges Conference, San Jose, California, February, 1997, Proc. SPIE, V. 2988, 1997.
5. Kare, J. T., "A Laser Orbital Maneuvering System With Aerostat Relays," in *Proceedings of the Beamed Energy Transportation, Annual Technology Workshop*, ed. by E. E. Montgomery, NASA Marshall Space Flight Center, MSFC, AL, October, 1996.
6   Dickinson, R. M., Interstellar Beamer Engineering, in *Proceedings, Space Tech. and Applic. Internat. Forum 2001*, M.S. El-Genk, ed., AIP CP552 (2001), pp. 565-570
7. Contract issued to TRW for observatory fabrication and integration, not including launch, announced Sept. 10, 2002. See, e.g., <http://ngst.gsfc.nasa.gov/FastFacts.htm>
8. Hyde, R. A. and Dixit, S., "Large-aperture Diffractive Optics for Space-Based Lasers," UCRL JC-139446, Lawrence Livermore National Laboratory, 2000.
9. Early, J., "Fresnel Optics For A Laser Driven Lightsail Interstellar Flyby Mission", these Proceedings.
10. Kare, J. T., "Pulsed Laser Thermal Propulsion For Interstellar Precursor Missions," *in Proceedings, Space Tech. and Applic. Internat. Forum 2000*, M. S. El-Genk, ed., AIP (2000).

The Energy Tanker Concept

Edward E. Montgomery IV

TD15
NASA Marshall Space Flight Center
Huntsville, Alabama 35812

Abstract. The energy tanker concept is based on the observation that solar flux increases as the square of the close distance to the sun. Such flux-rich environs will provide both a more plentiful supply of radiant energy and photon momentum. Robotic mission concepts to deliver an energy collection facility in the inner solar system, then transport energy from that facility to LEO and other destinations benefiting Human and Robotic Exploration and Space Science is proposed. Potential destinations are discussed. A list of alternative energy transport technologies is compared. The sensitivity of the performance of these systems as a function of proximity to the sun is discussed. Basing strategies are developed, taking advantage of non-keplerian orbit capabilities of solar sails. A primary trade-off is beam transmission of energy versus physical transport of energy storage containers. The impact on science and exploration in LEO/GEO space, interplanetary missions, and finally interstellar missions is addressed.

POWER AND PROPULSION NEED

Much of our scientific exploration of the universe beyond the confines of the planet Earth can be characterized as energy consumption at points vast distances away. Energy is consumed when payloads traverse space and planetary terrains, retrieve samples, create images, measure electrical and magnetic fields, store and transmit information, and often function or act to survive in harsh environments. Even the most passive space experiment at least requires energy for transportation.

The last few decades of the space program have demonstrated that two keys, power and propulsion, are the driving and limiting factors in what can be achieved in space.[1] This is evident in a number of ways. A mass summary of a typical commercial satellite mission is shown in figure 1). The non-power/propulsion mass is less than a fifth of the total mass inserted into the transfer orbit to a geostationary altitude. If the weight of the insertion stage and the launch vehicle were included, the payload mass would shrink to an almost vanishing fraction. The largest portion of mission cost has is for the launch services, demonstrating its importance. The lack of on-board power also limits the number of transponders on geostationary Comsats and the quality of their transmission.[2] Transponders provide income and are the leverage on positive ROI.

After a decade of vigorous development in closed ecology life support technology, NASA was forced to abandon most of the advances in the face of meager power availability on-board the space station.[3] Even after significant advancements in photovoltaic efficiency, the array modules are the largest dimensions of the

CP664, *Beamed Energy Propulsion: First International Symposium on Beamed Energy Propulsion,*
edited by A. V. Pakhomov
2003 American Institute of Physics 0-7354-0126-8

International Space Station. Additionally, the Mars Rover could travel only a short distance from its base due to inadequacy of power available to communications.[4]

A third to a half of most GTO payloads is the apogee kick motor
Another 10-20% is in solar arrays and batteries

Satellites	Power	Propulsion	Payload
Communications (TDRSS)	20 %	32 %	18 %
Remote Sensing (GEO)	13 %	24 %	30 %
Intelsat VI	12 %	55 %	17 %
Intelsat V	10 %	60 %	13 %
DOMSAT	14 %	58 %	9 %
TV-SAT	13 %	55 %	11 %
	10-20 %	30-60 %	10-20 %

FIGURE 1. Leveragable Mass of Typical Spacecraft Power & Propulsion

The two keys of power and propulsion have become interconnected in the advent of propulsion systems based on non-chemical means. Space nuclear systems development initiatives may soon provide significant on-board power to achieve specific impulse in thousands of seconds rather than hundreds.[5] This technology could provide a robust future to exploration in the energy-starved reaches of the outer solar system and interstellar space.[6] For exploration of the inner solar system, however, nuclear system options are not as dominating when compared to alternatives. Close to the sun, where thermal control is already more challenging, the waste heat characteristically generated by nuclear power is not a positive feature. While nuclear thermal and electric systems are capable of generous power output, they bring the burden of significant plant mass for reactors, shielding, and radiators. Fluids and fluid handling systems typical of nuclear designs are inherently less reliable than solid-state mechanisms. While nuclear propulsion has higher specific impulse than conventional chemical stages, nuclear propulsion still uses significant quantities of propellant as reaction mass.

AVAILABILITY OF RESOURCES

In contrast, there is a class of propellantless propulsion alternatives that includes, at least, contenders for missions with small masses and slower trip times. These suggest an architecture alternative to nuclear powered outer solar system missions – one that looks first towards the inner solar system.

Solar Energy

Such a view identifies several relevant characteristics. Compelling is the dominance of the sun in the space and planetary environments. The energy available from the sun for space missions can be inferred directly from the solar flux density, which varies

675

with the square of the distance to the sun. At one AU, the flux density of the sun (1370 watts/m^2) is weakly diffused. Large solar arrays are needed to power habitats and experiments. Demand lacks supply to such an extent that major design limitations have been imposed that in turn have reduced the scope of mission accomplishments.[7]

Such is not the case in the vicinity of inner planets, and at perihelion of smaller bodies on eccentric orbits (e.g. asteroids Icarus 1566 and Apollo). Figure 2 shows the dramatic increase that occurs within the orbit of Venus. With the abundance of energy flux also comes the potential for solving the transportation problems as well. Specifically the solar sail transportation option becomes very attractive. Consider that sail performance (thrust, acceleration, payload, trip time) also improves by square of proximity to the sun.[8] Solar Sails are fundamentally simple mechanical devices with relatively few moving mechanical parts, no fluid storage and handling, no propellant management, no energetic exhaust, no combustion chambers and nozzles under extreme pressure and temperature, and no electrodes to vaporize in repetitive arcing.

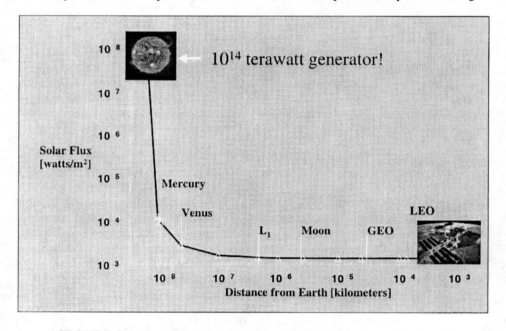

FIGURE 2. Available Solar Flux Increases Dramatically in the Inner Solar System

The inner solar system is certainly not a benign environment and is a concern of space mission designers. Early investigations of the effects of the space environment on thin films indicate radiation may not cause significant degradation in optical properties of the sail materials, even in the ultraviolet.[9] Significant degradation of the mechanical properties is a growing concern however. The sail technical community considers the high thermal input on orbits closer than 0.25 AU an unsolved problem. Although spacecraft have operated this close to the sun, they were made of structural members with significant thickness. Even though surface properties were degraded over time, these spacecraft relied on the underlying material bulk to maintain structural integrity through the life of the mission. There is no underlying bulk in the ultra-thin membrane of a solar sail. Furthermore, charge accumulation from the solar wind on the large, thin sail structure will require careful attention in design.

Transportation

Finally, the sun's gravitational field also varies with the inverse square of distance. Trajectories leaving the earth's orbit for the outer planets are adding energy at perihelion to raise aphelion. Although, the opposite is true for trajectories to the inner planets, significant energy is still required for interplanetary orbital transfers. About the same amount of energy is required for a Venus mission as is for a Mars mission and Mercury is about the same as Jupiter. Trip times for missions departing Earth for the inner solar system are much shorter and favorable trajectory windows occur more frequently.

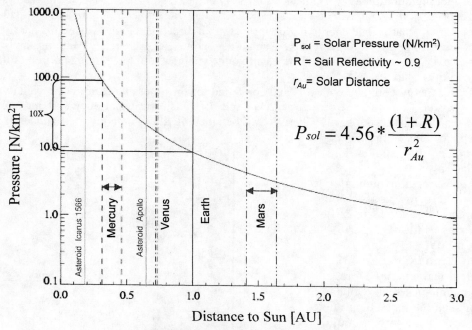

$$P_{sol} = 4.56 * \frac{(1+R)}{r_{Au}^2}$$

P_{sol} = Solar Pressure (N/km²)
R = Sail Reflectivity ~ 0.9
r_{Au} = Solar Distance

FIGURE 3. Solar Photon Sail Pressure

Solar Electric, Solar Thermal, and Solar Sails are all systems that will perform better with increased solar flux. To illustrate, figure 3 shows a plot of the dramatic increase in solar pressure available to a solar sail when operated at the inner solar system. For example, the same sail operating at the perihelion of Mercury (90 N/km²) would produce an order of magnitude more thrust and acceleration than it would in the vicinity of the earth. Sails are propellantless. Therefore, as long as the sun is producing photons, they are the ultimate reusable option.

A drawback of sails is that unreasonably large area sails are required to move payloads massing than a few hundred kilograms in the vicinity of earth.[16] A small robotic tanker concept may be feasible though. The most challenging leg of a transportation architecture connecting earth and the inner solar system would likely be thrusting at departure when a tanker might be fully loaded. Fortunately, that will also be when the tanker is deep in the solar system and can benefit from the higher thrust there. In this architecture, arrival of a full tanker at earth might be the most stressing. Perhaps aerobraking or a momentum exchange tether can be employed.

Energy Storage

If the transportation exists to move a commodity around frequently and efficiently between the earth and the inner solar system, the only question remaining is what shall that commodity be? It has already been established that plentiful, high quality power (or its accumulation as work, heat, or energy) is the next obstacle. Again, the sun is proposed as the source. The sun is a 4×10^{14} terawatt generator with lots of load capacity to spare. The energy is flowing out in all directions at once, which is hardly efficient for getting it to the relatively small region of space we occupy. But the supply is so overwhelmingly sufficient; we can do quite well just to intercept that portion of it that flows nearby.

The energy could be collected and converted to some storable form in the vicinity of the sun. Then it can be transported via a solar propulsion transportation system back to the vicinity of the earth. There it can be stockpiled and/or converted and used. Applications could include power for space and planetary science/research/mining/manufacturing bases or higher thrust/heavier lift transportation. Figure 4 provides a survey of conventional terrestrial electrical energy storage technologies. Photovoltaic or solar thermal engines in close solar orbit could charge these devices. Their energy would then be dumped into similar devices near the earth that have been discharged

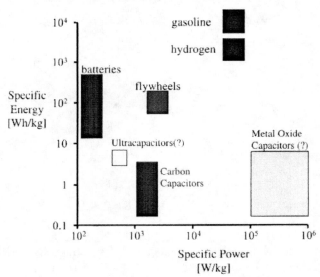

FIGURE 4. Electrical Power and Energy Storage Comparison [after 13]

The 78 kilowatt space station power system utilizes photovoltaic arrays while in sunlight and Nickel-Hydrogen batteries during the third of its orbit when the station is in eclipse. This means that tankers with payload capabilities in the range of 200-1000 kg can only bring enough energy to provide a few hours of operation at best. See figure 5. This is probably not worth the investment even if you were to employ a fleet of such systems with regular deliveries. It should be noted that data relates to technologies developed primarily for terrestrial applications. A design study specific to this application might produce more favorable results. Also, a momentum exchange tether could capture and store the momentum energy from an incoming sail mass in the form of potential energy by moving mass into a higher energy orbit.

Another option is to transmit the energy rather than transporting it. The photon is well suited for delivery of energy packets over great distances. The sun produces such a prodigious number of photons, it more than compensates for the relatively meager amount of energy in a single photon. Solar wind particles (protons primarily) are more massive and therefore produce a higher momentum exchange. But the natural flux of those particles is several orders of magnitude below radiated photonic energy. As early as the 1970's, NASA produced concepts for 100 megawatt orbital direct solar pumped laser power stations to take advantage of these abundant energies.[14]

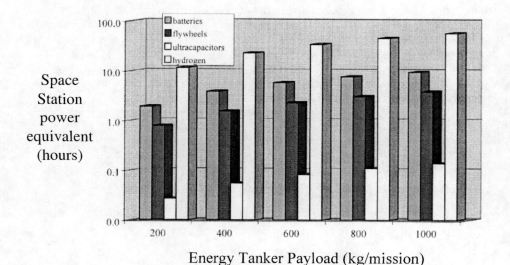

FIGURE 5. Transported Energy Trade Study

Over the last three decades the materials and structural design technologies for large lightweight apertures has progressed tremendously and many system concept studies have provided insight. It is now known with a high degree of confidence that the primary transmitting mirror is often the pivotal driver for the total mass of an on-orbit optical system larger that a few meters.[17] It is the aperture size of the transmitting solar pumped laser (or other?) station that also determines much of the character of a power transmission system connecting the inner solar system with earth. For a laser of wavelength λ, the projected spot diameter (d) at some distance R from transmitter of diameter D is given by the expression:

$$dD = 2.44\ \lambda R \tag{1}$$

For a selected wavelength and identical transmitter and receiver dimensions (i.e., d = D), the fundamental proportionality is revealed:

$$D = R^{1/2} \tag{2}$$

Figure 6 shows the total energy delivered to the earth of a pair of transmitters and receivers as function of transmitter proximity to the sun. Note that until the transmitter is within the orbit of Venus there is very little increase over the power available from a collector in low earth orbit. In contrast though, a 214 meter, diffraction limited receiver transmitter in the orbit of mercury would ideally deliver over 6 sols to LEO. This assumes only the full central lobe of an ideal gaussian intensity distribution in the beam is received and converted without loss from sunlight to laser. To correct for the latter effect, Landis has assessed the losses for conversion from sunlight to laser at the transmitter and then to electricity at the receiver.[15] Direct solar pumped conventional solid state lasers illuminating silicon photovoltaics should have a throughput of only about 5%. His concept for solar pumped semiconductor diode laser might have efficiency as high as 35%. The efficiency question is moot though, since even subject to the worst conceivable efficiency, there is still power to spare.

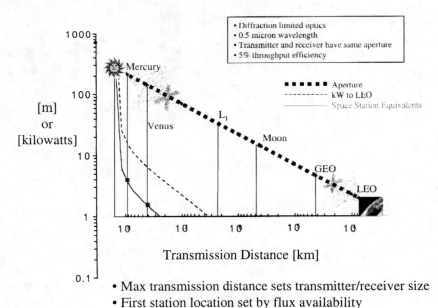

FIGURE 6. Transmitter/Receiver Aperture and Delivered Solar Power

Although it's not necessary, in this study the companion receiver to the transmitter mentioned above was chosen to be the same 214 meters in diameter. Such a collector would produce over 14 megawatts or 180 space station load equivalents! Even if the efficiencies reduce the performance and order of magnitude this is an interesting option. For transmitter orbits even closer to the sun, the energy delivered increases phenomenally. Smaller than diffraction limited apertures would still produce acceptable levels. Also, the receiver could not only collect the transmitted energy, but the normal ambient flux as well! This latter feature is a convenient graceful degradation path for a system. In the event the transmitting station fails, the power output of the receiver could still be significant.

The concluding observation is that it may be a better strategy to begin our expansion into the solar system by first going to a region of space where power and propulsion are in more abundant in-situ supply – toward the sun, and not away. An initial foray can be less expensive with smaller, simpler, robotic systems involved. Solar sails enable the benefits propellantless transportation and maybe necessary to enable non-Keplerian orbits required to keep a transmission system in alignment. Small payloads and relatively slow trip times will limit human participation for the near future. More robust, compact robotics should be sufficient until the inner solar system is better known and newer technologies can be developed to improve power and propulsion capabilities even further. The best means to bring energy from the sun out to the vicinity of earth seems to favor transmission over transportation, although development of tethers and or high temperature, long term, large quantity energy storage concepts specifically for this scenario could make that option more attractive. The option exists that the transmitted photons would be used for additional light pressure on a sail, forming a beamed energy transportation system rather than a power utility. Eventually this pipeline[10] can be an integral part of beamed energy transportation throughout the solar system and beyond.[11]

ACKNOWLEDGMENTS

The leverage of power and propulsion on space exploration was first illustrated to me by Dr. John D. G. Rather. From Dr. Glenn W. Zeiders, I learned the physics the laser propagation and the discipline of critical thought. It was L. Whitt Brantley's insight that space exploration is the projection of force to a far distant point and that large, thin films in space are possible. C. Les Johnson inspired me with the value of propellantless propulsion and gave me a job working on solar sails. John W. Cole provided support and guidance to the formation of the forum of revolutionary thought for which this work was prepared. A special acknowledgement is extended to the late Dr. Robert L. Forward, whose concept for a laser driven interstellar mission may someday spring from the foundation laid in the concepts presented here. All these men have paid a price for being champions of innovation and their courage deserves acknowledgement. Finally, I am not ashamed to thank Jesus Christ, for everything.

REFERENCES

1. J.D.G. Rather, "Power Beaming Options", proceedings of the *Second Beamed Space –Power Workshop*, NASA Conference Publication 3037, March 1989, Hampton, Virginia. pp. 21-40.
2. G. A. Landis and L. H. Westerlund, "Laser Beamed Power: Satellite Demonstration Applications," paper IAF-92-0600, presented at the 43rd Congress of the International Astronautical Federation, Washington, DC, 28 Aug.-3 Sept. 1992; also available as NASA Contractor Report CR-190793, 1992.
3. E. E. Montgomery , "Initial Assessment of Life Support Technology Evolution and Advanced Sensor Requirements", Systems Technology Group/SRS Technologies, Interim Technical Report, SRS/STG-TN91-03, NAS8-38781 Huntsville, AL, February 1991.
4. J. Matijevic, "Sojourner: The Mars Pathfinder microrover flight experiment", *Space Technology*, 17, No. 3/4, 143-149, 1997.
5. G. Langford, "Nuclear Systems Initiative, 38th Annual Joint Propulsion Conference, July 2002.
6. B. Farris, B. Eberle, G. Woodcock, and B. Negast, "Integrated In-Space Transportation Plan, Phase I Final Report", Gray Research, September 14, 2001.
7. Various authors, "From Sunlight to Power: International Space Station Solar Arrays", NASA Publication ET1988-07-003-HQ
8. R. L. Forward, " Grey Solar Sails", AIAA paper 89-2343, 25th Joint Propulsion Conference, Monterey California, July 10-14, 1989.
9. Dever/GRC, Semmel/Qualis Corp., Edwards/MSFC et al "RADIATION DURABILITY OF CANDIDATE POLYMER FILMS FOR THE NEXT GENERATION SPACE TELESCOPE SUNSHIELD", number AIAA-2002-1564, 43rd AIAA/ASME/ASCE/AHS/ASC Structures, Structural Dynamics, and Materials Conference, Denver, CO, 22-25 April 2002
10. E. E. Montgomery, "Solar Power Pipeline - Components for Beamed Energy Transportation", 1996 MSFC Research and Technology Report, Marshall Space Flight Center.
11. L. Johnson, "NASA charts course to sail to the stars on largest spacecraft ever built" Journal of Aerospace and Defense Industry News, May 16th, 2000
12. G. Garbe, "Solar Photon Sail, Basic Principles", presentation at Goddard Space Flight Center, September, 2002.
13. National Renewable Energy Laboratory, http://www.engineering.sdsu.edu/~hev/energy.html
14. J.D.G Rather; E. T. Gerry; and G.W Zeiders: "A Study to Survey NASA Laser Applications and Identify Suitable Lasers for Specific NASA Needs." W.J. Schafer Associates, Inc., February 1978.
15. G.. A. Landis., "Prospects for Solar Pumped Lasers", W.J.Schaffer Associates, NASA Lewis Research Center, 2121-09, 1993.
16. C. R. McInnes, *Solar Sailing: Technology, Dynamics and Mission Applications*, Praxis Publishing, Chichester, UK, 1999. "Solar Sails, " 200-100 kg in LEO
17 E. E. Montgomery IV, "Solar Power Pipeline - Components for Beamed Energy Transportation", 1996 MSFC Research and "Roadmap to Large Apertures in Space", NASA Marshall Space Flight Center, 1998 AIAA Defense and Civil Space Programs Conference and Exhibit, Von Braun Civic Center, Huntsville, Alabama, October 29, 1998.

Propulsion Systems Integration for a 'Tractor Beam' Mercury Lightcraft: Liftoff Engine

L.N. Myrabo

Department of Mechanical, Aerospace and Nuclear Engineering
Rensselaer Polytechnic Institute, Troy, NY

Abstract. Described herein is the concept and propulsion systems integration for a revolutionary beam-propelled shuttle called the "Mercury" lightcraft – emphasizing the liftoff engine mode. This one-person, ultra-energetic vehicle is designed to ride 'tractor beams' into space, transmitted from a future network of satellite solar power stations. The objective is to create a safe, very low cost (e.g., 1000X below chemical rockets) space transportation system for human life, one that is completely 'green' and independent of Earth's limited fossil fuel reserves. The lightcraft's airbreathing combined-cycle engine operates in a rotary pulsed detonation mode PDE for lift-offs and landings; at hypersonic speeds it transitions into a magnetohydrodynamic (MHD) slipstream accelerator mode. For the latter, the transatmospheric flight path is momentarily transformed into an extremely long, electromagnetic "mass-driver" channel with an effective 'fuel' specific impulse in the range of 6000 to 16,000 seconds. These future single-stage-to-orbit, highly-reusuable vehicles will ride "Highways of Light," accelerating at 3 Gs into space, with their throttles just barely beyond 'idle' power.

INTRODUCTION

Let's face it, we're smashed up against the hard physical limits of chemical-fueled airbreathing and rocket propulsion, and nothing revolutionary is likely to come from this familiar technology - at least until HEDM (High Energy Density Material) propellants, or antimatter becomes practical. What we desperately need is an efficient, non-polluting space transportation system that can propel us safely into orbit for the price of a common airline ticket. Otherwise, the dream of having thousands of us living and working in space, within our lifetimes, will continue to elude us.

The RPI Lightcraft Project proposes to exploit unconventional, ultra-energetic pathways to enable advanced propulsion/power systems and active control over hypersonic aerothermodynamics - for tomorrow's spacecraft. The obvious reason for choosing beamed energy propulsion is that historically, the future has rarely been changed in revolutionary ways by pursuing 'conventional' wisdom. It often takes novel and radical concepts to trigger such progress. Low cost launch systems will not magically appear just because some clever soul figures out how to increase a turbine's efficiency by 0.2%, or a chemical rocket's specific impulse by 10 seconds. However, if the power required for propulsion can be transmitted electromagnetically to the vehicle in flight, and the atmosphere is used for reaction mass (i.e., through momentum exchange with airbreathing engines), then the limits of the rocket equation can be circumvented altogether.

CP664, *Beamed Energy Propulsion: First International Symposium on Beamed Energy Propulsion*,
edited by A. V. Pakhomov
© 2003 American Institute of Physics 0-7354-0126-8/03/$20.00

The author's earliest concepts for laser powered shuttlecraft [1-4] were designed around orbital laser power plants (i.e., 'tractor-beam' configurations). Then in 1986, the Lightcraft Project was established at RPI under NASA Headquarters sponsorship, through an Advanced Design Program that was managed and directed by the Universities Space Research Association (USRA). The first two years of RPI's participation (academic years '86/'87 and '87/'88) in the USRA Advanced Design Program involved the conceptual design of a 5-person vehicle, the 'Apollo' Lightcraft [5-6] – named after the Lunar Command Module. In the '88/'89 timeframe, attention shifted to a near-term 1.4-m diameter microsatellite launcher called the Lightcraft Technology Demonstrator [7]. In the remaining years of the Advanced Design Program sponsorship, RPI's attention shifted to the single-place Mercury lightcraft [8-11] – i.e., the subject of this paper.

The Mercury Lightcraft Concept

The Mercury lightcraft relies upon beamed energy propulsion to greatly improve engine specific impulse (and thereby drastically cut launch mass) of futuristic manned missions to space. By definition, beamed energy propulsion must employ a remote source to project laser or microwave power beams directly to the craft in flight (i.e., a 'line-of-sight' wireless transmission link). The lightcraft is equipped with a receiving antenna that reflects this electromagnetic power beam into a propulsion energy converter (i.e., a unit that transforms the collected power into usable thrust).

One of the major assumptions underlying this Mercury lightcraft design effort, was that future satellite laser power stations could transmit any desired power level, repetitively-pulsed or continuous waveform, and number of beams – that the engine demanded. This freed up the Lightcraft design team to concentrate its attention exclusively upon creating the simplest and most efficient vehicle/engine/optics/ airframe configuration. The lightcraft is assumed linked to the orbital powerplant with a pre-programmed, interactive flight control system. The specific impulse of its hypersonic airbreathing engine is so high that only 10-20% of the gross liftoff mass must be dedicated to liquid expendables. For orbital insertion, circularization, or de-orbit, a brief laser-heated rocket mode is assumed, which would consume little propellant.

Figure 1 is the first 'cutaway-view' attempted for the Mercury lightcraft – pictured during a laser boost into orbit (by Ron Levan – artist) using its rotary pulsed detonation engine (PDE) mode. This initial conceptual design has subsequently evolved over the years, along with the dimensions of various engine/vehicle/airframe components (e.g., shroud, rocket-driven MHD generators, optics, etc.). Note that the vehicle is axisymmetric and that the repetitively-pulsed, laser-generated thrust acts upon the aft centerbody surface, which also serves as a plug nozzle and re-entry heat shield. The PDE thruster is employed for vertical liftoff and landings, as well as rapid accelerations through Mach 3. In this mode, the laser beam is brought in from above, then reflected by the large (off-axis parabolic) primary optics onto smaller secondary mirrors located under the engine cowl. The beam is then reflected into a tight focus across the vehicle afterbody, causing air breakdown and the formation of hemi-spherical blast waves that momentarily contain the high air plasma pressures.

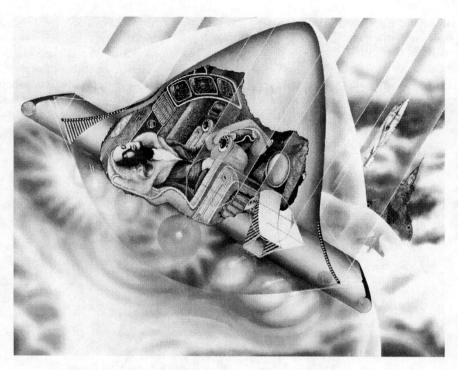

FIGURE 1. Artist concept for Mercury lightcraft at Mach 3 in rotary PDE mode

Several of the lightcraft's interior features are also visible in Fig. 1. The occupant is accommodated in a reclined seating position, for high G tolerance. Finger- and palm-actuated controls (e.g., trackballs) are integrated into the chair armrests and flat-screen color displays (or mini-display goggles) provide a visual link for the occupant. Together, they facilitate a 'user-friendly' interface to the lightcraft flight computers, and the outside world – perhaps interpreted through a virtual reality environment. Direct viewing of the outside flight environment (i.e., through one-way, 'electronically closeable' windows) is allowed only when the propulsive laser beam is shut off. The flat screen display can be linked into a 'super vision' system after the lightcraft has finished its transatmospheric flight and the power beam is terminated. When the vehicle is coasting through space, the 'super vision' system can engage the 2.2m-diameter primary optics, numerous smaller optical elements, fiber-optic cables, and electronic image processing – to function much like a powerful multiple-mirror telescope. With this high resolution optical telescope, the occupant can make astronomical observations, examine a destination spaceport (e.g., before deorbit), or send messages into deep space (laser communications).

Luggage is shown stowed in twin compartments under the armrests (Fig. 1), and the boarding/de-boarding steps retract flush with the interior wall (much like jet aircraft today). The Mercury lightcraft is equipped with tripod robotic landing gear (with each leg individually articulated) – the minimum-mass penalty for a vehicle that can take off and land vertically (VTOL). The simple gear design was inspired by "back-hoe" excavation machines, and was selected for its ability to accurately tilt and point the lightcraft's receiving antenna – at a low altitude laser relay satellite – for precision lock-on and tracking, just prior to lift-off. Finally, the entire vehicle is designed to 'kneel' in the tripod gear, when the occupant desires to board or exit the vehicle. Altogether, such features attempt to create a comfortable, ergonomic environment for the occupant during the short ascent (e.g., 3-8 minutes), and brief travel time - at most 45 minutes to any point on the globe. A small liquid hydrogen propellant tank is placed beneath the pilot/cockpit, enclosed by the lightcraft aftbody.

Rotary Pulse Detonation Engine Mode

Since it is likely that the MHD slipstream accelerator mode cannot be engaged below Mach 3 and perhaps 3-km altitude, a separate laser-thermal thruster (i.e., non-electric mode) is required for the initial portion of the launch. Figure 2 shows the

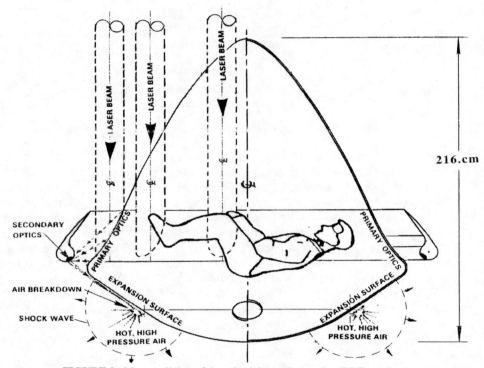

FIGURE 2. Mercury lightcraft in pulsed detonation engine (PDE) mode.

Mercury lightcraft in its rotary pulsed detonation engine (PDE) cycle [12], which develops laser-generated thrust upon the aft centerbody surface. For this propulsion mode, the shroud is translated forward to position "a" in Fig. 3 -- to engage special

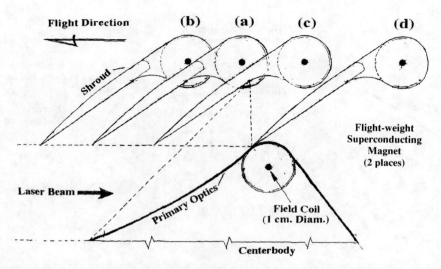

FIGURE 3. Translating shroud positions for key engine modes. Note two 'flight weight' superconducting magnets in a Helmholtz coil arrangement – for 2 Tesla field.

focusing mirrors (called SO, or secondary optics) mounted on the inner lower edge of the shroud, and to capture a sufficient air mass flow rate into the annular inlet of the PDE thruster. Note that the primary receptive optics (also called 'PO,' or primary mirror), represents about half of the vehicle's frontal area in Fig. 4, and is designed to receive pulsed laser beams at twenty-four (24) specific PO mirror locations. Each 22-cm circular laser beam has a uniform 'top-hat' intensity distribution. Several detectors around each primary mirror will monitor the laser intensity at the beam's edge, and will be linked to an active feedback system to provide accurate beam pointing and tracking.

After reflecting from the PO surface in Fig. 2, each beam is focused onto a smaller secondary mirror that covers the shroud's aft interior surface. This secondary mirror (i.e., 'SO' - which has a near cylindrical geometry), then reflects the beam into a tight focus across the vehicle aft-body, triggering air breakdown and a hemi-spherical blast wave. Laser energy is absorbed into air by the inverse Bremsstrahlung process, and ignites a high-pressure Laser Supported Detonation (LSD) wave; this 'plasma bubble' deteriorates into an unpowered blast wave after the laser pulse terminates. The airbreathing PDE engine develops thrust as these repetitively pulsed, high-pressure blast waves expand over the aft vehicle surface.

The laser-PDE thruster doesn't have to be an inherently noisy engine, and one potential 'noise abatement' procedure might even enable near silent-running PDE operations. For example, the laser firing sequence might be timed to detonate all 24 of the blast waves simultaneously *at sub-audible frequencies* (e.g., below 16 HZ). Alternately, the 24 beams could be pulsed *sequentially*, to precess in some rapidly-rotating pattern at *above audible frequencies* (i.e., above 20 kHZ) – as shown in Fig. 4. The latter mode embodies the concept for a rotary pulsed detonation engine (or rotary PDE). Clearly, additional blast waves could be inserted between the fundamental 24 sites (which are

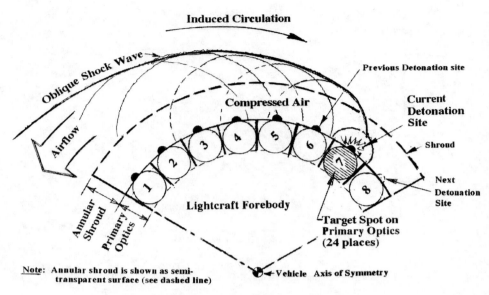

FIGURE 4. Top view of rotary PDE thruster concept

determined by the number of PO and SO mirrors), by invoking adaptive optics and/or vehicle rotation. Through proper programming, the PDE engine can directly manipulate vehicle thrust level & acceleration, flight attitude, and direction through *active thrust vector control*. In this manner, the PDE thruster can generate all the pitch and roll moments necessary for controlled flight of a lightcraft.

Above Mach 1, significant phase distortions due to aero-optical phenomena [13] may be produced in each 22-cm diameter laser beam as it traverses the bow shock wave and encounters a compressed wedge of high-pressure air (i.e., trapped between the bow shock and each PO surface). These effects are dominant mostly in the lower supersonic flight regime. To maintain a tight focus at each PDE breakdown site, all primary mirrors must have an adaptive surface that is actively controlled throughout the acceleration run (Fig. 5). Since the phase distortions are a direct function of flight Mach number (as well as the aerodynamic pressure across the fore-body), they can be sensed and corrected by manipulating the PO surface contour. The requisite adaptive optics technology is already here, and SOA reflectivities of high power laser mirrors are entirely adequate for the job.

Single pulse, proof-of-concept experiments for this PDE 'lift-off' engine were successfully demonstrated in Feb. 1991 at the Naval Research Laboratory, using the 1 um Pharos III laser [14]. As shown in Fig. 6, the tests involved a full-scale engine segment (30 cm diameter flat thruster plate) mounted onto a pendulum-type impulse measurement device. The beam was delivered at grazing incidence with a beam half-angle of 5 degrees - to a line focus measuring 3.5 to 4-cm long, positioned just 3-mm above the flat target surface. With a 0.5 Tesla permanent magnet in place, an impulse of 0.036 N-s was measured by the pendulum, for a laser pulse energy of 200 J (absorbed into the plasma); the laser pulse duration was 2 ns. It was surprising to discover that the momentum Coupling Coefficient (CC) fell from ~185 N/MW to ~120 n/MW when the magnet was removed.

These NRL tests proved conclusively that Tesla-level magnetic fields applied perpendicular to laser-PDE surfaces can improve CC performance by affecting a "magnetic nozzle" [14]. Incidentally, at 200 N/MW, the magnet-equipped NRL laser-PDE demo performed as well as the 1944 German V1 "Buzz-Bomb", and the first experimental kerosene-powered turbojet engines (i.e., time-average thrust produced per input thermal power level, or specific fuel consumption). Note however, that laser PDE thrust-to-weight ratios will be astronomically higher than today's best turbomachines.

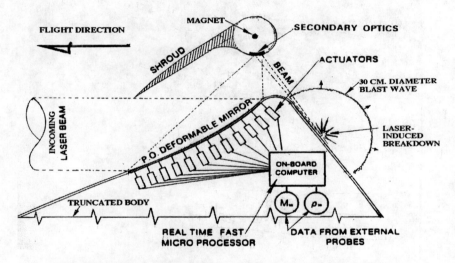

FIGURE 5. Adaptive primary optics (PO) used in rotary PDE thruster mode.

The first flight demonstration of a "tractor-beam' lightcraft was carried out by Sasoh [15] - using a repetitively pulsed CO2 TEA laser having a nominal maximum energy of 5 J/pulse at 100 Hz. Sasoh performed these "laser-driven in-tube accelerator" (LITA) experiments with a 2.7-cm diameter, 2.2 gram projectile and measured a momentum coupling coefficient of ~290 N/MW. The success of Sasoh's tests conclusively proved that the 'tractor-beam' PDE concept does indeed work.

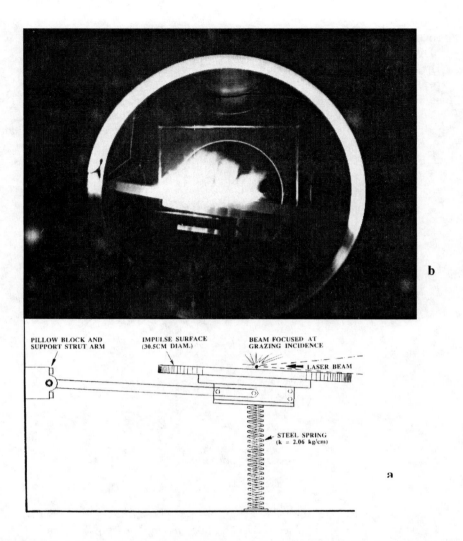

FIGURE 6. Impulse pendulum apparatus for PDE experiment (a), shown with open-shutter photo (b) of air plasma formation (1-um wavelength, Feb. 1991, NRL, Washington, D.C.).

Note that other U.S. [16] and Russian researchers have measured even higher momentum coupling coefficients - up to 580 N/MW - with simple parabolic laser PDE thrusters (single-pulse tests with no magnets, and mostly at 10.6 um).

For maximum acceleration performance at zero forward velocity, the beam-powered PDE must first 'supercharge' the inlet air, to elevate its initial static pressure and density above ambient atmospheric levels. As indicated in Fig. 4, the laser-PDE engine can accomplish this by driving an oblique rotating detonation wave around the vehicle base, to convect shocked, (i.e., pre-compressed) "fresh" air into the next detonation site. This inherent 'supercharging' function can provide very high thrust levels for lift-off, with only minor penalties paid in momentum CC efficiency.

Using 10.6 um laser radiation, the Paschen curve indicates a minimum breakdown threshold at 15-20 atm. pressure - which might be an efficient engine operating condition for that wavelength. At 15 atm, a rotating thrust area of only 22.6 sq. cm. is sufficient to levitate a 350 kg lightcraft for hover! Note that at 600 N/MW, a *single* rotating 10 MW beam can boost a 350 Kg lightcraft at 1.7 Gs; all 24 beams could yield over 40 Gs, when pulsed simultaneously (assuming the PDE can refresh itself adequately between pulses). This certainly represents *exceedingly high* levels of laser-PDE acceleration potential.   Even at 200 N/MW, two 10 MW laser power beams - rapidly pulsed (say, 560 Hz) and rotating - could easily hover a 350 kg lightcraft, whereas six active beams would accelerate it away at 3.4 Gs.

MHD Slipstream Accelerator Mode

TransAtmospheric vehicles (TAV) designed around advanced airbreathing engines must exchange momentum with a hypersonic airstream. One might be tempted ask the following: "How long are we going to cling to the concept of dumping raw chemical energy into a scramjet combustor, and just **hoping** to get positive net thrust at up to orbital velocities - skimming the upper atmosphere? If our TAV power plant of the future can employ electrical rather than thermal energy to do useful work, it has numerous advantages over the surface-applied pressure forces of existing aerospace propulsion systems. The Mercury lightcraft's hypersonic engine is designed around an alternative concept of exploiting 'high grade' electricity in a MHD slipstream accelerator, to reach out with 'action-at-a-distance' forces and interact with the surrounding airflow. The following briefly introduces this radical engine concept and attempts to explain why MHD should enable superior engine performance throughout the hypersonic regime - for future manned lightcraft.

Perhaps the most critical element in the Mercury lightcraft's propulsion system is its hypersonic airbreathing accelerator that boosts the vehicle to orbital and perhaps escape velocities – within the rarefied upper atmosphere. The basic idea is to exploit purely electromagnetic forces to move air and provide lift and maneuvering thrust. The original five-place Apollo lightcraft [5-8,17] used a shrouded electric airturborocket engine, based on the 'MHD Fanjet cycle proposed by Rosa [18-20] for an on-board nuclear-electric powerplant. The Mercury lightcraft employs an unshrouded version of this MHD engine to reduce the thermal management problems associated with cooling the inner and outer shroud surfaces, as well as the support struts.

The MHD slipstream accelerator engine should be superior to today's "accelerator-class" engines because it has the potential for operating at high efficiency over a very wide range of flight regimes. At lower supersonic speeds (e.g., Mach 3) and altitudes, it is desirable for a thrust-producing system to ingest a larger mass flow rate of air and expel it at a low velocity. On the other hand, for high-speed flight at high altitudes, it is desirable to ingest a small mass flow rate of air and expel it at a high velocity. With surface pressures developed by rotating airfoils made from solid materials (such as fan and compressor blades), it is very difficult to fashion a device that will do both jobs and still be light, reliable and efficient. Furthermore, designing efficient and versatile ramjets that can transition into scramjets that are capable of flight beyond Mach 12 (or so) has proven to be an exceedingly difficult task; note that these engines must derive propulsive thrust from surface pressures acting on inlets, combustors and nozzles. Electricity, on the other hand, is noted for its flexibility, adaptability, and ease of control.

A cross-sectional view of the MHD slipstream accelerator concept is shown in Fig. 7 for which the principal components include: a) laser-induced "airspike" to externally

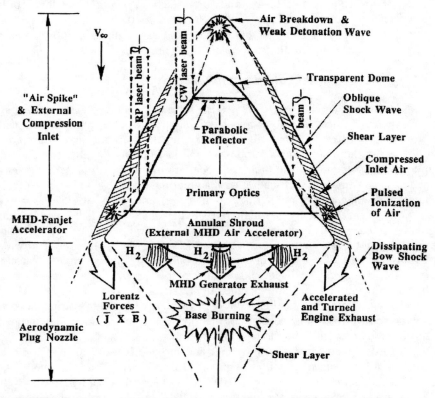

FIGURE 7. Hypersonic MHD slipstream accelerator mode for the Mercury lightcraft.

pre-compresses the inlet air [21];   b) annular MHD air accelerator, inclusive of electrodes and magnets [22];   c) set of four laser-heated, hydrogen-fueled MHD generators that are rocket-driven and open-cycle [23,24];   and,   d) a solid state electronic power system (i.e., for switching and load management) designed to link the four, 130 MWe MHD generators to the 520 MWe MHD air accelerator [22]. The conceptual design, physics and performance estimates for several critical MHD engine components are addressed in [24].   Contained in [25] is a study of optimized airbreathing launch trajectories for an Apollo Lightcraft with an MHD fanjet engine exhibiting specific impulse ranging from 6000 to 16,000 seconds (depending on the Mach number and altitude).

The ideal inlet for this airbreathing MHD engine is a novel device that enables active control of a lightcraft's external aerothermodynamics: i.e., a laser-induced airspike, or just 'airspike' for short. The physics of this novel device was first quantified by Raizer and Myrabo [21]. The principal function of an airspike is to replace the traditional sharp and massive conical forebody (normally proposed for streamlining a TAV or aerospaceplane and for precompressing the inlet air captured by its hypersonic airbreathing engine) – with a weightless alternative. The airspike accomplishes the same objectives by substituting directed energy, for mass - i.e., the mass of a sharp-nosed structure. The airspike power is varied 'upon demand' to maintain the desired bow shock wave geometry (Fig. 7) throughout the entire transatmospheric flight path. Both heat transfer and drag are greatly reduced, and the bow shock wave can never be 'swallowed,' because there is no need for an external shroud in a MHD slipstream accelerator.

Why do this? Because in ultra-energetic spacecraft, power is cheap and mass is the archenemy. Active control of external hypersonic aerodynamics enables a lightcraft airframe to be optimized under a completely different set of rules, and may even permit the use of ultra-light, pressurized tensile structures - in the pursuit of a minimum airframe mass solution. Hence lightcraft aeroshells are more likely to be blunt/rounded bodies (i.e., paraboloids of revolution, spheres, oblate spheroids, cylinders, etc.) – which present the minimum surface area to the severe hypersonic aerothermodynamic environment – thereby reducing thermal management problems. In hypersonic flight, such blunt bodies would normally produce a strong normal shock wave across their bows, and experience high aerodynamic drag and heat transfer. However, with the laser-induced airpike, this normal shock wave is transformed into a weaker oblique shock, tilted strongly aft with respect to the spacecraft forebody. Its shape can be varied at will to achieve the desired inlet pressure recovery for the lightcraft's MHD slipstream accelerator.

SUMMARY

This paper presents an overview of the propulsion systems integration for the 'tractor-beam' Mercury lightcraft concept – emphasizing the lift-off thruster mode which is a rotary pulsed detonation engine (PDE). A detailed description of the Mercury lightcraft's hypersonic MHD slipstream accelerator may be forthcoming at the next ISBEP gathering in Japan on 21 October 2003.

REFERENCES

1. Myrabo, L.N., "Solar-Powered Global Air Transportation," Paper 78-698, AIAA/DGLR 13th International Electric Propulsion Conference, San Diego, CA, 25-27 April 1978.
2. Myrabo, L.N., "A Concept for Light-Powered Flight," AIAA/SAE/ASME 18th Joint Propulsion Conference, Cleveland, OH, 21-23 June 1982.
3. Myrabo. L. and Ing, D., *The Future of Flight*, Baen Books, publisher, distributed by Simon & Schuster, New York, 1985.

4. Myrabo, L.N., "Advanced Beamed-Energy and Field Propulsion Concepts," BDM Corp. publication BDM/W-83-225-TR, Final Report for the California Institute of Technology and Jet Propulsion Laboratory under NASA Contract NAS7-100, Task Order No. RE-156, dated 31 May 1983.
5. Myrabo, L.N., et al., "Apollo Lightcraft Project," Final Report, prepared for the NASA/USRA Advanced Design Program, 3rd Annual Summer Conference, Washington, D.C., 17-19 June 1987.
6. Myrabo, L.N., et al, "Apollo Lightcraft Project, Final Report, prepared for the NASA/USRA Advanced Design Program, 4th Annual Summer Conference, Kennedy SFC, FL, 13-17 June 1988.
7. Myrabo, L.N., et al., "Lightcraft Technology Demonstrator," Final Technical Report, prepared under Contract No. 2073803 for Lawrence Livermore National Laboratory and the SDIO Laser Propulsion Program, dated 30 June 1989; See also, *Proceedings of the 5th Annual Summer Conference, NASA/USRA University Advanced Design Program*, June 1989.
8. Myrabo, L.N., et al., "Investigations into a Potential "Laser-NASP" Transport Technology, *Proceedings of the 6th Annual Summer Conference, NASA/USRA University Advanced Design Program*, held at NASA Lewis Research Center, 11-15 June 1990.
9. Myrabo, L.N., et al., "The Lightcraft Project," *Proceedings of the 7th Annual Summer Conference, NASA/USRA University Advanced Design Program*, held at Kennedy Space Center, 17-21 June 1991.
10. Myrabo, L.N., "The Lightcraft Project: Flight Technology for Affordable Space Transportation," *SSI Update -The High Frontier Newsletter*, Nov./Dec. 1992 issue, Space Studies Institute, Princeton, NJ.
11. Myrabo, L.N., "The Mercury Lightcraft Concept," *Proceedings of the SDIO/DARPA Workshop on Laser Propulsion, Livermore, CA, 7-18 July 1986*; LLNL-CONF-860778, Vol. 2, edited by J.T. Kare, T, Lawrence Livermore National Laboratory, California, 13-21 April 1987.
12. Strayer, T.D. and Myrabo, L.N., "Analysis of Laser-Supported Detonation Waves for Application to Airbreathing Pulsejet Engines," Paper No.95-2893, AIAA Joint Propulsion Conference, San Diego, CA, July 10-12, 1995.
13. Minucci, M.A.S., and Myrabo, L.N., "Phase Distortion in a Propulsive Laser Beam Due to Aero-Optical Phenomena," *Journal of Propulsion and Power*, 6, 416-425 (1990).
14. Lyons, P.W., Myrabo, L.N., et al., "Experimental Investigation of a Unique Airbreathing Pulsed Laser Propulsion Concept," Paper 91-1922, AIAA/SAE/ASME 27th Joint Propulsion Conference, Sacramento, CA, 24-26 June 1991.
15. Sasoh, A. "Laser-propelled ram accelerator," *J. Phys. IV France* 10, 41-47 (2000).
16. Myrabo, L.N., Libeau, M.A. Meloney, E.D., Bracken, R. and Knowles, T., "Pulsed Laser Propulsion Performance of 11-cm Parabolic Engines Within the Atmosphere," Paper 2002-2206, 32nd Lasers and Plasma Dynamics Conference, Hawaii, 20-23 May 2002
17. Myrabo, L.N., Moder, J.P., Blandino, J.S., and Fraizer S., "Laser-Energized MHD Generator for Hypersonic Electric Air-Turborockets," Paper 87-1816, AIAA/SAE/ASME/ASEE 24th Joint Propulsion Conference, Boston, MA, 11-13 July 1988.
18. Rosa, R.J., "Propulsion System Using a Cavity Reactor and Magnetohydrodynamic Generator," **ARS Journal**, July , 884-885 (1961).
19. Rosa, R.J., "The Application of Magnetohydrodynamic Generators in Nuclear Rocket Propulsion", AERL Research Report 111, AFBSD-TR-61-58, Contract No. AF 04(647)-278, Avco Everett Research Laboratory, Everett, MA, Aug. 1961.
20. Rosa, R.J., "Magnetohydrodynamic Generators and Nuclear Propulsion," *ARS Journal*, Aug., 1221-1230 (1962).
21. Myrabo, L.N., and Raizer, Yu.P., "Laser-Induced Air Spike for Advanced Transatmospheric Vehicles," AIAA Paper 94-2451, 25th AIAA Plasmadynamics and Lasers Conference, Colorado Springs, CO, June 1994.
22. Myrabo, L.N., D.R. Head, D.R., Yu.P. Raizer, S. Surzhikov, and R.J. Rosa, "Hypersonic MHD Propulsion System Integration for a Manned Laser-Boosted Transatmospheric Aerospacecraft," Paper No. 95-2575, AIAA, Wash. D.C. 1995.
23. Moder, J.P., Myrabo, L.N., and Kaminski, D.A., "Analysis and Design of an UltraHigh Temperature, Hydrogen-Fueled MHD Generator," *Journal of Propulsion and Power* 9 739-748 (1993).
24. Myrabo, L.N., Raizer, Yu.P., and Surzhikov, S.T., "Laser Combustion Waves in Laval Nozzles", *High Temperature* 3, 11-20 (1995).
25. Frazier, S.R., "Trajectory Analysis of a Laser-Energized Transatmospheric Vehicle," Master's Thesis, Rensselaer Polytechnic Institute, Troy, New York, August 1987

HIGH-POWER BEP USER

TESTING FACILITIES

Pulsed laser facilities operating from UV to IR at the Gas Laser Lab of the Lebedev Institute

Andrei Ionin, Igor Kholin, Boris Vasil'ev,
and Vladimir Zvorykin

Lebedev Physics Institute of the Russian Academy of Sciences,
53 Leninsky pr., 119991 Moscow, Russia
Phone/Fax: (095)132 0425; e-mail: aion@mail1.lebedev.ru

TO THE MEMORY OF N. BASOV

Abstract. Pulsed laser facilities developed at the Gas Lasers Lab of the Lebedev Physics Institute and their applications for different laser-matter interactions are discussed. The lasers operating from UV to mid-IR spectral region are as follows: e-beam pumped KrF laser (λ= 0.248 μm) with output energy 100 J; e-beam sustained discharge CO_2 (10.6 μm) and fundamental band CO (5-6 μm) lasers with output energy up to ~1 kJ; overtone CO laser (2.5-4.2 μm) with output energy ~ 50 J and N_2O laser (10.9 μm) with output energy of 100 J; optically pumped NH_3 laser (11-14 μm). Special attention is paid to an e-beam sustained discharge Ar-Xe laser (1.73 μm; ~ 100 J) as a potential candidate for a laser-propulsion facility. The high energy laser facilities are used for interaction of laser radiation with polymer materials, metals, graphite, rocks, etc.

INTRODUCTION

To solve the problem of launching a laser-propulsed satellite of 1-10 kg mass into the Earth orbit, one need to have a high-energy repetitively-pulsed (RP) MW class laser [1]. There are no so many RP lasers which can be considered as a MW class laser, and, hence, as a laser launcher for laser propulsion. One of such lasers is an e-beam sustained discharge (EBSD) CO_2 laser (λ=10.6 μm), that can produce IR radiation pulses with duration τ= 10^{-10}-10^{-3} s. Average output power up to 750 kW was reported [2]. Just EBSD RP CO_2 lasers with average output power of 10-30 kW is being used now for impressive experiments on laser propulsion in the USA [3], Germany[1], and Russia[4]. However, propagation of CO_2 laser radiation through the atmosphere is not good enough because of water vapor continuum and atmospheric CO_2, an absorption coefficient being rather high ~0.3 km^{-1}.

Another laser launcher can be a fundamental band EBSD RP CO laser (λ ~5 μm; τ =10^{-7}-10^{-3} s). Average output power up to ~100 kW is obtained for such a laser (see, for instance, a review [5]). However, this laser operates on a large number of wavelengths, some of which have very strong absorption of ~1.0 km^{-1}, but other ones have a very weak absorption of ~0.1 km^{-1}. A special procedure must be undertaken to match fundamental band CO laser spectrum with the atmospheric transparency window. An overtone CO laser (λ ~3-4 μm; τ=10^{-6}-10^{-3} s) [6] having the same design as a fundamental band CO laser, except for the optics, can be considered as a laser

CP664, *Beamed Energy Propulsion: First International Symposium on Beamed Energy Propulsion,*
edited by A. V. Pakhomov

launcher in future. The overtone CO laser can radiate within 3.3 – 4.1 μm atmospheric transparency window, the absorption coefficient being less than~0.1 km^{-1}.

A chemical HF(DF) laser (λ =2.7-4.0 μm) is a MW class laser (see, for instance, [7]). However, it is a continuous-waves (CW) one and uses quite a toxic and dangerous substance as an active medium. Non-chain reaction HF(DF) laser operates in RP mode, but still remains an experimental facility[8].

A chemical oxygen-iodine laser (COIL) (λ =1.315 μm) being a MW class laser [9] has the same drawbacks as an HF(DF) laser- CW mode of operation and toxic medium. COIL can operate in RP mode by using Q-switching and other procedures but its average output power is not high yet [10]. Another potential way of a RP COIL development in future is using a RP electric discharge generator of singlet delta oxygen on the base of EBSD [11].

Solid-state RP Nd-glass laser (λ=1.06 μm) of 100 kW average power can be used as a launcher, however it has a complicated design and very high cost of production. A free-electron laser looks also as a very expensive one.

Excimer lasers such as KrF laser (λ =0.248 μm; τ=10^{-7}-10^{-6} s)); XeF (λ =0.351 μm) and XeCl (λ =0.308 μm) are very promising as high-power lasers and can be MW class lasers, in principle. The design of "EMERALD" and "ELECTRA" laser facilities can be used as a prototype for such a laser. However, an absorption of UV radiation of these lasers in the atmosphere is extremely high. Raman frequency shifting to the visible spectral range is needed for their application as a laser launcher.

One of the promising candidate as a MW class laser launcher can be an Ar-Xe laser (λ -1.73 μm ; τ =10^{-6}-10^{-4} s). Its design can be something between RP EBSD CO$_2$ laser and "ELECTRA" facility. More detailed information about an Ar-Xe laser see in Part 4.

Although the most impressive up-to-date experiments on laser propulsion were carried out with a RP CO$_2$ laser, it does not follow from obtained results that a CO$_2$ laser is the best candidate as a laser launcher considering a coupling coefficient, angular divergence, and propagation of its radiation through the atmosphere. Experiments on laser-matter interaction must be done with lasers considered as a potential candidates for laser propulsion: fundamental and overtone CO laser, Ar-Xe and KrF laser. Numerous laser facilities developed at the Gas Laser lab of the Lebedev Physics Institute are able to deliver such an opportunity. The characteristics of these laser facilities and different sorts of their applications are discussed in this paper.

1. GAS LASER LABORATORY

The Gas Laser lab belongs to the Basov Quantum Radiophysics Institute which is a part of the Lebedev Physics Institute. The lab was founded by the Nobel Prize winner Nikolai Basov in 1983. The main objective of the lab was the study of high-power gas lasers for industry and other applications. Together with other Russian institutions the lab has taken part in R&D of RP and CW CO$_2$ and CO lasers with average power up to ~ 100 kW. The lab consists of four groups engaged in activity with molecular gas lasers, excimer lasers and lasers on atomic transitions of rare gases, especially, in activity with lasers pumped by e-beam and EBSD. Some of the laser facilities developed at the Gas Lasers Lab one can see in Figs. 1,6, and 9. The research projects of the lab have been supported by the Russian Foundation of Basic Research, NATO, European Office of Aerospace Research and Development (EOARD), International Scientific Technological Center (ISTC), and INTAS. The lab has taken part in international European Projects EUREKA 113 (CO Eurolaser), EUREKA 1390 (Ultralas), and in joint research with DuPont (USA), University of New Mexico (USA), Air Force Research Lab (USA), Naval Research Lab (USA), Directorate of Applied Technology (USA), German Aerospace Establishment , Colorado School of Mines, Rome University (Italy), and other research centers.

2. MOLECULAR LASERS AND THEIR APPLICATIONS

There are two groups at the lab engaged in molecular lasers study: EBSD molecular lasers and ammonia laser groups. Various EBSD molecular lasers were launched by using EBSD facilities with active volume of ~10 liters (Fig.1) and ~4 liters. Maximum output energy came up to 800 J for fundamental band CO laser ($\lambda \sim 5$-6 μm), ~500 J for CO_2 laser (λ =10.6 μm);~ 100 J for N_2O laser(λ =10.9 μm) and ~50 J for overtone CO laser ($\lambda \sim 3$ μm).

FIGURE 1. Electron beam sustained discharge molecular laser with active volume ~10 liters.

The experimental and theoretical research of a pulsed EBSD first-overtone CO laser was done together with the Troitsk Institute of Innovation and Thermonuclear Research (TRINITI) (Russia), AFRL and the University of New Mexico. The research demonstrated the overtone CO laser to operate within the spectral range of 2.5 - 4.2 μm that entirely covers up the atmospheric transparency window (~3.3 - 4.1 μm) (Fig2). The laser can be considered as an electric analog of a chemical HF(DF) laser. The experimentally obtained laser efficiency comes up to 11%, theoretical one being up to 20%. The frequency selective overtone CO laser operates on more than 400 spectral lines [6]. Small-signal gain on highly excited vibrational transitions is up to 0.4 m^{-1}. The experimental and theoretical results on the overtone CO laser has given us a real scientific basis for launching an RF excited supersonic CO laser at the AFRL [12]. Qualitative and quantitative analysis of spectral characteristics of the overtone CO laser and HF(DF) lasers demonstrates that the overtone CO laser has the best parameters of sensitivity and selectivity, if one is going to use the overtone CO laser as a radiation source for the spectral analysis of multicomponent gas mixtures.

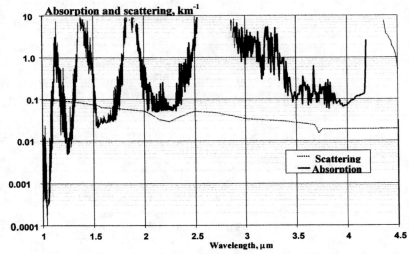

FIGURE 2. Dependencies of atmospheric absorption and scattering on wavelength [14].

Nonequilibrium vibrational kinetics of CO laser active medium was experimentally studied together with TRINITI and the High Temperature Institute (Russia). The kinetic model taking into account multiquantum exchange processes was developed on the basis of modern point of view about physical processes going in multilevel active medium of a CO laser. The experimental procedure using a double Q-switching of a frequency selective CO laser was applied. The procedure enabled us to produce a short disturbance of a vibrational distribution function (VDF) on selected ro-vibrational transition and to measure the time behavior of the VDF relaxation. The calculations made for these experiments demonstrated that, when modeling an active medium of a CO laser including an overtone CO laser operating on highly excited (V>15) vibrational transitions, one have to apply the multiquantum exchange model.

The procedure of study of nonlinear optical properties of a laser active medium was experimentally and theoretically developed. The procedure is based on detection and analysis of a phase- conjugation signal time behavior at intracavity degenerate four-wave mixing of laser radiation in the own active medium. Nonequilibrium kinetic processes taking place in the active medium manifest themselves in transient time behavior of the phase-conjugation signal. Experimental and theoretical study made together with the Institute of Applied Laser-Information Technologies (IALIT), "Astrophysics" Company(Russia) and the Directorate of Applied Technology (USA) demonstrated that the formation of a phase-conjugation signal in active medium of a pulsed EBSD CO_2 laser is related to two nonlinear mechanisms- resonant and thermal ones. The high energetic efficiency up to 20% for the intracavity phase conjugation in an active medium of the CO_2 laser is due to the properties of inertial thermal nonlinearity, energy loaded into an electric discharge and gas pressure being the most influential factors. The same method was for the first time applied together with

TRINITI to the study of nonlinear properties of an active medium of a pulsed CO laser having quite different mechanism of inversion formation and, accordingly, quite a different influence of resonant and thermal mechanisms of nonlinearity. The results obtained can be used for developing high-power laser systems and improving their angular divergence. The procedure itself can be applied for the research of nonlinear optical properties of various laser media. For instances, the study of a turbulent diffusion in an active medium of a fast axial flow CO_2 laser is under way right now at the IALIT together with the Gas Laser Lab participation.

Pulsed ammonia NH_3 laser optically pumped by CO_2 laser radiation was developed at the lab. The laser operates within ~11-14 µm spectral range, to which absorption spectra of many harmful freons belong. The design of a double frequency lidar using frequency selected CO_2 l;aser radiation as a reference beam and a frequency tuned NH_3 laser radiation as a measuring beam was developed. It was demonstrated that such a lidar operating within 9-14 µm spectral range can be used for remote sensing of freons and other atmospheric pollutants with concentration of ~1 ppt. The lidar can be used as a mobile one located in a mini-van. The weigth and size of such a device are 300 kg and 2.5x1.7x1.3 m^3.

Experimental and theoretical study of a possibility of molecular singlet delta oxygen production (SDO) in different sorts of electric discharge including EBSD with a yield adequate for developing electric discharge oxygen-iodine laser has been started at the Gas Laser Lab together with the Chemical Laser Lab and TRINITI. [11]. As a first step of the program, the study of electric breakdown and volt-ampere characteristics of a glow discharge in SDO was completed. SDO and the ground state oxygen turned out to have quite different electric properties.

Pulsed CO and CO_2 lasers developed at the lab were used in various experiments on laser-matter interaction. The experiments on laser-rock interaction were carried out together with the Colorado School of Mines, Solution Engineering (USA) and Moscow State Mining University for rocks typical for oil fields. Different types of sandstones, limestone, shale and granite saturated by water or mineral oil, or dry were irradiated by laser radiation. The energy density and intensity of laser radiation on a rock sample came up to 10^3 J/cm^2 and ~ 10^7 W/cm^2. For various rocks the dependencies of specific laser energy needed for excavation of volumetric unit of rock upon the energy density, the number of laser pulses, rock surface conditions, etc. were experimentally obtained. Measurements of a specific mechanical momentum (Fig.3) by a ballistic pendulum, which enabled us to get a coupling coefficient, and application of high-speed photography (Fig.4), gave us an opportunity to estimate an ablation pressure in interaction area and a velocity of expanded plasma plume for different gas dynamic regimes. The analysis of absorption and reflection IR spectrum made for rocks before and after the laser treatment demonstrated a surface crystal structure modification. Three mechanisms of rock destruction were demonstrated to take place in the experiments with dry and wet rocks: a) dry rock ablation; b) water boiling in a surface layer of porous rock saturated by water and following explosive water vapor expansion leading to the efficient rock destruction; c) rock destruction by a shock wave arising in rock when screening laser plasma by a thin layer of liquid transparent for the IR radiation.

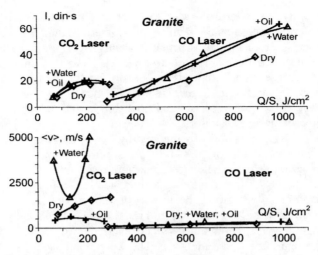

FIGURE 3. Momentum I and average velocity <v> vs. energy density on the rock sample

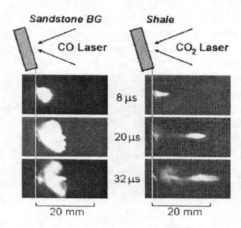

FIGURE 4. Frame records of CO and CO_2 laser -rock interaction plasma plume.

The experiments on interaction of frequency selective pulsed EBSD CO laser radiation with polymer materials such as nylon and other one having intense absorption bands near the wavelength ~6 μm were carried out together with DuPont (USA). The objective of the research was a formation of periodic structures on polymer fiber surface. The interaction parameters such as a geometry and a procedure of irradiation of polymer surface by laser radiation, energetic, spectral and temporal characteristics of laser radiation corresponding to different sorts of microstructures on fibers of treated polymer were determined. The connection between these microstructures and macrochanges of polymer surface such as a threshold of destruction, elimination of surface brilliance, etc. was analyzed. It was demonstrated that the procedure of using laser radiation for the treatment of polymer fabric consisted of fibers ~10-20 μm in diameter enabled us to form various periodic structures with a period of 3-5 μm (Fig.5) The procedure can be used for changing surface properties of polymer fabrics and production of optical fibers with surface periodic structures.

FIGURE 5. Periodic structure on polymer fibers.

3. HIGH-ENERGY KrF LASER AND ITS APPLICATIONS

The e-beam pumped excimer KrF laser facility named "Garpun" was developed at the lab (Fig.6). The laser with an active volume of ~25 l is pumped by two counter-propagating e-beams, specific pump power being 0.8 MW/cm^3 and the total pump energy ~2kJ. The laser operates in UV spectral range on the wavelength 0.248 μm. Its output energy is 100 J. Pulse duration is 100 ns. The laser facility "Garpun" has been actively used for laser-matter interaction experiments (see, for instance, [13])

FIGURE 6. Electron beam pumped KrF laser with active volume ~25 liters.

A set of experiments has been carried out on the interaction of UV laser pulses with planar targets. An ablation pressure as high as 5 Mbar was measured for laser intensities up to 5×10^{12} W/cm^2 on irradiated spot of 150-μm diameter. The effect of anomalous fast penetration of focused laser beam through a target was observed being explained by radial squeezing out of condensed matter by cone-shaped shock wave (Fig.7). It propagated in self-regulated manner together with an ablation front, both moving with supersonic velocity of 5-8 km/s. Fast martencitic phase transformation of pyrolitic graphite (being initially in quasi-single crystalline form) into diamond-like phase was for the first time observed in laser-driven shocks. Long-duration sample loading (if compared with phase transition time of few tens nanoseconds) by megabar pressure and strong shear stresses in conical shock wave created favorable conditions

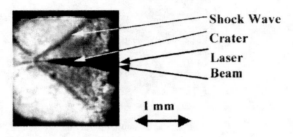

FIGURE 7. Conical shock wave and crater configuration in Plexiglas

for such graphite-diamond transformation. Hydrodynamics of plasma corona and its temperature were investigated: plasma stream velocity being measured by a streak camera reached ~50 km/s; its electron temperature being determined by absorption of soft X-ray emission in beryllium foils was ~100 eV. Emission spectra of plasmas for different target materials in extreme UV range 120-250 Å were measured by means of a compact spectrograph based on $MoSi_2$-Si sliced multilayer grating.

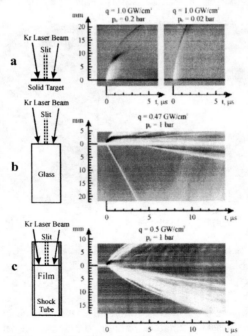

FIGURE 8. Slit scanning images of laser interaction with solid (a), transparent (b), and foil targets (c).

The design of miniature "laser-driven shock tube" has been proposed for generation hypersonic shock waves in gases with Mach number 10-20 and compression waves in liquids with pressure amplitude ~10 kbar. This novel experimental technique might be applied for studies of hydrodynamic instabilities at contact interfaces between different liquids and gases, hypersonic gas flow around the bodies, effects of strong shock wave refraction and cumulation in time scale of several microseconds and space scale of ten millimeters. To verify the concept shock wave time behavior was studied in KrF laser interaction with solid and thin-foil targets at moderate intensities 10^8-10^9 W/cm^2 and uniformly irradiated large spots of ~1-cm size (Fig.8).

A new approach to the fast-ignition inertial confinement fusion scheme has been proposed based on simultaneous amplification in the same large-scale electron-beam-pumped drivers long laser pulses of 5-100 ns for implosion of shell targets and short pulses of 10-100 ps for rapid heating of compressed fuel.

The problems of long-life, high transparency pressure foils (to isolate vacuum in the electron-beam diode from a working gas in the laser chamber) and of durable, stable optical windows has been studied. Both problems are important for the development of prototype of repetition-rate KrF laser driver for Inertial Fusion Energy program. The attractive foil based on aluminum-beryllium heterogeneous alloy was demonstrated to possess high transmittance for electrons, mechanical and chemical stability, fine heat conductivity. For a number of perspective UV optical materials (fused silica, calcium fluoride, magnesium fluoride and leucosapphire), which are used as laser windows, high-reflecting or anti-reflecting coatings, various optical properties have been studied. They were nonlinear absorption, transient (short-lived) and residual (long-lived) absorption induced by scattered fast electrons or bremsstrahlung X-ray emission of decelerating electron beams. It was shown that in the aggregate of properties the best choice for windows might be calcium fluoride and for coatings-leucosapphire.

The transport of high-current large cross section electron beams through titanium and aluminum-beryllium foils and their scattering in large-aperture KrF amplifiers have been measured and compared with numerical simulations on the base of Monte Carlo algorithm. It was demonstrated that guiding magnetic field prevents e-beam pinching, increase the efficiency of laser excitation, and eliminates the escape of scattered electrons onto laser windows thus decreasing induced absorption. In these conditions the only reason for degradation of optical windows transmittance would be accompanying bremsstrahlung X-ray emission. The distribution of X-ray sources inside laser chamber has been measured by means of scintillation detector. Absorption curves in different materials such as aluminum, copper and lead were measured in dependence on thickness. These attenuation curves were used further to reconstruct the X-ray spectrum. Regularizing algorithm has been developed to solve the incorrect inverse problem of spectrum reconstruction. With the help of thermoluminescence dosimeters energy fluences of X-ray radiation and distribution of absorbed doses in optical materials have been measured and scaling for larger KrF laser modules has been done.

The laser facility "Garpun" can be used together with a high-energy CO_2 laser of the Gas Laser Lab for combined UV & IR laser irradiation of a target which could increase a momentum transfer coupling coefficient. For a pulsed CO_2 laser-target interaction a coupling coefficient for the momentum transfer typically decreases at reduced pressure of surrounding air. Absorption coefficients of KrF laser radiation in evaporated target material as well in ambient air at temperatures higher then 8-10 kK are 3 to 4 orders less then for CO_2 laser radiation. Target irradiation by UV laser alone in intensity range up to 10^9 W/cm^2 would produce larger evaporated mass due to low absorption of radiation in ambient air. A subsequent IR laser pulse being absorbed in the dense cold vapor plume would give rise to a rapid ionization and plasma blow up. In combined UV&IR target irradiation a momentum transfer might increase significantly because of higher ablated mass and plasma acceleration to higher velocity. Of course, experiments should be performed to verify this idea.

4.HIGH-PRESSURE LASERS ON ATOMIC TRANSITIONS OF RARE GASES: Ar-Xe LASER.

High-pressure near-IR gas lasers can be quite important in controlling chemical processes, spectroscopy, atmospheric remote sensing, medicine, ranging, and communication. These lasers, which operate on rare gas transitions, are among the most extensively studied laser systems. The first such gas laser operating with helium-neon mixture was developed in 1961 and remains the most widely used. A variety of low-pressure glow-discharge pumped lasers operating on several hundreds transitions of neon, argon, krypton, and xenon in the visible and infrared spectral ranges have been developed. Despite their low efficiency ($\sim 10^{-5}$) and low output power, these lasers are in wide use because of their large wavelength range, chemical inertness, and long stable operation with a single gas filling. Rare gas lasers with higher energy were developed by increasing a gas pressure. Following the results obtained in the early 1970s, electric-discharge pumped lasers were developed with increased output energies from fractions of mJ up to several tens of mJ and efficiencies from $\sim 10^{-2}$ % up to $\sim 1\%$.

FIGURE 9. Electron beam sustained discharge and electron beam pumped laser on atomic transitions of rare gases with active volume ~10 liters.

Further progress in developing these lasers was achieved in a set of studies (see, for instance [15-17]) devoted to the e-beam and EBSD pumping, which were performed at the Gas Laser Lab. Various rare gas lasers were launched at the lab: Ar-Xe laser(~100

J, λ=1.73 µm), He–Ar laser(4.5 J, λ = 1.27 µm), He–Kr laser(4 J, λ= 2.52 µm) and He–Kr–Ne laser (1.5 J, λ = 0.73 µm) (Fig.9). The best laser turned out to be the Ar-Xe laser. The laser can operate with both e-beam and EBSD pumping. The Ar-Xe laser can operate on many IR wavelengths but about 90% of the laser energy is concentrated on $5d[3/2]_1^o - 6p[6p[5/2]_2$ transition with wavelength λ = 1.73 µm. The best output parameters of the high-pressure Ar-Xe laser were attained at the Gas Laser lab with the e-beam sustained discharge pumping of Ar–Xe working mixtures at pressure of 3.5 bar, consisting of lasing gas Xe (~1%) and buffer gas Ar, on 5d–6p transition of Xe atom with wavelength of λ = 1.73 µm. The laser efficiency was up to 5%, output energy being up to ~100 J. Specific volumetric output as high as ~10 J/liter with radiation beam divergence of 3×10^{-5} was obtained. This laser is of greatest interest for applications. Its wavelength coincides with the atmospheric transparency window with a very low absorption less than ~0.1 km^{-1} (Fig.2). It was shown that, based on these lasers, the industrial pulse-repetitive laser facilities with average radiation power of up to 1.0 kW/l and efficiency 3.2 % can be developed

Relatively short wavelength of λ = 1.73 µm and small divergence of the laser radiation allow Ar–Xe laser to compete with the most widespread CO_2 laser in radiance. To date, the high-pressure EBSD CO_2 laser surpass the Ar–Xe laser in specific output energy by a factor of no more than 5. Hence, even if the divergence of output radiation from CO_2 and Ar–Xe lasers differ from their diffraction limits to the same extent, the brightness of Ar–Xe laser is higher than that of CO_2 laser by a factor of $(\lambda_{CO_2}/\lambda)^2/5 \sim 6$. The wavelength λ = 1.73 µm corresponds to one of the most transparent atmospheric window whereas the wavelength λ = 10.6 µm is absorbed strongly enough, especially by water vapor. Contrary to radiation at λ = 10.6 µm, the radiation at λ = 1.73 µm can be transmitted with low losses via common glass fiber. In the material processing with the focused high-power laser beams, plasma screening of the treated material's surface occurs. A plasma heated by 10.6-µm radiation has a critical density of ~10^{19} cm^{-3}. Hence, the single-ionized air strongly reflects this radiation. For the radiation with λ = 1.73 µm, the plasma critical density is ~4×10^{20} cm^{-3}, which is much higher than density of the earth's atmosphere. Thus, the low-temperature laser plasma weakly screens the treated material's surface. As the active medium of the laser discussed is a mixture of rare gases that do not participate in the undesirable chemical reactions, one could use lasers with sealed-off gas-dynamic loops. The design of Ar-Xe repetitively- pulsed laser is the same as for repulsed CO_2 laser systems (for instance, produced earlier by AVCO (Textron)) except a design of high-pressure (2—4 atm) laser chamber and gas duct. It should be pointed out that repulsed "ELECTRA" laser facility (NRL) operates at such pressures. Therefore, Ar-Xe laser design combines that of repulsed CO_2 and KrF lasers and the laser can be considered as a potential high-power laser facility for laser propulsion.

CONCLUSIONS

The pulsed laser facillities developed at the Gas Laser Lab of the Lebedev Institute operate from UV to IR. They are situated in one place and give a unique opportunity to compare effects taking place at laser-matter interaction for different wavelengths and for their combination. Coupling coefficients which values are very important for the laser propulsion have to be compared for various materials and various laser wavelengths from UV to IR. The experiments should be performed to verify the idea of combined UV&IR laser irradiation of a target, which can enhance a momentum transfer coupling coefficient. High-power Ar-Xe laser operating on 1.73 μm wavelength coinciding with the atmospheric transparency window looks like a very promising candidate for a laser propulsion facility.

ACKNOWLEDGMENTS

Authors hold the memory of academician N.Basov who founded the Gas Laser Lab. They are very grateful to V.Danilychev who was the first head of the lab from 1983 till 1987 and hold the memory of A.Suchkov the next head of the lab from 1987 till 1996. Authors are very grateful to all members of the Gas Lasers Lab for their great contribution into developing laser facilities and their experimental and theoretical activity at the Gas Laser Lab.

REFERENCES

1. W.Bohn, Laser lightcraft perfomance, Proc SPIE, **3885**, 48 (2000)
2. T.Roberts, CO_2 lasers development, Poster presentation at the XII All-Union Conf on Nonlinear and Coherent Optics, 26-29 Aug 1985, Moscow, USSR
3. L.Myrabo, Flight experiments and evolutionary development of a laser propelled, trans-atmospheric vehicle, Proc SPIE, **3343**, 560 (1998)
4. Yu.Rezunkov, From development of lasers to the development of laser propulsion engines: the review of Russian beamed energy propulsion researches, First Int Symp Beamed Energy Propulsion, 5-7 Nov 2002, Huntsville, AL, USA
5. A.Ionin, Carbon monoxide lasers: problems of physics and engineering, Proc SPIE **3889**, 424 (2000)
6. N.Basov, G.Hager, A.Ionin et al, Efficient pulsed first-overtone CO laser operating within the spectral range of 2.5-4.2 μm, IEEE J Quant Electron, **36**,810 (2000)
7. R.M.Graves, D.G.O'Brien, "Star Wars laser technology applied to drilling and completing gas wells", Society of Petroleum Engineers Annual Technical Conference, New Orleans, USA, 27-30 Sept 1998, Paper SPE 49259.
8. B.Lacour, High average power HF/DF lasers, Proc SPIE, **4071**, 9 (2000)
9. S.E. Lambertson, The airborn laser, Proc.SPIE, **4760**,25,(2002)
10. N.Yuryshev, Pulsed mode of COIL, Proc.SPIE, **4760**,515 (2002)
11. A.Ionin,N.Napartovich, N.Yuryshev, Problems of development of oxygen-iodine laser with electric discharge production of singlet delta oxygen, , Proc.SPIE, **4760**,506 (2002)
12. J.McCord, R.Tate, G.Hager, A.Ionin, L.Seleznev, Supersonic RF discharge CO laser: multiline and single line spectral characterictics, Proc Int Conf LASERS 2000, Dec 4-8, 2000, ed. by V.Corcoran and T.Corcoran, STS Press, McLean,VA, 2001
13. V.Zvorykin, Comparative analysis of gasdynamic regimes of high-power UV and IR gas lasers interactions with solids in atmosphere, Proc.SPIE, **4065**, 128 (2000)
14. J.Cook, Propagation of high-power IR laser beam in maritime atmosphere, Int Workshop on laser pumped by chemical reactions or explosives, ISL, Saint-Louis, Nov 13-14, 2001
15. N.G. Basov, A.Yu. Chugunov, V.A. Danilychev, I.V. Kholin, and N.N. Ustinovsky, A powerful electroionization laser using Xe infrared atomic transition, IEEE J. Quantum Electron, **19** , 12(1983)
16. N.G. Basov, V.V. Baranov, A.Yu. Chugunov, V.A. Danilychev, A.Yu. Dudin, I.V. Kholin, N.N. Ustinovskii, and D.A. Zayarnyi, A 60 J quasistationary electroionization laser using Xe atomic metastable transitions, IEEE J. Quantum Electron. **21** 1756 (1985);
17. A.Yu. Dudin, D.A. Zayarnyi, L.V. Semenova, N.N. Ustinovskii, I.V. Kholin, A.Yu. Chugunov, Gain and lasing dynamics of Ar-Xe mixture e-beam pumped lasers, Quantum Electron. **23** (1993)

Facilities to Support Beamed Energy Launch Testing at the Laser Hardened Materials Evaluation Laboratory (LHMEL)

Michael L. Lander

Anteon Corporation
P.O. Box 33647
Wright-Patterson AFB, Ohio 45433 USA

Abstract. The Laser Hardened Materials Evaluation Laboratory (LHMEL) has been characterizing material responses to laser energy in support of national defense programs and the aerospace industry for the past 26 years. This paper reviews the overall resources available at LHMEL to support fundamental materials testing relating to impulse coupling measurement and to explore beamed energy launch concepts.

Located at Wright-Patterson Air Force Base, Ohio, LHMEL is managed by the Air Force Research Laboratory Materials Directorate AFRL/MLPJ and operated by Anteon Corporation. The facility's advanced hardware is centered around carbon dioxide lasers producing output power up to 135kW and neodymium glass lasers producing up to 10 kilojoules of repetitively pulsed output. The specific capabilities of each laser device and related optical systems are discussed. Materials testing capabilities coupled with the laser systems are also described including laser output and test specimen response diagnostics. Environmental simulation capabilities including wind tunnels and large-volume vacuum chambers relevant to beamed energy propulsion are also discussed. This paper concludes with a summary of the procedures and methods by which the facility can be accessed.

BACKGROUND

The Laser Hardened Materials Evaluation Laboratory (LHMEL) was established in 1976 at Wright Patterson Air Force Base, Ohio. This laboratory contains high power laser resources matched with environmental and diagnostic equipment dedicated to performing testing characterizing materials interaction with laser irradiation. Research is conducted in the LHMEL facility under the guidance of the Hardened Materials Branch of the Air Force Research Laboratory Materials Directorate (AFRL/MLPJ).

The LHMEL facility has been used to support several recent tests producing materials impulse coupling measurement and momentum coupling measurement data. Dr. Claude Phipps performed a test series in 1999 characterizing the impulse coupling for various materials and resulting in the confirmation of coupling coefficients as large as 113 dyne-s/J.[1,2] Dr. Leik Myrabo conducted several test series in 1999 and 2000 measuring momentum coupling and establishing vertical momentum on special lightweight sail materials.[3]

CP664, Beamed Energy Propulsion: First International Symposium on Beamed Energy Propulsion,
edited by A. V. Pakhomov
© 2003 American Institute of Physics 0-7354-0126-8/03/$20.00

Laser Systems

The laser systems described below can be combined with the environmental simulation and diagnostics systems described in the subsequent sections. The carbon dioxide (CO_2) and Nd:Glass laser systems available at the LHMEL facility are capable of producing high power output beams that have temporally stable, radially symmetric intensity distributions. The capabilities of each laser are summarized below.

LHMEL I 15kW Laser

The LHMEL I laser system, shown below, produces 15kW of output power for test durations up to 100 seconds. The laser utilizes a multi-mode resonator to produce an output beam having a uniform spatial intensity distribution (±17%). Optics combined with this laser can form beam diameters ranging from 0.75cm to 11cm, maintaining this spatial uniformity throughout the entire range. LHMEL I typically supports high volume testing and have produced up to 70 data producing tests in one day. This economically favorable device typically can achieve 25 laser tests per day and 125 laser tests per week.

FIGURE 1. LHMEL I 15kW Laser

LHMEL II 135kW Laser

The LHMEL II carbon dioxide (CO_2) laser system produces 135 kilowatts of output power in a beam that is temporally stable and has a uniform spatial intensity distribution.[4] This laser provides highly repeatable and economically favorable operating characteristics with a between-firing turn-around time on the order of minutes. The laser system is a continuous wave CO_2 electric discharge coaxial laser operating at a wavelength of 10.6μm. The laser produces multimode output beam having a flat top spatial intensity distribution uniform to within ±17% and can maintain power levels on target to within an average variation of ±5%. This stability is maintained over the entire target irradiation time lasting up to 100 seconds at maximum power. Longer shot durations are achieved when the lasers are operated at lower power levels. These reliable and repeatable lasers allow measurements of thermal properties to be made consistently, accurately and in a way that is easily modeled and analyzed.

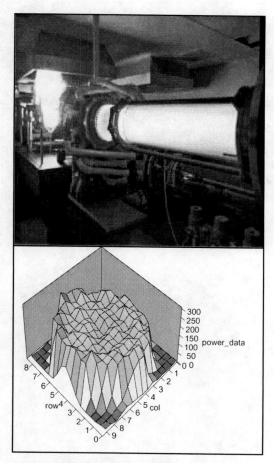

FIGURE 2. LHMEL II 135kW Laser

TABLE 1. LHMEL II 135kW Laser Parameters

Parameter	LHMEL II
Wavelength	10.6 μm
Operating Mode	Continuous Wave
Maximum Power Output (On Target)	135 kW
Spatial Intensity Distribution	Flat Top
Beam Uniformity	±17%
Minimum Spot Size (Diameter)	0.5 cm
Irradiance	0.01 to 690 kW/cm^2
Sample Exposure Time at Maximum Power	0.2 to 100 sec
Turnaround Interval	~20 min
Available Daily Test Time at Maximum Power Output	600 sec
Nominal Number of Shots/Day (Simple/Air)	10

Nd:Glass Laser

The Nd:Glass Laser system is made up of four laser heads.[5] Cavity optics can form two heads in series (operating independently or simultaneously depending on the desired pulse format) or as a four head resonator in a "U" configuration. The laser can operate in dump mode, Q-Switched Mode (with the insertion of Pockel Cell Q-switches and associated optics) or as a Master Oscillator Power Amplifier (MOPA). Each head contains a 6.4-cm diameter by approximately 670-cm long laser rod made of Nd-doped phosphate or silicate glass. The lasing wavelength of each type of glass is 1.054 mm and 1.06 mm. The Nd:Glass laser has flexibility in pulse width, repetition rates, energy per pulse and temporal distribution. The LHMEL Nd:Glass Laser has been used extensively for research applications including laser/materials interaction phenomenology and effects testing, nuclear fusion exploration, X-ray generation, laser shock processing of materials, and, ablation/coupling coefficient examination.[6,7]

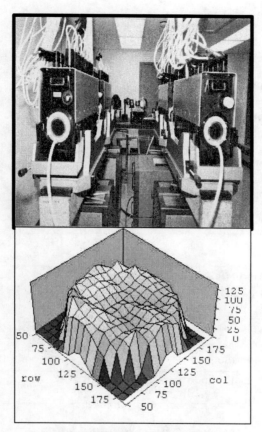

FIGURE 3. Nd: Glass Laser

TABLE 2. Nd:Glass Parameters

Nd:glass Parameters
Multimode beam: 6 cm dia. Can be focused to 0.5 cm diameter.

WAVELENGTH: 1.054 (PHOSPHATE)/1.060 MM (SILICATE)

PARAMETER	DUMP MODE	Q-SWITCHED MODE	MOPA
Pulse width	0.5 & 5 ms	60-300 ns	10 - 20 ns
Energy/pulse	100J - 10 kJ	5J - 100J	5J - 80 J
Power	4 MW	1GW	5 GW
Run duration	80 ms max	≤ 10kHz	20 ns
Beam Profile	Flat Top	Maltese Cross	Gaussian

713

Environmental Simulation

The LHMEL facility contains vacuum vessels and wind tunnels allowing specific environmental conditions to be achieved local to a test sample. The following sections describe these systems.

Vacuum Chambers

Three vacuum chambers are available for use at the LHMEL facility. The operating characteristics of these vacuum chambers are summarized below.

Table 3. LHMEL Vacuum Chambers

Parameter	30-Inch	7x9-Foot	22-Foot
Routine Vacuum	$1x10^{-5}$ torr	$1x10^{-6}$ torr	$1x10^{-6}$ torr
Maximum Test Article Dimension	20 cm	120 cm	3 m
Minimum Laser Spot Size on Target (Diameter)	1.0 cm	1.0 cm	1.0 cm
Maximum Laser Spot Size on Target (Diameter)	16 cm	100 cm	100 cm
Average Pumpdown time	30 min	30 min	120 min
Maximum Number of Pumpdown Sequences per Day	10	5	2

30" x 30" Vacuum Chamber

The smallest vacuum chamber at the LHMEL facility is a horizontal right cylinder with an inside diameter of 30 in. and a length of 40 in. A spherical end is connected to the vacuum system and also a flat end with the 5-1/4 in. ID laser entry window. It is connected to a Varian Turbo-V450 turbo-molecular vacuum pumping system and can simulate high-altitude environments to $1x10^{-5}$ torr within 20 to 25 minutes after the start of pump-down. It is equipped with six 8-1/2 in. ID ports for specimen handling. Signals are passed through the chamber walls via 2-3/4 in. OD high vacuum feedthroughs (Varian Conflat or equivalent). There is also one 5 in. ID and limited 1-1/2 in. ID viewing ports for pyrometers, video cameras and motion picture cameras. Special user-furnished equipment will also be accommodated if possible. The chamber has been designed for rapid turnaround time between experiments.

During a test the pumping system can maintain a pressure within the $1x10^{-3}$ torr range with a small amount of sample outgassing. The system cannot tolerate sample degradation and vacuum chamber contamination and maintain pressures within this range. The contamination of optical elements due to spallation from materials undergoing laser irradiation in a vacuum environment has been a matter of considerable concern. A technique using an internal cone placed against the laser entry window provides the best contamination control. This is accomplished by reducing the contaminant entrance area to a 5-cm diameter.

Figure 4. 30" x 30" Vacuum Chamber

7' x 9' Vacuum Chamber

The 7x9-Foot Space Environment Test Chamber is a horizontal right cylinder with an inside diameter of 7 ft. and a length of 9 ft. between the centers of the spherical ends. It incorporates at various locations nine 5-1/2 in. inside diameter (ID) ports that can be used for view ports or instrumentation ports. A line of seven 2-3/4 in. outside diameter (OD) high vacuum feedthroughs (Varian Conflat or equivalent) are on the rear end for instrumentation and coolant lines. The front end has a 46-inch-diameter port to which is mounted the Beam Entry Sub-System (BESS), two 5-1/2 in. ID ports, and one 9 in. ID port. All ports are situated such that the central region of the chamber is within the field of view through the port. The vacuum chamber is capable of accepting large test articles up to the full inside diameter of the chamber; however, the nominal maximum test article dimension is about 120 cm. The chamber is fabricated of stainless steel and incorporates a 10,000-liters-per-second oil-vapor diffusion pump and forepump. The chamber is capable of evacuation to 1×10^{-6} torr with LN_2 in the cold trap of the diffusion pump and the cryogenerator operating at -130° C.

Figure 5. 7' x 9' Vacuum Chamber

22' Diameter Vacuum Chamber

The 22-Foot Space Environment Test Chamber is an upright cylindrical volume with an inside diameter of 22 ft. and a height of 15 ft. It incorporates seven ports at various locations for viewing or instrumentation ranging in size from 5-1/2 in. to 10 in. diameter (ID). In addition, multiple instrumentation feedthroughs permit the monitoring of electrical and electronic signals internal to the chamber during testing. This chamber contains a 15-ft tall liquid nitrogen cryo-wall around the 22-ft diameter perimeter. The chamber can be configured with a Beam Entry Window allowing LHMEL facility lasers to access test samples inside the chamber. The vacuum chamber is equipped with a 10- by 8-foot access door allowing large test articles or systems to be assembled or placed inside the chamber. The chamber is capable of evacuation to 1×10^{-6} torr during test operation.

Figure 6. 22' Vacuum Chamber

Wind Tunnels

The LHMEL wind tunnels operate as regulated blow-down devices relying upon a high-pressure supply, which exhaust into a test region open to the atmosphere. The wind tunnels provide nozzles with exit dimensions of 2.5 x 7.6 cm and test Mach numbers ranging from 0.1 to 0.9 in standard configuration. They can also be fitted with a supersonic nozzle allowing velocities up to 2.3M to be attained at reduced pressure (< 1Atm) local to the sample. Test specimens may be positioned in the center of the air stream in a free jet, inserted into an extension of a three-wall channel, or inserted alongside a flat-plate test region. In addition, the optical delivery system for both lasers can be adjusted for incidence angles up to 60 degrees. Exhaust ducts are provided to remove sample debris and wind tunnel gasses. These capabilities are summarized below:

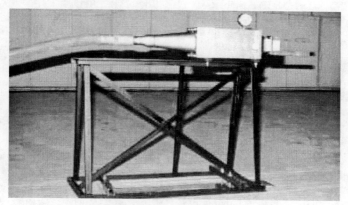

Figure 7. Wind Tunnel

Table 4. LHMEL II Wind Tunnel Parameters

Parameter	LHMEL II Wind Tunnel
Mach Number	0.1 to 2.3 M
Nozzle Exit Sizes	2.5 cm x 7.6 cm
Operating Medium	Air or Nitrogen
Operating Mode	Blow-down; Bottled Compressed Gas
Test Section Type	Free Jet, Channel or Flat Plate
Velocity	100 to 1000 ft/sec at 1.0 Atm
Mass Transfer Coefficient	0.02 to 0.13 lbm/ft^2 sec
Aerodynamic Shear Stress	0.001 to 0.04 lbf/in^2
Nominal Number of Tests/Day	10

SUMMARY

The Laser Hardened Materials Evaluation Laboratory has supported impulse coupling and momentum coupling experiments relevant to beamed energy propulsion considerations. The facility is available to support testing ranging from small-scale fundamental concept exploration to larger-scale materials response and limited flight demonstrations. The facility can be accessed via U.S. Government transfer of funds or through a standard purchase order via the Cooperative Research and Development Agreement (CRADA) established between Anteon Corporation and AFRL/MLPJ.

REFERENCES

1. Phipps, C., Lander, M. L., Seibert, D. B., Royse, R., Reilly, J. P., "Impulse Coupling Measurements for an Orion Demonstration." Accepted for presentation Space Technology and Applications International Forum (STAIF-99), 31 January – 4 February 1999, Albuquerque, NM. Accepted for publication in Conference Proceedings, STAIF 99, American Institute of Physics, Woodbury, NY 1999.
2. Phipps, C., Seibert, D., Royse, R., King, G., Campbell, J., "Very High Coupling Coefficients at Low laser Fluence with a Structured Target" High Power Laser Ablation 2000, Santa Fe, NM, 23-28 April 2000.
3. Myrabo, L. N., Knowles, T. R., Bagford, J. O., Seiber, D. B., Harris, H. M., "Laser-Boosted Light Sail Experiments with the 150kW LHMEL II CO2 Laser," *SPIE 2002*, #4670-100, High Power Laser Ablation IV, April 2002.
4. Lander, M. L., Maxwell, K. J., Reilly, J. P., Hull, R. J., "Continuous-Wave Carbon Dioxide Laser System Producing Output Power up to 135 Kilowatts," Gas and Chemical Lasers/High Power Lasers Conference Proceedings, Edinburgh, Scotland, UK, 25-30 August 1996.
5. Lander, M. L., Royse, R. W., Seibert, D. B., Eric, J. J., "Pulsed-laser Capabilities at the Laser Hardened Materials Evaluation Laboratory (LHMEL)" *High-Power Laser Ablation III,* edited by C. R. Phipps, Proceedings of *SPIE* Vol. 4065, 24-27 April 2000.
6. Walters, C. T., Clauer, A. H., Campbell, B. E., "Laser Shock Effects on Stressed Structural Materials - Experimental Results," Proceedings of the 6th DOD Conference on DEW Vulnerability, Survivability, and Effects, Gaithersburg, MD, 12-15 May 1987.
7. Walters, C. T., "Laser Generation of 100-kbar Shock Waves in Solids," *Shock Compression of Condensed Matter-1991*, edited by Schmidt, S.C., Dick, R.D., Forbes, J.W., Tasker, D.G., *Elsevier Science Publications,* B.V. 1992, pp. 797-800.

ADDENDUM

The following papers were included in the Symposium program, but have not been presented (withdrawn) or submitted for the publication:

From Development of Lasers to the Development of Laser Propulsion Engines: The Review of Russian Beamed-Energy Propulsion Researches (*withdrawn*)
  Y. A. Rezunkov

Towards a Microwave Beamed-Energy Rocket (*withdrawn*)
  K. Parkin

The Possibilities of Laser Application in Magnetic Fusion Devices (*withdrawn*)
  A. F. Nastoyashchii

Nuclear Waste Disposal as a Laser Launch Application (*unavailable for publication*)
  J. T. Kare

Unipolar Optical Pulse Generation and Beamed Optical Power for Lift-to-Orbit (*unavailable for publication*)
  M. Burleson and R. L. Fork

Laser Application in Correction of Satellite Orbits (*withdrawn*)
  A. F. Nastoyashchii

Laser Beam Control Algorithms for Precise Beaming of a Flying Vehicle in the Atmosphere (*withdrawn*)
  Y. A. Rezunkov

An Economic and Performance Analysis of Laser-Beamed Energy In-Space Propulsion Systems (*unavailable for publication*)
  G. Woodcock

System Requirements and Economics of a Laser Launch System (*unavailable for publication*)
  J. T. Kare

High Power Lasers for Industrial and Ecological Use (*withdrawn*)
  V. V. Apollonov and A. M. Prokhorov

Sherstobitov, V. E., 620
Shiho, M., 475
Singh, J. P., 251
Smalley, L., 509
Soms, L. N., 620
Stepanov, V. V., 149

T

Tabibi, B. M., 608
Taniguchi, K., 185, 535
Taylor, T. S., 369, 382
Thompson, M. S., 194, 206
Torikai, H., 454
Toro, P. G. P., 497

U

Uchida, N., 411
Uchida, S., 214
Urabe, N., 105, 454

V

Vasil'ev, B., 697

W

Wang, T.-S., 138

Y

Yabe, T., 185, 475, 535, 545, 557
Yamaguchi, M., 185, 535, 557
Yueh, F.-Y., 251

Z

Zeiders, G. W., 390
Zvorykin, V., 697